INTRODUCTION

This volume forms Volume 3 Part 1 of Grierson & Long's *Flora of Bhutan* and covers the monocotyledons with the exception of grasses (Poaceae) and orchids (Orchidaceae), which will form Volume 3 Part 2 and Volume 3 Part 3 respectively.

The treatment and arrangement of families largely follows Dahlgren, Clifford & Yeo (1985), which subdivides the unwieldy and heterogeneous families Liliaceae and Amaryllidaceae into smaller, more natural, groups. Whilst some may disagree with the delimitation of certain of these segregate families, and much further work and refinement remains to be done, there seems little doubt that the treatment represents a considerable advancement and, for this reason, it has gained wide acceptance.

Many groups covered in the present volume have been under-collected in the area for various reasons: some are thought obscure (e.g. *Juncus* and *Carex*), while others such as *Musa* and Arecaceae are difficult to collect adequately as herbarium specimens. Modern collections are needed for these groups and many interesting discoveries lie in store. For these, and also for difficult genera such as *Polygonatum* and *Ophiopogon*, the present treatments claim to be no more than provisional and summarise the position as deduced from specimens in British herbaria, together with a limited amount of field-experience in Yunnan and the E Himalaya.

In contrast to the perceived obscurity of some of the groups covered, many ornamental species are included, some of which are native to the region and widely grown in gardens worldwide (e.g. *Lilium, Iris*). On the other hand many exotic species of widely differing origins are cultivated in the gardens of the region, especially in Darjeeling and Sikkim, no doubt a legacy of British influence in these areas. An attempt has been made to cover the more conspicuous of these cultivated taxa, but an exhaustive treatment has not been possible (for example species of *Hosta, Cordyline* and *Aloe* are known to be grown, but have not been collected or identified). In addition to plants of horticultural value, some of the plants covered in this volume are of use as foods or medicines (e.g. *Dioscorea*, Arecaceae) or are valuable genetic reserves for crop plants (e.g. *Musa*). Much further work remains to be done on these aspects and collection of recent information on uses made by the native inhabitants is essential. It is hoped that this preliminary documentation will act as a stimulus for such exploration and research.

ACKNOWLEDGEMENTS

The author is extremely grateful to the many people who have contributed in diverse ways to this volume, including:

The Overseas Development Administration, London, who employed the author for three years.

Rosemary M. Smith, William T. Stearn, Sally J. Rae and David G. Long, who have contributed taxonomic accounts.

G.C.G. Argent (Musaceae), J. Dransfield (Arecaceae), R.B. Faden (Commelinaceae), P. Boyce (Araceae), I. Kukkonen, A.O. Chater, D.J. Mabberley, and D.A. Simpson (Cyperaceae), who have commented on the manuscripts of particular families.

D.G. Long, R.R. Mill, M.F. Watson, A.G. Miller and B.L. Burtt for immense quantities of general botanical help and discussion and Sally Rae for technical support.

R.J.D. McBeath and the staff of the Herbaceous and Alpine Department, for maintaining the excellent collection of living Sino-Himalayan monocotyledons at the Royal Botanic Garden Edinburgh.

The curators of the following herbaria for lending specimens: the Natural History Museum, London (BM); the Royal Botanic Gardens, Kew (K); the Central National Herbarium, Howrah, Calcutta (CAL); the National Museums & Galleries on Merseyside, Liverpool (LIV).

The Royal Botanic Gardens, Kew and the Natural History Museum, London for extensive use of facilities and valuable discussion with colleagues.

John Wood for hospitality, extensive co-operation, discussion and for making his collections available, and Chris Parker for discussion, information and collections of Bhutanese weeds.

The Edinburgh Botanic Garden (Sibbald) Trust, the William Steel Trust, the Friends of the Royal Botanic Garden Edinburgh and the Gordon Fraser Charitable Trust for sponsoring fieldwork in Yunnan and the eastern Himalaya.

Mary Bates, Sally Rae, Glenn Rodrigues and Mary Benstead for providing the illustrations.

Finally, the author wishes to thank the Publications Section of the Royal Botanic Garden Edinburgh, particularly Erica Schwarz, for preparing the manuscript for publication and for liaising with the printers, The Charlesworth Group.

2

FLORA OF BHUTAN

INCLUDING A RECORD OF PLANTS FROM SIKKIM AND DARJEELING

VOLUME 3 PART 1

H.J. NOLTIE

ROYAL BOTANIC GARDEN EDINBURGH
1994

This volume is dedicated to the memory of
Henry Robert Noltie, 1903–1985

Published by the Royal Botanic Garden Edinburgh, Inverleith Row, Edinburgh
EH3 5LR, UK

© Royal Botanic Garden Edinburgh 1994

ISBN 1 872291 11 2

Typeset, Printed and Bound by The Charlesworth Group, Huddersfield, UK, 01484 517077

CONTENTS

ANGIOSPERMAE continued (Monocotyledons)

COLOUR PLATES (between pp. 234 and 235)

FIGURES

ABBREVIATIONS

For reasons of conciseness the following standard works used for synonymy and localities are abbreviated as follows (*Flora of Bhutan* Bibliography numbers given in parentheses):

F.B.I.: Hooker, J.D. (ed.) (1890–1894). *The Flora of British India. Vol. 6. Orchideae to Cyperaceae*. London. (80).

M.F.B.: Subramanyam, K. (ed.) (1973). Materials for the Flora of Bhutan. *Rec. Bot. Surv. India* 20: 1–278. (117).

E.F.N.: Hara, H., Stearn, W.T. & Williams, L.H.J. (1978). *An Enumeration of Flowering Plants of Nepal. Vol. 1*. London. (72).

F.E.H.1: Hara, H. (ed.) (1967). *The Flora of Eastern Himalaya*. Tokyo. (69).

F.E.H.2: Hara, H. (ed.) (1971). *Flora of Eastern Himalaya. Second Report*. Forming Bulletin no. 2 of the University Museum, The University of Tokyo. Tokyo. (71).

F.E.H.3: Ohashi, H. (ed.) (1975). *Flora of Eastern Himalaya. Third Report*. Forming Bulletin no. 8 of the University Museum, The University of Tokyo. Tokyo. (101).

In the lists of localities, literature records are separated from those based on seen specimens by a semi-colon. For other literature references see Bibliography.

Abbreviations for languages and dialects of common names of plants used in this volume are:

Dz: Dzongkha
Sha: Shachop
Med: Bhutanese medicinal name

Lep: Lepcha
Eng: English
Nep: Nepali

Other abbreviations:

incl.: including
excl.: excluding
infl(s).: inflorescence(s)

fl.: flowering period
fr.: fruiting period

For remaining abbreviations, e.g. botanical authorities, see Volume 1 Part 1, p. 34.

Family 188. DIOSCOREACEAE

Commonly herbaceous, twining perennials arising from tubers. Leaves alternate or opposite, petiolate, simple or palmate. Infls. axillary, commonly spikes or racemes. Flowers usually unisexual, tepals 6 in two similar or dissimilar whorls, ± free. Stamens 6, or 3 + 3 staminodes, attached to base of tepals, anthers with 2 longitudinally dehiscing locules. Ovary inferior, 3-loculed, ovules axile. Fruit a capsule.

1. DIOSCOREA L.

Tubers stalked or not, elongate to globose, bearing superficial rootlets (sometimes armed). Stems twining to left or right, with or without recurved prickles, axillary bulbils often present. Dioecious. Male flowers borne singly or in clusters on spikes; spikes aggregated along ± leafless axes, axes axillary and terminal, or in axillary fascicles; flowers globose to cup shaped, commonly sessile, subtended by a bract and a bracteole; anthers dehiscing laterally; pistillode minute, commonly 3-pointed. Female flowers sessile on axillary spikes; ovary 3-loculed, 3-angled, narrowly ellipsoid, locules each with 2 ovules, style short, stout; stigmatic ridges 3, sometimes ± falcate. Capsule broadly 3-winged, shortly stipitate, dehiscing from apex, sometimes reflexed. Seeds 2 per locule, compressed, winged — wing basal and oblong or ± surrounding seed.

Yams (Eng) are very important medicinally and as a food source (tubers and bulbils, which often have to be carefully prepared to destroy poisons). Prain & Burkill (1936, 1939) collected much information on local names and uses in Sikkim and Darjeeling. Unfortunately very little similar information is available for Bhutan, where the genus is seriously under-collected. Nakao & Nishioka (1984) document the collection of wild yams and illustrate tubers of at least three species from Shemgang (Tongsa district), but show no leaves or flowers so identification is impossible. They mention a long-tubered, bulbiliferous species known as *shabrang kyi* which they tentatively identify as

---

FIG. 1. **Dioscoreaceae** I (leaves of **Dioscorea** spp.).
a, **D. alata**: from main stem. b–c, **D. hamiltonii**: b, from main stem; c, from lateral shoot. d, **D. bulbifera**: from main stem. e, **D. esculenta**: from main stem. f–g, **D. belophylla**: f, from main stem; g, from lateral shoot. h–i, **D. glabra**: h, from main stem; i, from lateral shoot. j–k, **D. deltoidea**: j, from main stem; k, from lateral shoot. l, **D. prazeri**: from main stem. m, **D. pubera**: from main stem. All × ¼. Drawn by Sally Rae.

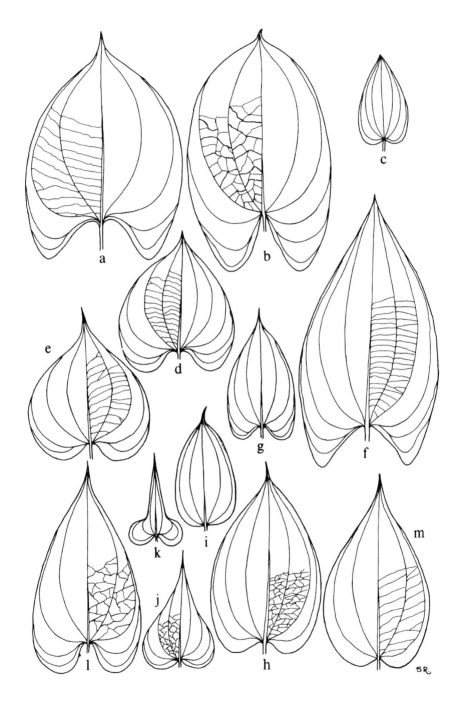

D. hamiltonii (or relative), and a round-tubered species known as *jyukpang*. In addition the following local names have been recorded for species of *Dioscorea* in Bhutan: *gitta lahara* (Nep), *gong-jogtang* and *borang-jogtang-tsala* (Sha) for species with reddish tubers, *borang-jogtang-ballingmeen* (Sha) for a species with a white tuber and *faantang* (Sha) for a species with a round tuber.

Tubers not seen of most species; descriptions taken from Prain & Burkill (1936, 1939). Local names marked P & B are taken from the same work; others come from annotations on specimens. Many additional Sikkim/Darjeeling localities are given in Prain & Burkill: localities from this work are only given if no specimens have been seen from the area.

1. Plant twining to right (Fig. 3b); leaves simple; capsules not reflexed against spike axis; seeds surrounded by ± equal, circular wing 2
+   Plant twining to left (Fig. 3a); leaves simple or compound; capsules reflexed; wing oblong or circular (when usually very unequal) 6

2. Stems angled or winged 3
+   Stems terete 4

3. Stems conspicuously winged, drying yellowish-brown; cultivated; male flowers never seen; capsule wings over 2cm wide .... **9. D. alata**
+   Stems ridged to narrowly winged, drying reddish-brown; wild; male flowers on zigzag axis; capsule wings under 2cm wide   **11. D. hamiltonii**

4. Plant (especially infl.) pubescent .... **10. D. pubera**
+   Plant glabrous 5

5. Leaves thin-textured; secondary veins not prominent on underside, not forming ladder-like pattern about midrib; upper (small) leaves truncate, no leaves sagittate .... **12. D. glabra**
+   Leaves becoming coriaceous; secondary veins prominent on underside, forming ladder-like pattern; upper leaves usually shallowly cordate, lower ones ± sagittate with angled basal lobes .... **13. D. belophylla**

6. Leaves compound 7
+   Leaves simple 10

7. Leaflets always 3, the central 3-costate; capsules over 4cm; male spikes very short and compact about equalling their peduncles ... **8. D. hispida**
+   Leaflets usually 5, if 3 then central leaflet with 1 midrib with pinnate venation; capsules under 3cm; male spikes laxer, exceeding their peduncle 8

8. Stem glabrous, prickles absent; small bulbils abundant; leaflets usually under 1.5cm wide, apex piliferous .... **7. D. melanophyma**
+ Stem hairy at least when young, prickles present or absent; bulbils sparse; leaflets usually over 2.5cm wide; apex cuspidate 9

9. Infl. greyish-white pubescent; flowers small; stems prickly; male spikes arranged along a lateral, inflorescence axis .... **5. D. pentaphylla**
+ Infl. creamy-white pubescent; flowers larger; stems lacking prickles; male spikes often axillary on main stem .... **6. D. kamoonensis**

10. Plant densely hairy .... **3. D. esculenta**
+ Plant glabrous 11

11. Petals and sepals linear, whitish, over 2mm, flowers scented; capsules oblong in outline; seeds with basal, oblong wing .... **4. D. bulbifera**
+ Petals and sepals inconspicuous, greenish, under 2mm; capsules obovate in outline; seeds surrounded (unequally) by wing 12

12. Leaves hastate, at least when young; male infl. very slender; leaves usually hispid on veins beneath .... **2. D. deltoidea**
+ Leaves ovate; male infl. stouter; leaves glabrous beneath .. **1. D. prazeri**

1. D. prazeri Prain & Burkill; *D. sikkimensis* Prain & Burkill. Lep: *kuchyong, kunching, kancheong, cumchong, chanchiyang*; Nep (Sikkim): *kukur toral.* Fig. 1(1), Fig. 2q–r.

[Tuber much-branched, surface rootlets many, flesh white]. Stem twining to left, lacking prickles, bulbils few or absent. Plant glabrous. Stem leaves alternate, triangular-ovate, abruptly acuminate to cuspidate, base cordate, 10–22 × 6–15.5cm, rather glossy, thick-textured; infl.-branch leaves smaller; petioles shorter than blades. Male infl.: spikes in 2s and 3s, arranged along leafless branches or in axillary groups; spikes 3–17cm, axis sharply angled, bearing 2–4-flowered clusters throughout length; bracts minute. Male flower: petals and sepals similar, obliquely spreading, oblanceolate, apex rounded, pale brown, margins hyaline, 1.3–1.5 × 0.8–1mm; pedicel short (under 1mm); stamens 6, filaments 0.5–0.7mm, outwardly curved, anthers 0.1 × 0.2–0.3mm, transversely elongate, 2-lobed, downwardly directed. Female spikes borne singly in axils. Female flower: petals and sepals similar to male but larger (1.6–2 × 0.9–1mm); style 0.7–0.9mm, stout. Capsules reflexed, widely oblong-elliptic to oblong-obovate in outline, truncate or sometimes shallowly retuse, 1.8–2.5cm, wings 1.1–1.3cm wide. Seed c.0.6 × 0.4cm, ± oblong, surrounded by whitish wing, unequally wide (to 0.5cm) ± elliptic in outline.

7

Bhutan: S — Chukka district (Marichong); **C** — Mongar district (Ngasam); **Terai** (Sivoke, Naksaibari, Jalpaiguri Duars); **Darjeeling** (Little Rangit, Tukvar, Rishap, Chunbati, Lebong, Birick, Bakrikot, Kalimpong, Ribong, Tista, Lanki, Rongo, Mongpu, Geille, Birick, Dasogadhara); **Sikkim** (Dikeeling, Rinchinpong to Pemionchi, Rishi to Rinchinpong (P & B)). *Shorea* forest; among *Artemisia* scrub, 180–1220(–1525)m. Fl. June–October; fr. July–February.

Root poisonous and not eaten by Lepchas who use it as a lice-killing soap for washing hair; also used as a fish poison. A note by K. Biswas on a sheet at Kew comments that the diosgenin content is high and of use in contraceptive pills.

2. D. deltoidea Wall. ex Kunth. Fig. 1j–k, Fig. 2o–p.

Similar to *D. prazeri* in flower and fruit, differing as follows: leaves usually smaller, thinner-textured, distinctly hastate with basal lobes present at least when young, finally ovate (i.e. sides curved rather than ± straight), usually hispid on veins beneath. Male spikes: flower clusters rather distant, absent from lower part of the filiform axis.

Bhutan: C — Mongar district (Lhuntse Dzong); **Sikkim** (Dikeeling; Yoksum (F.E.H.1)). Low forest, 1220–1800m. Fl. April–May.

A species of drier areas and commoner in the W Himalaya, where used as a soap.

3. D. esculenta (Loureiro) Burkill; *D. sativa* L.; *D. spinosa* sensu F.B.I. Fig. 1e.

[Tubers several to many, stalked, shapes various, flesh white; superficial roots sometimes spiny]. Plant softly hairy, hairs T-shaped. Stems twining to left, softly woolly-pubescent, prickly especially below, lacking bulbils. Leaves alternate, broadly ovate, shortly acuminate, base cordate, often rather shallowly, 8–15.5 × 6.5–14cm, subglabrous above, densely woolly-hairy beneath, sometimes becoming subglabrous; petiole about equalling blade, subtended by 2 downwardly directed prickles. Male infl. (not seen from our area): spikes borne singly in axils, axis to 20cm, woolly, bearing distant, sessile flowers in upper ¾. Male flowers: broadly funnel-shaped, petals and sepals similar, lanceolate, acute, hairy on outside, fused into cup below, c.2.5 × 0.9mm; stamens 6, anthers c.0.5 × 0.4mm, shorter than filaments. Female flowers and fruit very rare, not seen.

Terai (Jalpaiguri, P & B); **Darjeeling** (into hills at least to Tista Bridge, P & B); **Sikkim** (unlocalised Hooker specimen, possibly from Darjeeling).

A widely cultivated species probably originating in Indo-China. Tubers eaten boiled, or dried and ground into a meal, slightly sweet-tasting.

4. D. bulbifera L.; *D. sativa* auct. non L. Nep (Bhutan): *gittha*; Nep (Sikkim):
gita, githa lahara, kumchong simpot; Lep: *kachem, kuching*. Fig. 1d, Fig. 3c–e.
Tubers — see below. Plant glabrous. Stems twining to left, sharply angled
(sometimes narrowly winged), lacking prickles; bulbils warty, many. Leaves
alternate; upper ovate, finely acuminate, base deeply cordate, 8–24 ×
5.5–17cm; lower larger, broader (to 31 × 30cm); petiole about equalling blade,
narrowly winged above, auriculate at base. Male infl.: spikes borne in fascicles
(sometimes subtended by small leaves) on elongate axillary axes, sometimes in
groups of 4–6 in axils of main stem; spikes long (4–15cm), slender, flowers
borne singly, sessile, adjacent in upper ⅘; bracts lanceolate, acuminate, shorter
than flowers, membranous. Male flowers: sepals and petals similar, erect,
linear-lanceolate, subacute, 2.1–5 × 0.6–0.9mm, whitish in bud becoming
green, brown or purple; stamens 6, anthers 0.4 × 0.2mm, about equalling
filaments. Female spikes axillary in groups of 2–5, flowers overlapping, inserted
on upper ⅔ of axis. Female flowers: petals and sepals similar to male, white,
sweet-smelling; stigma 0.8mm. Capsules reflexed, 1.8–2.4cm, oblong-elliptic in
outline, apex subtruncate, wings 0.6–0.7cm wide. Seeds with basal, oblong
wing c.0.8 × 0.4cm.

Bhutan: S — Samchi (Tamangdhanra Forest), Phuntsholing (Phuntsho-
ling, below Ganglakha), Chukka (Chukka Dzong, Marichong), Sarbhang
(17km from Sarbhang on Chirang road) and Gaylegphug (Sham Khara)
districts; **C** — Punakha (Tsarza La to Samtengang, between Lobesa and
Tinlegang) and Tashigang (Dangme Chu Valley) districts; **Terai** (Sukna,
Debiganji, Apalchand Forest); **Darjeeling** (Tukvar, Punkabari, Muckeyebarry,
Budumthan, Pomong, Nahore, Tista Valley, Kalimpong, Farsing, Kalijhora,
Great Rangit opposite Manjitar, Little Rangit above Singla Bazaar, Lankoh,
Labdah, Mongpu; Happy Valley (F.E.H.1)); **Sikkim** (Yoksum, N of Gangtok,
14km W of Singtam; Martam, Gangtok (F.E.H.2)). Subtropical forest; *Shorea*
forest; riverside scrub, 90–1900(–2150)m. Fl. June–September; fr. September.

Prain & Burkill described four varieties (not distinguishable from herbarium specimens
and not separated above) from Darjeeling and Sikkim:

(A) Wild varieties with small, acrid tubers

var. **vera**: male sepals to 2mm. Darjeeling and Terai to 914m.

var. **simbha** (Lep: *simbha*): male sepals to 4mm; tubers globose, flesh when cut pale
yellow, slightly tinted salmon; stem smooth. Common in Sikkim Himalaya 610–1524m.
Eaten in times of need by Lepchas but must be washed for 2 days before cooking.

(B) Cultivated varieties with improved tubers

var. **kacheo** (Lep: *kacheo*; Nep (Sikkim): *ghar gita*): tubers large, depressed-globose,
shortly stalked, sparingly covered with rootlets, flesh when cut yellow near outside,
salmon-coloured at top and centre; stem winged. Commonly cultivated by Lepchas and
running wild in Darjeeling and Sikkim to 1524m.

188. DIOSCOREACEAE

var. **suavior**: cultivated for its bulbils — specimens at K cultivated in Singapore collected from **Darjeeling**: Mongpu (local name: *singulbok*) and Silake (local name: *kunchong*).

Two types are readily distinguishable in the field in Bhutan and Sikkim:

(a) Leaves lanceolate, stems terete and reddish — possibly var. *simbha*.

Bhutan: Phuntsholing (below Ganglakha), Chukka (above Chukka Colony) and Punakha (E of Chusutsa) districts; **Sikkim** (N of Gangtok, Yoksum); higher altitudes (1640–1940m).

(b) Leaves suborbicular, stems winged — possibly var. *kacheo* or var. *suavior*.

Bhutan: Phuntsholing (below Kamji) and Punakha (above Lobesa) districts; **Darjeeling** (Great Rangit opposite Manjitar, Little Rangit above Singla Bazaar); **Sikkim** (14km W of Singtam); lower altitudes (450–1640m).

5. D. pentaphylla L. Lep: *sili kussok, kussok*; Nep (Sikkim): *bhegur, behagur*. Fig. 2e–g.

Tubers — see below. Plant hairy, hairs simple, greyish-white. Stem sometimes prickly, densely appressed hairy, or subglabrous, bulbils usually present. Leaves (3- or) 5-foliolate, rugose in life; leaflets, middle largest with 1 strong midrib, narrowly elliptic to oblanceolate, cuspidate, narrowed to sessile base, 7–13 × 2.5–4.3cm, densely white-bristly above and beneath or larger 8.2–19 × 3.3–6.8cm, subglabrous beneath, distinctly petiolulate (0.3–0.5cm); petiole 5–6cm. Male infl.: spikes borne singly or in pairs on branched axes 15–49cm, axes hairy, sometimes bearing small leaves, borne singly in axils; spikes dense, greyish-white hairy, 1.2–3 × 0.2cm, exceeding peduncles even in bud; bracts ovate, cuspidate, clasping, brown, hairy, exceeding flowers. Male flowers: globose, usually sessile, borne singly, sepals lanceolate, acute, 1–1.3 × 0.5–0.8mm, brown, hairy; petals narrowly oblanceolate, rounded, 0.9–1.2 × 0.8–0.9mm, brown, glabrous; anthers 3, c.0.3mm, exceeding filaments; staminodes 3; pistillode columnar, exceeding stamens. Female spikes in 2s or 3s, slender, bearing flowers in upper part. Female flowers: sepals lanceolate, acute, c.2 × 0.9mm, with scattered bristles; petals glabrous; ovary densely appressed reddish- or whitish-hairy. Capsules reflexed, oblong in outline, subtruncate, 2–2.4cm; wings 0.6–0.7cm wide; seeds 0.3 × 0.4cm, wing 1–1.3 × 0.5–0.6cm, oblong, basal.

FIG. 2. Dioscoreaceae II (leaves and fruits of **Dioscorea** spp.). a–c, **D. hispida** var. **daemona**: a, leaf from main stem; b, fruit; c, seed. d, **D. melanophyma**: leaf from main stem. e–g, **D. pentaphylla**: e, leaf from main stem; f, fruit; g, seed. h–i, **D. glabra**: h, fruit; i, seed. j, **D. alata**: fruit. k–l, **D. pubera**: k, fruit; l, seed. m, **D. hamiltonii**: fruit. n, **D. belophylla**: fruit. o–p, **D. deltoidea**: o, fruit; p, seed. q–r, **D. prazeri**: q, fruit; r, seed. Leaves × ¼; fruits and seeds × ½. Drawn by Sally Rae.

10

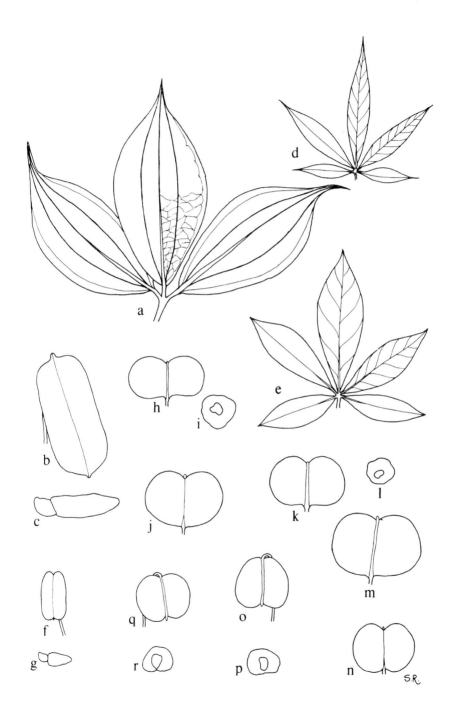

a

d

e

b

c

f

g

h

i

j

k

l

m

n

o

p

q

r

S.R.

Bhutan: S — Chukka district (Chukka, Gedu, Chasilakha, Putlibhir); **Terai** (Lower Tondu Forest); **Darjeeling** (Tukvar, Bakharikot, Tista, Lebong, Kodabari, Mongpu, Kalimpong, Mongpu to Sureil; Jalapahar (Mukherjee, 1988)); **Sikkim** (Dentam to Sangachelling, S slope of Pemionchi Hill (P & B)). 180–1520m. Fl. July–September; fr. December.

The specimen from Chukka (*Cooper* 1217, E, BM) is atypical, having long-pedicelled male flowers, glabrous sepals, small bracts and lax, elongate male spikes.

According to Prain & Burkill, Lepchas in Sikkim distinguish three varieties, all with tubers that are eaten: *suli bok* (Nep: *begur*), *kussok zyok* and *kussok ding*. The first they described as var. **suli** with an elongate tuber bearing few rootlets and white flesh, and white-hairy stems; the latter two they united under var. **kussok** having an ovoid-globose tuber bearing many surface rootlets and pale yellow flesh, and rusty red-pubescent stems.

Two forms appear to be distinguishable in the herbarium:

(1) Leaflets sessile, small, narrow, densely white hairy beneath.
(2) Leaflets petiolulate, very large, broad, subglabrous beneath.

From determinations by Prain & Burkill the former seem chiefly referable to var. *suli* and the latter to var. *kussok*. Intermediates occur and considerable confusion exists over the native names on the specimens (e.g. one of the type sheets of var. *suli* bears the name 'kussok' and one of the Gamble specimens labelled *sili kasok* has tubers conforming to var. *kussok*). It is not known whether this confusion is due to mixing of labels or genuine complexities in the wild. Records above have therefore been amalgamated. Further fieldwork is clearly needed bearing in mind the great variability of the species over its range.

The distinction between this and the following species is by no means clearcut.

6. D. kamoonensis Kunth ('*kumaonensis*'). Lep: *tukjhok* (P & B).
Differs from *D. pentaphylla* as follows: stems lacking prickles; male infls. usually of short (to 7cm), 3–5-clustered, axillary spikes; spikes and bracts with cream hairs, male flowers larger, sepals subglabrous.
Bhutan: C — Thimphu (Paro (M.F.B.)) and Tashigang (Tashi Yangtsi Valley) districts; **Terai** (Rungnoo); **Darjeeling** (Sureil, Pashok, Darjeeling to Simenbong (P & B)); **Sikkim** (Lachung Valley, Choongtam, Rungnoo Valley (P & B)). 'Hedges'; lower levels of temperate forest, 1070–2500m. Fl. August–September; fr. October.

The (male) flowers on the only Bhutanese specimen seen are heavily galled. Seldom eaten, according to Prain & Burkill.

7. D. melanophyma Prain & Burkill. Fig. 2d.

Differs from the previous two species as follows: stems glabrous; leaflets more narrowly elliptic, usually smaller (4–12 × 0.9–1.5(–2.2)cm), gradually narrowed to piliferous apex and the more abundant, small (pea-sized) bulbils.

Bhutan: C — Thimphu (Tsalimaphe, Paro) and Punakha (Punakha) districts. Apparently occurring in dry areas, 1350–3350m. Fl. August–September; fr. October.

Recorded for Sikkim in Naithani (1990), but no records in any of his cited references.

8. D. hispida Dennstedt var. **daemona** (Roxb.) Prain & Burkill; *D. daemona* Roxb. Lep: *rukloo, roklu, raklu-rick*. Fig. 2a–c.

[Tuber depressed-globose, often lobed, sparsely covered with small roots, skin straw-coloured to grey, flesh white to pale yellow]. Stem twining to left, lacking bulbils, sometimes prickly, shortly pubescent especially when young. Leaves trifoliolate, leaflets narrowly to broadly elliptic, rhombic or obovate, acuminate or cuspidate, 7.5–20.5 × 4–15cm, outer asymmetric, middle symmetric, 3-costate, glabrous or hairy above, sparsely hairy especially on veins beneath; petiolules 0.5–2cm, pubescent; petiole 5–19cm, pubescent, sometimes prickly. Male infls.: spikes borne in groups of 2–4 on pubescent axes 6–41cm, axes borne singly in axils, sometimes with short, basal branch; spikes 0.5–1 × 0.2cm, extremely dense, about equalling their peduncles; bracts brown, hairy, exceeding flowers. Male flowers globose, sepals c.0.8 × 0.7mm, membranous, hairy; petals subacute, 0.6–1 × 1mm, stout, glabrous, green; stamens 6, anthers c.0.2mm, slightly exceeding filaments. Female spikes borne singly or in pairs bearing flowers on upper part. Female flowers: sepals subacute, hairy, c.1.3mm, exceeding hairy petals; ovary densely pubescent. Capsules reflexed, oblong in outline, apex subtruncate, 4–4.7cm, glabrous, wings 1–1.3cm wide; seeds to 1cm, wing to 3cm, oblong, basal.

Bhutan: C — Punakha (Tinlegang) and Mongar (Shongar Chu above Lingmethang) districts; **Terai** (Garidura, Tondu Forest (W Duars)); **Darjeeling** (Great Rangit, Punkabari, Rungnoo, Bamunpokri, Lebong, Tista Valley, Rishap, Geille Ghora; Singtam Tea Estate (Mukherjee, 1988)); **Sikkim** (Pakyong to Rhenock (F.E.H.1), S of Pemionchi (P & B)). 150–1980m. Fl. July–November; fr. August–March.

Poisonous; eaten by Lepchas but must be soaked in running water for 7 days before cooking.

Prain & Burkill cite a Griffith specimen from Dewangiri (Deothang) but the specimen bears an 'E Himalaya' label and probably comes from Darjeeling. Only one *Dioscorea* is listed in Griffith (1848) and this probably refers to HEIC 5548 (a mixture of *D. pubera* and *D. alata*) which has a 'Bootan' label.

Small specimens are similar to *D. pentaphylla* — male plants distinguished by having 6 stamens; female by larger capsules; vegetatively by the 3-costate middle leaflet.

9. D. alata L. Lep: *bok*; *pem bok, pirieh bok* (P & B); Nep: *toral*. Fig. 1a, Fig. 2j.

Tubers — see below. Plant glabrous. Stems twining to right, commonly 4-angled, angles conspicuously wavy-winged, lacking prickles, bulbils apparently sometimes produced. Stem leaves mainly opposite, ovate or triangular-ovate, cuspidate, base cordate, basal lobes sometimes slightly angled, 10–19 × 6–13cm, subcoriaceous, glossy; leaves on flowering branches smaller, bases sometimes subtruncate; petioles about equalling blades, narrowly winged. Male infls. seldom produced. Female spikes axillary, single or in pairs on lateral branches, axis 12–34cm, stout, angled, flowers distant, borne almost to base. Female flower: sepals and petals similar, narrowly ovate, rounded, 1.8–2 × 1–1.5mm, brownish, margins hyaline; stigma c.0.7mm, stout. Capsules forwardly directed, apex shallowly retuse, 2–2.4cm; wings 1.5–1.8cm wide, half-obovate. Seeds not fully developed on specimens seen, surrounded by ± circular wing.

Bhutan: S — Phuntsholing district (Phuntsholing). [A Griffith Bhutan sheet bears this species and *D. pubera* and perhaps comes from Deothang]; **Darjeeling** (Mongpu, Peshok, below Barnesbeg); **Sikkim** (S of Legship, Singtam). Cultivated near houses, 300–750m.

Widely cultivated for its tubers in Old and New World tropics, probably originating in mainland SE Asia. Two forms occur in Sikkim according to Prain & Burkill: at lower altitudes (to 1830m) a cultivated form with short, globose tubers called *pem bok* by Lepchas; at higher altitudes (914–2743m) a wild form called *pirieh* (*piriyeh*) *bok* with long, thin, deeply buried tubers with a magenta-tinted superficial layer and unpleasant odour when cooked — this is also eaten and unusually for the species is evidently frost tolerant.

10. D. pubera Blume; *D. anguina* Roxb. Lep: *tusung buk, suong buk, zo-um-buk*; Nep (Sikkim): *panglang, pangla torul* (P & B). Fig. 1m, Fig. 2k–l.

[Tubers 1 or 2, narrowly cylindric, rootlets few, flesh lemon yellow]. Plant shortly pubescent throughout. Stem twining to right, lacking prickles, bulbils apparently sometimes produced. Leaves alternate to subopposite, ovate (occasionally suborbicular), cuspidate to shortly caudate, base shallowly cordate, margins cartilaginous, 8–16.5 × 5.2–11.5(–17)cm, persistently pubescent beneath, scattered hairy becoming glabrous above; petiole usually shorter than blade. Male spikes in groups of (1–)6–8 on leafless axes; axes 4.5–38cm, sometimes branched below, borne singly or in unequal pairs in axils; spikes 1–2.5cm, very dense, flowers borne to base; bracts minute. Male flowers brownish, globose, sessile; sepals ovate, subacute, keeled, c.1.2 × 0.8mm, hairy on outside; petals slightly smaller, glabrous; stamens 6; anthers c.0.3 × 0.2mm, slightly exceeding filaments. Female spikes single or paired, axillary or widely divergent from short, axillary axis; spikes bearing distant flowers throughout. Female flowers: sepals and petals shorter and wider than male; ovary densely pubescent; style stout, c.0.4mm. Capsules forwardly directed, apex retuse, 1.8–2cm, wings suborbicular, 1.5–1.8cm wide, pubescent in angles. Seeds surrounded by brown wing to 0.8cm wide, ± circular in outline.

Bhutan: S — Samchi district (Dhoankhola), also an unlocalised Griffith specimen possibly corresponding to the record from Dewangiri (Deothang (Griffith, 1848)); **Terai** (Dumdumura Jhar, Chenga, Phansidoora); **Darjeeling** (Great Rangit, Peshok, Darjeeling, Riang, Mongpu). 150–1070m. Fl. August–December; fr. December–January.

Lower (white) part of root eaten by Lepchas.

11. D. hamiltonii Hook.f. Lep: *pa-sok-book*; *pu-um-bok* (P & B); Nep: *jat torul*. Fig. 1b–c, Fig. 2m.

[Tuber deeply buried, long-stalked, narrowly cylindric (?c.2cm diameter), skin dark or black, flesh white]. Plant glabrous. Stem twining to right, 6-angled, drying reddish-brown, lacking prickles, bulbils apparently abundant. Leaves mostly opposite, lower ones ovate, cuspidate, base deeply cordate, to 21 × 12.5cm, upper and lateral ones deltoid, acuminate, cordate, basal lobes slightly angled, 6–11 × 3.5–6cm; petioles shorter than to equalling blades, angled. Male infl.: spikes borne in groups of 2–6 on leafless, unbranched axes; axes 7–20cm, borne singly or in unequal pairs, angled; spikes 1–2cm, axis zigzag, filiform, flowers distant. Male flowers globose; sepals ovate, concave, subacute to rounded, c.1 × 0.8–1mm, thick-textured, reddish-brown; petals smaller, oblanceolate, fused into cup below; anthers c.0.3 × 0.2mm, longer than filaments. Female spikes single or paired in axils, rather few-flowered. Capsules erect, apex retuse, c.2.9cm, pale brown; wings c.2.2cm wide, semi-circular, margin thickened. Seed surrounded by circular wing.

Bhutan: S — Gaylegphug district (Sham Khara and Maorey forests); **Terai** (Siliguri); **Darjeeling** (Mongpu, Riang). 200–1650m. Fl. August–October; fr. December.

Tuber collected from wild and highly prized by Lepchas.

12. D. glabra Roxb. Lep: *thamring book*; *chimeo bok, shimeo bok* (P & B). Fig. 1h–i, Fig. 2h–i.

[Tuber stalked, elongate, to 50 × 4cm, flesh white]. Plant glabrous. Stems terete, apparently prickly near base, bulbils not produced. Leaves opposite, upper oblong-elliptic to narrowly ovate, cuspidate or acuminate, base truncate or very shallowly cordate, 6–12 × 3.2–7cm, thin-textured, secondary veins inconspicuous, not raised beneath; lower larger, cordate with rounded basal lobes, to 16.5 × 9.5cm; petioles shorter than to almost equalling blade. Male infls.: spikes in groups of 4–7 on slender axes; axes single or paired in axils, 9–31cm; spikes slender, to 2cm, flowers distant, sessile; bracts minute. Male flowers: petals and sepals erect, sepals triangular-ovate, 1.2–1.5 × 1.1–1.2mm, brown, margins narrowly scarious; petals oblanceolate, 1.3–1.5 × 0.7mm, thick-textured; anthers c.0.3mm, about equalling filaments. Female spikes single or paired, bearing flowers throughout. Female flowers like male; style very short (c.0.2mm). Capsules spreading to erect, apex shallowly retuse, 1.7–2cm; wings transversely elongate, 1.6–1.8cm. Seeds surrounded by brown wing.

Bhutan: S — Phuntsholing district (Phuntsholing (M.F.B.)); **Terai** (Muraghat Forest, Jalpaiguri, Puhar Goomgoomia); **Darjeeling** (Tista, Birick, Kalimpong, Rebong, Kuman, ?Kurthjnem). 305–610m. Fl. August–November; fr. November–December.

13. D. belophylla Voigt ex Haines; *D. glabra* sensu F.B.I. p.p.; *D. lepcharum* Prain & Burkill (at least p.p.). Lep: *singol bok, kacheo bok; pazok bok* (P & B); Nep (Sikkim): *panu torul, ghita torul* (P & B); Bhutea: *nachray kyn* (P & B). Fig. 1f–g, Fig. 2n.

Differs from *D. glabra* in having at least the larger leaves sagittate with angled basal lobes, upper leaves lanceolate, leaves all thicker textured, with secondary veins prominent beneath forming a ladder-like pattern about the midrib; bulbils apparently numerous; male infl. stouter and often more contracted (axis sometimes not developed so spikes axillary), male flowers larger; capsule with narrower wings so more obovate in outline.

Bhutan: S — Phuntsholing (Phuntsholing, Sorchen) and Chukka (Chukka Colony to Taktichu) districts; **Terai** (Jalpaiguri Duars (P & B)); **Darjeeling** (Little Rangit, Mongpu, Silake, Riang, Labdah); **Sikkim** (Kabi; Rishi to Rinchinping (P & B)). *Shorea* plantation; scrub in disturbed subtropical forest, 200–1900m. Fl. October–December; fr. to January.

Tuber good to eat, but deeply buried. Probably only a variety of *D. glabra* — the local names also suggest that species 12 and 13 are not always distinguished.

Doubtful and doubtfully recorded species:

D. lepcharum Prain & Burkill
The distinctness of this taxon is doubtful: many syntypes are given in the original description most of which can be referred to *D. belophylla*, though some to *D. glabra*. In the type cover at Kew are three specimens from Sikkim

(one of which — *Gage* 34207 — selected as lectotype by Jayasuryia, though this seems not yet to be published) all of which are *D. belophylla*. One specimen from the Abor Hills (*Burkill* 35905, labelled by Burkill '*D. lepcharum* var. *bhamoica*') is *D. glabra*. The two specimens in the type cover of var. *bhamoica* are probably *D. belophylla*. It is hard to understand why Prain & Burkill described this species since vernacular names on at least the Sikkim specimens agree with those they published for *D. belophylla*; none bear the Lepcha name *chimeo tendeo bok* which they give for *D. lepcharum*. Jayasuryia (on determinavit slips) misidentified all the sheets mentioned above as *D. hamiltonii* but none of them have the characteristic stems and zigzag infl. of that species.

D. wattii Prain & Burkill
   Characterised by having very coriaceous leaves with distinctive thickened cartilaginous margins; upper leaves narrowly lanceolate, bases cuneate to rounded; capsules larger than any of above species having wings 2.6–2.8cm wide, c.3cm long and a truncate apex. Fl. March–April. No specimens seen but recorded from Darjeeling (Tassiding and Rishap Jhora) by Prain & Burkill who give the Lepcha name *palam bok*, which they say is also used for *D. hamiltonii*.

Family 189. TACCACEAE

S.J. Rae

Perennial, rhizomatous herbs. Leaves in rosette at apex of rhizome, with entire or pinnately dissected, pinnately veined, glabrous blade and long, erect petiole with sheathing base. Infl. umbellate, borne on a leafless scape, subtended by (2–)4(–12) showy, herbaceous bracts, usually in 2 whorls. Floral bracts long, filiform, falling after anthesis. Flowers actinomorphic, bisexual, campanulate with 6 perianth lobes in 2 whorls, united into tube below. Stamens 6, lower part of filaments fused to perianth tube, upper part free, helmet-shaped, with anthers adnate to inner face. Ovary inferior, obpyramidal, 6-ribbed, 1-celled, ovules numerous, on 3 parietal placentae; style short, stigma peltate, with 3 obcordate lobes. Fruit berry-like, many-seeded, disintegrating irregularly.

1. TACCA J.R. & G. Forster

Description as for family.

1. T. integrifolia Ker Gawler; *T. laevis* Roxb. Fig. 3f.
   Rhizome cylindric, to 3cm diameter. Leaves 2–13, blades commonly oblong or lanceolate, (7.5–)40–72 × (3–)11–32cm, acuminate to acute apex, base

17

attenuate; petiole (4.5–)18–57cm. Scape (9–)24–48(–100)cm. Infl. (1–)5–12(–30)-flowered. Outer involucral bracts 2, opposite, sessile, commonly elliptic, acute, (1.5–)6.5–9.5(–14) × (0.5–)2.5–4(–7)cm, yellowish-green to purple with darker veins; inner bracts 2, inserted in axil of one of outer bracts, blade ovate, acuminate to 13.5 × 7cm, narrowed below into petiole-like base 2.5–5.5cm, cream flushed purplish at base, veins purplish. Floral bracts to 38 × 0.1cm, yellowish-green, darker at base. Pedicels (0.5–)2–4cm. Flowers blackish-purple; perianth lobes reflexed during anthesis, to 1.4cm, outer elliptic, to 0.8cm wide, inner broadly obovate, to 1.2cm wide; perianth tube subglobose, to 5(–8) × 9(–15)mm. Lower part of filaments to c.2mm, upper part to c.3mm. Ovary to 8(–15) × 3(–7)mm; style to c.2mm; stigma to c.3.5mm diameter. Fruit 2.5–5 × 1–2.5cm, triangular to circular in section. Seeds 3.5–6mm, ovoid, ribbed.

Bhutan: S — Deothang district (Deo River), apparently quite frequent in S Bhutan (teste U.C. Pradhan); **C** — Tongsa district (near Mamoun, Shemgang (Nakao & Nishioka, 1984, as *T. chantrieri* André)). Stream and river banks in subtropical forest, 300–500m. Fl. February–September.

Family 190. TRILLIACEAE

Rhizomatous perennials. Stem unbranched bearing a whorl of 3–6 simple leaves. Flowers terminal, solitary, 3–10-merous, tepals in two whorls, all similar or outer sepaloid and inner either petaloid or filiform. Stamens equal in number to tepals; anthers basifixed, dehiscing inwards or outwards; connective sometimes prolonged upwards. Ovary superior, of 3–6(–8) locules or ± unilocular, apex sometimes with thickened rim; style divided at apex into stigmatic lobes. Fruit a berry or capsule. Seeds sometimes with fleshy sarcotesta.

FIG. 3. **Dioscoreaceae, Taccaceae, Trilliaceae, Smilacaceae.**
a, **Dioscorea sp.**: stem twining to left. b, **Dioscorea sp.**: stem twining to right. c–e, **D. bulbifera**: c, habit with male infl. (× ⅓); d, female half flower (× 5); e, male half flower (× 5). f, **Tacca integrifolia**: habit (× ⅙). g, **Trillium tschonoskii** var. **himalaicum**: half flower (× 1.5). h, **Trillidium govanianum**: half flower (× 1.5). i, **Paris polyphylla** var. **appendiculata**: stamen (× 2). j–k, **P. violacea**: j, habit (× ½); k, stamen (× 2). l, **Smilax lanceifolia**: junction of leaf and peduncle showing prophyllate infl. (× 1). m–p, **S. menispermoidea**: m, habit (× ⅓); n, junction of leaf and peduncle showing eprophyllate infl (× 1); o, female half flower (× 5); p, male half flower (× 5). q–r, **Heterosmilax japonica**: q, male flower; r, male half flower (both × 5). Drawn by Mary Bates.

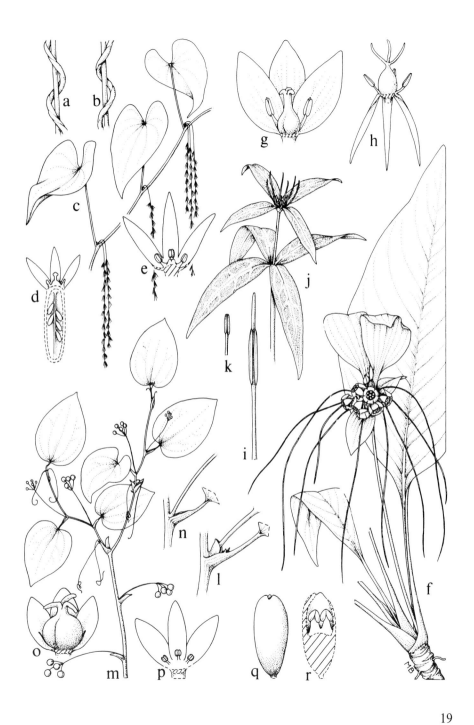

1. Leaves more than 3 per whorl; sepals 4 or more, herbaceous; petals filiform, equalling number of sepals; fruit a capsule; seeds with scarlet sarcotesta .... **3. Paris**
+ Leaves 3 per whorl; sepals 3; petals 3, similar or different from sepals; fruit a berry; seeds without scarlet sarcotesta 2

2. Sepals and petals ± similar, linear, reddish-purple; leaves lanceolate to ovate .... **2. Trillidium**
+ Sepals herbaceous; petals white or purplish, elliptic; leaves rounded-rhombic .... **1. Trillium**

1. TRILLIUM L.

Perennial herbs with thick, horizontal rhizomes. Stems simple, 1 or more produced from apex of rhizome, with basal, sheathing, membranous scale leaves. Leaves in a whorl of 3 near top of stem, simple, usually sessile, with 3–5 main veins which branch reticulately. Flower single, terminal, sessile or pedicellate. Sepals 3, green, persistent. Petals 3, commonly white or purplish. Stamens 6, attached at base of tepals; anthers basifixed, linear, dehiscing inwards, connective prolonged upwards. Ovary superior, 3-loculed, each locule with many ovules; stigma lobes 3. Fruit a globose berry.

1. T. tschonoskii Maximowicz var. **himalaicum** Hara. Fig. 3g.
Stem 16–32cm, stout, erect; scale leaves pale brown. Leaves sessile, rounded-rhombic, acuminate, 6.5–11.5 × 4.8–12.5cm, main veins 3. Pedicel at flowering 3.5–5mm, curved, lengthening in fruit to 1.6cm. Sepals narrowly elliptic, slightly contracted below subacute apex, margins very narrowly membranous, 1.4–2.6 × 0.5–0.9cm. Petals persistent (becoming damaged), elliptic, slightly contracted below subacute apex, slightly shorter than or equalling sepals, 1.5–2.2 × 0.7–1cm, white or purplish. Filaments 3.5–4.9mm, flattened; anthers 3–4mm, minutely apiculate. Ovary 5–7mm diameter, ovoid, narrowed above, with 6 longitudinal ridges; stigma lobes c.3mm, recurved. Berry 0.8–1.2cm.
Bhutan: C — Ha (Batte Dzong), Thimphu/Punakha (Dochu La), Tongsa (Pele La, Lamse La), Bumthang (above Lami Gompa), Mongar (Pimi, below Sengor, Donga La — W side) and Tashigang (Donga La — E side) districts; **Sikkim** (Lachen, Rookah, Lachung). Shady forest (incl. *Picea*, *Tsuga*, mixed and deciduous), 2290–3200m. Fl. April–May.

The record of *T. obovatum* Pursh (Chung Thang to Lachen (Rao, 1964b)) almost certainly refers to the white-flowered form of *T. tschonoskii*.

2. TRILLIDIUM Kunth

Differs from *Trillium* in having 6 ± similar, linear, reddish-purple tepals; anthers dehiscing outwards and stigma lobes erect, filiform.

1. T. govanianum (Wall. ex D.Don) Kunth; *Trillium govanianum* Wall. ex D. Don. Fig. 3h.

Stem 5.5–24cm, erect; scale leaves pale brown. Leaves lanceolate to ovate, acuminate, base truncate to shallowly cordate, 4.1–7.1 × 2.6–4.1cm, main veins 3; petioles 0.2–1cm. Pedicel 0.6–1.2cm, ± erect, straight, not greatly lengthening in fruit. Tepals reflexed, ± similar (outer 3 sometimes slightly wider), linear, acute, 0.9–1.5cm × 0.9–3.1mm, greenish or reddish- to blackish-purple. Filaments 1.5–2.2mm; anthers 1.5–3mm, apiculate. Ovary narrowly ellipsoid to narrowly ovoid, smooth, 3.1–6.5 × 2–3.5mm; stigma lobes 3–6.5mm. Berry ellipsoid, c.1.1 × 0.8cm, dark purple. Seeds c.2mm.

Bhutan: C — Ha (Tseli La), Thimphu (Olaka; Barshong to Nala (F.E.H.2)), Tongsa (Tunle La), Mongar (below Sengor) and Tashigang (E side of Donga La) districts; **N** — Upper Mo Chu district (near Kohina; Gasa to Pari La (F.E.H.2)); **Darjeeling** (Sandakphu; Phalut to Chia Bhanjang (F.E.H.1)); **Sikkim** (Lachen; Migothang to Nayathang (F.E.H.1)). Banks and streamsides in forest (*Abies*, *Tsuga*, rhododendron, bamboo), 2900–3960m. Fl. April–June.

3. PARIS L.

Perennial herbs; rhizomes thick, bearing annular scars. Stem erect, with basal chaffy scale leaves and whorl of 4–12 narrow, petiolate or sessile leaves in upper part. Flower single, terminal, pedicellate. Number of floral parts inconstant, even in a single population. Sepals 4–10, herbaceous, persistent. Petals usually same number as sepals, filiform, yellowish, greenish or purplish. Stamens usually twice number of sepals; filaments attached at base of sepals and petals; anthers with two lateral locules, connective often prolonged as an apiculus. Ovary superior, oblong-globose, angled, truncate with thickened rim at apex, ovules many, parietal; style thick, divided into stigmatic lobes at apex. Fruit a fleshy loculicidal capsule, seeds with fleshy, scarlet sarcotesta.

The *Paris polyphylla* complex is extremely difficult to deal with taxonomically. The following account largely follows Hara (1969), who was aware of the difficulties and knew the plants in the field. Although Takhtajan (1983) was probably correct in separating *Daiswa* from *Paris*, his account is not satisfactory and fails to deal adequately with the variation observed even within single populations, misleadingly dividing the genus into a series of species, which have reality only on paper. Varietal rank seems far more appropriate for the most distinct nodes of variation.

1. Leaves without white markings along veins, margins not wavy; anthers over 3.8mm, connective developed into apiculus at least 0.4mm
                                                                  1. P. polyphylla s.l.
+   Leaves variegated with whitish markings along veins, margins wavy; anthers under 2.5mm, connective not developed .... **2. P. violacea**

1. P. polyphylla Smith var. **polyphylla**. ?Hindi: *patar ko*; Nep: *satuwa*.
Flower stem 16–56cm (probably sometimes longer). Leaves (5–)6–9(–13), narrowly oblanceolate to narrowly elliptic, finely acuminate, base cuneate, 7–13.7 × (1.1–)1.4–3.5cm, glabrous; petioles 0.3–1.5(–2)cm. Pedicel 0.9–5.5cm (to 7.4cm in fruit), erect. Sepals (3–)4(–6), lanceolate, acuminate, 3.6–6.5 × 0.8–1.6(–2.4)cm. Petals equalling number of sepals, (½–)⅔ to longer than sepals, filiform, yellowish or greenish. Stamens ± twice number of sepals; filaments 3.8–6mm; anthers 3.8–8mm, connective apiculus 0.4–1mm. Ovary c.5 × 6mm; style 1.7–4.5mm; stigma lobes usually 4, 2–3mm, recurved at tips.
Bhutan: C — Ha, Thimphu, Punakha, Tongsa, Bumthang and Mongar districts; **N** — Upper Mo Chu district; **Sikkim** (Lachen, Chungtham); **Darjeeling** (Senchal, Tonglu, Ghum, Kalapokri). Broad-leaved (incl. *Quercus*) and *Tsuga* forest and among scrub, 1300–3960m. Fl. April–June.

The above description refers to plants matching the type (from Nepal); the plant is, however, very variable, with local populations (and plants within a population, to judge from herbarium sheets) differing greatly in morphology.

The following are the chief variants:

var. **stenophylla** Franchet; *P. lancifolia* Hayata.
    Differs from var. *polyphylla* in its more numerous (10–15), sessile, very narrow (under 1.3cm wide) leaves.
    Bhutan: C — Mongar district (Sengor, 2590m, Donga La — W side, 2286m); **N** — Upper Mo Chu district (Tamji to Gasa, 2200–2500m (F.E.H.2)); **Sikkim** (Praig Chu, 2500m (F.E.H.2)).

It does not seem possible to maintain this variety: there seems to be little correlation between leaf number and width, for example *Grierson & Long* 4482 (E) consists of two specimens, one typical var. *polyphylla*, the other tending towards var. *stenophylla* in having 10 leaves, 1.2cm wide; *Ludlow, Sherriff & Hicks* 20177 — the specimen at E fits the description, but a duplicate at BM has 10–11 leaves to 1.9cm wide; *Griffith* 915 at K bears two specimens on a single sheet — one is typical var. *polyphylla*, the other has 11 leaves 0.87–1.25cm wide.

var. **wallichii** Hara. Plate 1.

Differs from var. *polyphylla* in its mature leaves being much wider (3.5–6cm), oblong-oblanceolate, with more truncate bases and petioles much longer (1.2–4cm) with respect to the lamina.

Bhutan: S — Chukka district (Bunakha); **C** — Mongar district (Sengor); **Darjeeling** (Patasi, Ghumpahar, Gohr, Rungbool); **Sikkim** (Relli Chu; Yoksum (F.E.H.3)). 1520–2590m.

Extremely large plants (to 1.2m with leaves to 8.5cm wide) from Darjeeling (Rhikisum, 1950m) and Bhutan (Gaylegphug district: Gale Chu, 1675m) resemble large Chinese forms vegetatively but differ from var. *chinensis* (Franchet) Hara in having petals much longer than sepals and are possibly just forms of var. *wallichii*.

var. **pubescens** Handel-Mazzetti.

Differs from var. *polyphylla* in the papillose-ciliate margins of the leaves and sepals, and the veins on the undersides of the leaves and sepals being densely papillose-hairy.

Bhutan: C — Thimphu/Punakha district (Jato La, 3048m).

Previously only recorded from China. This single gathering from Bhutan consists of two specimens, one of which is very large with 10 leaves and 6 enormous (10.3 × 2cm) sepals; the other resembles var. *polyphylla* except for its hairiness. The duplicate at BM, however, is not hairy and is referable to var. *polyphylla*.

var. **appendiculata** Hara. Fig. 3i.

Differs from var. *polyphylla* in having long apiculi to the anther connectives (at least 3.5mm and often equalling or even exceeding the anthers).

Bhutan: C — Thimphu district (Thimphu to Chimakothi, 2200–2400m (F.E.H.2)); **Sikkim** (above Tsoka Village; Bakkim to Dzongri, Migothang to Nayathang, 2500–4000m (F.E.H.1)).

Included by Takhtajan under *Daiswa thibetica* (Franchet) Takhtajan, though with a query and not included by him in the distribution of this species, which is more eastern and has very much longer ((6–)11–17.5mm) anther apiculi.

var. *brachystemon* Franchet seems to be part of var. *polyphylla* as treated by Hara latterly (E.F.N.). Takhtajan, however, places it under *D. lancifolia*.

2. P. violacea Léveillé; *Paris marmorata* Stearn; *Paris polyphylla* Smith subsp. *marmorata* (Stearn) Hara. Fig. 3j–k.

Differs from *P. polyphylla* in being usually smaller (8–15cm), with fewer (4–6), smaller (5.5–7.4 × 1–2.1cm), variegated leaves with whitish markings along the veins, margins wavy, upper surface dark green, lower surface dark purplish; flowers with 3–4 smaller (2–3.3 × 0.5–1cm) sepals, shorter filaments (2.5–3mm), smaller anthers (1.5–2.5mm) and connective not produced into apiculus.

Bhutan: C — Thimphu (Drugye Dzong, Dechhenphu, Dechencholing to Punakha), Thimphu/Punakha (Sinchu La), Punakha (E of Dochu La) and Punakha/Tongsa (Pele La (F.E.H.2)) districts; **N** — Upper Mo Chu district

(Gasa to Pari La to Chamsa (F.E.H.2)); **Darjeeling** (Ramam; Rimbick to Ramam (F.E.H.2)). Damp mossy forest (incl. broad-leaved, evergreen and blue pine), 1900–3500m. Fl. April–May.

Family 191. SMILACACEAE

Dioecious shrubs or semi-woody climbers. Roots tuberous. Stems often with stout recurved prickles. Leaves alternate, simple, often coriaceous, with prominent parallel veins (costae) linked by weaker reticulate veins. Petioles commonly persistent, winged in lower part, with tendrils arising from apex of wings. Infls. of pedunculate umbels, borne singly or in racemes in axils of leaves, sometimes subtended at base by a prophyll between base of peduncle and main stem. Pedicels arising from swollen 'receptacle' at end of peduncle which may bear bracteoles. Flowers unisexual, of 2 slightly differentiated whorls each of 3 free tepals, or tubular with 3 apical lobes. Male flowers usually with 6 stamens, filaments free or fused. Ovary superior, trilocular. Fruit a berry.

1. Flowers with 6 free tepals; if peduncle compressed then combination of leaf characters not as below .... **1. Smilax**
+ Flower tubular with 3 apical lobes; peduncle compressed; mature leaves large (over 15cm), semi-herbaceous, slightly shining, petiole ± terete, c.2mm diameter, over 3cm long .... **2. Heterosmilax**

1. SMILAX L.

Description as for family: flowers with 6 tepals free to base; male flowers with filaments free to base.

The name *kukoor-zoo* (Sha) refers to a species of *Smilax*.

Fɪɢ. 4. **Smilacaceae** I (leaves and petioles of **Smilax** spp.).
a–b, **S. prolifera**: a, leaf from main stem; b, petiole. c–d, **S. perfoliata**: c, leaf from main stem; d, petiole. e–f, **S. orthoptera**: e, leaf from main stem; f, petiole. g–h, **S. wallichii**: g, leaf from main stem; h, petiole. i–j, **S. aspericaulis**: i, leaf from main stem; j, petiole. k–l, **S. ovalifolia**: k, leaf from main stem; l, petiole. m–n, **S. lanceifolia**: m, leaf from main stem; n, petiole. All × ¼. Drawn by Sally Rae.

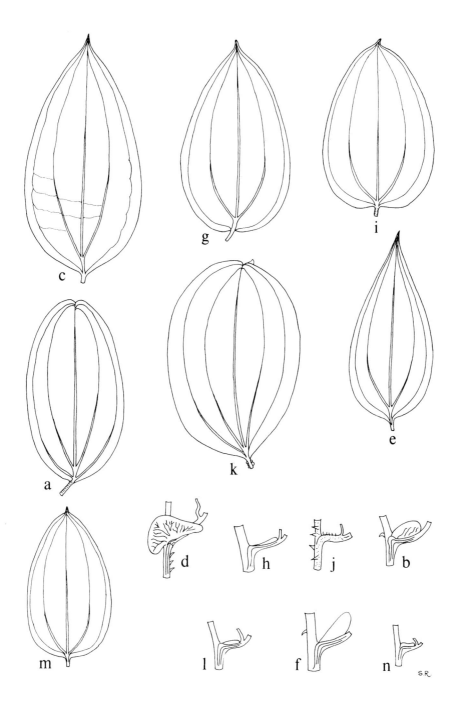

c

g

i

a

k

e

m

d

h

j

b

l

f

n

S.R.

25

1. Infl. a raceme of sessile umbels; leaves usually triangular or hastate
with rounded basal lobes .... **1. S. aspera**
+ Infl. of 1 or more peduncled umbels; leaves without expanded basal
lobes 2

2. Prophyll (bract-like structure) present between base of infl. axis and
stem (Fig. 3l) 3
+ Prophyll not present at base of infl. axis (Fig. 3n), though a scale
(bract) sometimes present between infl. axis and petiole of subtending leaf 9

3. Wings of petiole conspicuously expanded, bases clasping stem 4
+ Wings of petiole not clasping at base 5

4. Infl. of more than 3 umbels arranged in whorls of 2 or 3 along an
elongate axis .... **2. S. prolifera**
+ Infl. of 3 or fewer umbels inserted singly along infl. axis   **3. S. perfoliata**

5. Stems densely granulate with many small warts and bristles, recurved
prickles present or absent .... **6. S. aspericaulis**
+ Stems not granulate, smooth or furrowed, recurved prickles usually
present 6

6. Petiole wings erect, triangular .... **4. S. orthoptera**
+ Petiole wings very narrow, oblong 7

7. Usually more than 1 umbel per infl. axis; leaves broadly ovate-elliptic
7. S. ovalifolia
+ Umbels borne singly; leaves narrowly ovate 8

8. Peduncle longer than subtending petiole; tepals over 8mm; leaves
drying greyish .... **5. S. wallichii**
+ Peduncle shorter than or equalling subtending petiole; tepals under
5mm; leaves drying dark green .... **8. S. lanceifolia**

9. Leaves ± sessile, crowded, ovate to orbicular, small (under 3cm);
spiny shrub .... **9. S. myrtillus** var. **rigida**
+ Leaves petiolate, variously shaped, if under 3cm then a non-spiny
climber 10

10. Peduncle strongly compressed, shorter than or just exceeding subtending petiole; receptacle with conspicuous, lanceolate bracteoles c.3mm .... **13. S. elegans** subsp. **subrecta**
+ Peduncle less obviously compressed, conspicuously exceeding subtending petiole; bracteoles filiform, under 1mm 11

11. Petiole wings pale, semi-elliptic; plant strongly armed; leaves immature at flowering .... **10. S. ferox**
+ Petiole wings very narrowly oblong or gradually tapered into petiole above; plant unarmed; leaves mature at flowering 12

12. Plant without tendrils, even on old persistent petioles; point of leaf abscission between top of petiole wing and base of blade; apex of petiole straight, not expanded .... **11. S. minutiflora**
+ Plant with tendrils (which may be extremely small on young leaves, but conspicuous on old persistent petioles); leaf abscission from base of blade; apex of petiole often hooked, sometimes expanded 13

13. Petiole wings gradually tapered from base upwards; leaves ovate, not glaucous beneath; lateral branches not strongly zigzag
.... **12. S. menispermoidea**
+ Petiole wings narrowly oblong; leaves lanceolate, glaucous beneath; lateral branches strongly zigzag .... **13. S. elegans** subsp. **elegans**

1. S. aspera L. Fig. 5a–b.

Climber; stems zigzag, ridged, prickly (sometimes sparsely). Leaves triangular to hastate with rounded basal lobes (lobes occasionally not developed so leaves lanceolate or ovate), acute or mucronate, margins sometimes minutely spiny, base truncate to weakly cordate, 6–9(–12) × 2.2–5.3(–7.5)cm, costae 5–7(–9), midrib sometimes prickly on underside, coriaceous. Petioles 1.6–2.5cm, sometimes prickly, extreme basal part (to 5mm) with very narrow wing; tendrils strongly developed; abscission from base of blade. Infl. a zigzag raceme 4–9cm of up to 8(–12) alternate, sessile umbels, racemes arising from prophylls in leaf axils of lateral branches terminated by an infl. Female umbels 2–3-flowered; receptacles globose, c.2mm diameter; bracteoles very short, brown; pedicels 3–4mm; flowers smaller than male, ovary ellipsoid, c.1.5mm long, stigmas 3, very short, staminodes 6. Male umbels 8–16-flowered; pedicels to 6.5mm; flower buds clavate, tepals reflexed at maturity, cream, outer narrowly oblanceolate, 3.9–5.2 × 0.8–1.4mm, inner linear 0.4–1mm wide, filaments 1.5–2.8mm, anthers oblong to 1.4 × 0.7mm. Berries red, 0.5–0.7cm diameter, 3-seeded; seeds c.3.5mm diameter.

Bhutan: C — ?Thimphu (Thimphu to Tinlegang), Punakha (Lometsawa, Punakha, Wangdi Phodrang to Samtengang, Tinlegang; Bhotoka to Rinchu (F.E.H.2)) and Tashigang (Tashigang, Jangphu) districts; **Sikkim** (Tingling). Among shrubs on dry hillsides, 1220–2100m. Fl. October–November; fr. April–August.

2. S. prolifera Roxb.; *S. perfoliata* sensu F.E.H.1 non Loureiro. Nep: *kukurdainyi*. Fig. 4a–b.
Large climber; stems weakly striate, prickly. Leaves elliptic or oblong-ovate, usually blunt or retuse (due to damage?), sometimes shortly cuspidate, base rounded, 11–17.5 × 6.5–11.5cm, costae 5 + 2 weak marginals, middle 3 fused at base for 1.5–2cm, coriaceous. Petiole 2–4.4cm, winged for lower ⅓–½, wings oblong to half-ovate, 0.35–0.7cm wide, brown, papery, weakly reticulate-veined, slightly clasping to amplexicaul at base, disintegrating in older leaves, upper part of petiole transversely wrinkled on upper surface at point of abscission; tendrils usually developed. Infl. to 11cm with up to 6 whorls of 2 or 3 long-peduncled (to 4cm) umbels, each whorl with a small brown ovate bract, infl. arising from brown, coriaceous, ovate prophyll c.1cm in axils of leaves of lateral branches. Female umbels 14–18-flowered; tepals similar to male, ovary 1.8mm; stigmas 3, c.2.5mm, spirally recurved, staminodes 3. Male umbels 23–26-flowered; receptacle globose, c.2.5mm diameter; bracteoles short, brown, mucronate; flower buds clavate, tepals reflexed at maturity, outer oblong c.4.7 × 1.4mm, inner linear c.0.5mm wide, filaments 3.2–3.6mm, anthers lanceolate, cuneate at base c.1.9 × 0.6mm. Berry red, c.0.5cm diameter.
Terai (Titalya, Apalachaud and Muraghat Forests); **Darjeeling** (Mahanadi, Tukvar, Sivoke, Raja Bhat Khawa, Datiwan, Tista Valley; Peshok (F.E.H.1)). Mixed Plains, Lower and Middle Hill Forest, 300–1830m. Fl. February–May; fr. September–November.

3. S. perfoliata Loureiro; *S. roxburghiana* Wall. ex Hook.f.; *S. ocreata* A.DC. Fig. 4c–d.
Climber; stems usually prickly. Leaves lanceolate, cuspidate (occasionally damaged and appearing retuse), base rounded to cuneate, occasionally shallowly cordate, 11.5–18.5 × 6–7.5(–14)cm, costae 5 + 2 weak marginals, 3 central fused at base for 0.8–1cm, coriaceous. Petiole 1.9–3.1cm, winged in lower half, wing massive (0.5–1.5cm wide), half-ovate, brown, coriaceous, reticulately veined, strongly amplexicaul at base; tendrils usually produced. Infl. of 1–2(–3) peduncled, bracteate umbels, axis sometimes developing vegetatively above, infl. arising from brown, coriaceous, ovate, keeled, mucronate prophyll c.0.9cm in axils of leaves of lateral branches. Female umbels 10–24-flowered; receptacle globose or elongate, 3–5mm diameter; without bracteoles at fruiting; flowers smaller than male, ovary c.2.5mm, stigmas short,

staminodes 3. Male umbels more than 30-flowered; receptacle globose, c.3.5mm diameter; bracteoles linear, brown; pedicels c.0.5cm; flower buds oblong, tepals pale yellow, reflexed at maturity, outer 5.8(–8.0) × 1.8(–2.1)mm, inner c.0.8mm wide, filaments c.4mm, anthers c.1.2 × 0.6mm. Berries c.0.9mm diameter, green when young.

Bhutan: C — Tongsa (S of Shamgong, Tama) and Mongar (below Mongar) districts; **Darjeeling** (Rongsong, Darjeeling, Kalimpong, Rangit, Ging, Passan); **Sikkim** (Yoksum; Gangtok (F.E.H.2)). Broad-leaved (incl. *Quercus griffithii*) forest, 300–2000m. Fl. March–May; fr. June–August.

4. S. orthoptera A.DC. Nep: *kukurdina*. Fig. 4e–f.

Differs from *S. perfoliata* in its erect, triangular petiole wings which are not clasping at the base.

Bhutan: S — Sarbhang district (W of Singi Khola); **Duars** (Santrawari Road, Buxa); **Darjeeling** (Dumsong, Rungdung, Peshok). Subtropical forest, 150–900m. Fl. March–May.

Further work is needed on variability of development of the petiole wing of the previous two species to ascertain whether or not they are specifically distinct.

5. S. wallichii Kunth. Fig. 4g–h.

Differs from *S. perfoliata* in its non-clasping petiole wings. Differs from *S. orthoptera* in its narrower (to 4mm wide), oblong to half-lanceolate petiole wings; infls. always borne singly; peduncles stouter and longer; receptacles elongate; umbels larger (female 35–80-flowered; male c.45+ flowered); flowers larger (female: outer tepals c.6 × 2mm, inner c.1mm wide, ovary c.4 × 2.2mm; male: outer tepals 8.5–9.7 × 2.1–2.5mm, inner tepals c.1mm wide, filaments longer, c.9.1–9.3mm).

Darjeeling (Rangirun, Kurseong, Ambyok); **Sikkim** (Mamring). Subtropical forest, 610–1830m. Fl. February–April.

6. S. aspericaulis Wall. ex A.DC.; *S. bracteata* Presl subsp. *verruculosa* (Merrill) Koyama. Nep: *dathun*. Fig. 4i–j.

Climber; stems densely granulate with rounded warts and extremely short (under 0.5mm) bristles, recurved prickles also sometimes present. Leaves usually oblong-lanceolate (occasionally ovate), cuspidate, margins narrowly cartilaginous, base rounded, 11–19.5 × 3.5–10cm, costae (3–)5, free from base or fused for lower 0.8cm. Petiole 1.1–2cm, lower ⅓ verrucose and bearing very narrow (under 1mm) oblong wing; tendrils usually strong; abscission between apex of wing and base of blade. Infl. of 1–2(–3) peduncled, bracteate umbels arising from small (3–6mm) ovate, apiculate prophylls in axils of mature leaves on lateral branches. Female umbels 19–22-flowered; receptacle globose, c.2mm diameter; without bracteoles; pedicels c.0.7cm, flowers smaller than male, outer tepals c.5 × 1.5mm, ovary ellipsoid, c.2.5 × 2mm, stigma lobes erect and recurved, staminodes 3. Male umbels 33–42-flowered; receptacle globose, c.1.5mm diameter; bracteoles usually not conspicuous; pedicels c.0.8cm, flower buds oblong, tepals reflexed at maturity, outer 6.4–7 × 1.1–1.7mm, reddish, inner 0.5–0.7mm wide, cream, filaments 4–5.7(–6.5)mm, anthers 1.5–1.9 × 0.5–0.6mm.

Bhutan: S — Samchi (Changda Chungli Hill), Phuntsholing (Kamji (M.F.B.)) and Chukka (Gedu to Ganglakha) districts; **Duars** (top of Leesh Block, British Bhutan); **Darjeeling** (Takdah, Birch Hill, Kurseong, Rungbee, Sureil, Mumsong, Rimbick to Palmajua, Sittong, Kalimpong, Gopaldara, Mangwa, W of Mongpu); **Sikkim** (Gangtok, Yoksum to Mintok Khola). Evergreen oak forest, 300–2130m. Fl. October–December; fr. May.

7. S. ovalifolia Roxb.; *S. macrophylla* Roxb. Nep: *kukur dyoti*. Fig. 4k–l.

Climber; stems terete or slightly ridged, sparsely prickly. Mature leaves ovate to orbicular, abruptly contracted to cuspidate apex, base cuneate to rounded, old leaves sometimes slightly cordate, 14–18 × 9.5–18cm, costae 5–7 + 2 weak marginals, sometimes sharply raised on underside, free from base, coriaceous. Petiole 2.3–2.7cm, upper part often with margins waved on upper surface, winged for lower ⅓, wing oblong, narrow, c.0.5mm; tendrils strong; abscission from just above tip of wings. Infl. of 2(–3) bracteate, pedunculate umbels, arising from brown, ovate prophylls to 0.7cm. Female umbels c.14-flowered; peduncles stout (especially in fruit); pedicels c.1.2cm; receptacle globose, large (to 5mm diameter) with conspicuous bracteoles to 1.3mm, flowers not seen. Male umbels c.20-flowered; pedicels c.4mm; receptacles globose, c.3mm diameter, with brown bracteoles, flower buds oblong, tepals reflexed at maturity, outer hooded 5.7–6 × 1.5–1.9mm, inner 0.6–0.8mm wide, filaments 5.3–5.4mm, anthers linear 1.2–1.5 × 0.3–0.6mm. Berry red, c.0.7cm diameter.

Terai (Mohurgong, Sukna); **Darjeeling** (Lusuguri, near Rangit below Badamtam, Peshok to Tista Bridge, Glen Cathcart). Lower Hill Forest, 300–1220m. Fl. March–August; fr. December.

Records of *S. wightii* A.DC. (F.E.H.1) probably refer to this species.

Fig. 5. Smilacaceae II (leaves and petioles of **Smilax** spp.).
a–b, **S. aspera**: a, leaf from main stem; b, petiole. c–d, **S. myrtillus** var. **rigida**: c, leaf from main stem; d, petiole. e–f, **S. ferox**: e, leaf from main stem; f, petiole. g–i, **S. minutiflora**: g, leaf from main stem; h, leaf from young shoot; i, petiole. j–k, **S. menisperoidea**: j, leaf from main stem; k, petiole. l–n, **S. elegans** subsp. **elegans**: l, leaf from main stem; m, leaf from young shoot; n, petiole. All × ⅔. Drawn by Sally Rae.

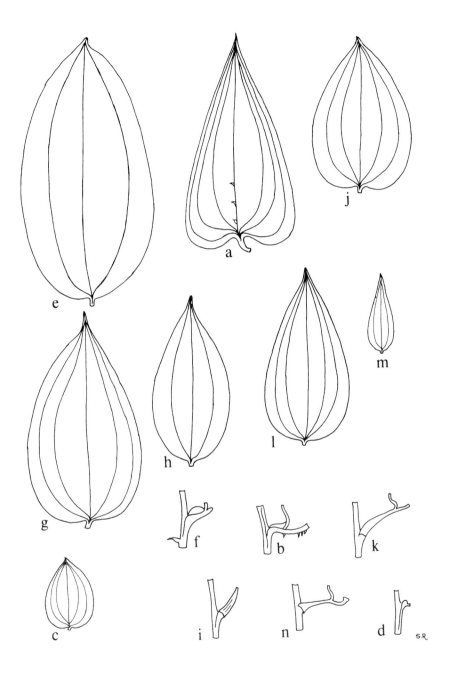

191. SMILACACEAE

8. S. lanceifolia Roxb. Fig. 3l, Fig. 4m–n.

Climber; stems terete or slightly ridged, not prickly or sometimes with prickles on older wood. Leaves oblong-lanceolate, cuspidate, base rounded, 9.5–16 × 5–9cm, costae 3–5 + 2–4 very weak marginals, free from base or middle 3 joined for up to 0.4cm, coriaceous. Petiole 1.4–2cm, winged for lower ⅓, wing narrow (0.8–1.3mm) oblong; tendrils developed. Infl. always of a single ebracteate umbel, peduncle usually shorter than adjacent petiole, prophyll c.0.5cm, brown, ovate with curved apiculate tip. Female umbel 27–48-flowered; pedicels 0.6–1.9cm; receptacle globose or elongated 2–3.5mm diameter; bracteoles very short; outer tepals c.2.9 × 0.9mm, inner c.0.5mm wide, ovary c.2.3mm, stigmas very short, staminodes 6. Male umbel 20–30-flowered; receptacle globose, c.2.5mm diameter; flower buds clavate, tepals reflexed at maturity, outer 4–4.5 × 1.1–1.3mm, inner 0.9–1.1mm wide, filaments 2.9–4.2mm, anthers oblong c.1.1 × 0.5mm. Berries c.9mm diameter, blackish.

Bhutan: S — Samchi (Changda Chungli Hill), Phuntsholing (Ganglakha to below Suntlakha), Chukka (Sinchu La) and Gaylegphug (Chabley Khola) districts; **C** — Punakha (Bhotokha to Rinchu, Mishichen to Khosa) and Tongsa (S of Shamgong) districts; **Darjeeling** (Rishap, Labdah, Sureil, Rungbee, Mangion, above Mongpu); **Sikkim** (Kulhait, Yoksum, Burmiok, Rishi to Rinchinpong, Rhenok). Subtropical and *Quercus griffithii* forest, 910–2130m. Fl. October–March; fr. November–June.

9. S. myrtillus A.DC. var. **rigida** Noltie; *S. rigida* Wall. ex Kunth non Russell. Fig. 5c–d.

Shrub to 2.5m; stems with strong ridges decurrent from leaf bases; spines numerous, ± straight, to 1cm. Leaves almost sessile, ovate (older occasionally orbicular), shortly mucronate, base rounded becoming shallowly cordate, 1.4–2.8 × 0.9–2.2cm, costae 3, free from base, coriaceous. Petioles extremely short with a pair of minute brown filiform, ciliate, free stipular scales at base; tendrils absent. Infl. a single 2–5-flowered umbel on peduncle 0.4–0.9cm, borne in axil of lowest leaf of immature lateral shoots which develop vegetatively after flowering. Female flowers: outer tepals c.3.8 × 1.7mm, inner c.1.5mm wide, ovary c.1.2mm, stigmas 3, erect, c.0.9mm, staminodes 3. Male flowers: bud ovoid, flower campanulate, outer tepals 2.2–2.9 × 1.1–1.3mm, inner tepals c.1mm wide, anthers c.0.3 × 0.3mm, equalling filaments. Berries c.0.5cm, black.

Bhutan: S — Chukka district (Chukka); **C** — Punakha (Dobemen Chu; Ratsoo to Samtengang, Samtengang to Ritang (F.E.H.2)), Tongsa (Longte Chu, Changkha to Chendebi; Tongsa to Tratang (F.E.H.2)) and Tashigang (Balfi) districts; **N** — Upper Mo Chu (Kohina; Khosa to Tamji (F.E.H.2)),

Upper Kuru Chu (Nashima) and Upper Kulong Chu (Tobrang) districts; **Darjeeling** (Tiger Hill, Sonada, Sandakphu, Ghum, Rungbi, Tonglu, Senchal, Neebay; Rimbick to Ramam (F.E.H.1)); **Sikkim** (Yoksum to Bakkim, Bakkim to Dzongri). Broad-leaved (incl. oak) forest; moss on wet cliff-face, 1680–3400m. Fl. May; fr. September–October.

Var. *myrtillus*, which differs in its rhombic-ovate, more herbaceous leaves with acute bases and fewer spines, has also been recorded from Bhutan (Tamji to Gasa (F.E.H.2)), but I have seen no specimens.

10. S. ferox Wall. ex Kunth. Fig. 5e–f.

Climber (shrubby when young); lateral branches strongly ridged, surrounded at base by involucre-like persistent petiole-base and brown coriaceous prophyll (to 1cm) with reflexed tip, old stems with strong recurved prickles. Leaves at flowering lanceolate, mucronate, base cuneate, very thin, drying brown. Mature leaves lanceolate to oblong, mucronate, base rounded, 5.2–8.5(–10.5) × 1.8–2.5(–5.5)cm, costae 3 + 2 very weak marginals free almost from base, coriaceous, glaucous beneath. Petiole 3.5–8mm, conspicuously winged for almost whole length, wings 0.7–2mm wide half-obovate to half-elliptic, membranous, pale; tendrils only occasionally developed on older wood; abscission from apex of petiole. Infl. a single 14–21-flowered umbel on peduncle of c.2cm borne in axil of lowest leaf of a lateral shoot which develops vegetatively after flowering; receptacle elongate (to 6mm in female); bracteoles conspicuous, brown, lanceolate (shorter and deciduous by fruiting in female). Female flowers: ovary c.2.4mm, stigmas 3, c.1.3mm, spreading, staminodes 2–3. Male flowers campanulate, outer tepals 3–5 × 1.9–2.3mm, inner 0.9–2mm wide, filaments 1.4–2.2mm, anthers oblong, 0.6–1 × 0.4mm. Berries red, with glaucous bloom, c.1.1cm diameter; seeds reddish-brown, c.5.5mm.

Bhutan: S — Chukka district; **C** — Thimphu, Punakha, Tongsa, Bumthang and Mongar districts; **N** — Upper Mo Chu, Upper Kulong Chu and Upper Kuru Chu districts; **Darjeeling** (Jorepokri, Senchal, Labdah, Rungirun, Palmajua, Sonada, Ghumpahar, Rungbul, Alubarie, Batasi, Sukia Pokhri to Manibhanjang). Broad-leaved (incl. oak) forest, often among disturbed scrub, 1450–2950m. Fl. April–May; fr. May–December.

11. S. minutiflora A.DC. Fig. 5g–i.

Shrub, sometimes epiphytic; stems terete or slightly furrowed, without prickles, lateral branches subtended by persistent petioles and keeled, folded, ovate, blunt prophylls (to 1.2cm). Leaves variable in shape, ones on old wood at least sometimes narrowly ovate, to 17 × 10cm, base cordate, costae 7; upper lanceolate (occasionally ovate), gradually acuminate to subacute, sometimes shortly caudate, base rounded, 3.1–12.7 × 1.2–3.5cm, costae 5 (middle

3 strong), free from base, herbaceous at first, becoming coriaceous after flowering, dull above, glaucous beneath. Petiole 0.7–1.3(–2)cm, winged for about ¾ length, wing narrow, c.1.5mm wide at base, gradually tapered into petiole above; tendrils absent; abscission between base of blade and top of wing. Old petioles to 1(–2.3)cm, apex not recurved or expanded, without tendrils. Infl. of single umbels inserted in axils of lowest 1 or 2 leaves of lateral shoots, a conspicuous straw-coloured, ovate bract (to 4mm) present between base of peduncle and subtending leaf. Female umbel 2–7-flowered; peduncle 1.5–4cm; receptacle not expanded; bracteoles absent; pedicels c.0.6cm; outer tepals 2–2.4 × 0.9–1mm, inner tepals 0.6–1.1mm wide, ovary globose c.1.3mm, stigma lobes 0.5–1mm, recurved, staminodes 0–4. Male umbel 6–10(–22)-flowered; peduncle to 6cm, slender, greatly exceeding petiole, sometimes recurved; receptacle scarcely expanded; bracteoles absent; pedicels unequal, 0.4–1.2cm; flower bud oblong-ovoid, tepals incurved, outer 1.9–2.5 × 0.9–1.2mm, inner c.0.9mm wide, filaments c.0.4mm, anthers c.0.4 × 0.4mm. Berry black, c.0.8cm diameter, 2–3-seeded.

Bhutan: S — Chukka (between Jumudag and Tala) and Gaylegphug (Chabley Chola above Sham Khara) districts; **C** — Thimphu district (Thimphu to Dochu La); **Darjeeling** (Ghum to Tiger Hill, Senchal, Tonglu, Sonada, Rungbool, Sandakphu to Garibans, Kurseong, Ging, Punkabari); **Sikkim** (Chola, Prek Chu to Bakkim; Chiabanjan to Dentam, Bakkim to Dzongri (F.E.H.1)). Shady, moist forest (incl. *Castanopsis/Quercus*), 1800–3050m. Fl. May–July; fr. August–February.

S. elegans subsp. *osmastonii* (Wang & Tang) Noltie possibly occurs in the area. It is a shrubby species differing from *S. minutiflora* in having coriaceous leaves, with densely reticulate veins raised on upper surface when dry, and undersurface papillose. A Hooker specimen bearing a printed 'Sikkim' label (1524–2134m) has been seen, but bears no field label and could perhaps be a Khasian specimen. Specimens recorded as *S. osmastonii* from Bhutan (F.E.H.2) have been re-determined as *S. minutiflora*.

12. S. menispermoidea A.DC.; *S. elegans* sensu F.B.I. non Wall. ex Kunth. Fig. 3m–p, Fig. 5j–k.

Climber; stems terete, without prickles, lateral branches subtended by persistent petiole (to 3.2cm) and elongate, folded prophyll (0.8–1.8cm) with mucronate often recurved tip. Leaves ovate, subacute, mucronate (occasionally damaged and retuse), base shallowly cordate (occasionally truncate), 4–10.5(–14) × 3.2–6.5(–10.5)cm, costae 5, free from base, thin, herbaceous at flowering, thickening with age, drying brownish-green. Petioles 1–1.4cm, winged for ⅔ length, wing gradually tapered from base to insertion of tendrils, apex sometimes hooked; tendrils always present, initially minute, filiform; abscission from between base of blade and apex of wing. Infl. a single umbel borne in axil of lowest leaf of lateral shoots with mature leaves at time of flowering. Female umbel 4–9-flowered; peduncle 1.9–3cm; bracteoles very small, deciduous; outer tepals (2.3–)3.3 × (0.9–)1.4mm, inner tepals c.1mm wide, ovary globose 1.2–1.6mm, stigmas 0.9–1.3mm, recurved, staminodes 2–3. Male umbel 12–17-flowered; peduncle c.4cm; receptacle scarcely expanded; bracteoles membranous, lanceolate, acuminate; flower buds globose, tepals purple, outer 3–4 × 1.5–1.9mm, inner 1–1.4mm wide, anthers c.0.5 × 0.7mm, filaments 0.3–0.7mm. Berries black with glaucous bloom, 0.7–1cm; seeds 1–3, reddish-brown, c.4mm.

Bhutan: S — Chukka district; **C** — Ha, Thimphu, Punakha, Tongsa, Bumthang, Mongar and Tashigang districts; **N** — Upper Mo Chu and Upper Bumthang Chu districts; **Darjeeling** (Tonglu, Sandakphu, Manibanjan; Garibans (F.E.H.1)); **Sikkim** (Lachen, Karponang, Kalapokri, Chungthang, Lachung, Dzongri, Yakla). Among shrubs (*Pyrus, Acer, Prunus*) in oak and conifer (*Abies/Rhododendron, Pinus/Tsuga*) forest, 2130–4270m. Fl. May–August; fr. May–November.

13. S. elegans Wall. ex Kunth subsp. **elegans**; *S. parvifolia* sensu F.B.I. p.p.; *S. glaucophylla* sensu F.E.H.1 non Klotzsch. Fig. 5l–n.

Differs from *S. menispermoidea* in its very zigzag branches; its smaller, lanceolate leaves, rounded at base, often glaucous beneath, drying greyish-green; petiole wings oblong, with free triangular tips, tip of petiole expanded and often recurved; abscission from joint at base of blade; male peduncle shorter (1–1.8cm); smaller flowers — outer tepals c.2 × 1mm (female), 2–2.4 × 0.7–1.1mm (male); staminodes 3–6.

Bhutan: S — Chukka district (below Chimakothi); **C** — Thimphu (Chimakothi to Thimphu, below Lobnakha) and Punakha (Ritang to Ratsoo) districts; **N** — Upper Kuru Chu district (Denchung). Also an unlocalised Griffith specimen (*Griffith* 2641, K, BM). *Quercus/Pinus* forest (shady cliffs; scrub on wet slopes), 2130–2530m. Fl. April–June.

Records of *S. glaucophylla* Klotzsch from Sikkim (Rimbick to Ramam and Ramam to Phalut (F.E.H.1)) and Punakha district (Samtengang to Ritang (F.E.H.2)) almost certainly refer to this taxon.

subsp. **subrecta** Noltie; *S. elegans* var. *major* A.DC.; *S. longebracteolata* Hook.f.

Climber; stems smooth. Lateral branches subtended at base by persistent petioles to 2.3cm and triangular, keeled prophyll to 1cm. Leaves narrowly ovate, abruptly contracted beneath acute triangular apex, base cuneate to rounded, 5–8.8 × 2.8–4.5cm, costae 5 + 2, free to base, coriaceous, drying greyish. Petiole 1–1.5cm, winged for lower ½–⅔, wings oblong, to 1mm wide, with free acute apices; tendrils developed, especially on old growth; abscission from slightly expanded petiole apex. Female umbels borne singly in axils of mature leaves all along side branches, without prophyll or bract-like scale; peduncle compressed, equalling or very slightly exceeding petiole; receptacle hemispheric, bearing conspicuous lanceolate bracteoles to 3.5mm; pedicels c.0.7cm. Berries blackish, c.4.5cm diameter.

Bhutan: C — Mongar (Baga La) and Tashigang (Tashi Yangtsi Dzong) districts. 1770–1830m. Fr. August.

Doubtfully recorded species:

S. quadrata A.DC. has been recorded from Darjeeling (Kurseong (Matthew, 1967)) but most probably in error.

2. HETEROSMILAX Kunth

Differs from *Smilax* chiefly in its bottle-shaped flowers with 3, free apical lobes; male flowers with (2–)3(–4) stamens (6 or 9–12 in some non-Indian species), filaments fused at base to varying degrees.

1. H. japonica Kunth; *H. indica* A.DC. Fig. 3q–r.
Climber; stems terete, spineless. Mature leaves ovate (young leaves oblong-lanceolate), contracted below tip to cuspidate apex, base cuneate to cordate, c.18.5–21 × 9.5–13.5cm, costae 5–7 + 2 weak marginals, free from base, herbaceous, both surfaces slightly shining. Petiole to 5cm × c.2mm, extremely narrowly winged for lower 0.5cm; tendrils strongly developed; abscission c.⅓ way between blade and attachment of tendrils. Umbels borne singly; peduncles 3–6cm, flattened, naked at base. Female umbel c.10-flowered; receptacle globose c.2.5–4mm diameter. Male umbel (from Khasian specimens) c.25-flowered; pedicels curved, unequal; flowers 3.5–4.5 × 1.7–2mm, narrowly obovoid, attenuate to base, apical lobes triangular, very short; stamens 3, about equalling perianth, filaments flattened, fused for less than ⅓ length. Berries c.1cm diameter, ?blackish; seeds pale orange-brown, c.5.5mm diameter.

Bhutan: S — Samchi district (Khagra Valley near Gokti); **Darjeeling** (Rungdung). Subtropical forest, 300–550m. Fr. March–April.

Family 192. CONVALLARIACEAE

Perennial, usually rhizomatous herbs. Leaves in basal rosettes (sometimes distichous), or spiral, opposite or whorled along a stem, linear or differentiated into blade and petiole. Infl. a spike-like raceme on a leafless scape, or in a terminal panicle or raceme or borne in axillary clusters on a leafy stem. Flowers bisexual, actinomorphic, hypogynous or epigynous; tepals usually 6, commonly all similar and tubular below. Stamens usually 6, filaments free or fused below, commonly inserted at apex of tube, anthers basifixed or dorsifixed dehiscing longitudinally, inwards. Ovary usually 3-loculed, superior or partly inferior, style simple, stigma capitate or 3-lobed, ovules 2–several per locule, basal or axile. Fruit a berry or irregularly rupturing capsule; seeds sometimes with a showy sarcotesta.

Young shoots of several species of *Polygonatum* and *Maianthemum* used as vegetables in Sikkim.

1. Plant with stout, decumbent to semi-erect stems; leaves petiolate, oblan-ceolate, cuspidate; infl. an apparently terminal raceme; pedicels jointed; ovary inferior, rupturing early in development of fruit, so seeds exposed for most of development; seeds covered with fleshy, pale blue sarcotesta .... **(4) Ophiopogon dracaenoides**
+ Plant not as above 2

2. Leaves alternate, opposite or whorled on erect stems; infl. a terminal raceme or panicle, or flowers axillary 3
+ Leaves all basal; infl. a terminal spike-like raceme on a leafless scape .. 4

3. Infl. a terminal raceme or panicle; flowers with 6 free tepals
   2. Maianthemum
+ Flowers in few-flowered axillary clusters; flowers tubular with short, free, apical lobes .... **1. Polygonatum**

4. Leaves linear, under 1.1cm wide; flowers white, pinkish or violet; bracts shorter than to slightly exceeding flowers 5
+ Leaves over 1.1cm wide, usually with distinct petiole and expanded blade; if ± linear and not differentiated into blade and petiole, then flowers orange to greenish and bracts greatly exceeding flowers 6

5. Ovary superior; pedicel not jointed; fruit a many-seeded globose berry
   3. Theropogon
+ Ovary inferior; pedicel jointed about the middle; fruit as for *Ophiopogon dracaenoides* (see above) .... **4. Ophiopogon**

6. Filaments fused into a corona; ovary at least partly inferior (Fig. 61); fruits as for *Ophiopogon* .... **5. Peliosanthes**
+ Filaments not fused into a corona; ovary superior (Fig. 6o); fruit a berry .... **6. Tupistra**

1. POLYGONATUM Miller

Rhizomatous herbs. Stems simple; scale leaves present at base; lower part commonly bearing 1 or more intermediate leaves. Stem leaves alternate, in opposite pairs or whorls of 3 or more. Flowers pedicellate, 1 or more on a common (often recurved) peduncle in leaf axils. Perianth tubular, with 6 short apical lobes, whitish to purplish, lobes often darker. Stamens 6, included within tube, filaments short, attached above middle of tube; anthers linear, dorsifixed. Ovary of 3 locules, each with 2 or more ovules; style slender; stigma capitate or with 3 very short lobes. Fruit a few- to many-seeded, globose berry.

The following account is provisional. Revisions of the many Sino-Himalayan taxa based solely on morphological characters have severe limitations. A revision of the genus is badly needed using molecular and cytological techniques, which might elucidate the complex pattern of variation observed and demonstrate whether or not this complexity is partly the result of hybridisation.

1. Dwarf plant usually under 5cm; corolla purple, lobes elliptic, over 5.6mm .... **1. P. hookeri**
+ Plant more robust, over 5cm; corolla cream or purple, lobes triangular, under 5.6mm 2

2. Leaves coriaceous, slightly shining when dry; corolla tube urceolate, whitish flushed pinkish .... **2. P. punctatum**
+ Leaves not coriaceous, not shining when dry; corolla tube ± cylindric, yellowish, whitish, greenish or purplish 3

3. Leaves strictly in opposite pairs 4
+ Leaves alternate or in whorls of 3 or more 6

4. Peduncles stout, bearing 5–10 flowers; leaves with conspicuous cross-veinlets when dry .... **3. P. oppositifolium** var. **oppositifolium**
+ Peduncles slender, 2-flowered; leaves usually without conspicuous cross-veinlets 5

FIG. 6. **Convallariaceae.**
a–b, **Polygonatum cathcartii**: a, habit ($\times$ 1/6); b, stamen showing swollen filament apex ($\times$ 5). c–d, **Maianthemum fuscum**: c, habit ($\times$ 1/3); d, flower ($\times$ 3). e, **M. purpureum**: flower ($\times$ 2). f–g, **Theropogon pallidus**: f, half flower ($\times$ 3); g, fruit ($\times$ 2). h, **Ophiopogon wallichianus**: fruit ($\times$ 3). i–j, **O.** cf. **bodinieri**: i, habit showing stolons ($\times$ 1/2); j, half flower ($\times$ 3). k–l, **Peliosanthes macrophylla**: k, habit ($\times$ 1/6); l, half flower ($\times$ 5). m–o, **Tupistra aurantiaca**: m, habit ($\times$ 1/6); n, half flower ($\times$ 3); o, half flower with ovary removed ($\times$ 3). Drawn by Mary Bates.

5. Filaments expanded at apex (Fig. 6b) .... **4. P. cathcartii**
+   Filaments not expanded at apex .... **7. P. singalilense**

6. Leaves strictly alternate 7
+   At least some leaves in whorls of 3 or more 8

7. Peduncles over 4.5cm .... **5. P. nervulosum**
+   Peduncles under 1.5cm .... **6. P. brevistylum**

8. Filaments swollen at apex (Fig. 6b) .... **4. P. cathcartii**
+   Filaments not swollen at apex 9

9. Leaves coiled at tips, often 5 or more per whorl; flowers sometimes more than 2 per peduncle 10
+   Leaves not coiled at tips, though young ones may be hooked, never more than 4 per whorl; flowers never more than 2 per peduncle 11

10. Leaves linear-lanceolate, over 4.2mm wide; flowers usually flushed purplish; stems usually spotted purplish .... **8. P. sibiricum**
+   Leaves linear, under 3.8mm wide; flowers white; stems unspotted
    9. P. cirrhifolium

11. Flowers purple; plant usually under 40cm at flowering; leaves under 1cm wide .... **12. P. kansuense**
+   Flowers whitish to cream, with greenish tips; if plant under 40cm then leaves over 1cm wide 12

12. At least some leaves in opposite pairs the rest in 3s, leaves over 1cm wide; plant under 45cm .... **7. P. singalilense**
+   Leaves in whorls of 3s and 4s; plant usually over 60cm (if less then leaves under 1cm wide) 13

13. Leaves usually over 1cm wide, apex subacute; plant stout
    10. P. verticillatum
+   Leaves usually under 1cm wide, apex very acute; plant slender
    11. P. leptophyllum

1. P. hookeri Baker.

Rhizome slender, under 3mm diameter. Stem 3–5(–16)cm. Scale leaves several, transparent, whitish, veined. Leaves alternate to subopposite but very crowded, lanceolate to linear-lanceolate, blunt, gradually narrowed to base,

1.7–4 × 0.4–0.8cm. Flowers lilac to deep pink, fragrant, erect, borne singly (occasionally paired) in lower axils, corolla tube (7.8–)9–14 × 2–3.1mm, lobes narrowly elliptic, 5.6–7.5(–9)mm; peduncle 4–5mm. Anthers c.2(–2.8)mm, sessile. Ovary ellipsoid to subglobose, 2.5–3 × 2–2.5mm; style 1.4–2mm. Berry c.1cm, deep pink.

Bhutan: C — Ha (Chelai La, Sharithang), Thimphu (Mem La; Shodu to Barshong (F.E.H.2)) and Punakha (Dunshinggang) districts; **N** — Upper Mo Chu (Lingshi to Chebesa, above Laya; Yabuthang to Laya (F.E.H.2)), Upper Bumthang Chu (Mon La Karchung, Lhabja, Pangotang, Chamka) and Upper Kulong Chu (Shingbe) districts; **Sikkim** (Lachen, Beeroom, Yeumtang, Lhonak, Samiti, above Thangshing, Chaunrikiang); **Chumbi**. Dry grassy hillsides, sometimes among shrubs and glades in conifer forest, 3050–4880(–5790)m. Fl. April–July.

Roots eaten by Monal Pheasant.

2. P. punctatum Royle ex Kunth.

Rhizome 0.8–1.2cm diameter; roots thick. Stem 20–38cm. Scale leaves several, short, membranous (often flushed pink). Intermediate leaves 1 or more, oblong. Leaves alternate, subopposite or in irregular whorls of 3, narrowly elliptic, contracted below blunt apex, margins narrowly cartilaginous, gradually narrowed to base, 4.9–7.5 × 1–1.8cm, coriaceous, slightly shining when dry. Infls. borne all along stem, 2 flowers per peduncle; peduncles 4–11mm, with narrow cartilaginous wings; pedicels 2–4.5mm. Flowers white with pinkish or purplish flushing, urceolate, corolla tube 5.4–7.7 × 2.7–4mm, lobes 1.3–1.6mm. Filaments 0.3–1mm; anthers 1–1.4mm. Ovary ellipsoid to ovoid, 2.8–3.3 × 1.8–2.3mm; style 1.5–2.2mm. Berry c.5.5mm.

Bhutan: C — Tongsa (E of Pele La) and Tashigang (Tashi Yangtsi Dzong) districts; **N** — Upper Kuru Chu (Dengchung) and Upper Kulong Chu (Tobrang) districts; **Darjeeling** (Srikola; Batasi, Palmajua to Rimbick, Ramam (F.E.H.1)); **Sikkim** (Jorpokree, Chola, Latong, Lachen). Epiphyte — on mossy trees in wet forest, 2000–2440m. Fl. April–May.

3. P. oppositifolium (Wall.) Royle var. **oppositifolium**. Sikkim name: *ruk khirola*.

Rhizome c.1.2cm diameter. Stem 40–95(–130)cm. Scale leaves triangular to oblong, cream, membranous, striate. Intermediate leaf sometimes present. Leaves strictly opposite (lower sometimes slightly staggered), lanceolate to narrowly elliptic, acuminate to caudate, narrowed below into short (0.3–1cm) petiole, 5.6–15 × 2.1–5cm, thick-textured, short cross-veinlets usually conspicuous when dry. Peduncles 1.3–2.6cm, very thick, striate, flattened, sometimes reflexed, (3–)5–8(–10)-flowered; pedicels short (0.9–1.8cm), stiff.

Flowers white, tubular, tube 8.5–12.2 × c.4mm, lobes 3.8–4.8mm. Filaments 2.2–4mm; anthers acute, bases long, slightly divergent, 4.2–4.7mm. Ovary cylindric, 4–4.8 × 1.7–2.2mm; style 5.5–7.3mm. Berry 0.5–1cm, scarlet.

Bhutan: S — Samchi (Changda Chungi Hill), Phuntsholing (Kamji) and Chukka (E of Jumudag) districts; **C** — Tongsa district (Dakpai); **Darjeeling** (Rissum, Mahalderam, Jhandihill, Lopchu, Takdah, Tumlong, Damsing, Ghumpahar; Batasi, Kurseong (F.E.H.1)); **Sikkim** (Relli Chu, Dikeeling, Yoksum — cultivated in pot; Dentam (F.E.H.1)). Epiphyte — on trees in evergreen oak forest, 1200–2130m. Fl. April–May; fr. August–April.

Eaten as vegetable in Sikkim (Rao, 1964b).

var. **decipiens** Baker.

Stem 27–33cm. Leaves mainly alternate but with some in subopposite pairs, narrowly lanceolate, finely acuminate, 8–10.3 × 1.2–2.2cm, fairly thick textured, cross-veinlets just visible. Petioles 0.4–0.8cm, sometimes deflexed. Peduncles 0.6–0.9cm, erect; pedicels c.0.6cm (fl.), 1–1.5cm (fr.). Flowers 1.1 × 0.27cm, two per peduncle.

Sikkim (Chola). 2438m.

New collections are required of this plant to ascertain its true status. It might well prove to be a distinct species or hybrid. At present it would almost certainly be better placed under *P. brevistylum*, having short, 2-flowered peduncles.

4. P. cathcartii Baker. Fig. 6a–b.

Rhizome 0.9–1.2cm diameter, beaded; roots fleshy. Stem arching, 25–80cm, usually unmarked, naked for lower third. Scale leaves apparently deciduous. Leaves opposite or in 3s, lanceolate, acuminate, base rounded, 5.2–11.5 × 2.1–3.6cm, upper narrower and more acuminate, glossy above, glaucous beneath; petiole short (3–5mm). Infls. borne in lower half of leafy part of stem; peduncles 1.8–5(–7)cm, very slender, about half length of subtending leaf, 2–3(–4)-flowered; pedicels 0.4–1.9cm. Flowers greenish-white to yellow, widely cylindric, tube 8.2–10.4 × 5.3–6.5mm, lobes 1.8–2.6mm. Filaments 1.4–2.2mm, greatly expanded at junction with anther; anthers 2.1–3.1mm. Ovary ellipsoid, 4–5 × 1.8–3mm; style 3.5–4.5mm. Berry c.7mm, 9–15-seeded, red.

Bhutan: C — Punakha (Mara Chu Valley) and Mongar (below Sengor) districts; **Darjeeling** (Senchal to Tiger Hill (F.E.H.1)); **Sikkim** (Sakkargong, Lachen, Bakkim to Tsoka). Epiphytic and on moss-covered rocks in *Tsuga/Quercus* forest, 2440–3040m. Fl. May–August.

5. P. nervulosum Baker.

Stem 30–59cm, conspicuously zigzag in upper part, leafless for lower third. Scale leaves whitish, oblong, persisting as fibres. Leaves alternate, lanceolate to elliptic, abruptly caudate, base rounded, 8.2–11.5 × 1.5–4.2cm, cross-veinlets prominent; petiole 0.5–0.9cm. Infls. borne all along stem; peduncles 4–8cm (more than half length of subtending leaf), often slender, (1–)2–4-flowered; pedicels 1–3cm, sometimes with minute bracteoles. Flowers yellow, cylindric, tube 10–12.6 × 3.8–5mm, lobes 2–2.3mm. Filaments 1–1.5mm, expanded at junction with anther; anthers apiculate, linear, 3.2–4.3mm. Ovary cylindric, 3.6–5.1 × 1.7–2mm; style 3.6–5.6mm.

Bhutan: S — Chukka (Bunakha) and Deothang (S of Riserboo, Raidong) districts; **C** — Mongar district (near Lhuntse); **Sikkim** (Lachung). Moist, broad-leaved forest, 2130–3050m. Fl. June–July.

Specimens from Gaylegphug district (W bank of Chabley Khola, 2060m, *Grierson & Long* 4072) and Sikkim (Phedup, *Ribu* 297) probably belong here but are atypical in having very narrow linear-lanceolate leaves, some in subopposite pairs. They are also very similar to *P. oppositifolium* var. *decipiens* from which they differ in having long peduncles.

6. P. brevistylum Baker.

Stem 25–44cm, naked for lower third. Scale leaves apparently deciduous. Leaves elliptic to lanceolate, gradually narrowed to subacute apex, base rounded, 6.8–10.4 × 2.4–4cm; petiole very short (2–4mm). Infls. borne in lower ⅔ of leafy part of stem; peduncles 0.9–1.5cm, about ¼ length of subtending leaf, 2–4(–6)-flowered; pedicels 0.7–1.5cm, with minute caducous bracteoles. Flowers greenish-yellow, cylindric, slightly contracted at middle, tube 9.4–10.9 × 4–4.5mm, lobes c.4mm. Filaments 3–4.2mm, not expanded; anthers 3–3.8mm. Ovary 6.2–7.5 × 2.5–2.9mm, narrowly ellipsoid, tapered into style (3.5–4mm). Berry c.8.5mm, c.13-seeded, dark red.

Sikkim (Bakkim); **Darjeeling** (Rungbool, Takdah, Ghumpahar, Singalila; Tonglu to Chitrey, Tiger Hill to Senchal (F.E.H.2)). Epiphyte — on mossy tree trunks in oak forest, 1830–3050m. Fl. May; fr. July–October.

7. P. singalilense Hara. Bhutanese name: *tema*.

Rhizome to c.1cm diameter, beaded. Stem erect, 28–45cm. Intermediate leaf 1, oblong, clasping, to 3.5cm, membranous. Leaves opposite or in 3s above, lowest usually single, lanceolate, slightly contracted below blunt apex, narrowed to sessile base, 3.5–6.5 × 1.2–2.1cm, glaucous beneath. Infls. restricted to lowest 2(–3) nodes; peduncles 0.7–1.6cm, recurved, 2-flowered; pedicels 0.4–0.7cm. Flowers yellowish to greenish-white, slightly contracted at middle, tube 7.5–8 × 3.4–4mm, lobes 3–3.1mm. Anthers 1.9–2.5mm sessile or with short filaments to 0.7mm, not expanded. Ovary narrowly ovoid, 2.5–3.6 × 2–2.6mm; style 2–3.8mm.

Bhutan: C — Ha district (Kale La); **N** — Upper Kulong Chu district (Shingbe); **Sikkim** (Laghep, Lampokri, Dzongri, Lingtoo, Jamlinghang to Bikbari, Thangshing to Lam Pokhri; Migothang to Nayathang (F.E.H.1)). Marshy meadows; dwarf rhododendron scrub, 2740–4270m. Fl. June–August.

Leaves cooked as vegetables according to Bedi.

This species requires further investigation. Specimens seen from E Nepal and Thimphu district (Pajoding, *Noltie* 59, E) with wider leaves (to 2.8cm), very glossy above and in regular opposite pairs with purple striped stems perhaps belong to this species. A specimen from Bumthang district (above Gortsam, *Noltie* 139, E) also perhaps belongs here, being similar but having leaves in 3s. This variability in leaf arrangement shows the artificiality of the division of the genus on this character; it could perhaps be partly explained by hybridisation. For example, plants seen at Bikbari (Sikkim) growing with both species were intermediate in appearance between *P. singalilense* and *P. sibiricum*.

8. P. sibiricum F. Delaroche s.l.

Rhizome 0.7–2.5cm diameter, cylindric, with transverse, circular scars. Stem 43–122cm, densely purple-spotted. Intermediate leaves 2. Leaves in whorls of 3–9, linear-lanceolate, strongly coiled at tips, 4.2–11.6mm wide. Infls. borne in lower half of leafy part of stem; peduncles 0.8–1.3cm, strongly recurved, 2–4-flowered; pedicels often shorter than peduncle, bearing minute bracteoles. Flowers usually tinged purplish, 7–8.6 × 2–3.5mm, slightly constricted below middle, tube 5.5–6.1mm, lobes 1.7–2.5mm. Filaments 0.3–1mm; anthers 1.5–2.2mm. Ovary narrowly ellipsoid, 1.4–3.1 × 0.9–2.6mm; style 0.9–2mm.

Bhutan: C — Thimphu (Drugye Dzong, Dotena, Tsalimaphe, above Ragyo, Dochu La, above Motithang Hotel) and Bumthang (above Dhur) districts; **N** — Upper Mo Chu district (Gasa); **Sikkim** (Laghep, Lachen, Lachung, Yumthang, Latong, Dzongri, Pemiongchi, Chungthang, Yakthang, Tookit, Tangsap, Jamlinghang to Bikbari, Thangshing to Lam Pokhri); **Chumbi**. Rhododendron scrub; meadows; pine forest, 1830–4200m. Fl. May–July.

The juvenile form of the above has been seen in dry habitats (pine forest and scrubby stream banks) at Motithang and Atsho Chhubar (both Thimphu district); the leaves are single or in whorls of up to 3, semi-coriaceous and with a distinct central pale stripe. At first sight these appeared to represent a distinct species until the progression to the mature form was observed.

9. P. cirrhifolium (Wall.) Royle. Bhutanese name: *caidu*.

Differs from *P. sibiricum* in its linear leaves (0.6–3.9mm wide), and always cream flowers.

Bhutan: S — Chukka district (near Chukka); **C** — Ha (Ha), Thimphu (above Bongde Farm, below Chang Gang Kha temple, Thimphu, Tsalimaphe to Hinglai La, Motithang Guest House, Paro) and Tongsa (Kinga Rapden) districts; **Sikkim** (Choongtam, Zemu Valley). Dense conifer forest; shingle by stream; disturbed shrubby thickets; cliff ledge; moist shrubbery, 1670–4200m. Fl. April–June.

As in the *P. verticillatum* group much work is required to resolve the complex of forms with cirrhose leaf tips. *P. cirrhifolium* with its extremely narrow leaves merely seems to form one end of the range; the commonest form occurring in E Himalaya and SW China has broader leaves. These broad-leaved forms with small flowers are probably all best placed under *P. sibiricum* meantime. Many species have been described in this complex and our plant is certainly synonymous with *P. fuscum* Hua and *P. trinerve* Hua from SW China, and with *P. fargesii* Hua which was described in part from Sikkim specimens.

10. P. verticillatum (L.) Allioni.

Stem 60–93cm (or more?), stout. Intermediate leaf 1 (or more?). Leaves in whorls of 3–4, except for lowest two nodes which sometimes have a single leaf or a subopposite pair, lanceolate, subacute, never hooked or coiled at tip (even when young), narrowed to sessile base, 7.5–14.5 × 0.8–1.8(–3)cm. Infls. borne almost to top of stem; peduncles 1.7–3.5cm, always 2-flowered; pedicels very short (c.⅕ length of peduncle), sometimes with minute bracteoles. Flowers creamy-white or pale green, with darker lobes, cylindric to urceolate, tube 6.5–8mm, lobes 2–3.8mm. Filaments 0.3–1mm; anthers 1.5–2.2mm. Ovary narrowly ellipsoid, 2–3.9 × 2–2.2mm; style 1.5–3.1mm. Berry c.1cm, red, c.10-seeded, seeds large.

Bhutan: C — Ha (Ha), Thimphu (above Ragyo, Dotena to Barshong, Chi La to Paro) and Mongar/Tashigang (Donga La) districts; **N** — Upper Mo Chu (Gasa to Koina) and Upper Pho Chu (Foomay) districts; **Darjeeling** (Tonglu; Phalut (F.E.H.2)); **Sikkim** (Jamlinghang). *Abies* forest (in clearings and at edges); alpine meadows, 3000–3920m. Fl. June–July.

11. P. leptophyllum (D. Don) Royle.

Differs from *P. verticillatum* s.s. in being usually more slender, with narrower leaves (3.3–)4.1–12mm wide, gradually tapered to very fine, acute apex, which may appear hooked when young; peduncles shorter (0.7–2.2cm); pedicels c.½ length of peduncle; berries smaller (to c.0.7cm, 1–8-seeded).

Bhutan: C — Ha (below Saga La, Sele La), Thimphu (Simkarap, Pajoding, above Ragyo), Bumthang (above Gortsam) and Tashigang (Dib La) districts; **N** — Upper Mo Chu district (Gasa to Pari La); **Darjeeling** (Sandakphu, Tonglu); **Sikkim** (Lachen, Yampung, Zemu Valley, Yakla, Natong, Tumbok).

Conifer (*Tsuga, Rhododendron, Abies*) forest; juniper scrub, 2440–4270m. Fl. June–July.

12. P. kansuense Maximowicz ex Batalin; *P. erythrocarpum* Hua. Med: *rangey*. Plate 2.

Differs from *P. verticillatum* and *P. leptophyllum* chiefly in its purplish flowers, commonly restricted to the lowest few nodes, leaves 1–4 per whorl, often inconstant within same plant; anthers sessile and larger (1.7–2.9mm). Plant relatively small (19–39(–53)cm) at least at flowering, leaves narrow (2.9–10mm wide), hooked when young (hence often misidentified as *P. cirrhifolium* in herbaria).

Bhutan: C — Ha (Ha, Tare La, Kale La), Thimphu (Pajoding to Motithang, Cheka, Dochu La, above Ragyo), Punakha (Lao La) and Bumthang (E side of Ura La) districts; **N** — Upper Bumthang Chu (Waitang) and Upper Kulong Chu (Shingbe) districts; **Darjeeling** (Sandakphu); **Sikkim** (Dzongri, above Sherabthang, above Lamtung, Laghep, Kailee, Bikbari). Clearings and margins of conifer (*Picea/Tsuga, Abies*) forest; open grassy slopes; juniper and rhododendron scrub, 2740–4270m. Fl. May–July.

Hara's record (F.E.H.2) of *P. graminifolium* later re-determined by him as the Chinese *P. curvistylum* (F.E.H.3) presumably refers to this taxon.

According to Mikage & Suzuki (1993) the rhizomes of *P. verticillatum* and *P. cirrhifolium* are used in Tibetan medicine as *Ra-mNye, Ra-Mo-Shag* and *Lug-Mo-Shag*.

2. MAIANTHEMUM G.H. Weber

Rhizomatous herbs. Stems simple; scale leaves present at base; lower part commonly bearing 1 or more intermediate leaves. Stem leaves alternate, simple, usually subsessile, sometimes petiolate. Infl. a terminal raceme or panicle. Flowers shortly pedicellate, subtended by minute bracteoles; tepals 6, free to base, ± similar, or outer 3 smaller than inner 3. Stamens 6; filaments linear or swollen; attached at base of tepals; anthers small, dorsifixed. Ovary superior, of 3 locules each with 2 ovules; style with 3 often recurved stigmatic lobes. Fruit a 1–3-seeded berry.

1. Petioles over 1cm; inner and outer tepals strongly differentiated (outer shorter, more rounded; inner longer, narrower: Fig. 6d); filaments fleshy, swollen .... **1. M. fuscum**
+ Leaves sessile or petiole under 1cm; inner and outer tepals ± similar; filaments linear 2

2. Tepals greenish; infl. axis and pedicels glabrous .... **2. M. tatsiense**
+ Tepals purplish or whitish; infl. axis and pedicels hairy 3

3. Leaf margins ciliate; inner tepals usually under 5mm; anthers $\pm$ round, under 1mm; style under 1mm; infl. a cylindric raceme (sometimes with 1 or 2 erect basal branches) .... **3. M. purpureum**
+   Leaf margins not ciliate; inner tepals usually over 5mm; anthers oblong, over 1mm; style over 1mm; infl. a panicle with widely spreading branches .... **4. M. oleraceum**

1. M. fuscum (Wall.) La Frankie; *Smilacina fusca* Wall. Fig. 6c–d.

Stem 16–60cm, glabrous, or occasionally densely hairy in upper part. Scale leaves usually 2, whitish, membranous. Intermediate leaf usually 1, linear, partly green. Leaves 3–6, blades ovate (sometimes narrowly elliptic), acuminate to caudate, margins narrowly membranous, minutely serrulate, sometimes ciliate, base cordate (sometimes cuneate), 5.5–15.5(–18) × 3–8.7cm, glabrous or sometimes with short white hairs above and beneath; petioles 1–4cm. Panicle 5–17cm, spreading irregularly; pedicels and branches glabrous, slender; bracteoles minute, lanceolate, brownish. Outer tepals greenish, orbicular, blunt, concave, 2–3.5 × 1.5–3mm; inner tepals edged purple, orbicular to elliptic, larger, 2.2–4.5 × 2–3.5mm. Filaments fleshy, flattened-globose, 0.5–1.1mm diameter; anthers attached along whole of back to tip of filaments, facing upwards, round, 0.4–0.6 × 0.4–0.7mm. Ovary obscurely trigonous, depressed, 1–1.7mm diameter; stigma lobes extremely short, erect, arising directly from ovary. Berry red, 1 (or more?)-seeded.

Bhutan: C — Thimphu (Dotena, Dochu La, Olaka), Punakha (SW of Wangdi Phodrang, Phobsikha), Tongsa (W of Chendebi, Shing Jakpa La), Mongar (W side of Donga La) and Tashigang (Tashi Yangtsi Dzong) districts; **N** — Upper Mo Chu (Tamji to Gasa (F.E.H.2)) and Upper Kuru Chu (Denchung) districts; **Darjeeling** (Rechi La, Sonada, Srikola, Sukia, Rhikisum, Mahalderam, Takdah); **Sikkim** (Lachen, Tendong, Tumlong). Shady banks in wet (incl. bamboo and *Tsuga*) forest, 1500–3660m. Fl. April–June.

Plants with a sparse covering of short white hairs on leaves, leaf margins, infl. axis and branches were described as *S. fusca* var. *pilosa* Hara, but intermediate states occur (e.g. with hairs only on the infl.) and Hara's *S. fusca* f. *papillosa* probably also represents an intermediate condition. I have therefore included these in the above description and distribution.

2. M. tatsiense (Franchet) La Frankie; *Smilacina tatsiensis* (Franchet) Wehrhahn; *S. yunnanensis* (Franchet) Handel-Mazzetti.

Stem (9–)21–43cm, glabrous, slightly zigzag above. Scale leaves 0–2, whitish or brownish, membranous. Stem leaves 5–9, elliptic (sometimes narrowly), shortly acuminate or contracted below rounded apex, base rounded and slightly clasping or very shortly petiolate, 6–9 × 1.7–4cm, glabrous, veins beneath minutely papillose. Infl. 1.5–6cm, a short raceme, sometimes with several ascending branches from base, axis slightly zigzag, glabrous, very narrowly membranous winged; pedicels to 3.5mm; bracteoles oblong, greenish membranous, conspicuous. Tepals greenish or purplish-green, lanceolate, blunt, margins narrowly membranous, 2.8–3.9 × 0.8–1.1mm. Filaments linear, c.0.3mm; anthers ± round, c.0.6 × 0.5mm. Ovary widely ovoid, depressed, c.1 × 1.4mm; stigma lobes arising directly from ovary, erect, c.0.4mm. Berry orange, c.1.2cm diameter, 2-seeded.

Bhutan: S — Phuntsholing district (Sorchen to Gedu); **C** — Ha (Damthang), Thimphu (Dotena; Tzatogang to Dotanang, Nala to Tzatogang (F.E.H.2)), Punakha (Wangdi Phodrang), Tongsa (Longte Chu) and Mongar (Lhuntse) districts; **N** — Upper Mo Chu (above Gasa Dzong) and Upper Kulong Chu (Lao) districts; **Chumbi** (Do-ree-chu). Moist (incl. broad-leaved and *Tsuga*) forest; juniper/rhododendron scrub, 1700–3280m. Fl. May–June.

According to Hara (1968a) this species is functionally dioecious. I have only dissected one flower and it had apparently functional stamens and ovary. Specimens at BM also seem to be bisexual.

3. M. purpureum (Wall.) La Frankie; *Smilacina purpurea* Wall.; *S. oligophylla* (Baker) Hook.f.; *S. pallida* Royle. Fig. 6e.

Stem 14–76cm, hairy above. Scale leaves 1–3, broad, whitish, membranous. Stem leaves 5–8(–10), lower ovate, upper narrowly elliptic, acuminate to rounded apex, margin ciliate, suddenly contracted to slightly clasping base, 4–16.3 × 1.5–7cm, glabrous or occasionally sparsely hairy above, short white bristly hairs present especially on veins beneath; petiole absent or under 3mm. Infl. 3.5–15cm, a cylindric raceme (occasionally with a few basal branches); pedicels 0.4–1.3cm, infl. axis and pedicels densely hairy; bracteoles small, lanceolate, brownish, membranous. Flowers dark red, purplish or reddish-brown outside, whitish or pale greenish inside, occasionally entirely white, outer tepals oblong-elliptic, blunt, 4–6.5 × 1.8–3.8mm, inner tepals lanceolate to oblong-elliptic, blunt, 4–6.4 × 1.3–4.3mm. Filaments 0.7–1.8mm, widened towards connate bases; anthers round 0.4–0.8mm. Ovary ovoid to subglobose, obscurely 3-lobed, 1–2 × 1.1–2mm; style 0.2–0.7mm or sometimes absent; stigma lobes 3, recurved, c.0.3mm. Berry c.1cm.

Bhutan: C — Ha (Ha, Sele La), Thimphu (above Pajoding, above Ragyo; Cheka (F.E.H.2)), Bumthang (above Gortsam) and Mongar (W side of Thrumse La, Pung La) districts; **N** — Upper Mo Chu (above Naha, Kangla Karchu La), Upper Pho Chu (Loona), Upper Bumthang Chu (Kantanang) and Upper Kulong Chu (Shingbe) districts; **Darjeeling** (Sandakphu, Tonglu; Phalut (F.E.H.1)); **Sikkim** (Changu, Megu, Sakkargong, Lingtu, Teumtong,

Dzongri, etc.); **Chumbi** (Yatung, Pit za La). *Abies* and juniper forest; rhododendron scrub, 3500–4270m. Fl. May–July.

Young shoots eaten.

White-flowered forms predominate in the W Himalaya but are rare in our area; they have been recognised as *Smilacina pallida* Royle (though this name seems not to have been validly published) and were treated by Hara (1987) as *S. purpurea* f. *albiflora* (Wall.) Hara. Small plants with up to 5 stem leaves and short racemes are not worth recognising; they were treated by Hara as *S. purpurea* f. *oligophylla* (Baker) Hara, but the species decreases in stature with altitude, with no significant correlated morphological characters. The above description includes both pale-flowered and dwarf variants.

4. M. oleraceum (Baker) La Frankie var. **oleraceum**; *Smilacina oleracea* (Baker) Hook.f. ?Lep: *chokli-bi*.

Stem (21–)52–152cm, sometimes slightly zigzag, hairy above. Stem leaves 8–12, lanceolate or broadly to narrowly elliptic, gradually acuminate or caudate, margins not ciliate, base rounded to truncate, occasionally cuneate, (11–)14–25.5 × (4.4–)5.3–12.5cm, usually glabrous above, with tiny appressed hairs beneath, especially on veins; petiole 0.5–0.7cm. Infl. 2–26 × 4–23cm, a many-branched panicle, branches spreading to ascending, infl. axis, branches and pedicels densely hairy; bracteoles tiny, filiform. Flowers white, sometimes flushed pink or mauve, outer tepals oblanceolate to elliptic, 2.3–4mm wide, inner tepals elliptic to broadly elliptic (occasionally obscurely rhombic), 4.5–7.5 × 3.2–5mm. Filaments linear 1–3mm; anthers oblong, 1–1.7 × 0.5–0.7mm. Ovary globose, obscurely 3-lobed, 1.4–2mm diameter; style 1.5–3mm, 3-lobed to varying degrees (to half length of style), stigma lobes sometimes recurved. Berry pale mulberry-coloured, 1–2-seeded, to 8mm diameter.
    Bhutan: S — Chukka district (Chimakothi to Putlibhir (F.E.H.2)); C — Ha (Ha, Jaccha below Sale La), Thimphu (Zado La, Tsalimaphe, Pumo La, above Ragyo), Punakha (Dochu La, Pele La, Mara Chu valley), Bumthang (above Lami Gompa, Yuto La), Tongsa (Chendebi) and Sakden (Sakden) districts; N — Upper Mo Chu (between Laya and Gasa, Gasa to Kohina), Upper Kuru Chu (Singhi Dzong) and Upper Kulong Chu (Lao) districts; **Darjeeling** (Sonada, Ghumpahar, Tonglu, Rungbool, Jorepokri Bungalow, Senchal, etc.); **Sikkim** (Lingtu, Zemu Valley, Laghep, Tsoka to Yoksum, Dzongri, Samding, etc.). *Abies* and rhododendron forest (edges and clearings, ravines, streamsides), 2130–3350(–3700)m. Fl. May–August.

Eaten as a vegetable (Hooker, 1854, 2: 48).

var. **acuminatum** (Wang & Tang) Noltie.
Flowers deep wine-red to blackish; infl. laxer, axis zigzag, pedicels longer.
Bhutan: C — Thimphu (Pumo La, Pyemitangka), Punakha (Maru, Lao La, Tang Chu) and Tongsa (Sephu) districts; **N** — Upper Kulong Chu district (Shingbe); **Sikkim** (Tsoka to Jamlinghang, Lingtu); **Chumbi** (Yatung, Rinchinging). *Abies* forest — at higher altitudes than type variety, (2740–)3050–3700m.

Although the two varieties are very distinct in the field and were found at different altitudes in W Sikkim, they are more difficult to distinguish in the herbarium and intermediates seem to occur, for example forms with open, flexuous panicles but white flowers (Lachen, Tangu). Records of fruiting and otherwise uncertain specimens are placed under the type var. above.

3. THEROPOGON Maximowicz

Tufted perennial herbs. Roots thickened. Leaves all basal, linear, distichous with overlapping sheathing bases; bladeless sheaths present, decreasing in size to base. Infl. a terminal raceme on a flattened, leafless scape; pedicels not jointed, petaloid; flowers borne singly, subtended by a bract and a bracteole. Tepals 6, in two whorls, free to base. Stamens 6, inserted at base of tepals, filaments wide, flattened, anthers basifixed. Ovary superior. Style simple, filiform. Fruit a several-seeded berry.

1. T. pallidus (Kunth) Maximowicz. Fig. 6f–g.
Leaves grass-like, channelled, acute, 15–56 × 0.3–1.2cm, midrib prominent, sheathing bases papery. Scape 15–35cm, almost equalling leaves, usually several-angled or narrowly winged in upper part. Infl. 3–18-flowered, somewhat flexuous, 1.5–10cm; bracts and bracteoles linear with membranous margins, bracts shorter than to exceeding flowers; pedicels curved downwards, 0.3–1.6cm. Flowers subglobose, white sometimes tinged pinkish or purplish; tepals 6.5–7.7mm, apex slightly hooded, thick textured, outer lanceolate, 2.3–3.7mm wide, inner ovate, 3.5–4.3mm wide. Filaments 1–1.7mm; anthers oblong to triangular, apex acute, lobes divergent to cordate base, 2–4 × 0.9–2mm. Ovary globose, 1.8–2.8mm; style 4.3–5mm. Berry c.7mm, up to 12-seeded. Seeds c.2mm, round, pale orange-brown.
Bhutan: C — Thimphu (Drugye Dzong, Tashi Cho Dzong, Dotena; Gunisawa to Cheka (F.E.H.2)), Punakha (Ritang) and Tongsa (Chendebi) districts; **Sikkim** (Lachen, Mempup, below Tangu, Chungtam). Dry rocky hillsides; shady banks and cliffs in forest (incl. oak), 1830–3970m. Fl. May–August.

4. OPHIOPOGON Ker Gawler

Perennials with thick rootstocks, roots often tuber-bearing; sometimes with underground stolons. Stout leaf bearing stem sometimes developed or leaves in basal rosettes. Leaves linear, or differentiated into blade and petiole. Infls. spike-like racemes borne on often flattened, leafless scapes, flowers borne singly or in small groups subtended by a bract and a bracteole. Pedicels jointed, lower part herbaceous, upper part tepaloid. Flowers epigynous, campanulate, whitish or purplish, tepals 6, free. Stamens 6, inserted opposite tepals; anthers basifixed, oblong to lanceolate, bases cordate. Ovary inferior, of 3 locules, each with usually 2 basal ovules; style about equalling tepals, slightly toothed at apex. Top of ovary breaking irregularly after flowering exposing developing seeds. Seeds large, with bluish-black fleshy sarcotesta, resembling berries.

Measurements for pedicels are from scape to base of ovary; those of flower size from base of ovary to tip of tepals (flower diameter being impossible to measure on herbarium specimens).

1. Leaves with distinct petiole and oblanceolate blade .. **1. O. dracaenoides**
+ Leaves linear, grass-like 2

2. Plant not stoloniferous (though sometimes with thick, elongate rootstock) 3
+ Plant stoloniferous 4

3. Infl. usually rigidly curved, less than ¼ length of scape, fewer than 14-flowered; flowers funnel-shaped, small (4–6.7mm), usually shorter than bracts .... **2. O. wallichianus**
+ Infl. straight or flexuous, ⅓ to ½ scape, 15–50-flowered; flowers campanulate, to 10mm, usually exceeding bracts .... **4. O. intermedius**

4. Scape exceeding leaves; leaves 0.7–2.8mm wide; infl. 3–9-flowered; flowers to 8.5mm, always violet; anthers to 3.5mm; stolons very slender .... **3. O. cf. bodinieri**
+ Scape shorter than leaves; leaves over (3–)4mm wide; infl. 4–13-flowered; flowers to 11mm, usually white; anthers to 5.1mm; stolons stout .... **5. O. clarkei**

1. O. dracaenoides (Baker) Hook.f.
Stem decumbent to semi-erect, 2–3.5mm diameter, bearing membranous, encircling scales. Leaves borne along stem and in apical rosettes, blades oblanceolate, cuspidate, blunt, narrowed below, to 12 × 4cm, longitudinal veins

many, strong; transverse veinlets weak, irregular; petiole to 8.5cm. Scape c.5cm; infl. to 20-flowered, narrowly cylindric, to 7cm; bracts shorter than flowers, lanceolate, margins scarious. Flowers white or pale blue, borne singly or lower in pairs, c.6.2mm, tepals narrowly lanceolate, acute, c.5.5mm; pedicels c.5mm. Seeds oblong-ellipsoid, c.9 × 7mm; sarcotesta pale blue.

Sikkim (Hee); **Darjeeling** (below Kurseong, Sureil, Labdah). 1220–1830m. Fl. July.

This interesting species has apparently not been collected recently.

O. intermedius agg. (spp. 2–4)
Several distinct forms of this extremely difficult group occur in Bhutan and Sikkim, but their nomenclature is far from being resolved and the treatment is provisional. Much work remains to be done and more collections are required with more care in collecting underground parts and fuller annotations on ecology and flowering time. No literature records have been accepted.

2. O. wallichianus (Kunth) Hook.f.; *O. parviflorus* (Hook.f.) Hara; *O. intermedius* var. *parviflorus* Hook.f. Fig. 6h.

Rootstock thick, knobbly; some roots bearing oblong-ovoid tubers to 2 × 0.7cm, tapered into root at both ends; apparently never stoloniferous. Leaves in dense tufts, sometimes swollen at base. Leaves linear, stiff, curved, rough with margins and some veins minutely serrate, bases with expanded, papery sheathing margins, 17–59cm × 2.3–5(–6.5)mm. Scapes flattened, angles slightly winged, usually $\frac{2}{3}$ length to slightly exceeding leaves, (10–)16–37cm. Infl. (2–)4–14-flowered, stiffly curved, (1–)2.5–7cm ($\frac{1}{10}$–$\frac{1}{4}$ length of scape); bracts linear-lanceolate, usually exceeding flowers. Flowers usually bluish or violet, sometimes white, scarcely overlapping, mostly borne singly or lowest ones paired, funnel-shaped, 4–6.7mm; tepals ± erect, elliptic, blunt, 3–5.4mm; pedicels 2.3–4.6mm. Filaments 0.2–0.3(–0.5)mm; anthers 1.2–3.8 × 0.4–0.9mm. Seeds ellipsoid, c.6 × 5mm, translucent, brownish; sarcotesta bluish-black.

Bhutan: S — Phuntsholing, Chukka and Gaylegphug districts; **C** — Thimphu, Tongsa, Bumthang and Mongar districts; **N** — Upper Mo Chu and Upper Bumthang Chu districts; **Darjeeling** (Senchal, Palmajua to Batasi, Labdah, Ryang Bridge, etc.); **Sikkim** (Lachen, Phedang to Choka, Karponang, Burmiok, Kabi, Yoksum to Bakkim, etc.). Wet shady places usually in forest (bamboo, *Abies*, *Tsuga*/deciduous, broad-leaved evergreen (incl. oak)); rhododendron/juniper scrub, sometimes epiphytic, (270–)910–3960m. Fl. (June–)July–August.

Small-flowered forms are not worthy of recognition and grade imperceptibly into typical *O. wallichianus.*

The following specimens are atypical but agree with each other in having large flowers, but short infls. on long scapes, and are probably best placed under this species: Punakha (Wangdi Phodrang, *Ludlow & Sherriff* 206, BM), Tongsa (Mangde Chu, *Bowes Lyon* 15062, BM) and Tashigang (Gamri Chu, *Ludlow & Sherriff* 579, BM; *Bowes Lyon* 9132, E) districts.

3. O. cf. bodinieri Léveillé. Fig. 6i–j.

Plant with slender, spreading stolons. Leaves very narrow, often recurved, 7–38cm × 0.7–2.8mm. Scape overtopping leaves, 7.5–26.5cm. Infl. 3–9-flowered, cylindric, straight, 1.5–3.5cm; bracts about equalling pedicels. Flowers always violet, usually borne singly throughout, pendent, very widely campanulate, 6.1–8.5mm; tepals 5.2–7mm; pedicels 3.4–6.3mm. Filaments 0.2–1.1mm; anthers 2.3–3.5 × 0.6–1mm. Seeds c.6 × 5mm; sarcotesta black.

Bhutan: C — Ha (Ha Dzong), Thimphu (Dotena, Pumo La, Drugye Dzong, Taba, Tsalimaphe, Paro, hill above Thimphu Hospital) and Punakha (Ritang, Tasakha) districts. Grassy banks, often among blue pine forest; river beds and banks; marshes, 1830–3810m. Fl. May–August.

This distinctive species, characterised by its slender stolons spreading just below the soil surface (hence easily uprooted), has until now been misidentified as *O. intermedius* or (by Hara) as *O. clarkei*. Similar specimens were seen from Tibet and Yunnan, but until the Chinese species are revised the application of the name is uncertain.

4. O. intermedius D. Don; *O. wallichianus* sensu Hook.f. p.p., non Kunth; *O. intermedius* vars. *macrantha* Hook.f. and *vulgaris* Hook.f. (*nom. nud.*) p.p.

Rootstock stout, elongate, root tubers large (to 6cm); not stoloniferous. Leaves 3.6–10mm wide. Scapes shorter than leaves, 20–33cm. Infl. 15–c.50-flowered, cylindric, ± straight, dense, c.⅓–½ (occasionally more) length of scape, 7–19cm; bracts lanceolate, shorter than flowers. Flowers white or purplish, overlapping, many in groups of 2 or more, cup-shaped to campanulate, 6.2–10mm; tepals spreading, 4.5–6.6mm; pedicels 5.2–9mm. Anthers 2.8–4.3mm.

Bhutan: S — Chukka district (Gedu to Ganglakha); **C** — Punakha district (Mara Chu); **Darjeeling** (Sittong, Sureil, Rungbee, Simsibong, Mehalderam, Selimbong, Senchal, below Tiger Hill, Baghghora); **Sikkim** (Yoksum to Bakkim, Tsoka to Yoksum, above Gangtok, Chumthang, Chiabanjan, Talung). Dense broad-leaved (e.g. *Lithocarpus*/dipterocarp) forest, 1520–2750(–3140)m. Fl. May–June(–July).

There has been much confusion over the application of this name and I have followed the neotypification of Bailey (1929). The taxon in this sense is characterised by early flowering, long, many-flowered spikes of large flowers on scapes shorter than leaves and is apparently rare in Bhutan, being chiefly W Himalayan.

5. O. clarkei Hook.f.

Roots sometimes bearing small tubers (to 2 × 0.6cm). Stoloniferous, stolons initially slender (c.1.5mm diameter), covered with pale brown papery scales, later becoming thickened (to 5.5mm). Leaves borne in tufts, often swollen at base, 22–38cm × (3.1–)4–7mm, initially minutely serrate on margins, veins usually smooth. Scapes 12–20(–27)cm, usually less than half length (occasionally equalling) leaves, strongly curved below infl. Infl. 4–13-flowered, rather lax, 2–6cm; bracts conspicuous, with broad membranous margins, almost equalling flowers. Flowers usually pure white (occasionally flushed pink or lilac), spreading to erect, borne singly, cup-shaped, 7–11.1mm; tepals 6–8.5mm; pedicels 3–5.9mm. Filaments 1–2mm; anthers 3.7–5.1 × 0.9–1.6mm. Seeds 6–7.5 × 4.5–6.5mm; sarcotesta blue.

Bhutan: C — Tongsa (Tongsa), Mongar (above Namning) and Mongar/ Tashigang (Donga La) districts; **N** — Upper Kulong Chu district (Lao); **Darjeeling** (Senchal, Rungbee, between Mount Tonglu and Sandakphu, Kurseong, Darjeeling, Sittong, Palmajua to Batasi, Sukia Pokhri to Manibhanjang, below Tiger Hill); **Sikkim** (Bakkim to Tsoka, Chakung Chu). Banks and mossy rocks in wet broad-leaved forest, 1520–3050m. Fl. July–October.

Additional cultivated ornamental:

Liriope muscari (Decaisne) L.H. Bailey var. **variegata** L.H. Bailey superficially resembles *Ophiopogon* but differs in having flowers with a superior ovary and borne in clusters. Grown in Gangtok.

5. PELIOSANTHES Andrews

Perennial, rhizomatous herbs. Roots thickened, densely hairy. Leaves and papery scales scattered along rhizome, or mainly in apical rosette. Leaves with expanded blade and petiole. Infl. a cylindric raceme on a scape bearing several small bracts. Flowers borne singly or in small groups subtended by a bract and a bracteole. Flowers greenish or purplish, thick textured; tepals 6, in two whorls, tubular below and fused to ovary to varying degrees. Filaments united to form a corona attached at base of tepals; anthers 6, attached by backs to inside face of corona. Ovary inferior to semi-inferior, of 3 locules, each locule with 2 or more basal ovules, apex of ovary rupturing as in *Ophiopogon*; style stout, divided at apex into 3 very short stigmatic lobes. Seeds with bluish fleshy sarcotesta.

1. Bracts all single-flowered 2
+ At least lower bracts subtending more than one flower .... **3. P. teta**

2. Leaf blade over 30cm, thick textured; longitudinal veins more than 50,
unequal in thickness; cross-veinlets not visible or if so then irregularly
spaced and oblique; flowers over 1cm diameter .... **1. P. macrophylla**
+ Leaf blade under 30cm, thin textured; longitudinal veins fewer than
25, equal in thickness; cross-veinlets conspicuous, closely spaced, horizontal; flowers under 0.5cm .... **2. P. griffithii**

1. P. macrophylla Wall. ex Baker. Nep: *chille dhotisara*. Fig. 6k–l.
   Rhizome stout (c.1cm diameter), oblique. Scale leaves to 20cm. Rosettes
2–4-leaved. Leaf blade narrowly elliptic, cuspidate, margins sometimes crimped,
gradually tapered to base, (28–)35–56 × (5.3–)8–13cm, longitudinal veins over
50, unequal in thickness, cross-veinlets usually not visible, if visible then weak,
oblique, irregularly spaced, subcoriaceous; petiole (24–)37–67cm. Scape bearing
several flowerless bracts throughout length, stout (3.5–4mm diameter), 14–37cm.
Raceme 30–60-flowered, stout, 14–31cm (at maturity); bracts each bearing a
single flower, lanceolate to linear-lanceolate, shorter than to exceeding flower,
brownish, membranous. Flowers greenish or greenish-purple, c.1–1.2cm diameter, tepals blunt, 3.5–6mm, outer ovate, 3–4.1mm wide, inner 2.5–2.7mm wide;
pedicels recurved, stout, 3–4mm. Anthers oblong, 1.5–2 × 0.8mm. Ovary
c.1.7–2.5mm diameter; stigma 1–2mm. Seeds oblong c.10 × 5mm.
   Bhutan: S — Phuntsholing (Kamji) and Gaylegphug (c.15km N of
Gaylegphug, Karai Khola above Aie bridge) districts; **Terai** (Jalpaiguri,
Mahanadi); **Darjeeling** (Labdah, Kurseong, Rongbe, Punkabari, Rishap,
Ryang, Brik, Mongpu, Sureil, Kalighora, Tista, Lal, Latpanchor Forest).
Subtropical forest (shaded bank above stream), 300–1830m. Fl.
February–May.

2. P. griffithii Baker; *P. bakeri* Hook.f.; *P. violacea* Wall. ex Baker var. *minor*
Baker.
   Differs from *P. macrophylla* as follows: scale leaves shorter (to 7cm);
rosettes 4–10-leaved; petioles shorter (5–25cm); leaf blades smaller (9.5–20 ×
2.5–6.1cm), more widely elliptic, apex subcaudate, thinner textured, parallel
veins fewer (9–22), all similar in thickness, cross-veinlets conspicuous, evenly
spaced and horizontal; scape very short (under 5cm), concealed by scale leaves;
raceme 9–17cm (lengthening at maturity to c.30cm), very slender; flowers
yellow, smaller (c.4.5mm diameter, outer tepals c.2.5 × 1.6mm, anthers c.0.4
× 0.5mm, ovary c.1.2mm diameter, stigma c.0.9mm); pedicels very slender;
seeds fatter (c.9.5 × 6.5mm).

Terai (Mal Forest); **Darjeeling** (Brip, Rayang Valley, Little Rangit, Selim, Rishap, Labdah, Mongpu, Lankoh; Takdah to Rayang (F.E.H.1)). 240–1520m. Fl. September–November.

3. P. teta Andrews.
Differs from both the above in having more than 1 flower per bract.
I have seen a single specimen ex Mongpu (Darjeeling) and Jessop (1976) cites a specimen from 'Sikkim'.

6. TUPISTRA Ker Gawler

Herbs with thick, creeping rhizomes. Leaves in rosette at apex of rhizome, narrowly elliptic to oblanceolate, commonly narrowed into petiole-like base. Scape stout, leafless, bearing a dense terminal spike, with sessile flowers each subtended by a bract. Flowers fleshy, campanulate, with 6(–8) free perianth lobes fused into broad tube below. Stamens 6(–8) inserted on tube below lobes; anthers dorsifixed; filaments with lower parts expanded and fused to tube, upper part free, sometimes very short, sometimes deflexed. Ovary superior, globose, slightly lobed, of 3(–4) locules, each with 2 axile ovules; style stout, sometimes very short; stigma peltate or of 3(–4) ridge-like lobes. Fruit a globose berry, commonly 1-seeded.

The retention of *Campylandra* Baker as a separate genus seems unjustified in view of the range of variation in filament (length and posture) and stigma (lobed or peltate) characters of Chinese species.

1. Stigma peltate; perianth lobes not papillose on either surface; bracts blunt, shorter than flowers .... **1. T. nutans**
+ Stigma not peltate; perianth lobes papillose on margins and one face; bracts acute, longer than flowers 2

2. Leaves narrowly elliptic to oblanceolate, usually over 3.7cm wide, narrowed into petiole below; fused (lower) part of filaments with indistinct, smooth margins .... **2. T. wattii**
+ Leaves ± parallel-sided, under 2.8cm wide, petiole scarcely distinct; fused part of filaments with distinct, irregularly crested margins (Fig. 6o) .... **3. T. aurantiaca**

1. T. nutans Wall.; incl. *T. clarkei* Hook.f. Nep: *shan khuley*, *nakima*.
Scale leaves persistent, oblong, acute, bases encircling rhizome, to 34 × 3cm, pale brown, membranous. Rosettes 2–6-leaved; blades narrowly oblance-olate, asymmetric about midrib, contracted below apex, 36–139 × 3.3–11.5cm,

midrib thickened on underside, narrowed below into distinct petiole 13–39cm. Scape 8–13cm. Spike densely cylindric, curved, (3–)4.5–8cm; bracts short, blunt, membranous. Flowers brownish-green streaked purple, fleshy, sessile; perianth tube broadly campanulate, 5.5–10mm; lobes triangular, not papillose, 7–9 × 5.5–7mm. Free part of filaments extremely short, so anthers subsessile; lower parts expanded, fused to tube; anthers oblong, 1–2mm. Ovary globose, c.3mm, gradually narrowed upwards into style 3.5–7mm; stigma fleshy, peltate, whitish, c.3–3.5mm diameter, usually closing mouth of tube.

Bhutan: S — Samchi (Sangura) and Gaylegphug (Betni) districts; **Terai** (Buxa to Bhutan road 31 miles from Jalpaiguri); **Darjeeling** (Darjeeling, Sureil); **Sikkim** (Tinding, Yoksum, Phodong to Kabi, N of Rangit, Tumlong, Luybik, Rungji, Bakkim, Rungbee, head of Rayang Valley, Dikeeling; Pamianchi to Tingling Bridge (F.E.H.3)). Broad-leaved forest, often on rocky slopes; under *Alnus* in cardamom plantation, 370–2440m. Fl. September–December; fr. July.

Perhaps occurring wild but also cultivated in gardens for its flowering spikes which are used as a vegetable in Gangtok and Samchi; fruits edible; Hooker records the smoking of petioles in hookahs.

T. clarkei is supposed to have larger leaves, a longer style and smaller stigma (not closing tube mouth) than *T. nutans*, but this distinction seems not to be valid from the very inadequate specimens seen. New collections are badly needed. Many of the old specimens are labelled *T. squalida* Ker Gawler, but this appears to be a different species described from cultivated material of uncertain origin.

2. T. wattii (C.B. Clarke) Hook.f.; *Campylandra wattii* C.B. Clarke. Nep: *nakshima*.

Rhizome 0.8–1.5cm diameter, cylindric, with circular scars; roots thick, felted. Scale leaves not persistent, to 10.5 × 0.8cm, membranous. Rosettes 4–5-leaved; leaves distichous, bases overlapping, sheathing, blades narrowly elliptic to oblanceolate, gradually acuminate to very acute apex, gradually narrowed into petiole-like base, 28–60 × (2.7–)3.7–8cm, apparently thin-textured. Scape 2.5–9cm. Spike with apical tuft of bracts, densely cylindric, 2.5–4cm; bracts lanceolate, margins minutely ciliate near apex, exceeding flowers, to 1.8(–2.2)cm, largely membranous. Flowers green becoming pale orange; perianth tube c.4–5.5mm; lobes triangular, c.2.5 × 3mm, fleshy. Upper part of filaments 1.5–2.5mm, rigidly recurved so anthers sunk into depression in ridge running down centre of lower (fused) part of filaments, divergent edges of lower filaments indistinct, smooth; anthers c.1.1mm. Ovary globose, obscurely 3-lobed, 2.2–2.4mm; style to 0.3mm, with 3 stigmatic ridges at apex. Berry 0.85–1.8cm, scarlet, globose or two-lobed, 1–2-seeded.

Bhutan: S — Gaylegphug district (N of Jumudag); **C** — Punakha (SW of Wangdi Phodrang, Samtengang to Tsarza La), Tongsa (above Dakpai, Kinga Rapden) and Tashigang (Balfi) districts. Evergreen oak forest; rocky slope under scrub in dry valley, 1800–2780m. Fl. April–June; fr. February.

3. T. aurantiaca (Wall. ex Baker) Hook.f.; *Campylandra aurantiaca* Wall. ex Baker. Fig. 6m–o.
Differs from *T. wattii* in its narrower (1.1–2.8(–3.9)cm), more parallel-sided leaves; longer (to 3.7cm) more herbaceous bracts; larger perianth lobes (4–6 × 3.5–5mm); free (upper) parts of filaments less strongly recurved, divergent edges of fused (lower) parts conspicuously toothed.
Darjeeling (Darjeeling, Mahalderam, near Senchal, Kurseong; Victoria Fall, Batasi (F.E.H.1)); **Sikkim** (Yoksum to Bakkim, Rungji Ghora, Permaghora; Damthang to Temi, Damthang to Rabong La (F.E.H.1)). Evergreen oak forest, 1220–2440m. Fl. October–May; fr. July.

All Bhutanese specimens seen labelled as this have been re-determined as *T. wattii* — to which the Japanese records (F.E.H.2) from Punakha and Tongsa districts should probably also be referred. Further fieldwork and collections are required to see if the two species are really distinct or possibly geographical races, with the eastern form being broader-leaved.

Family 193. ASPARAGACEAE

Rhizomatous, shrubby or climbing perennials; stems herbaceous or woody. Leaves reduced to scales which may become spiny. Photosynthetic organs are linear, leaf-like cladodes (modified stems). Flowers solitary or in clusters or racemes, radially symmetric, unisexual or bisexual, tepals 6, tubular at base. Stamens 6, reduced in functionally female flowers; filaments free; anthers dorsifixed, dehiscing inwards. Ovary superior, of 3 locules, ovules axile; style divided at apex into 3 stigmatic lobes. Fruit a berry.

FIG. 7. **Asparagaceae, Dracaenaceae, Agavaceae, Hypoxidaceae, Phormiaceae.**
a, **Asparagus racemosus** var. **racemosus**: habit (× ½). b–d, **Dracaena angustifolia**: b, habit (× ⅙); c, half flower (× 1.5); d, fruit (× 1). e–g, **Agave lurida**: e, habit (× ¹⁄₃₆); f, fruit (× ⅓); g, half flower (× ¼). h, **Furcraea selloa**: half flower (× ⅓). i–j, **Hypoxis aurea**: i, habit (× ⅔); j, seed (× 12). k–n, **Molineria capitulata**: k, habit (× ¹⁄₁₅); l, half flower (× 1.5); m, stamen (× 3); n, anther t.s. (× 6). o–q, **Curculigo orchioides**: o, half flower (× 2); p, stamen (× 5); q, anther t.s. (× 10). r–t, **Dianella ensifolia**: r, habit (× ¹⁄₁₅); s, stamen (× 5); t, leaf base (× 1). Drawn by Mary Bates.

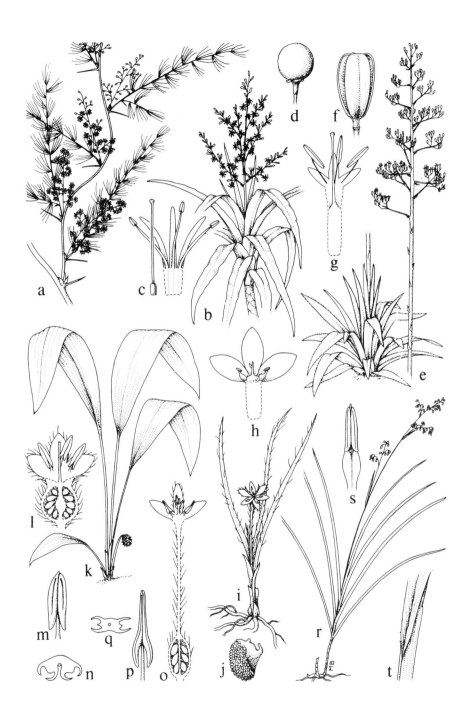

1. ASPARAGUS L.

Herbaceous or woody, rhizomatous perennials (herbs, shrubs or climbers). Roots usually tuberous. True leaves reduced, scale-like, bases often developed as spines. Cladodes linear, borne singly or in whorls in axils of scale leaves. Flowers white or greenish, hermaphrodite or functionally unisexual, borne singly, in clusters, racemes or occasionally panicles. Pedicels jointed. Perianth segments 6, similar, fused into short, broad tube at base. Stamens 6, attached at base of free part of perianth segments; anthers dorsifixed. Ovary of 3 locules each with 2 or more axile ovules. Style short, divided at apex into 3 stigmatic lobes. Berry red or black, 1–3-seeded.

Roots of several species are of medicinal importance in SE Asia (Satyavati *et al.*, 1976).

The name *nga-la-nyom* (Sha) refers to a species of *Asparagus*.

1. Flowers in racemes or panicles; cladodes 3-angled; stems woody
    1. A. racemosus
 + Flowers borne singly or in 2s or 3s; cladodes distinctly flattened or needle-like; stems herbaceous 2

2. Cladodes needle-like in whorls of usually >10; cultivated
    3. A. officinalis
 + Cladodes flattened, in whorls of 3–6; wild 3

3. Peduncles exceeding cladodes .... **2. A. filicinus** var. **giraldii**
 + Peduncles shorter than cladodes 4

4. Cladodes usually under 1mm wide, inserted in 4s and 5s; flowers usually under 3.5mm .... **2. A. filicinus** var. **filicinus**
 + Cladodes over 1.2mm wide, inserted in 3s; flowers c.4mm
    2. A. filicinus var. **lycopodineus**

1. A. racemosus Willdenow var. **racemosus**. Nep: *satamuli, satavali*; Sha: *malakma*. Fig. 7a.

Densely branched climber. Old stems woody, with ± straight spines to 1.5cm. Cladodes in whorls of 5–8(–11), linear, slightly curved, V-shaped in cross-section, upper surface channelled, 5–22 × 0.5–0.6mm. Flowers in racemes or sometimes panicles inserted at base of branches or branchlets, racemes 1.8–9.5cm (panicles to 18cm); pedicels subtended by minute membranous bracteoles, upper part of pedicel 2.2–3mm, longer than persistent lower part (1.5–2.5mm). Flowers white, fragrant, 4–4.5mm, tepals narrowly oblanceolate, rounded, 3–4mm. Filaments 2.1–2.8mm; anthers round, c.0.5mm. Ovary obovoid, narrowed to base, 1.7–2.3 × 1.1–1.7mm; style 0.3–0.5mm; stigma lobes recurved. Berry black, 0.5–0.8cm, globose or lobed, 1–2-seeded.

Bhutan: S — Phuntsholing district (Kamji); **C** — Punakha (Wangdi Phodrang), Tongsa (N of Shamgong) and Tashigang (Ghunkarah, Tashigang) districts; **Darjeeling** (Peshok, Rangit, Balasun Valley); **Sikkim** (N of Naya Bazaar). Dry rocky hillsides; open banks in scrub jungle; by streams, 300–1830m. Fl. September–November; fr. April.

Roots eaten in Sikkim.

var. **subacerosus** Baker. Lep: *machalchampalet*.

Differs from var. *racemosus* as follows: cladodes in whorls of 1–3, longer (0.7–)1.6–3.4cm, ± equally triquetrous in cross-section; spines stouter, more strongly down-curved; racemes much more slender and shorter 0.7–2(–3)cm (to 5cm in fruit); pedicels very slender, lower part longer (2.2–)2.4–3.5mm and exceeding the upper part (1.5–1.7mm); flowers smaller, tepals 2.4–3mm.

Terai (Dulkajhar, Meshbash, Jalpaiguri Duars, Bamunpokri). 150–1220m. Fl. July–September; fr. October–December.

Almost certainly worth recognising at specific rank, but modern collections are required.

2. A. filicinus Buchanan-Hamilton ex D. Don var. **giraldii** C.H. Wright; var. *filicinus* sensu F.B.I. Bhutanese name: *nakhachu*; Nep: *satavara*. Plate 3.

Stem 38–122(–200)cm, herbaceous, without spines. Cladodes in whorls of 4–6, whorls not densely crowded, flat, curved, unequal in length within whorl, very narrow, 4.5–20 × 0.6–1mm. Flowers borne singly or in pairs in axils of cladode whorls in middle and upper parts of main stem and ultimate branches; pedicels very slender, 0.7–1.4cm (to 2.1cm in fruit), lower part 5.5–8mm, slightly exceeding upper part. Flowers white or greenish, 2.6–3.5mm, tepals narrowly oblanceolate, rounded, 2.7–3.5mm. Filaments 1.7–2.1mm; anthers round, 0.6–1 × 0.3–0.6mm. Ovary obovoid to globose, 0.8–1.3 × 0.7–1mm; style c.0.2mm. Berry black, globose, c.7.5mm.

Bhutan: C — Ha (Ha), Thimphu (Paro Chu, Taba, Paga, Drugye Dzong, below Dotena, Motithang Guest House, hill above Thimphu Hospital; Geptemhill (M.F.B.), Thimphu to Chimakothi (F.E.H.2)) and Punakha (Naki) districts. *Picea* forest; dry places among boulders and shrubs, 1830–2980m. Fl. April–June; fr. July–October.

var. **filicinus**; var. *brevipes* Baker.

Differs from var. *giraldii* in having much more closely spaced cladode whorls; flowers inserted only at base of ultimate branchlets on lateral branches; pedicels very short (lower part c.1mm, upper part c.1.5mm); flowers ?unisexual.

Sikkim (Yoksum). 1800m. Fl. May.

var. **lycopodineus** Baker.
Differs from var. *filicinus* and var. *giraldii* in its broader cladodes (1.2–1.5mm), constantly 3 per whorl; larger flowers (c.4mm); larger ovary (c.1.5 × 1.4mm) gradually tapered into longer style (c.1mm), and from var. *giraldii* in its shorter pedicels (2–3mm; lower part c.0.5mm). Position of flowers as in var. *filicinus*.

Bhutan: C — Tongsa (9km SW of Tongsa) and Mongar (above Shersing Thang) districts; **Darjeeling** (below Darjeeling, Rangirun). Mixed broad-leaved forest, 1680–2100m. Fl. May; fr. July–August.

3. A. officinalis L. Nep: *kurilo*.
Differs from the previous species in having long (to 2.5cm), very fine (c.0.2mm wide), needle-like cladodes borne in groups of usually more than 10. Flowers borne in 1–3s at base of ultimate branches; lower part of pedicel c.5mm, slightly exceeding upper. Berry red.
Cultivated in Thimphu and Paro valleys, the young shoots being eaten as a vegetable.

The following S African species are cultivated worldwide for their ornamental foliage:

A. setaceus (Kunth) Jessop. Eng: *asparagus fern*.
Scrambler; cladodes very short (under 1cm) in groups of 8–20; flowers solitary.
Bhutan (Phuntsholing, as pot plant).

A. virgatus Baker.
Dwarf, clump-forming shrub to c.60cm. Stems erect, sharply angled, bearing many suberect branches; cladodes to 9mm, 3-angled, in groups of 3; flowers in groups of up to 6.
Darjeeling (Darjeeling Town and Sonada, in gardens).

Family 194. DRACAENACEAE

Evergreen herbs, shrubs or dwarf trees; stout trunk-like stem developed or not. Leaves with parallel venation, linear to narrowly ovate, sometimes succulent, usually restricted to ends of stems or rhizomes. Infl. axillary, pedunculate, racemose or paniculate. Flowers borne on jointed pedicels, bisexual, actinomorphic, tubular below with 6 similar lobes. Stamens 6, attached at base of corolla lobes; anthers dorsifixed, dehiscing inwards. Ovary superior, of 3 locules, each with a single ovule; style slender; stigma capitate or lobed. Fruit a 1–3-seeded berry.

1. Plant with stout stem to 3m; leaves not variegated, borne at end of stems .... **1. Dracaena**
+ Plant stemless; leaves variegated, erect in a basal rosette   **2. Sansevieria**

1. DRACAENA Vandelli ex L.

Thick trunk-like stem developed. Leaves restricted to ends of stems. Panicle usually broad.

1. D. angustifolia Roxb. Nep: *makai pate*. Fig. 7b–d.
Pachycaul herb. Stem stout, simple or sometimes forked, 0.3–3m. Leaves linear-lanceolate, gradually tapered to acute apex, margins wavy, bases encircling stem leaving V-shaped scars, 38–65 × 2–3.9cm, diagonal veinlets visible between the many fine longitudinal veins, midrib thickened on undersurface. Panicle 14–24cm, with up to 8 branches finally spreading at a wide angle; branches subtended by small leaf-like bracts and bearing many evenly spaced, usually 3-flowered clusters. Flowers cream, corolla lobes oblong, subacute, hooded, 8–9.5 × 1.2–1.5mm, tube c.4mm; pedicels persistent, slender, to 0.7cm, jointed at apex, subtended by small membranous bracteoles. Filaments 5.5–7.3mm; anthers 1.9–2.2mm, locules linear, divergent. Ovary narrowly ellipsoid, 2.1–2.5 × 1–1.4mm; style 10.5–12mm; stigma capitate. Berry 1–1.6mm diameter, spherical or 2-lobed, 1–2-seeded.
Bhutan: S — Samchi (Sibachang to Bhainkhola) and Gaylegphug (Karai Khola above Aie bridge) districts; **Terai/Duars** (Assarbari, Gorumara, Rajabhathawa); **Darjeeling** (Kalimpong, Dalingkota). Subtropical forest, 305–760m. Fl. December; fr. March.

2. SANSEVIERIA Thunberg

Rhizomatous, stem not developed. Leaves commonly variegated, stiffly erect, often succulent, borne in rosette at end of rhizome. Panicle narrow.

1. S. trifasciata Prain. Eng: *Mother-in-law's tongue*.
Leaves erect, c.5 per rosette, linear-lanceolate, flat, channelled towards base, acute, 30–120 × 2.5–7cm, coriaceous, margins cream-coloured, central area mottled in shades of green.
Native to tropical W Africa, but widely cultivated as a pot plant or in gardens in warm areas for its ornamental leaves. Occasionally grown in **Bhutan** (Tashigang Dzong) and **Sikkim** (Gangtok, Namli) to 1800m.

Family 195. AGAVACEAE

S.J. Rae & D.G. Long

Short-stemmed succulent rosette perennials, often suckering, commonly dying after flowering. Leaves linear, thickly coriaceous, fibrous within, spine-tipped, margins prickly. Infl. a spike, raceme or panicle on a long, leafless scape. Flowers bracteate, in umbellate clusters. Perianth of 6 petal-like lobes, united into a tube at base. Stamens 6, inserted at base of perianth lobes; anthers versatile. Ovary inferior, 3-loculed, with numerous axile ovules; style simple, slender, with 3 stigmatic lobes. Fruit a 3-valved, loculicidal capsule. Seeds black, flattened.

1. Trunk not produced; flowers cylindric, filaments and style not swollen at base (Fig. 7g) .... **1. Agave**
+   Stout trunk produced; flowers bell-shaped, filaments and style swollen at base (Fig. 7h) .... **2. Furcraea**

1. AGAVE L.

Flowers with long tube; anthers exserted; filaments and style not swollen at base; style terete; trunk not developed (in Bhutanese species); otherwise as for family.

The name *menchha-na* (Sha) refers to a species of *Agave*.

1. Leaves 12–18cm wide, never variegated, marginal prickles 4–6mm, straight or curved; panicles 6–7m; perianth lobes erect, 18–24mm; capsules oblong, c.5 × 2.5cm .... **1. A. lurida**
+   Leaves 3.5–10cm wide, sometimes variegated, marginal prickles 2–5mm, often hooked; panicles 3–5m; perianth lobes reflexed when dry, 25–30mm; capsules broadly ellipsoid, c.3.5 × 2.5cm
                                                        2. A. angustifolia

1. A. lurida Aiton; *A. vera-cruz* sensu Drummond & Prain non Miller. Nep: *hatibar*; Sha: *shom por shing*. Fig. 7e–g.
    Plant not suckering. Leaves linear-lanceolate, deeply concave, 110–150 × 12–18cm, dull green to glaucous, never variegated, margins nearly straight, prickles regularly and closely (1–2cm) spaced; apical spine 3–4.5cm, 6–8mm wide at base, shallowly grooved; prickles flattened, triangular, straight or curved, (2–)4–6mm, blackish. Panicles 6–7m, with 20 or more ascending

branches in upper ½ to ⅓. Flowers greenish-yellow, 58–65mm, perianth lobes erect, incurved at tip, 18–24mm; pedicels slender. Capsules oblong, c.5 × 2.5cm. Seeds rounded-triangular, narrowly flanged, c.1 × 0.8cm, black.
   Bhutan: C — Tashigang district (Tashigang Dzong). Cultivated and naturalised on dry hillsides, 1220–1330m.

What is probably this species observed in **Sikkim** — extensively cultivated or naturalised on hillsides near Jorethang (500m) and more sparsely in W Sikkim to 1800m.

Native of Central America, cultivated for fibre.

2. A. angustifolia Haworth.
   Similar to *A. lurida* but plant suckering; leaves 60–120 × 3.5–10cm, light green to glaucous, sometimes with wide cream margins, margins straight to undulate, prickles more widely spaced; apical spine 1.5–3.5cm, flat or shallowly grooved; prickles 2–5mm, tips often hooked; panicles 3–5m, sometimes viviparous, with 10–20 horizontally spreading branches; flowers yellowish-green, 50–65mm, perianth lobes reflexed when dry, 25–30mm; capsule broadly ellipsoid, c.3.5 × 2.5cm.
   Bhutan: S — Sarbhang district (Sarbhang). Cultivated in garden, 460m.

Native of Central America, cultivated as an ornamental. The above record refers to the variegated var. *marginata* Gentry — the single specimen is sterile and might perhaps refer to *Furcraea selloa*.

2. FURCRAEA Ventenat

Differs from *Agave* in having bell-shaped flowers with a very short tube; anthers included, filaments swollen at base; style swollen at base, prominently 3-angled.

1. F. selloa K. Koch 'Marginata'. Fig. 7h.
   Besides differing in its flowers, it differs from the above species of *Agave* in developing a stout trunk to 1.5m.
   Darjeeling (Mongpu) and **Sikkim** (S of Ranipul, Jorethang). Cultivated as an ornamental up to 500m.

Family 196. HYPOXIDACEAE

Rhizomatous or cormose perennial herbs. Leaves basal, linear or with lanceolate, sometimes pleated blade, sometimes petiolate. Infl. on a leafless scape. Flowers bracteate, borne singly, in few-flowered umbels, or in racemes which may be condensed and capitate. Flowers actinomorphic. Perianth segments 6, outer 3 usually hairy on outside, free or fused into tube below. Stamens 6, attached to base of segments, filaments free, anthers linear. Ovary inferior, 3-loculed, sometimes continued upwards into solid beak. Ovules axile. Fruit a capsule or berry, often with perianth persistent on top. Seeds usually black, with protruding raphe.

1. Slender cormose herb to 24cm; leaves linear; scapes 1- or 2-flowered
    1. Hypoxis
+   More robust rhizomatous herbs to 2m; leaves lanceolate, narrowed below into petiole; scapes more than 2-flowered, flowers in dense heads, racemes or spikes (scape sometimes short and hidden in leaf bases) ... 2

2. Ovary not prolonged upwards into beak; perianth tube wide, extremely short; anthers with rounded back, both locules dehiscing to adaxial face (Fig. 7n) .... **2. Molineria**
+   Ovary prolonged upwards into beak; perianth tube filiform, over 2cm; anthers with locules dehiscing laterally (Fig. 7q) .... **3. Curculigo**

1. HYPOXIS L.

Rootstock a corm with fibrous tunics. Leaves linear. Scapes slender, flowers single or in few-flowered umbels (or racemes). Tepals free to base. Anthers with bases sagittate, locules dehiscing laterally. Style short, stout. Ovary clavate. Fruit a capsule eventually dehiscing longitudinally (or sometimes by a lid). Seeds ovoid, papillose, with prominent raphe.

1. H. aurea Loureiro. Fig. 7i–j.
Plant slender. Corm 0.6–1cm, blackish; roots fleshy. Leaves 5–24 × 0.1–0.6cm, usually with sparse, long white hairs. Scapes 1(–2)-flowered, to 10cm; flowers pedicellate, subtended by a filiform bract. Tepals oblong, 4.8–7mm, outer three yellow on inside, greenish and hairy on outside, inner three narrower, yellow on inside, outside glabrous sometimes with central greenish stripe. Capsule 0.5–1cm, oblong-ellipsoid, crowned by persistent perianth. Seeds dark brown.
Bhutan: S — Chukka and Deothang districts; **C** — Ha, Thimphu, Punakha, Tongsa, Bumthang and Mongar districts; **N** — Upper Mo Chu district; **Darjeeling** (Senchal, Nagree, Dumsing, etc.); **Sikkim** (Ratong, Yoksum, Chungthang, etc.). Among grass in open situations and in blue pine and oak (*Quercus griffithii*) forest, 1000–3660m. Fl. April–September.

2. MOLINERIA Colla

Perennial herbs with stout rhizomes. Leaves in basal rosettes, large, lanceo-late, pleated, narrowed below into petiole. Scapes thick, flattened, villous, sometimes very short. Infl. racemose, sometimes very condensed and capitate. Flowers bracteate, yellow; perianth lobes usually united at base to form a tube (sometimes extremely short). Filaments short; anthers rounded on back, both locules dehiscing to adaxial face. Filaments short. Ovary oblong or ovoid, sometimes elongated above into a solid beak. Fruit a berry crowned with remains of perianth; seeds smooth or striate.

1. Leaves densely tomentose on whole undersurface ...... **4. M. crassifolia**
+   Leaves tomentose only on veins on undersurface 2

2. Infl. a very condensed, ovoid, capitate raceme (under 5cm); perianth lobes under 9mm .... **1. M. capitulata**
+   Infl. laxer (at maturity), more elongate (over 5cm), sometimes cylindric; perianth lobes over 10mm 3

3. Perianth lobes under 12mm; infl. c.20-flowered; lower pedicels under 1cm .... **2. M. gracilis**
+   Perianth lobes over 12mm; infl. massive, more than 40-flowered; pedi-cels of lower flowers over 1cm .... **3. M. prainiana**

1. M. capitulata (Loureiro) Herbert; *Curculigo recurvata* Dryander; *C. capitulata* (Loureiro) Kuntze. Nep: *dhoti sara, wurdo lago*. Fig. 7k–n.
   Leaf blade: apex very acute (becoming torn), to 90 × 19cm, upper surface glabrous, underside usually with dense appressed hairs on veins, with scattered spreading hairs on lamina (var. *araneosa*), or subglabrous; petiole to 65cm, tomentose in lower part, or sometimes throughout. Scape to 23cm, occasionally very short (under 5cm), densely brownish tomentose. Infl. a condensed ovoid, deflexed, capitate raceme to 6 × 4cm; bracts lanceolate, exceeding flowers (ones in lower part of head sometimes to 5 × 0.7cm), hairy beneath, glabrous above. Flowers c.1.5cm diameter; perianth lobes yellow inside, oblong-ovate, to 7.5(–9)mm, outer 3 hairy on back; tube funnel-shaped, under 3mm; pedicels very short. Filaments very short; anthers coherent around style. Style slender to 7mm long; stigma capitate. Ovary ovoid, narrowed above, 3–7mm. Berry ovoid-globose, c.9mm, hairy. Seeds black, c.1.5mm diameter, obovoid, the flattened top beaked and raphe slightly protruding laterally, visibly longitudi-nally ridged under 10 × lens.
   Bhutan: S — Samchi (Sangura), Phuntsholing (below Ganglakha,

Phuntsholing), Chukka (Chukka Colony to Taktichu, W of Gedu) and Deothang (Riserboo) districts; **C** — Punakha (21km E of Wangdi Phodrang) and Mongar (Lhuntse Dzong) districts; **Duars** (Buxa–Bhutan border Jalpaiguri); **Darjeeling** (Rongbe, Rishap, Kurseong, Tukvar, Darjeeling, Tista, Riang; Rimbick-Ramam (F.E.H.1)); **Sikkim** (Yoksum; Gangtok (F.E.H.1)). Subtropical forest; banks and scrub in disturbed forest, 200–2400m. Fl. April–June.

2. M. gracilis Kurz; *Curculigo gracilis* (Kurz) Hook.f.
Differs from *M. capitulata* as follows: raceme laxer, more elongate, cylindric; flowers larger, over 2cm diameter; perianth lobes narrower, longer, 10–12mm; ovary oblong, not contracted above.
Darjeeling (Rongsong, near Algarah, Pomong); **Sikkim** (Samdong). 610–1830m. Fl. April–May.

3. M. prainiana Deb. Nep: *doti soro.*
Leaves very large (to 2m); blade 90–103(+?) × 11–19cm, tomentose only on veins beneath; petiole c.27cm, tomentose. Scape 13–17cm, stout (0.7–1cm diameter), brownish tomentose. Infl. more than 40-flowered, erect and capitate at first elongating and becoming ± recurved later, to 8 × 8.5cm; pedicels elongating, becoming deflexed with age, lower ones 1.2–2cm, tomentose. Bracts linear-lanceolate to lanceolate, acute, 3–5 × 0.75–1cm, finally shorter than flowers, tomentose beneath. Flowers 2.4–3.2cm diameter; perianth lobes yellow, oblong to ovate, acute, 1.2–1.7 × 0.5–0.95cm, tomentose on outside. Filaments 1–2mm; anthers 6–7mm, coherent around style. Ovary slightly curved, cylindric, not or slightly contracted above, 0.7–1.3cm. Style 1–1.35cm, long exserted; stigma club-shaped.
Bhutan: S — Samchi (Chepuwa Khola); **Sikkim/Darjeeling** (unlocalised Hooker specimen). Ravine in subtropical forest, 550m. Fl. March.

4. M. crassifolia Baker; *Curculigo crassifolia* (Baker) Hook.f.
Differs from *M. capitulata* as follows: leaves persistently, densely pale-brown tomentose beneath; scape stouter, to 8mm diameter; infl. larger, to 8 × 6cm, sometimes erect; bracts broadly ovate tapering to acute apex, more or less glabrous beneath; flowers larger, to c.2.7cm diameter, perianth lobes c.12mm, inner broadly ovate, wider than outer; style and stigma stouter; filaments longer, c.2mm; anthers longer, c.5mm, not coherent around style; ovary oblong, not contracted above; berry oblong, c.1.5cm; seeds slightly narrower, not visibly ridged under 10 × lens.
Bhutan: S — Chukka district (Chukka Colony to Gedu); **C** — Mongar district (Sengor); **Darjeeling** (Manibhanjan to Tonglu); **Sikkim** (Mintok to

Paha Kholas). Wet, shady banks in broad-leaved (incl. oak) forest, 1980–2360m. Fl. May–June.

Occasionally grown in gardens as at Sukia Pokhri and in Darjeeling Town.

3. CURCULIGO Gaertner

Differs from *Molineria* in its tuberous rootstock and especially in its anthers with locules dehiscing laterally. In our species ovary prolonged upwards into a long, slender beak.

1. C. orchioides Gaertner. Fig. 7o–q.
Tuber vertical, with bundle of fleshy roots at base. Leaf blade narrowly to broadly lanceolate, 8–38 × 1–3cm, usually glabrous, petiolate. Sometimes flowering before leaves developed. Scape hidden in leaf bases, to 6cm, hairy, hairs appressed, silky, white. Infl. a spike, upper flowers male only, lower bisexual; bracts narrowly lanceolate, membranous. Perianth lobes yellow, lanceolate, c.1 × 0.2cm, outer 3 slightly wider and hairy on outside; tube to 4cm, usually hairy. Capsule sessile, hidden in leaf bases, irregularly oblong-ovoid, tapered into beak. Seeds few, smooth, shiny black with hook-like projection.
Terai (Bhutan Ghat Jalpaiguri); **Darjeeling** (Mongpu, Lebong, Rangit, Viewpoint Tista River, Sukna Forest, Geille; Kurseong (Matthew, 1967)). Subtropical forest; dry grassy slopes, 180–910m. Fl. June–July.

Family 197. PHORMIACEAE

Rhizomatous herbs or subshrubs. Leaves distichous, linear, parallel-veined, often rigid and fibrous, laterally compressed at least at base. Infl. a terminal panicle. Flowers bisexual, actinomorphic; tepals 6, free or shortly fused at base, similar or outer 3 slightly differentiated from inner 3. Stamens 6; filaments free or fused at base, sometimes expanded at apex; anthers usually basifixed, dehiscing by apical pores or longitudinal slits. Ovary superior, 3-loculed, each locule with few to many axile ovules; style slender. Fruit a capsule or berry.

1. DIANELLA Lamarck

Rhizomatous perennials producing stout, erect stems bearing rigid, ensiform, distichous leaves, especially near apex. Panicle borne on elongate scape. Flowers with 6 similar tepals, white or bluish. Stamens 6, inserted at base of tepals; anthers basifixed, opening by terminal pores; filaments wide, expanded below junction with anther. Ovary superior, globose, of 3 locules each with several axile ovules; style filiform; stigma minutely capitate. Fruit a few-seeded berry.

1. D. ensifolia (L.) DC. Fig. 7r–t.

Plant to 1(–2)m; leafy part of stem to 45cm. Leaves gradually tapered to very acute apex, encircling stem at base, to 67 × 2.9cm, midrib keeled, often minutely serrate on underside and margins. Scape 20–30cm. Panicle to 34cm, primary branches in groups of 1–3, subtended by leaf-like bracts; primary branches, sometimes again branched; ultimate partial infls. condensed, bearing flowers in clusters of up to 7; pedicels often deflexed, persistent, jointed at apex. Tepals white or yellowish, narrowly lanceolate, slightly hooded at apex, 6–8mm. Filaments: lower part parallel-sided, c.1.4 × 1mm; upper part swollen, with rounded sides, c.1mm; anthers linear, c.2.9mm. Ovary globose, c.2mm; style c.2.7mm. Berry bluish-purple, 2–4-seeded, c.6.5mm. Seeds black, shiny, oblong-ovoid, slightly angled, c.4 × 2.6mm.

Bhutan: C — Tongsa district (above Dakpai near Shamgong, Berthi to Tama); **Darjeeling** (Labdah); **Sikkim** (Silak). Rocky slope under scrub in dry valley, 500–1800m. Fl. April–July.

Additional cultivated ornamentals:

D. caerulea Sims.

Differs from *D. ensifolia* in having ultimate partial infls. spike-like, bearing more than 20 flowers on an elongate axis.

Native of Australia widely grown as an ornamental; a variegated cultivar is grown at Namli (Sikkim).

Phormium tenax J.R. & G. Forster. Eng: *New Zealand flax*.

Massive, trunkless, plant with habit of an *Agave*. Leaves to 3m, rigidly coriaceous. Scape to 5m. Flowers to 5cm, tepals fused into short tube at base; filaments not expanded at apex, anthers dorsifixed, dehiscing longitudinally. Capsule to 10cm, oblong-trigonous.

FIG. 8. **Hemerocallidaceae, Anthericaceae, Alliaceae.**
a, **Hemerocallis fulva**: habit (× ⅕). b–c, **Chlorophytum khasianum**: b, habit (× ⅕); c, half flower (× 2). d, **C. arundinaceum**: fruit (× 3). e–g, **Allium prattii**: e, habit (× ⅕); f, half flower (× 3); g, l.s. of ovary showing gynobasic style (× 6). h, **Milula spicata**: habit (× ⅓). i, **Nothoscordum borbonicum**: half flower, showing terminal style (× 3). Drawn by Mary Bates.

71

Native of New Zealand cultivated worldwide as a fibre crop; grown occasionally for ornament in Darjeeling Town.

Family 198. HEMEROCALLIDACEAE

Glabrous herbs with fascicle of swollen roots. Leaves basal, linear, with many parallel veins. Infl. terminal on a scape, dichotomously branched into 2 cymose partial infls. Flowers orange or yellow, funnel-shaped, of 6 ± similar segments fused below into a distinct tube. Stamens 6; filaments free, attached at top of tube; anthers dorsifixed, dehiscing longitudinally. Ovary 3-loculed, each locule with numerous ovules; style long, curved; stigma punctiform. Fruit a 3-valved, loculicidal capsule.

1. HEMEROCALLIS L.

Description as for family.

1. H. fulva (L.) L. Fig. 8a.
Leaves (27–)44–60(–100) × (1–)1.4–1.6(–2.2)cm. Scape terete, (28–)30(–67)cm, bearing a lanceolate leaf-like bract (to 5.5cm) below infl. Infl. (4–)6–9-flowered, with small, partly scarious bracteoles at base of pedicels and infl. branches. Flowers dull orange; perianth lobes c.7cm, outer narrower than inner; tube 1.9–2.1(–2.7)cm. Filaments c.4.5(–5.7)cm, shorter than perianth lobes; anthers c.8 × 2mm. Ovary cylindric, c.8 × 2.5mm; style 7.7(–9)cm, almost equalling perianth lobes.
Bhutan: C — Thimphu (Taba) and Bumthang (Kandana, Kamephu) districts; **Darjeeling** (Selimbong, Simbong, Kurseong). Cultivated, possibly occasionally escaping, 1520–3660m. Fl. May–July.

The herbarium specimens all represent a smaller plant than the commonly cultivated European forms of *H. fulva*, but the leaves are too wide for them to be referred to var. *angustifolia* Baker. Showy cultivars were observed in Indian Army gardens along the road below Chapcha (Bhutan) and in Gangtok (Sikkim). However, the commonly cultivated form in Darjeeling (and to a lesser extent in Sikkim) is a double-flowered form — probably var. *kwanso* Regel.

Family 199. ASPHODELACEAE

1. KNIPHOFIA Moench

Herbaceous perennials with thick rhizomes. Leaves usually all basal, in several ranks, linear, long (to 2m), tapering to acute apex, often keeled beneath. Infl. a dense raceme on a stout scape. Flowers with short pedicels, subtended by scarious bracteoles, usually opening from base of raceme upwards, usually spreading to pendent at maturity; buds usually red, mature flowers commonly yellowish, so racemes 2-coloured. Perianth to 4.5cm, tubular, with 6 short (to 5mm), free, terminal lobes. Stamens 6, of 2 lengths; anthers dorsifixed, dehiscing inwards, filaments inserted at base of tube. Ovary superior, 3-loculed, ovules numerous. Style slender, eventually longer than stamens, apex minutely toothed. Fruit a globose to ovoid, loculicidal capsule.

Red-hot pokers (Eng) are well known as garden plants worldwide, but originate mainly from S and E Africa. Many of the cultivated forms are derived, by hybridisation, from *K. uvaria* (L.) Hooker. A specimen from Thimphu district (garden at Taba) probably involves that species (which it resembles in its large (3.5cm) flowers that are yellow at maturity) and *K. ensifolia* Baker (having acuminate, acute bracteoles, minutely serrate leaf margins and exserted stamens). Other, more showy species/cultivars are grown, however, as observed in Bhutan (Gedu) and Sikkim (Yoksum, Rabong La to Singtam).

Family 200. ANTHERICACEAE

Perennial herbs, roots often swollen or tuber-bearing. Leaves in basal rosette, bases sheathing. Infl. terminal on leafless scape, racemose, simple or compound. Flowers radially symmetric or slightly zygomorphic; tepals 6, free or tubular at base; pedicels sometimes jointed. Stamens 6 (sometimes 3); filaments free; anthers basifixed or dorsifixed, dehiscing longitudinally, inwards. Ovary superior, 3-loculed, each locule with several axile ovules; style simple; stigma simple or 3-lobed. Fruit a loculicidal capsule, often 3-angled.

1. CHLOROPHYTUM Ker Gawler

Leaves linear or oblanceolate. Raceme usually branched, sometimes very condensed. Flowers borne in small groups subtended by a bract; pedicel jointed, upper part tepaloid and falling with fruit. Flowers white, tepals 6, ± similar, free. Stamens 6, inserted opposite tepals; anthers lanceolate with cordate base, basifixed, dehiscing longitudinally. Style filiform, not thickened or divided at apex. Fruit an emarginate, 3-lobed, sometimes slightly winged capsule. Seeds flattened.

1. Infl. very lax, flowers spreading to deflexed, widely spaced; lower bracts ± parallel-sided, with broad blunt tips; upper pedicel longer than lower .... **1. C. nepalense**
+ Infl. denser, flowers erect, overlapping; bracts lanceolate, tapered to acute apex; upper pedicel longer or shorter than upper 2

2. Tepals over 1cm; upper pedicel shorter than lower; bracts herbaceous;
capsule longer than wide; leaves under 1cm wide, parallel-sided
                                                            2. C. khasianum
+ Tepals under 1cm; upper pedicel longer or equalling lower; bracts
whitish, membranous; capsule wider than long; leaves usually over
1.3cm wide, oblanceolate .... **3. C. arundinaceum**

1. C. nepalense (Lindley) Baker; *C. undulatum* Wall. ex Hook.f. Plate 4.
    Roots not bearing tubers. Leaves 6–11 per rosette, blades ± parallel-sided
for most of length, gradually narrowed to subacute apex, margins slightly
wavy, 14–65 × 0.8–2.2(–3)cm, longitudinal veins 11–23, transverse veins not
visible. Scape 21–100cm, naked or bearing one bract-like leaf at or above
middle. Infl. 13–49(–64)cm, simple (in small specimens) or with up to 5 rather
long, widely spreading branches from lower nodes of raceme, branches and
flower-groups subtended by herbaceous bracts, lowest bract parallel-sided with
broad, blunt tip, 1.5–4(–13)cm. Flowers usually in groups of 3, usually spreading to deflexed, widely spaced (internode between 2 lowest groups over 3cm);
upper pedicel 3–6.7(–9)mm, usually longer than lower part, elongating in fruit.
Tepals 0.9–1.5cm. Filaments 3–4.3mm; anthers 5–8.2(–8.8)mm. Ovary oblong-
ovoid, small; style filiform, 6.5–11.5mm. Capsule deeply emarginate, 3.8–6 ×
5.5–7mm, wider than long. Seeds blackish, flattened, c.3.5mm diameter.
    Bhutan: S — Chukka (Chukka, Chukka Colony to Taktichu) and
Deothang (Wamrung (M.F.B.)) districts; **C** — Tashigang district (Kori La,
Tashi Yangtsi, Shapang); **Darjeeling** (Lloyd Botanic Garden, Dumsong,
Tonglu, Manibhanjan); **Sikkim** (Yoksum, Rishee, Namchi, Lachen, Tsumtang
to Lachung). Broad-leaved forest, often among rocks; grassy banks at edge of
scrub jungle, 1370–3048m. Fl. July–November.

2. C. khasianum Hook.f. Fig. 8b–c.
    Differs from *C. nepalense* in its wider (at least in Bhutan specimens) leaves
(5.4–8.2mm); denser racemes, with erect, overlapping flowers (lowest floral
internode under 3cm); shorter, more erect infl. branches; lanceolate bracts,
tapered to very fine apex; flowers in pairs; upper pedicel shorter (1.8–4mm),
usually shorter than the lower part; longer filaments ± equalling the anthers
and larger capsule, (c.9.5 × 7.5mm) longer than wide.
    Bhutan: C — Mongar district (Lhuntse, Takila). Dry hillsides, in open or
among scattered *Pinus roxburghii*, 1830–2440m. Fl. July.

3. C. arundinaceum Baker. Fig. 8d.
    Roots bearing narrow, elongate tubers to 2.5cm. Leaves c.6 per rosette,
blades oblanceolate, gradually acuminate to very acute apex, narrowed below

into pseudo-petiole, 16–68 × (0.6–)1.3–5cm, longitudinal veins (19–)25–35, short transverse veinlets visible when dry. Scape 14–44cm, about half length to just exceeding leaves, naked. Raceme 4.2–20.5cm, usually simple, sometimes with a short, erect, basal branch, very dense, lowest floral internode under 1cm. Bracts lanceolate, acute, lowest 0.7–3.5cm, whitish, membranous. Flowers in groups of 3, erect. Upper pedicel 2–4mm, longer than lower part. Tepals 8.5–9.5mm. Anthers 4.1–5.5mm, longer than filaments (1.9–4.2mm). Capsule 5.5–8 × 7–9.5mm, only slightly wider than long.

Bhutan: S — Chukka district (Marichong); **Terai** (Tista, Balasun, Gohna, Sukna, Gajaldoba, Siliguri, Garidoora). 120–910m. Fl. September–May.

The Bhutan specimen is atypical in its narrower, more parallel-sided leaves, but agrees with the Terai plant in other respects. It is doubtful whether *C. arundinaceum* is distinct from (the earlier) *C. breviscapum* Dalzell from Bombay and S India.

'Sikkim' specimens referred to in F.B.I. under *C. breviscapum* are probably merely a form of *C. arundinaceum*; they differ from the description above in having very short scapes (less than ⅓ length of leaves); infl. bract leaf-like, half length to equalling the very condensed raceme; tepals longer (c.1.35cm); anthers longer (4.6–5.9mm). Specimens of this form have been seen from the Terai (Garidoora, Bamunpokri, Sukna, Tista; Hara's record (F.E.H.1) of *C. breviscapum* from near Siliguri, later re-determined (F.E.H.3) as *C. arundinaceum*, probably also belongs here). Modern collections are required, and it might prove worthy of some form of taxonomic recognition.

Additional cultivated ornamental:

C. comosum (Thunberg) Jacques '**Variegatum**'. Eng: *spider plant.*
A widely cultivated plant originating in S Africa, characterised by its variegated leaves and arching scapes bearing plantlets.
Sikkim (Legship, Yoksum, Tashiding).

Family 201. ALLIACEAE

W.T. Stearn

Perennial, bulbous or rhizomatous herbs, often strongly smelling. Leaves linear to ovate, basal or sheathing lower part of scape. Infl. a terminal umbel or spike on a leafless scape, enclosed in bud by a spathe composed of 1 or more valves which may persist in flower. Flowers radially symmetric, tepals 6, free or tubular at base. Stamens 6, inserted on tube or at base of tepals; filaments often flattened, anthers dehiscing longitudinally, inwards. Ovary superior, of 3 locules each with 2 or more ovules; style simple, arising from base, or apex of ovary; stigma simple or 3-lobed. Fruit a loculicidal capsule.

1. Infl. a dense cylindric spike .... **2. Milula**
+   Infl. a hemispheric to globose umbel 2

2. Plant smelling of garlic/onion; tepals ± free to base; style arising from
    base of ovary (Fig. 8g) .... **1. Allium**
+   Plant not smelling of garlic/onion; perianth tubular at base; style arising
    from apex of ovary (Fig. 8i) 3

3. Bulbous; flowers small (under 1.5cm), whitish .... **3. Nothoscordum**
+   Rootstock tuberous; flowers large (over 2cm), usually blue
    4. Agapanthus

1. ALLIUM L.

Bulbous perennials (rarely biennials), strongly smelling of onion/garlic.
Leaves basal or along lower part of stem, linear, tubular, or sometimes differentiated into blade and petiole; bases sheathing. Stem solid or hollow. Infl. a
few- to many-flowered umbel enclosed in bud by persistent or deciduous 1- or
several-valved spathe. Tepals free to base. Stamens inserted at base of tepals;
anthers dorsifixed. Ovary locules each with usually 2 ovules; style arising from
base of ovary, filiform, not usually divided at apex. Capsule loculicidal, subglobose or ± 3-lobed, with few black, angular or rounded seeds.

The name *rigog* (Med) refers to a species of *Allium*.

1. Leaves mostly elliptic, (0.8–)1.5–4.7cm wide, distinctly narrowed into
    a petiole; bulb tunic forming a long tube of densely reticulate fibres
    1. A. prattii
+   Leaves linear, under 2cm (usually under 0.7cm) wide, not narrowed
    into a petiole; bulb tunic membranous or with ± parallel fibres 2

2. Infl. composed of bulbils (flowers, if present, abortive) .. **10. A. sativum**
+   Infl. of normally developed flowers, bulbils not present 3

3. Tepals not reflexing, stamens ± concealed; flowers blue or purple 4
+   Tepals reflexing or spreading, stamens usually exposed; flowers purple,
    white or violet 5

4. Flowers rose-purple, tepals 8–11mm; umbels loose, pedicels 1–4cm
    4. A. macranthum
+   Flowers blue, tepals 4.5–10mm; umbels dense, pedicels under 0.5cm
    3. A. sikkimense

5.   Stem 7–9cm; pedicels c.2mm, much shorter than tepals; umbel 1.5–2cm diameter .... **2. A. phariense**
+   Stem over 10cm; pedicels usually exceeding tepals; umbel over 2cm diameter 6

6.   Stem inflated in lower part, to 3cm wide; bulb usually subglobose
                                                    6. A. cepa
+   Stem not inflated, to 1cm wide; bulb cylindric to narrowly ovoid 7

7.   Plant robust, 80–125cm; leaves sheathing lower ⅓ to ½ of stem; leaves smooth, hollow .... **5. A. rhabdotum**
+   Plant more slender, under 80cm; leaves sheathing for under ¼ of stem; leaves flat 8

8.   Flowers purple; leaves 7–20mm wide, keeled .... **7. A. wallichii**
+   Flowers white; leaves 1–9.2mm wide, not keeled 9

9.   Roots slender; anthers c.1.5mm .... **8. A. hookeri**
+   Roots tuberous; anthers c.0.7mm .... **9. A. fasciculatum**

1. A. prattii C.H. Wright; *A. victorialis* auct. non L. var. *angustifolium* Hook.f. Sikkim name: *kok-pa*. Fig. 8e–g.
    Bulb cylindric, 7–20mm wide; tunic forming a tube of densely matted, reticulate fibres. Leaves 2, basal, narrowly to very narrowly elliptic, acute, narrowed at base into short or long petiole, 9–30 × (0.8–)1.5–4.7cm, sometimes with whitish central stripe. Stem smooth, 21–45cm. Spathe 2-valved, persistent. Umbel hemispheric, few- to many-flowered, 2–4cm; pedicels 7–12mm. Perianth pink, cup-shaped; tepals narrowly elliptic, ascending, 4.3–5.6 × 1–1.6mm. Filaments 5–6mm; anthers rounded-oblong, exserted, 0.8–1.1mm. Ovary deeply emarginate, 1.4–2 × c.1.7mm; style 4.1–6.2mm, long exserted. Capsules rather flattened, c.5 × 2.5mm, 3-seeded.
    Bhutan: C — Ha (Tare La, Tremo La to Ha), Thimphu (Barshong, Dotena), Tongsa/Bumthang (Yuto La), Bumthang (Bumthang to Kyi Kyi La, Rudong La, above Gortsam) and Mongar (Ghijamchu, Ghijamchu to Thrumse La) districts; **N** — Upper Mangde Chu (Ju La) and Upper Kuru Chu (Singhi Dzong) districts; **Sikkim** (Yumthang, Phamanasa, Lhonak, Laghep, Bikbari, Dzongri, etc.); **Chumbi.** *Abies* forest; peaty meadows; among grass and shrubs by river, 2440–4570m. Fl. June–August.

Leaves used for seasoning curries in Sikkim.

2. A. phariense Rendle.
Bulb narrowly ovoid, c.0.5–1cm wide; outer tunics thickly membranous, dark brown. Leaves 2–4, almost basal, linear, flat, curved, 5–8 × 0.15–0.3cm. Stem 5–9cm. Spathe 2-valved, persistent. Umbel hemispheric to subglobose, dense, many-flowered, c.1.4–2cm; pedicels 2–5mm. Perianth white or reddish, cup-shaped; tepals elliptic, obtuse, c.5–6 × 1.5–2.3mm. Filaments c.6–8mm; anthers exserted, 0.6–0.8mm. Ovary 2.8–3 × 1.7–2mm; style 5–6mm.
Chumbi (Po-ting La 2 miles N of Phari, Yatung, Rama). 3050–4880m. Fl. July–August.

3. A. sikkimense Baker; *A. kansuense* Regel; *A. tibeticum* Rendle.
Bulbs densely clumped, narrowly cylindric, 3–5mm wide; outer tunics with slightly reticulate or almost parallel fibres. Leaves 2–3, basal, linear or filiform, 5–22 × 0.05–0.3cm. Stem smooth, 6–28cm. Spathe 1-valved, persistent. Umbel 4–15-flowered, dense, 1–2.5cm; pedicels 2–5mm. Perianth blue, cup-shaped; tepals elliptic, obtuse, 4.5–10 × 2.2–3.5mm. Filaments c.4–5mm; anthers round, included, c.0.5mm. Style included, 2–2.4mm. Capsule c.3.3 × 4.2mm.
Bhutan: C — Thimphu (Somana, Barshong to Dotena, Thimphu to Naro), Tongsa district (Thampe Tso); **N** — Upper Mo Chu (Phile La, below Chhew La) and Upper Bumthang Chu (Kurmathang, Pangotang) districts; **Sikkim** (Tarkarpo, Lhonak, Gayumchhona, Kangralamo, Cholamo, Chulong, Tangu, Kangra Lama); **Chumbi**. Grassy pasture, among shrubs by river; sandy soil and banks, 3050–5180m. Fl. July–October.

4. A. macranthum Baker; *A. oviflorum* Regel. Plate 5.
Bulb narrowly cylindric, 0.3–0.6(–1)cm wide; outer tunics membranous. Roots fleshy. Leaves 3–8, basal, linear, flat, to 32 × 0.2–0.6(–1.5)cm. Stem laterally compressed, ridged, 14.5–53cm. Spathe 2–3-valved, deciduous. Umbel 3–28(–60)-flowered, loose with some flowers pendulous, 2.8–7(–12)cm; pedicels unequal, 1–4cm. Perianth cup-shaped; tepals narrowly ovate, truncate-rounded, 8–11 × 3.5–5mm, purple, midrib green. Filaments slightly shorter than tepals; anthers very shortly exserted, c.0.9mm. Style long exserted, 1–1.3cm.
Bhutan: C — Ha district (Ha to Chelai La); **N** — Upper Mo Chu (SW of Lingshi Dzong, Gangyuel Chu) and Upper Pho Chu (Chojo Dzong, Thanza) districts; **Sikkim** (Taling, above Tangu, Lachen); **Chumbi**. Damp rocky ground; grassy hillside with scrub, 3670–4880m. Fl. July–September.

5. A. rhabdotum Stearn. Bhutanese name: *tunchu gop*.
Bulb narrowly cylindric, c.1.5–2cm wide; outer tunics reddish-brown, papery, breaking up into strips. Leaves 2–4, distichous, arranged in one plane,

smooth, hollow, 18–40 × 0.3–0.7cm, glaucous, sheaths clothing lower ⅓ to ½ stem. Stem 80–125cm, glaucous. Spathe 1-valved, ovate, quickly falling. Umbel globose, dense, many-flowered, 3–4cm; pedicels to 1.5cm. Perianth cup-shaped; tepals elliptic, obtuse, 9–10 × 3.5–5mm, whitish or mauve with dark median stripe. Filaments c.12mm, white; anthers long exserted, rounded, c.1.1mm, blue-grey. Ovary c.4 × 2.4mm; style included, 6–7.5mm.

Bhutan: C — Thimphu (Pajoding, Darkey Pang Tso, Bimelang Tso) and Punakha/Tongsa (Dunshinggang) districts; **N** — Upper Bumthang Chu (Kantanang) and Upper Kulong Chu (Shingbe, Me La to Chola) districts. Among rhododendron scrub or in open grassland, sometimes by streams, 3810–4270m. Fl. August.

Stems and leaves eaten.

6. A. cepa L. Sha: *kogpa*; Eng: *onion*.
Bulb very variable in size and shape, often almost globose but somewhat flattened above and below, to 10cm wide; tunics papery. Stem hollow, inflated in lower part, to 100cm. Leaves at first basal, later sheathing lower ⅙ of stem, hollow, semicircular in section. Spathe often 3-valved, persistent. Umbel subglobose to hemispheric, dense, many-flowered, 4–9cm; pedicels to 4cm. Perianth white, stellate; tepals ovate to oblong, obtuse or acute, 3–4.5 × 2–2.5mm. Filaments 4–5mm; anthers exserted.

Seen in markets and shops in Bhutan, Darjeeling and Sikkim but only observed growing at Thimphu and Yoksum.

Though unknown in a wild state, this widely cultivated species is probably derived from a Central Asiatic species, *A. oschaninii* O. Fedtschenko, but modified into a biennial during some 3000 years of cultivation.

7. A. wallichii Kunth. Name at Laya: *lagop*; Nep: *gopa*.
Bulb cylindric, c.1–1.5cm; outer tunics breaking up into papery strips or stout, erect, almost parallel fibres. Leaves 4–5, basal, flat, keeled beneath, to 51 × 0.7–2.5cm. Stem strongly 3–4-winged, 22–80cm. Umbel hemispheric (or fastigiate), loose, many-flowered, 3–8cm; pedicels ascending, 1.5–4cm. Spathe 1-valved, deciduous. Perianth red-purple, reflexing; tepals narrowly oblong-elliptic, acute, 5–10 × 1.5–3.5mm. Stamens erect; filaments 4.5–6.8mm; anthers 1.2–2mm. Ovary 2–4 × 2–5mm; style 2.5–5.4mm.

Bhutan: S — Chukka district (above Lobnakha); **C** — Ha (Ha to Chelai La, Ya La, Pya La, Gile La), Thimphu (Dechencholing to Punakha, above Paro Pass, Dochu La, Bela La, etc.), Punakha (Ritang), Bumthang (above Gortsam) and Tashigang (Tashigang) districts; **N** — Upper Mo Chu (Soe, above Laya), Upper Pho Chu (Foomay, Gyophu La), Upper Bumthang Chu

(Chamka) and Upper Kulong Chu (Shingbe) districts; **Darjeeling** (Tonglu); **Sikkim** (Laghep, Changu, Thango, Dzongri, etc.). Conifer (incl. juniper and *Tsuga*) forest; open wet cliff-ledges and hillsides; among scrub, 2670–4420m. Fl. July–October.

A specimen from Thimphu district (Barshong, 3962m, *Cooper* 1956, E, BM) probably belongs here. It differs from *A. wallichii* in its narrower leaves (2.8–5.4mm), smaller anthers (c.0.8mm) and reticulately fibrous tunics. In these it resembles the Chinese *A. bulleyanum* Diels (the original description of this species mentions a 'similar form' from Sikkim (Diels, 1912)) which, however, is probably only a form of *A. wallichii* as treated most recently by Wang & Tang (1980).

8. A. hookeri Thwaites. Name at Changkha: *bacha.*

Bulb narrowly cylindric, 0.5–1.2cm wide; outer tunics whitish, papery, with a few short parallel fibres at base. Leaves 4–5, linear, flat, to 30 × 0.58–0.92cm, sheathing lower ¼ of stem. Stem 30–92cm. Spathe quickly falling. Umbel hemispheric to globose, many-flowered, loose, 2.5–5cm; pedicels slender, c.1.5–2.5cm. Perianth white, stellate; tepals narrowly lanceolate, acute, 6–9 × 1–2mm. Filaments c.6mm; anthers linear, prominently exserted, c.1.5mm. Ovary c.2.5 × 1.6mm; style 3–4mm. Capsule subglobose with flattened top, c.3 × 3.4mm.

Bhutan: C — Bumthang (Bumthang) and Tashigang (Shapang) districts. Marshes, 1980–3050m. Fl. July–August.

A. hookeri was described from Ceylon and further work is needed to determine whether the extraordinary disjunction is real and whether or not the plants from SW China, Tibet, Bhutan and Khasia really belong to the same taxon (the type of which is certainly extremely similar, but has slightly narrower tepals).

A species commonly cultivated for its edible leaves is seen in markets at Thimphu, Gangtok and Darjeeling and appears to be this species. It has been seen growing in gardens in Chukka district (Changkha and Gedu) and Sikkim (Bitu N of Gangtok). There is no previous record of its cultivation in India, though it is recorded as being cultivated in China (Xu, 1990).

9. A. fasciculatum Rendle; *A. gageanum* W.W. Smith.

Bulb narrowly cylindric, c.2–5mm wide; outer tunics papery, with a clump of short, very stiff, rust-coloured fibres at base; roots tuberous, c.2–3mm thick. Leaves 3–5, sub-basal, linear, flat, 6–28 × 0.1–0.38cm. Stem smooth, 10–53cm. Spathe 1-valved, quickly falling. Umbel hemispheric to globose, many-flowered, 1.5–3cm; pedicels 0.5–2.5cm. Perianth white or greenish-yellow, stellate; tepals narrowly lanceolate, acute, 3.5–4.5 × c.1mm. Filaments c.5mm, white; anthers oblong, prominently exserted, c.0.7mm. Ovary c.2.7 × 2mm; style 3–4mm. Capsule segments obcordate, 2.8–3.6 × 3.2–3.8mm.

Bhutan: N — Upper Kulong Chu district (Me La); **Sikkim** (Llonakh, Natu La, Dong Dong); **Chumbi** (Lingshi La, Teling, near Phari). Among sand and gravel, scree, 3810–4880m. Fl. July–August.

Specimens from Upper Mo Chu district (between Gangyuel and Lingshi), Chumbi (Ka-po-op) and Sikkim (Tang ka La N of Zelep La) agree in having tuberous roots but differ in having narrower leaves, lacking persistent basal bristles; umbels fewer-flowered and flowers greenish-yellow. They perhaps represent an undescribed species.

10. A. sativum L. Sha: *lam*; Nep: *lahsun*; Eng: *garlic*.
Bulb ovoid, composed of many bulblets; tunics papery. Leaves flat, keeled beneath, c.0.7cm wide, sheaths covering lower ½ of stem. Stem solid, to 82cm (or more). Spathe 1-valved, persistent, to 37cm. Infl. composed of 3 or more bulbils (c.1cm diameter); flowers if present aborting early.
Bulbs frequently seen in markets and presumably cultivated in the area, but only one specimen seen (Barshong, Thimphu district).

Doubtfully recorded species:

A. chinense G. Don f. (syn. *A. bakeri* Regel) has been recorded as being sold in Darjeeling Bazaar (F.E.H.1); this seems unlikely and no specimens have been seen. The species has hollow, 3–5-angled leaves and a loose, hemispheric umbel with up to 18 open, cup-shaped, pale violet flowers.

2. MILULA Prain

Differs from *Allium* in its dense cylindric spike of flowers, subtended by a single persistent spathe and its flowers which are widely campanulate, lower half tubular, with 6 free lobes above; filaments in 2 whorls, the outer greatly expanded in lower part.

1. M. spicata Prain. Fig. 8h.
Bulb narrowly cylindric, 0.6–1.5cm wide; tunics densely reticulately fibrous. Leaves 4–7, basal, linear, margins minutely denticulate, 1.8–3mm wide. Stem 6–24cm. Spike 2–5cm. Perianth white or tinged pink, c.4–6mm; perianth lobes widely ovate, c.2 × 2.8mm. Filaments greatly exceeding perianth; anthers cordate c.0.7mm. Ovary 3-lobed, flattened, c.2.4mm diameter; style c.4.3mm; stigmatic lobes 2, reflexed, c.0.6mm.
Chumbi (Do-tho, Phari to Tuna, Sangang). 4120–4570m. Fl. August.

3. NOTHOSCORDUM Kunth

Differs from *Allium* in not smelling of onion or garlic; spathe valves 2, connate at base; tepals fused at base into short tube; lower parts of filaments adnate to perianth tube; style terminal.

1. N. borbonicum Kunth; *N. gracile* sensu Stearn, non (Aiton) Stearn; *N. inodorum* auct. Fig. 8i.

Bulb ovoid, c.1.5cm wide; outer tunics papery. Leaves c.5(–8), inserted near base of stem, linear, rounded, to 28(–40) × 0.4–0.6(–1)cm. Stem smooth, to 40(–60)cm. Spathe valves persistent. Umbel c.10(–15)-flowered, flowers ± erect; pedicels to 1.5(–6)cm, unequal. Perianth fragrant, white with brownish or pinkish streaks, funnel-shaped; tepals oblanceolate, blunt, narrowed to base, to 10(–15) × 4.5mm. Stamens shorter than tepals; filaments flattened; anthers linear, c.2.2mm. Ovary cylindric, obscurely 3-angled, c.3 × 2mm; style c.3.5mm. Capsule obovoid, c.7 × 4mm, pale brown. Seeds c.2.5mm, angled, shiny black.

Cultivated in Bhutan (as in garden at Chasilakha, Chukka district, fl. February). Native of S America, but widely cultivated and naturalised worldwide.

4. AGAPANTHUS L'Héritier

A. africanus (L.) Hoffmansegg (= *A. umbellatus* L'Héritier) has been recorded from Kurseong (Matthew, 1967). Members of this S African genus are commonly cultivated as garden ornamentals in Darjeeling and Sikkim but are difficult to name. They grow from a tuberous rootstock, have linear, basal leaves and an umbel of pedicelled flowers borne on a stout, leafless scape. Flowers usually blue, perianth tubular in lower part, with 6 partly spreading lobes.

Family 202. AMARYLLIDACEAE

W.T. Stearn

Perennial, usually bulbous herbs. Leaves commonly linear and in 2 rows, with sheathing bases which sometimes form a neck to the bulb. Infl. umbel-like, 1–many-flowered, terminal on a usually leafless scape, enclosed in bud by 2 or more, usually persistent spathe valves. Flowers radially or bilaterally symmetric, usually tubular at base, with 6 free lobes and also sometimes an inner, petaloid 'corona'. Stamens 6, inserted on tube or at base of lobes; filaments sometimes expanded and united into staminal corona; anthers usually dorsifixed, dehiscing longitudinally, inwards. Ovary inferior, 3-loculed, ovules axile or basal; style simple; stigma simple or 3-lobed. Fruit a capsule or berry.

1. Flowers solitary; spathe narrowly tubular; perianth lobes free almost to base .... **3. Zephyranthes**
+ Flowers in few- to many-flowered umbels; spathe divided to base into 2 valves; perianth tubular below 2

2. Filaments united below into a toothed, membranous, cup-like corona arising from top of perianth tube .... **4. Pancratium**
+ Filaments free to base 3

3. Perianth white, tube very slender, over 5cm .... **1. Crinum**
+ Perianth red, tube wide, c.2.5cm .... **2. Hippeastrum**

1. CRINUM L.

Bulbous perennials. Leaves linear to ensiform. Scape solid. Spathe with 2 free valves. Infl. a few- to many-flowered umbel; bracts filiform. Perianth with a long tube and 6, usually spreading, lobes. Stamens 6, inserted at throat of tube; anthers dorsifixed, versatile. Ovules numerous, axile. Style filiform; stigma entire, minute. Capsule subglobose breaking open irregularly. Seeds rounded, greenish.

1. C. amoenum Roxb. ex Ker Gawler. Fig. 9a.
Bulb subglobose, 5–10cm. Leaves c.5–9, spreading, ensiform, acute, with very narrow membranous margins, 45–52(–60) × 2.5–3.5(–5)cm. Scape 30–40(–90)cm, sometimes purplish. Umbel 3–10-flowered, flowers ± sessile. Perianth fragrant; tube reddish, 5.5–10(–13.5)cm; lobes white, spreading, linear-lanceolate, 6–9.5 × 0.7–1.3cm. Filaments purplish-red above, whitish below, curved, c.5–6.5cm (shorter than lobes); anthers linear, 0.9–2cm. Style exceeding stamens, curved, purplish.
Possibly native in Bhutan, Sikkim and Darjeeling; but the only possibly native specimens seen are old and from Darjeeling (below Punkabari, 1829m; Reinak, 914m).
The following recent records all refer to cultivated plants in gardens.
Bhutan (Tashigang); **Darjeeling** (Lloyd Botanic Garden); **Sikkim** (Yoksum; Sedonchen to Rongli (Rao, 1964b)). 1400–2070m. Fl. June–August.

Several other taxa are cultivated in Darjeeling and Gangtok including one with pink, campanulate flowers — possibly *C.* × *powellii* Baker.

2. HIPPEASTRUM Herbert

Bulbous perennials. Leaves linear to oblong. Scape hollow. Infl. a 2–10-flowered umbel, subtended by 2-valved, persistent spathe. Perianth zygomorphic, tubular at base, with 6 spreading lobes. Stamens 6, inserted at top of tube, declinate, curving upwards near apex; anthers dorsifixed. Ovules few to many, axile. Style declinate, curving upwards near apex; stigma obscurely to markedly 3-lobed. Fruit a capsule; seeds black.

1. H. puniceum (Lamarck) Voss; *H. equestre* (Aiton) Herbert; *Amaryllis belladonna* L. p.p.

Bulb subglobose, to 5cm wide. Leaves 3–8; developing fully after flowering, to 50 × 4.5cm. Scape 40–60cm. Spathe valves c.6cm. Umbel 2–4-flowered, flowers held horizontally, pedicels 3.5–4(–8)cm. Perianth bright orange-red, greenish-white or yellowish at base inside, trumpet-shaped, 9–12.5(–13)cm; tube c.2.5cm; lobes narrowly ovate, acuminate, 6.5–10cm. Filaments 7–11cm, shorter than lobes; anthers oblong 0.5–0.8cm. Ovary ± cylindric; style 8.5–12cm, slightly exceeding stamens.

Native of S America, but widely cultivated as an ornamental worldwide, as in Bhutan (Gaylegphug and Sarbhang, fl. March).

3. ZEPHYRANTHES Herbert

Low-growing bulbous perennials. Scape hollow. Spathe narrowly tubular below, apex usually bifid. Infl. one-flowered. Perianth funnel-shaped, erect or almost so, the lobes joined at base into short, narrow tube. Stamens erect, inserted near throat of tube. Style long, filiform; stigma distinctly 3-lobed. Capsule subglobose, with black, flattened seeds.

1. Flowers pink; stigma 3-lobed; leaves flaccid .... **1. Z. carinata**
+ Flowers white; stigma ± capitate; leaves narrow, erect ... **2. Z. candida**

1. Z. carinata Herbert; *Z. grandiflora* Lindley, *nom. illegit.* Fig. 9b.

Bulb ovoid, c.2cm wide, tunics dark brown. Leaves 3–6, linear, not fully developed at flowering, to 40cm long, 2–4.5mm wide. Scape 7–18.5cm. Spathe

FIG. 9. **Amaryllidaceae, Melanthiaceae, Burmanniaceae.**
a, **Crinum amoenum**: habit (× ¼). b, **Zephyranthes carinata**: habit (× ¼). c, **Hymenocallis littoralis**: half flower (× ⅓). d–e, **Tofieldia himalaica**: d, habit (× ½); e, fruit and pedicel (× 3). f, **Ypsilandra yunnanensis** var. **himalaica**: half flower (× 4). g–i, **Burmannia coelestis**: g, habit (× ⅔); h, flower opened up (× 3); i, stamen (× 18). Drawn by Mary Bates.

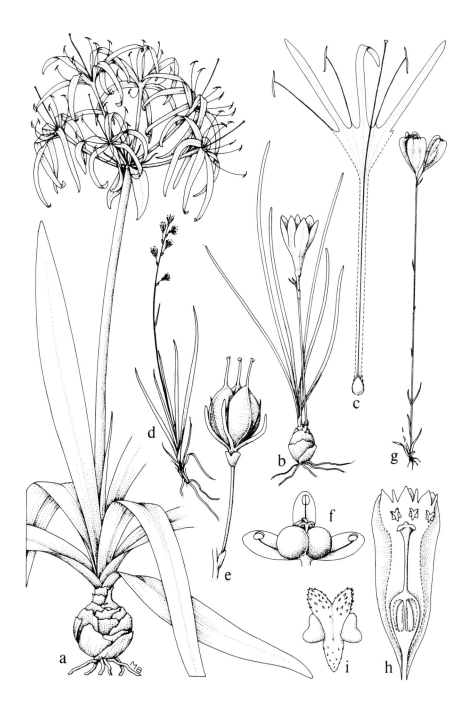

longer than pedicel, membranous, purplish. Pedicel 1.5–3.4cm. Perianth pink, 4.5–8.2cm (total length); tube to 1.3cm; lobes with oblong-elliptic, apiculate blade to 2cm wide, narrowed to claw-like base. Filaments c.3cm; anthers linear, versatile, over 1cm. Ovary ellipsoid, 0.5–0.7cm; style 5–6.5cm, longer than stamens; stigma lobes c.0.3cm, recurved. Apparently not setting seed.

Bhutan: C — Thimphu district (Thimphu); **Darjeeling** (Mongpu, Jorebungalow); **Sikkim** (Yoksum, Sambuk, Legship to Tindong). Cultivated in gardens; naturalised on waste ground, by paths and roads and edges of fields, 1200–2450m. Fl. April–August.

Native of Mexico, but widely cultivated.

2. Z. candida (Lindley) Herbert.

Differs from *Z. carinata* in its narrower, stiffly erect leaves; flowers white; anthers shorter (5–8mm), held erect; stigma ± capitate, style about equalling stamens.

Darjeeling (Mongpu); **Sikkim** (Gangtok). Cultivated in gardens.

Native of Argentina and Uruguay. The record of *Z. tubispatha* (L'Héritier) Herbert ex Traub from Puttabong, Darjeeling (Mukherjee, 1988) perhaps refers to this species.

4. PANCRATIUM L.

Bulbous perennials. Scape solid. Leaves linear to oblong. Spathe with 2 free valves. Infl. a 1- or few-flowered umbel. Perianth funnel-shaped, tubular at base with 6 narrow lobes. Stamens inserted at mouth of tube, filaments joined below by conspicuous membrane forming a funnel-shaped, toothed corona, free parts of filaments short; anthers dorsifixed. Ovules axile, numerous, in 1 or 2 columns per locule; style filiform; stigma entire, minute. Capsule subglobose or 3-angled, loculicidal. Seeds black, angled.

1. P. verecundum Aiton.

Bulb ovoid, c.5cm or more, elongated above into neck. Leaves 4–10, linear, gradually tapered to sheathing base, 35 × 1–3.5cm. Scape 10–30cm. Umbel 2–6-flowered; pedicels 5–10mm. Perianth tube greenish, slender, 6.5–10cm; lobes white, linear, 3.5–6cm. Corona white, 2–2.5cm, with 2 teeth between each filament. Anthers c.0.5cm. Style exceeding stamens.

Terai. No recent records.

Additional cultivated ornamentals:

Hymenocallis littoralis (Jacquin) Salisbury. Fig. 9c.
Like a large *Pancratium* — flowers white, with a corona; differs in the ovaries having few, basal ovules.
Native of Central and northern S America; cultivated as a garden plant in Sikkim (Namli).

Clivia miniata (Lindley) Bosse.
Infl. an umbel of 12 to 20, usually orange, funnel-shaped flowers borne on a flattened scape.
Native of S Africa but widely cultivated as a pot plant and seen in Darjeeling.

Family 203. MELANTHIACEAE

Rhizomatous perennials. Leaves mainly in a basal rosette, linear or oblanceolate, flat and grass-like or ensiform and distichous. Infl. a terminal, spike-like raceme on ± leafless scape. Flowers bisexual, actinomorphic, hypogynous or semi-epigynous, tepals 6, free or fused to ovary below. Stamens 6, filaments free, anthers basifixed, extrorse. Ovary superior or partly inferior, of 3 locules or 3 partly fused carpels; styles separate or fused. Ovules 2–many per locule, axile. Fruit a loculicidal capsule or 3 follicles. Seeds often with terminal projections.

1. Leaves oblanceolate; flowers commonly reddish-brown or pale blue
    (only occasionally whitish) .... **3. Ypsilandra**
+   Leaves linear (± parallel-sided); flowers whitish or greenish-white 2

2. Involucre of membranous bracts at base of flower; perianth lobes linear
    (under 1.2mm wide); styles 3, divergent .... **2. Tofieldia**
+   Involucre of bracts not present at base of flower; perianth lobes wider,
    not linear; style single .... **1. Aletris**

1. ALETRIS L.

Leaves linear, grass-like. Scape bearing several bract-like leaves. Infl. a spike or raceme. Perianth campanulate, tubular below, adnate to ovary to varying degrees, with 6 free lobes. Stamens 6, filaments inserted at base of perianth lobes, anthers basifixed, with cordate bases. Ovary appearing partly inferior due to adherence of perianth tube, of 3 carpels which may be fused or appressed; style apparently single, expanded slightly at apex into stigmatic lobes, splitting in fruit. Ovules many. Fruit a loculicidal capsule, or 3 adherent

203. MELANTHIACEAE

follicles. Seeds very small, longitudinally striate, sometimes with small terminal appendages.

1.   Upper part of scape and pedicels densely hairy .... **1. A. pauciflora**
 +   Upper part of scape and pedicels glabrous 2

2.   Plant under 20cm; stamens and style conspicuously exserted; perianth
      lobes reflexed; flowers not viscid .... **2. A. gracilis**
 +   Plant over 30cm; stamens and style enclosed by perianth; perianth
      lobes erect; flowers viscid .... **3. A. glabra**

1. A. pauciflora (Klotzsch) Handel-Mazzetti; *A. nepalensis* sensu F.B.I.
      Rhizomes bearing fleshy roots. Rosette leaves (4–)5–10, longest (2–)4–18.5
× (0.1–)0.33–0.5cm, veins 8 or more, fibrous remains of old leaves not usually
present. Scape (1.5–)5–16cm, upper part densely felted with short white hairs
(sometimes almost glabrous by fruiting); scape leaf 1. Infl. (2–)10–20-flowered,
densely pyramidal at first, elongating and becoming cylindric, 1–5cm; pedicels
spreading, becoming erect later, 1–3.5mm, hairy; flowers subtended by 2 bracteoles, lower usually much exceeding flower. Flowers pinkish or white, campanulate, enclosing stamens and style, fused to ovary for under half its length,
4.2–4.7mm; perianth lobes erect or recurved 1.6–2.1mm, c.half length of flower.
Filaments 0.1–0.3(–0.6)mm; anthers oblong-ovate c.0.6 × 0.5mm. Ovary
ovoid, 2.3–3.1 × 1.8–2.6mm; style 0.3–0.9mm. Capsule ovoid, acute, c.5 ×
3mm. Seeds slightly curved, without appendages, c.0.8mm, yellowish-brown.
      Bhutan: C — Ha, Thimphu, Punakha, Tongsa and Bumthang districts;
N — Upper Mo Chu, Upper Mangde Chu, Upper Bumthang Chu and Upper
Kulong Chu districts; **Darjeeling** (Tiger Hill); **Sikkim** (Changu, Dzongri,
Lachen, etc.); **Chumbi**. Alpine pasture, sometimes among shrubs (berberis,
rhododendron, juniper, etc.), 2290–4900m. Fl. May–August.

2. A. gracilis Rendle.
      Rhizome short, roots not fleshy. Rosette leaves 7–10, longest to 8(–13) ×
0.5cm, veins not conspicuous, fibrous remains of old leaves persistent. Scape
5–19cm, glabrous, bearing 1–3 bract-like leaves. Infl. 4–11-flowered, 5–3.5cm;
pedicels erect, 0.5–1.5mm, glabrous; bracteoles 2, partly scarious, both shorter
than flower, upper minute. Flowers pinkish-white, tube fused to ovary for at
least half length, 4.5–5.8mm; perianth lobes narrow, reflexed, so stamens and
upper part of ovary exserted, 2.4–3.6mm. Filaments 1.2–1.8mm; anthers
narrow, locules divergent, deeply cordate at base, 0.8–1.3 × 0.3–0.7mm. Ovary
ellipsoid, 2.6–3.5 × 1.5–2.1mm; style 0.9–1.1mm. Capsule ellipsoid, acute, c.5
× 3mm. Seeds c.0.7mm, reddish-brown, with tiny membranous appendages
at both ends and a longitudinal membranous ridge.

88

Bhutan: C — Punakha/Tongsa (Pele La (F.E.H.2)), Tongsa (Tongsa to Tratang (F.E.H.2)), Bumthang (Byakar, Kamephu, below Dhur) and Mongar (Sengor) districts; **Darjeeling** (Phalut (F.E.H.2)); **Sikkim** (Latong, Yakla); **Chumbi**. Bank in open blue pine forest; grassy hillsides and flushes, 1830–3800m. Fl. June–July.

3. A. glabra Bureau & Franchet; *A. sikkimensis* Hook.f. Plate 6.
Rosette leaves 6–12, longest to 18 × 1.2cm, with 7–11 prominent parallel veins, fibrous remains of old leaves persistent. Scape 34–46cm, glabrous, viscid, bearing 4–7 bract-like leaves. Infl. 22–46-flowered, very narrow, 12–14cm; pedicels spreading horizontally at flowering, lower ones becoming erect later, 0.5–2mm, glabrous; bracteoles 2, lower shorter than or equalling flower, upper minute. Flowers greenish-white or orange-green, viscid, covered with minute sessile glands, asymmetrically urceolate at base, completely enclosing stamens and ovary, fused to ovary for under half its length, 4.4–4.9mm; perianth lobes erect, narrow, 1.6–2.2mm. Filaments 0.3–0.8mm; anthers oblong, 0.4–0.7 × 0.4mm. Ovary broadly ovoid, apparently of 3 closely appressed carpels united only at base, 2.3–3.2 × 2–2.6mm; style 0.6–0.8mm. Fruit 5–8 × 3.5–4mm, of 3 appressed follicles (held together by persistent perianth tube), each splitting longitudinally on outer face. Seeds c.0.5mm, reddish-brown, linear, without terminal appendages.
Bhutan: S — Chukka district (above Lobnakha); **C** — Thimphu (Sinchu La, Tsalimaphe to Pumo La, Cangnana, hill above Thimphu Hospital) and Bumthang (N of Byakar Dzong) districts; **N** — Upper Kulong Chu district (Lao); **Sikkim** (Yumthang, Tromo Chu, Jakeynpyak, Lachen, Thangu, Takit, Zemu; Lhonak (Smith & Cave, 1911)). Wet meadows and flushes on open hillside; forest; shaded streamside, 2360–3960(–4270)m. Fl. June–July.

2. TOFIELDIA Hudson

Leaves in basal tufts, ensiform, distichous, with overlapping bases. Infls. terminal spikes or racemes on scapes bearing a few reduced leaves. Flowers small, usually whitish, tubular at extreme base, with 6 persistent perianth lobes. Stamens 6, inserted at base of perianth lobes; anthers basifixed. Ovary of 3 carpels fused at base, upper parts and styles free. Fruit of 3 basally fused follicles opening on inner face.

1. T. himalaica Baker. Fig. 9d–e.
Basal leaves with curved apiculus, membranous margins minutely serrate, 2–11.5 × 0.1–0.3cm, 3–5-veined. Scape 8–18.5cm (to 32cm in fruit), with 1–3 reduced, bract-like leaves. Raceme 4–22-flowered, cylindric; pedicels erect,

3–10mm (to 20mm in fruit), subtended by minute bracts. Flowers subtended by a membranous, irregularly 3-lobed involucre, 0.7–0.9mm. Flowers white, basal tube 0.7–0.9mm, lobes linear-oblanceolate, sometimes hooded at apex, 2.5–3.2 × 0.4–1.2mm. Filaments erect, stout, 3.1–4mm, exceeding perianth lobes; anthers ovate, c.0.5 × 0.5mm. Carpels 3.1–4.5 × 0.5–1mm, narrowed into style; stigmas small, capitate, exserted from perianth; upper parts of carpels and styles free, divergent. Fruit to 6 × 3.5mm, with 3 divergent apical stylar beaks to 1.5mm, surrounded by persistent perianth. Seeds to 1.5mm, linear, reddish-brown, with minute stalk-like projection at one end.

Bhutan: C — Thimphu (Nala to Tzatogang (F.E.H.2)) and Bumthang (Shabjetang, Chakong) districts; **N** — Upper Mo Chu (Laya; Chamsa to Kohina (F.E.H.2)) and Upper Kulong Chu (Me La) districts; **Darjeeling** (Phalut to Sandakphu (F.E.H.1)); **Sikkim** (Yakla, Changu to Karponang, Subarkum, Yeumting, Laghep, Lachung; Yumcho La, Zemu (Smith & Cave, 1911)). Sandy soil by stream; scree; among moss and peat, 3050–4570m. Fl. May–September.

3. YPSILANDRA Franchet

Leaves in basal rosette, oblanceolate. Infl. a raceme; scape bearing reduced leaves. Flowers with 6 persistent tepals, free to base. Stamens 6, inserted on base of tepals; anthers reniform, latrorse. Ovary superior, strongly 3-lobed; style very short with 3 recurved stigmatic lobes, or long with stigma lobes very short. Fruit a 3-lobed capsule. Seeds linear with long terminal projections.

1. Y. yunnanensis W.W. Smith & J.F. Jeffrey var. **himalaica** Hara. Fig. 9f.

Rosette leaves 6–9, oblanceolate, apiculate, gradually narrowed to base, to 8.5 × 1.2cm, relatively thick-textured, veins many, parallel. Scape (at flowering) (3–)5–16(–34)cm, stout, at first almost concealed by overlapping scape leaves, elongating later; scape leaves to 8(–11), oblong, apiculate, sometimes recurved, margins membranous. Infl. 5–10(–14)-flowered, densely cylindric, 0.8–1.5(–3)cm; pedicels 0.5–2mm, lowest 2 with minute bracteoles. Flowers pale blue, whitish or reddish-brown, unpleasantly scented; tepals persistent, lanceolate or elliptic, blunt, (2.5–)3–4.3 × 1.4–1.5mm. Stamens shorter than tepals; filaments 2.6–3mm; anthers c.0.6 × 0.7mm. Ovary of 3 ellipsoid, slightly inflated locules, deeply emarginate at apex, 1.8–2.4 × 2.4–4.2mm; stigma lobes recurved, divided almost to base. Capsule 3-lobed, each lobe c.5 × 2mm, over-topping stigmas, flattened, ellipsoid, splitting longitudinally. Seeds c.3mm, yellowish-brown, linear, with short stalk-like projection at base and thin, curved, acute membranous projection at apex.

Bhutan: N — Upper Kuru Chu/Upper Kulong Chu (Pang La) and Upper Kulong Chu (Me La) districts. Wet cliffs and hillsides; grassy screes and clearing, 4115–4270m. Fl. June–July.

Family 204. BURMANNIACEAE

Annual or perennial, sometimes saprotrophic herbs. Leaves linear or scale-like. Flowers bisexual, usually actinomorphic, borne singly or in cymes; perianth lobes usually 6, in 2 similar or differentiated whorls, fused into tube below, tube sometimes winged. Stamens 3 or 6, sessile, inserted at mouth of tube. Ovary inferior, 1- or 3-loculed, ovules parietal or axile; style filiform, usually 3-branched above. Fruit a capsule, dehiscing irregularly or transversely, often crowned with persistent perianth. Seeds minute.

1. BURMANNIA L.

Autotrophs or saprotrophs, usually annual. Leaves linear, in basal rosette and/or along stem or all reduced and scale-like. Perianth 6-lobed, 3 outer larger, 3 inner smaller (sometimes absent), tube 3-winged, wings continuing downwards to base of ovary. Stamens 3, with connective produced downwards into spur and upwards into 2 crests; anthers lateral. Ovary with 3 locules; style branched with 3 stigmas. Capsule dehiscing laterally, perianth and wings persisting in fruit.

1. Basal leaves short, narrow (under 12 × 2mm), stem leaves minute, usually few; flowers 1–4 .... **1. B. coelestis**
+   Basal leaves longer and wider (over 30 × 6mm), stem almost covered with leaves; flowers more than 4, usually in a bifid cyme
                                                                  2. B. disticha

1. B. coelestis D. Don. Fig. 9g–i.
    Slender annual, stems 5.5–24(–41)cm. Basal leaves scarcely forming a rosette, linear-lanceolate, finely acuminate, 5–12 × 1–2mm, thin-textured; stem leaves usually few (to 7), erect, clasping, to 1.5(–2.5)mm wide. Infl. of 1–2(–4) terminal flowers subtended by minute linear bracts, lateral, pedunculate flower occasionally present lower down stem. Flowers intense bluish-purple, oblong or occasionally elliptic in outline, 0.9–1.8cm from base of wings to tip of lobes; wings 1.5–3mm wide, oblong or outer margin sometimes curved; outer lobes triangular, c.2.3 × 1.5mm, thick-textured; inner lobes linear, c.1.7 × 0.3mm. Connective spur c.0.6mm. Style c.4.5mm; branches c.1.5mm; stigmas disc-like. Capsule narrowly obovoid, truncate, gradually narrowed below, 4.5–6 × 2.5–3mm, surmounted and surrounded by persistent perianth and wings, rupturing horizontally. Seeds minute (c.0.2mm), translucent.
    Terai (Siliguri, Sukna, Jalpaiguri; Dulkajhar, near Titalyah (Jonker, 1938));

Darjeeling (Darjeeling, Great Rangit Valley). [Damp grassy places], 75–305m. Fl. September–December.

2. B. disticha L.

Differs from *B. coelestis* in being much more robust, with a distinct rosette of much broader (over 6mm wide), thicker-textured basal leaves, more abruptly contracted to apiculate tip. Stem conspicuously leafy, with broader stem leaves. Flowers usually more than 4, commonly in a bifid cyme, bracts wider. Connective spur shorter (c.0.3mm), anthers larger.

Darjeeling (Darjeeling). [Open boggy pastures, 2440–3050m in W China].

Family 205. COLCHICACEAE

Twining perennial with stoloniferous, tuberous corm. Stem herbaceous bearing alternate, opposite or whorled leaves. Leaves simple, sessile, parallel-veined, apex cirrhose (i.e. forming a tendril). Infl. terminal, racemose, flowers borne singly in axils of leaf-like bracts; pedicels long, recurved at apex. Tepals 6, free, strongly reflexed. Stamens 6, filaments free, inserted at base of tepals; anthers linear, dorsifixed, extrorse. Ovary superior, narrowly ellipsoid, of 3 locules each with many axile ovules; style filiform, strongly reflexed at base, with 3 filiform stigmatic lobes. Fruit a coriaceous, septicidal capsule. Seeds globose.

This description applies to the single Bhutanese genus.

1. GLORIOSA L.

Description as for family.

1. G. superba L. Fig. 10a–b.

Stems extensive, to 6m. Leaves lanceolate, 10–20 × 1.2–4cm. Tepals red above, yellow below, linear-lanceolate, very acute, margins usually wavy, 4.5–7.5cm. Stamens shorter than tepals. Fruit c.6 × 3.5cm. Seeds c.0.3cm.

Darjeeling (Ramam, Punkabari); **?Sikkim** (No No La). 610m. Fl. August.

Only very inadequate specimens seen; perhaps sometimes cultivated.

Fig. 10. **Colchicaceae, Uvulariaceae.**
a–b, **Gloriosa superba**: a, habit (× ⅓); b, fruit (× ⅔). c–d, **Clintonia udensis** subsp. **alpina**: c, habit (× ⅓); d, fruiting infl. (× ⅓). e–f, **Streptopus simplex**: e, habit (× ⅓); f, fruit (× ⅔). g–h, **Tricyrtis maculata**: g, habit (× ⅓); h, half flower (× 1.5). Drawn by Mary Bates.

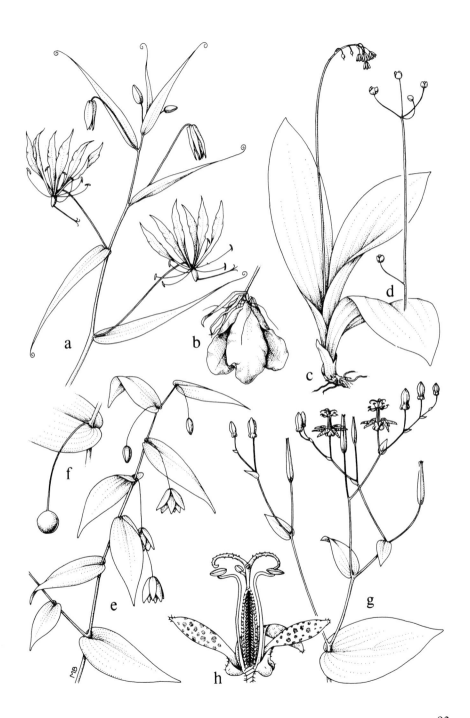

93

Family 206. UVULARIACEAE

Rhizomatous perennials. Leaves in basal rosettes or inserted spirally along a stem, commonly 'dicot-like' with elliptic blades and sometimes reticulate venation; scale leaves sometimes present at base of stem. Infl. a terminal raceme on a leafy stem or leafless scape, or consisting of 1–few-flowered axillary or terminal fascicles. Flowers actinomorphic, bisexual, hypogynous; tepals 6, free, all or outer 3 often with nectar spurs or sacs at base. Stamens 6, filaments free or connate at base, anthers dorsifixed or basifixed, extrorse. Ovary superior, 3-loculed, style slender, stigma usually 3-lobed. Fruit a septicidal capsule or berry.

1. Leaves all basal; infl. on a leafless scape .... **1. Clintonia**
+ Leaves arranged along stem; infl. axillary, terminal, or on short lateral shoots from main axis 2

2. Infl. a terminal raceme; flowers greenish-white, spotted purplish; stigma lobes very stout, spreading, bifid at apex; leaves hairy; fruit a linear capsule .... **4. Tricyrtis**
+ Flowers single, axillary, or in fascicles on short lateral shoots (occasionally terminal); flowers not spotted purplish; stigma lobes slender, not bifid; leaves glabrous; fruit a berry 3

3. Flowers in 2–9-flowered fascicles, terminal or on short lateral shoots; tepals saccate or spurred at base .... **2. Disporum**
+ Flowers borne singly, axillary; tepals not saccate or spurred at base
3. Streptopus

1. CLINTONIA Rafinesque

Leaves in basal rosette. Infl. a terminal raceme, (contracted and umbellate at first) borne on a leafless scape. Flowers campanulate, tepals 6, ± similar. Stamens 6. Ovary 3-loculed; stigma 3-lobed. Fruit a berry, 2–many-seeded.

1. C. udensis Trautvetter & C.A. Meyer subsp. **alpina** (Kunth ex Baker) Hara. Fig. 10c–d.
Leaves 2–5, blade at flowering, elliptic, cuspidate, narrowed to base, 9–16 × 4.5–7.5cm, afterwards elongating, finally oblanceolate to 24.5cm, longitudinal veins many, fine, linked by short transverse veinlets. Scape 17–42cm at flowering, with short, dense, appressed hairs especially above. Infl.

7–12-flowered, pedicels at first shorter than tepals, hairy; bracteoles deciduous, linear-lanceolate, to 5.5mm, hairy. Tepals whitish to pale blue, streaked mauve, oblanceolate, margins narrowly membranous, 6–10 × 3–3.3mm, appressed hairy outside. Filaments 4.5–6mm; anthers oblong, 1–1.7 × 0.5–1mm. Ovary narrowly ellipsoid, 3–4 × 1.5–2mm; style 3.5–4.5mm, stout; stigmatic lobes short. Berries black, c.1cm, many-seeded, erect, pedicels upward-curving. Seeds straw-coloured, ± ovoid, laterally compressed, c.3mm.

Bhutan: C — Ha, Thimphu, Punakha, Tongsa, Bumthang, Mongar and Tashigang districts; **N** — Upper Mo Chu, Upper Kuru Chu and Upper Kulong Chu districts; **Darjeeling** (Sandakphu); **Sikkim** (Lachen, Lachung, Kapoop, above Tsoka, etc.); **Chumbi**. Banks and streamsides in *Tsuga*, rhododendron and *Abies* forest, 2590–3810m. Fl. April–June.

2. DISPORUM Salisbury ex D. Don

Roots fleshy. Stems simple or dichotomously branched with sheathing membranous scale leaves on lower part and simple, alternate leaves on upper part. Infl. usually a fascicle of pedicellate flowers either terminal or on short lateral branchlets, so appearing lateral; occasionally flowers borne singly. Flowers campanulate, tepals 6, all similar, often saccate or spurred at base. Stamens 6, filaments often thickened below, attached to tepals above top of spur or sac; anthers dorsifixed. Ovary superior, 3-loculed; style divided into 3 stigmatic lobes at apex. Fruit a berry, often slightly 3-lobed, commonly black.

1. Stems unbranched; infl. a single terminal fascicle; pedicels and margins of tepals not papillose; flowers white .... **3. D. leucanthum**
+ Stems branched above, infls. numerous, apparently lateral; pedicels and margins of tepals minutely papillose; flowers usually greenish, cream or purplish 2

2. Tepals saccate at base, sacs under 1.5mm .... **1. D. cantoniense**
+ Tepals spurred at base, spurs over 4mm, often divergent from pedicel
    2. D. calcaratum

1. D. cantoniense (Loureiro) Merrill; *D. pullum* Salisbury. Plate 7.
    Stem 50–150cm, (2.2–)2.8–5mm diameter, dichotomously branched above, branches ascending, sometimes branched again. Scale leaves to 7 or more, loosely sheathing, brownish, membranous. Leaves lanceolate, finely acuminate, base cuneate into short (0.2–0.5cm) petiole, 5–14 × 1.8–3.7cm, main veins 5–9, raised and papillose beneath. Infl. of 4–9-flowered, apparently lateral, fascicles inserted opposite leaves on upper part of stems; pedicels papillose,

1.1–2.1cm, becoming rigid and deflexed (to 3.2cm) in fruit. Flowers greenish, cream or dull purple, tepals oblong to oblanceolate, widest ½ to ⅓ from apex, tapered or acuminate to apex, narrowed below, variable in size on same plant, largest 11.5–17.3(–20.3) × 2–4.7(–5.5)mm, papillose (visible on margins); basal sacs 0.9–1.5mm, rounded. Filaments 3.2–8.2(–10)mm, thickened below, papillose; anthers oblong, 2–4.5 × 1.2–2mm. Ovary narrowly obovoid, 2.7–4 × 2–3mm; style 5.5–10.2(–12.2)mm; stigma lobes extending from ¼ to almost whole length. Berry black, 3-seeded, 0.6–0.8cm diameter.

Bhutan: C — Thimphu, Punakha, Tongsa, Mongar and Tashigang districts; **N** — Upper Kuru Chu district; **Darjeeling** (Rayang, Aloobarry, Rungun, etc.); **Sikkim** (Latong, Chungthang, Singhik, etc.). Among shrubs and on banks in forest (mixed, *Quercus*, *Castanopsis*, *Pinus*); marshy ground at edge of paddy field, 1370–3350m. Fl. April–June.

The Himalayan plant has been called var. *parviflorum* (Wall.) Hara or f. *parviflorum* (Wall.) Hara.

The species is very variable and the only variety which is possibly worth recognising is var. *sikkimense* Hara, which has very large white flowers (tepals over 2cm and long-acuminate), longer filaments and a longer style with relatively very short stigma lobes; it is known only from Sikkim (Demthang Forest above Terni and Rechi La). The Bhutan specimens cited by Hara (1984, 1988) as being intermediate between var. *sikkimense* and var. *parviflorum* seem to fall within the range of variation of the latter.

2. D. calcaratum D. Don.

Differs from *D. cantoniense* chiefly in the basal tepal spurs which are 4.2–5mm long and sometimes deflexed from the pedicel. Leaves often not so finely acuminate.

Bhutan: S — Chukka district (Marichong); **Darjeeling** (Phubsering, Sureil, junction of Great and Little Rangit Rivers); **Sikkim** (Karponang (Smith, 1913)). 1070–2440m. Fl. July–August.

Probably only a form of *D. cantoniense*.

3. D. leucanthum Hara.

Plant with long, slender, creeping rhizomes. Stems 17–40cm, slender (1–2.1mm diameter), not branched. Scale leaves up to 5 on lower part of stem, closely sheathing, brownish, membranous. Leaves 3–5 on upper ⅓ to ⅛ stem, immature at flowering; mature leaves on non-flowering shoots narrowly ovate, acute to cuspidate, narrowed at base into short petiole (0.5–0.7cm), 6.8–8 × 2.5–2.8cm, subcoriaceous, with 3 prominent and many slightly weaker intermediate veins. Infl. terminal, 2–5-flowered; pedicels 1–2.5cm, not papillose.

Flowers white, more open than in *D. cantoniense*, tepals oblanceolate, acuminate, margins wavy, scarcely saccate at base, 13.5–19.3 × 3–4.8mm, not papillose. Filaments 8–12.5mm, thickened below; anthers linear, 3.8–5.4 × c.1mm. Ovary narrowly obovoid, 2.3–3.4 × 1.8–2.2mm; style 9.9–14mm; stigma lobes short to almost half length of style.

Darjeeling (Mahalderam, Rhikisum, Ghumpahar, Srikola; Kurseong (F.E.H.3)); **Sikkim** (Chungtam). 1530–2740m. Fl. April–May.

3. STREPTOPUS Michaux

Roots fleshy. Plants glabrous, stem usually branched above, bearing sheathing scale leaves below. Leaves alternate, simple, sessile with clasping bases. Flowers white, borne singly in axils of leaves. Stamens inserted at base of tepals; filaments flattened, expanded at base; anthers basifixed. Fruit a many-seeded berry.

1. Not stoloniferous; stem usually branched above; stigmatic lobes more than ⅓ length of style; filaments shorter than anthers ..... **1. S. simplex**
+   Stoloniferous; stem usually unbranched; stigmatic lobes less than ⅓ length of style; filaments equalling anthers .... **2. S. parasimplex**

1. S. simplex D. Don. Fig. 10e–f.
Stem (6–)19–80cm, glabrous, with several ascending branches above (except dwarf plants). Scale leaves up to 3, oblong, pale brown. Leaves borne on upper half of stem, lanceolate to elliptic, finely acuminate, base cordate, basal lobes clasping stem, (3.5–)6–10.5 × (1.3–)1.6–4.5cm, with many parallel veins. Flowers white, sometimes finely spotted with red, tinged greenish, yellowish or purplish at base, cup-shaped; tepals oblong to elliptic, 1–1.9cm, outer (3–)4–4.3mm wide, inner wider (4.3–)5.7–7.8mm; pedicels slender, shorter than subtending leaf. Filaments flattened, lanceolate, 1.3–2.5mm, shorter than anthers; anthers lanceolate, base cordate, 2.5–3 × 1–1.3mm. Ovary obovoid, c.1.7 × 1.6mm; style stout, 3.3–4mm; stigma lobes from ⅓ to almost equalling style. Berry red, c.1cm. Seeds c.2.5mm, pale brown, longitudinally wrinkled.

Bhutan: C — Ha, Thimphu, Punakha, Tongsa, Bumthang, Tashigang and Sakden districts; **N** — Upper Mo Chu, Upper Bumthang Chu and Upper Kulong Chu districts; **Darjeeling** (Sandakphu, Tonglu; Phalut (F.E.H.2)); **Sikkim** (Tsoka to Dzongri, Lingtu, Lachen, Gnatong, Zemu); **Chumbi**. Among moss and on banks in dense forest (*Abies*, rhododendron, mixed and juniper); occasionally epiphytic, 2900–3960m. Fl. May–July.

206. UVULARIACEAE

2. S. parasimplex Hara & Ohashi.
Plant under 22cm, usually unbranched. Differs from dwarf forms of *S. simplex* in its filiform, cream-coloured stolons (under 1mm diameter) and especially in its short stigma lobes (under ⅓ length of style), smaller anthers (under 2.5mm) and relatively longer, thinner filaments equalling the anthers.
Sikkim (Megu, Kapon near Dzongri, Pheonp; Migothang to Nayathang, Nayathang to Chia Bhanjang (F.E.H.3)). 3100–4270m. Fl. June–July.

4. TRICYRTIS Wall.

Stems simple or branched above. Leaves alternate, simple, sessile, with clasping bases. Infl. a terminal raceme, sometimes branched. Tepals in 2 strongly differentiated whorls; outer with conspicuous pouches near base. Filaments connivent at base, anthers dorsifixed. Ovules axile; stigmatic lobes 3, each bifid at apex. Fruit a linear-oblong, 3-angled, septicidal capsule.

1. T. maculata (D. Don) J.F. Macbride; *T. pilosa* Wall. Dz: *gentshay*. Fig. 10g–h.
Stem to 70cm, simple, shortly pilose. Leaves lanceolate to oblanceolate, long acuminate (to caudate) or suddenly contracted and cuspidate, margins ciliate, deeply cordate, sessile, 9–13.5 × 3.2–6.2cm, scattered hairy above and mainly on veins beneath. Infl. 3- to many-flowered, subcorymbose, sometimes dichotomously branched, occasionally a subsidiary infl. also present in axil of one of upper leaves; pedicels with small, herbaceous, deciduous bracteoles at base. Flowers greenish-white, spotted purple, tepals spreading, contracted to short, papillose, apex, to 2cm, outer oblong, with 2-lobed pouches near base, hairy on outside, inner narrowly lanceolate, lacking pouches, not hairy. Filaments shorter than tepals, strongly recurved near tips, wider below, hairy; anthers elliptic, c.2.6 × 1.4mm. Ovary narrow, c.1 × 0.3cm, gradually tapered into style; stigma lobes spreading, bifid apices deflexed. Capsule c.3 × 0.5cm. Seeds flattened, oblong-elliptic, c.2 × 1.4mm, dark brown, surface reticulately sculptured.
Bhutan: S — Chukka district (Chukka Colony to Taktichu); **C** — Punakha (Senphu; Mishichen to Khosa (F.E.H.2)), Tongsa (Tsanka, Tongsa Dzong) and Tashigang (Tashi Yangtsi) districts; **N** — Upper Mo Chu (Gasa, Kencho) and Upper Kuru Chu (Denchung, Kurted) districts; **Sikkim** (Lachen, Lachung, Chungtam). Dense, shady broad-leaved forest (incl. oak and bamboo); wet meadows and swamps, 1830–2440m. Fl. July–August.

Family 207. LILIACEAE

Perennial, bulbous herbs; bulbs composed of 1–many fleshy scales, enclosed by a membranous tunic or not. Stems erect, herbaceous; leaves basal or arranged (usually spirally) along stem, glabrous, linear to ovate, bases sheathing, apex sometimes developing into tendril (cirrhose). Infl. terminal, racemose, umbel-like, or reduced to a single flower. Flowers actinomorphic or weakly zygomorphic; tepals 6, usually all similar, free, with basal nectaries. Stamens 6; filaments free. Ovary superior, of 3 locules each with many axile ovules; style simple; stigma simple, 3-lobed or of 3 crests. Fruit a loculicidal capsule.

1. Flowers borne on leafy stem, leaves linear to ovate; tepals over 1.8cm . 2
+ Flowers borne on leafless scape (small leaf-like bracts sometimes present below infl.), leaves basal, filiform (under 5mm wide); tepals under 1.8cm 6

2. Leaves differentiated into petiole and ovate blade (over 10cm wide); flowering stems massive, to 4m .... **2. Cardiocrinum**
+ Leaves not differentiated into petiole and blade, linear to lanceolate (under 5cm wide); flowering stems under 1.5m 3

3. Infl. a raceme, sometimes single-flowered, flowers trumpet-shaped ($\pm$ zygomorphic), tepals reflexing 4
+ Flower single, terminal, bell-shaped (actinomorphic), tepals not reflexing 5

4. Stigma capitate; bulb not dying after flowering, of several fleshy scales, not covered with tunic .... **1. Lilium** (p.p.)
+ Stigma 3-lobed; bulb dying after flowering producing numerous bulbils, covered with brown papery tunic .... **3. Notholirion**

5. Leaves cirrhose, some whorled, or if inserted singly then non-cirrhose, elliptic and glaucous; flowers greenish, chequered darker inside; bulb scales 2, wide; inner margins of capsule lobes toothed ...... **4. Fritillaria**
+ Leaves not cirrhose, not glaucous, always inserted singly; flowers purplish or yellowish, if chequered inside then dark reddish-brown outside; bulb scales more than 7, narrow; inner margins of capsule lobes not toothed .... **1. Lilium** (p.p.)

6. Infl. a few-flowered raceme or single (erect or nodding) flower, not subtended by unequal pair of leafy bracts; tepals white or if yellow then not green-striped on outside .... **5. Lloydia**

+ Infl. umbel-like, subtended by unequal pair of leafy bracts, the longer overtopping infl., flowers erect; tepals yellow with green stripe on outside .... **6. Gagea**

1. LILIUM L.

Bulbs of many, fleshy, overlapping scales. Leaves usually linear to lanceolate, spirally arranged. Infl. a terminal raceme (sometimes reduced to a single flower) on a leafy stem. Flowers funnel-shaped, weakly zygomorphic or campanulate, actinomorphic; tepals often reflexed, papillose at apex. Anthers dorsifixed. Stigma ± capitate. Margins of capsule lobes not toothed. Seeds thin, flattened, densely stacked.

1. Tepals over 8cm 2
+ Tepals under 8cm 3

2. Flowers white (sometimes flushed yellow inside), base narrowly tubular; leaves linear (usually under 5mm wide) .... **1. L. wallichianum**
+ Flowers yellowish-green, lower half of tepals sometimes dark crimson brown inside, base widely funnel-shaped; leaves over 9mm wide
2. L. nepalense

3. Flowers dull reddish-brown with golden chequering on inside
3. L. sherriffiae
+ Flowers deep purple to lilac or pale yellow, with varying degrees of darker flecking inside .... **4. L. nanum**

1. L. wallichianum J.A. & J.H. Schultes var. **wallichianum**.
Bulb c.5 × 4cm; scales c.20, lanceolate, to 1.5cm wide, fawn to reddish-brown, margins scarious. Rooting stem short (5–15cm), erect. Flower stem 70–113cm. Leaves linear, finely tapered to apex, 10.5–20 × 0.45–0.65cm, veins

---

FIG. 11. **Liliaceae**.
a–c, **Lilium nepalense** var. **nepalense**: a, habit showing bulb of many narrow scales (× ¼); b, capsule showing toothless margins (× ½); c, seed (× ½). d–f, **Cardiocrinum giganteum**: d, habit (× ⅙); e, capsule showing toothed margins (× ½); f, seed (× ½).
g, **Notholirion bulbuliferum**: habit showing bulb disintegrating into bulbils (× ⅙).
h, **Fritillaria cirrhosa**: habit showing bulb composed of two broad scales (× ⅓).
i, **Lloydia yunnanensis**: habit (× ½). j, **L. flavonutans**: capsule (× 2). Drawn by Glenn Rodrigues.

1–3. Infl. a single (occasionally 2) terminal flower, pedicel to 3.5cm, subtended by leaf-like bract. Flower sweetly scented, white or yellowish-white, sometimes golden yellow inside, held horizontally, funnel-shaped, narrowly tubular at base, apex of tepals reflexed. Tepals 16–21cm; inner to 4.5cm wide; outer to 3.5cm wide, with narrowly elliptic blade, contracted at extreme apex to blunt tip, narrowed below into a claw for basal third. Filaments yellowish-green, to 12.5cm; anthers yellowish-green, to 2.5cm. Ovary very narrowly cylindric, to 4cm; style to 15cm; stigma to 8mm diameter. Capsule c.5.5 × 3.5cm, segments spathulate. Seeds irregularly triangular, broadly (unequally) winged, c.9 × 5mm, buff-coloured.

Bhutan: C — Punakha (Samtengang, Wangdi Phodrang), Tongsa (near Tongsa Bridge, Tongsa Dzong), Mongar (Khoma) and Tashigang (Jiri Chu, Yonpu La, Tashi Yangtse) districts; **Sikkim** (near Singhik). Open grassy hillsides among rocks, sometimes in open forest, 1220–2440m. Fl. June–September.

2. L. nepalense D. Don var. **nepalense**. Nep: *khiraule*. Fig. 11a–c.

Bulb c.4.5 × 4cm; scales c.7, oblong-ovate, c.2.5cm wide, purplish. Rooting stems spreading horizontally, bearing secondary bulbs. Flower stem 46–108cm. Leaves narrowly elliptic to oblong-lanceolate, narrowed to subacute apex, 4.5–14 × 0.9–2.7cm, veins 5(–7). Infl. usually a single terminal flower subtended by a whorl of 4–5 leaf-like bracts; pedicels 3–8(–16.5)cm. Flowers drooping, funnel-shaped, apex of tepals reflexed. Tepals yellowish-green above, dark crimson-brown below, lanceolate, contracted at extreme apex to minute, blunt tip, narrowed towards base, 8–15 × 1.7–3cm, inner broader than outer. Filaments 5.5–7.8cm; anthers yellow or brown, 0.8–1.2cm. Ovary cylindric, c.1.2cm; style to 8cm.

Bhutan: C — Punakha (Ritang) and Tongsa (Rukubji, near Chendebi) districts. On rocks on scrubby hillside; on cliffs and rocky hillsides among bracken and shrubs, 2130–2800m. Fl. June–July.

Occasionally cultivated in gardens as at Jorebungalow (Darjeeling).

var. **concolor** Cotton.

Differs from var. *nepalense* in having uniformly yellowish-green to pale yellow flowers without the reddish blotch.

Bhutan: S — Deothang district (Keri Gompa); **C** — Mongar district (Lhuntse); **N** — Upper Kulong Chu district (Tobrang). Shady cliff; bank of ravine in dense mixed forest; beside rocks among shrubs and bracken on open hillside, 1520–2590m. Fl. June–July.

Dasgupta & Deb (1984) place var. *concolor* under *L. primulinum* Baker var. *primulinum*. The Bhutan specimens certainly resemble the illustration of the type of *L. primulinum* (*Bot. Mag.* t. 7227), but it would be necessary to see fresh material of the Bhutan plant to determine tepal texture for certainty. Further work is in any case necessary on this group — for instance it should be noted that *L. primulinum* is supposed to have 3- (as opposed to 5–7)-veined leaves, whereas the illustration of the type clearly shows 5-veined leaves.

A fruiting specimen from Samtengang (*Cooper* 2352, E, BM) with clavate, truncate capsules, c.4 × 3cm, and oblong to D-shaped seeds, c.8 × 6mm, with unequal rim to 2mm is unusual in having a 3–4-flowered infl. This was determined by Dasgupta & Deb (1984) both as *L. primulinum* var. *primulinum* and var. *ochraceum*. It certainly does not belong to the latter, but flowering collections are necessary to ascertain its identity.

3. L. sherriffiae Stearn. Bhutanese name: *abecas.*
    Bulb c.2(–5) × 2cm; scales c.10, lanceolate, to c.0.7cm wide, yellowish-white. Rooting stem short (c.2.5cm), erect. Flower stem to 46(–90)cm. Leaves linear-lanceolate, gradually tapered to blunt apex, to 12 × 0.95cm, many-veined. Flower single (occasionally 2), campanulate, drooping, reddish-brown chequered with gold inside. Tepals elliptic, contracted below apiculate apex, not reflexed, c.6cm, outer c.1.6cm wide, inner c.2.3cm wide. Filaments c.1.9cm; anthers c.1.3cm. Ovary c.1.2cm; style c.2.8cm, thickened upwards. Capsule oblong in outline, 6-winged, 2–2.5 × 1.5cm, shortly stipitate.
    Bhutan: C — Bumthang district (Dhur Chu); **N** — Upper Kulong Chu (Lao). Sandy soil among willows by stream; rocky and sandy grassy hillsides; damp humus-rich slope in *Abies* forest, 2740–3680m. Fl. May–July.

4. L. nanum Klotzsch f. **nanum**. Med: *abbikha*; Bhutanese name: *cima.*
    Bulb 2.5–3cm; scales 10–20, lanceolate, brown or yellowish. Flower stem (2–)9–50cm. Leaves linear, blunt, 3–10.5 × 0.15–0.65cm, lower sometimes wider than upper. Flower single, drooping, campanulate, deep reddish-purple to lilac, inside usually paler, with varying degrees of darker flecking. Tepals narrowly elliptic, contracted at extreme tip, apex not reflexed, with tufts of branched hairs at base on both sides of basal nectar furrow, 1.7–3.1 × 0.7–1.4cm, inner usually wider and broader at apex. Filaments greenish, 0.6–1.4cm; anthers dull orange to dark purple or brown, 2–4.5mm. Ovary cylindric, 0.6–1cm, striped; style 1.5–8mm. Capsule pale with longitudinal purplish-brown stripes, oblong-ellipsoid, truncate, segments 1.5–2.5 × 0.9–1.3cm. Seeds pale brown, pear-shaped, narrowly (c.0.6mm) winged.
    Bhutan: C — Thimphu, Tongsa, Bumthang, Tashigang and Sakden districts; **N** — Upper Mo Chu, Upper Bumthang Chu and Upper Kulong Chu districts; **Sikkim** (Changu, Dzongri, Chemathang, Thangu, Chamnago, etc.)

207. LILIACEAE

Meadows and grassland; peaty hillsides; among rocks in open, by rivers or commonly among shrubs (rhododendron, juniper, potentilla etc.), 3350–4880m. Fl. June–August.

f. flavidum (Rendle) Hara.
Plant larger. Flowers creamy white streaked with yellow or green, sometimes faintly speckled with brown.
Sikkim (Thangshing, Jamlinghang, Dzongri, Tista Valley, Thangku, E of Bikbari, Samiti); **Chumbi** (Chumbithang, Koo-ma-py-a, Cho-leh-la, Yatung).

Very distinct in the field and no intermediates observed; less distinct in the herbarium, and impossible to identify in absence of colour-notes. Probably a distinct species, but further work required.

Additional cultivated ornamental:

L. lancifolium Thunberg (*L. tigrinum* Ker Gawler), a widely cultivated species originating in Japan with strongly reflexed, spotted, orange tepals, is commonly cultivated in Darjeeling and less so in Sikkim (N of Tashiding, Rumtek).

2. CARDIOCRINUM (Endlicher) Lindley

Differs from *Lilium* chiefly in its large, heart-shaped, petiolate leaves with dichotomously branched venation. Main bulb dying after flowering, producing many small bulbils. Capsule lobes with prominent, inward-pointing marginal teeth.

1. C. giganteum (Wall.) Makino; *Lilium giganteum* Wall. Sha: *loo-dhung*; name at Tongsa: *umdare*. Fig. 11d–f.
Bulb to 10cm diameter; scales few, broadly ovate. Stem 0.9–3(–4)m, massive, to 4.5cm diameter, hollow. Basal leaves in rosette, blade ovate, subacute, deeply cordate, to 33(–45) × 24(–40)cm, primary venation pinnate, lateral veins dichotomously branched; petiole to 25cm; stem leaves smaller, more acuminate, less deeply cordate. Infl. a 6–10(–25)-flowered raceme; bracteoles deciduous by flowering; pedicels short (under 1cm). Flowers white tinged greenish outside, streaked reddish or purplish inside (especially inner tepals), sweetly scented, horizontal or deflexed, trumpet-shaped. Tepals narrowly oblanceolate, apex rounded or subacute, reflexed, saccate at base, 11–14.5(–20) × 1.6–2.9cm, inner broader than outer. Filaments 6–7.4cm; anthers purplish-yellow, 0.6–1.5cm. Ovary cylindric, 2.3–3.2cm; style pale yellow, 4.5–5.9cm; stigma lobes 3, short, rounded. Capsule segments oblong, to 7 × 3.5cm; seeds reniform, 1–1.5cm, surrounded by very broad membranous wing.

104

2. CARDIOCRINUM

Bhutan: S — Chukka district (Chimakothi (F.E.H.2)); **C** — Ha (Ha), Thimphu (Jato La, Chapcha, Barshong to Dotena, Pumo La, Paro; Thimphu to Dochu La (F.E.H.2)), Punakha (Wangdi Phodrang to Pele La, Ztachu Pho Hot Springs, Mara Chu), Mongar (Mongar to Jakar) and Tashigang (Yonpu La) districts; **N** — Upper Mo Chu (Gasa Dzong) and Upper Bumthang Chu (Shimitang) districts; **Darjeeling** (Kalapokri, Tonglu, Siri); **Sikkim** (Samatek, Zemu Valley, Lachen, Bakkim; Lachung, Lower Chakung Chu (Smith, 1913)). Moist forest (incl. *Tsuga*), 1830–3660m. Fl. June–August.

3. NOTHOLIRION Wall. ex Boissier

Differs from *Lilium* in its few-scaled bulb covered with a brownish papery tunic, which dies after flowering producing numerous small bulbils. Linear basal leaves present during winter. Stigma with 3 recurved lobes. Margins of capsule lobes not toothed.

1. Tepals mauve-lilac, without green spot at tip; flowers 5 or fewer; plant under 30cm .... **1. N. macrophyllum**
+ Tepals mauve-lilac or crimson, with green spot at tip; flowers (6–)10–30; plant over 60cm 2

2. Tepals mauve-lilac, less than 4.4 × 1.3cm; flowers zygomorphic, spreading horizontally .... **2. N. bulbuliferum**
+ Tepals crimson, usually over 4.4 × 1.3cm; flowers campanulate, drooping .... **3. N. campanulatum**

1. N. macrophyllum (D. Don) Boissier; *Lilium roseum* sensu F.B.I.
Bulb scales c.4, lanceolate, whitish; tunic segments ovate, to 5 × 2cm, chocolate-brown, dull. Stem 15–40cm, 1.5–3.5mm diameter. Leaves 3–6 inserted evenly along stem, linear-lanceolate, tapering gradually from near base to blunt apex, 9–16 × 0.65–0.8cm, many-veined. Infl. a (1–)2–5-flowered raceme, flowers subtended by leaf-like bracts; lower pedicels to 2.5cm, longer than upper. Flowers held horizontally or slightly upturned, weakly zygomorphic, purple or mauve, sometimes darker at base inside, smell offensive. Tepals oblanceolate, blunt, lacking green tip, 2.7–5.7cm, outer 5.4–9.7mm wide; inner 9–18mm wide. Filaments 1.8–2.2cm; anthers 4–7mm. Ovary cylindric, 0.7–1cm; style 1.8–2.5cm. Capsules 1–1.8cm, segments oblong, truncate, c.0.9cm wide. Seeds irregularly triangular, with extremely narrow, equal rim, to 4 × 3.5mm, reddish-brown.
Bhutan: C — Ha (above Ha), Thimphu (Dotena, Chapcha, above Taba, Pajoding, Dochu La, Yanchengphug, Ragyo), Punakha (Tinlegang), Tongsa

(Chendebi, Rukubji), Bumthang (Ura, Dhur Chu, Bumthang) and Mongar (Rip La) districts; **N** — Upper Pho Chu district (Lhedi, Chojo Dzong); **Sikkim** (Lachung, Mempup, Tankra); **Chumbi** (Ho-ko-chu). Open grassy hillsides, rocky banks, sometimes in *Pinus* forest, 1980–3960m. Fl. June–July.

2. N. bulbuliferum (Lingelsheim) Stearn; *N. hyacinthum* (E.H. Wilson) Stapf. Fig. 11g.

Bulb disappearing by flowering, replaced by mass of tiny bulbils surrounded by membranous, brown tunic-scales. Stem 65–105(–135)cm, 0.6–1.2cm diameter. Leaves 7–17, linear lanceolate, tapered to subacute apex, 11–14(–25) × 1.4–2.4(–2.8)cm, many-veined. Raceme (6–)10–20(–30)-flowered; pedicels short (under 1cm), subequal, recurved immediately after flowering, each subtended by a linear-lanceolate, leaf-like bracteole. Flowers held horizontally, zygomorphic, lilac, mauve or pale purple, darker near base inside, scentless. Tepals oblanceolate, blunt, green tipped, 3.1–4.4 × 0.8–1.3cm. Filaments curving upwards, 2.7–3cm; anthers 5.5–7.4mm. Ovary cylindric, c.1cm; style curving upwards, c.2cm. Capsule oblong-trigonous, blunt, segments c.2 × 0.8cm.

Bhutan: C — Tongsa district (Tibdey La); **N** — Upper Mangde Chu (Dur Chutsen, Worthang), Upper Kuru Chu (Singhi Dzong) and Upper Kulong Chu (Me La) districts; **Sikkim** (Dikchu Valley, Chamnago); **Chumbi** (Langrang, Rinchinging). Open hillsides among grass and bushes; clearings in fir and spruce forest, 2740–3350m. Fl. July–September.

3. N. campanulatum Cotton & Stearn.

Similar to *N. bulbuliferum* (especially when dried) but differing in its flowers which are crimson, campanulate and pendent and its larger tepals (to 4.9 × 1.5cm in Bhutan specimen; to 5.5 × 1.8cm in Burmese specimens).

Bhutan: C — Bumthang district (W side of Rudong La). *Abies* forest, 3505m. Fl. July.

4. FRITILLARIA L.

Bulb subglobose, usually of 2 very thick scales sometimes with thin, papery tunic. Leaves linear to lanceolate, spirally arranged or whorled, sometimes cirrhose at tips. Flower single, campanulate, usually drooping. Tepals chequered inside, each with basal nectary. Anthers usually basifixed. Stigma obscurely 3-lobed. Capsule erect, sometimes winged, margins of lobes toothed.

1. Leaves elliptic to lanceolate, over 1cm wide, apex subacute, leaves single or paired; tepals persistent around capsule .... **1. F. delavayi**

+   Leaves linear, to 5.5mm wide, with filiform or cirrhose tips, at least
    some leaves in whorls of 3; tepals falling before capsule ripe
                                                              2. F. cirrhosa

1. F. delavayi Franchet; *F. bhutanica* Turrill. Bhutya (Sikkim): *yadok*; Med:
karpo chheek thup, abikha.
    Bulb 2.5–4 × 1.5–3cm; tunic yellowish-brown. Stem 12–31cm, mostly
underground, leafy part of stem short, 1.1–6.5cm. Leaves 3–5, subdistichous,
lower alternate, upper 2 opposite, narrowly elliptic or lanceolate, subacute,
base slightly amplexicaul, longest (usually lowest) 4–9 × 1.5–2.3cm, lowest
usually wider (sometimes shorter) than upper, veins not conspicuous, greenish-
to brownish-glaucous, coriaceous. Flower drooping, outside flushed tawny or
olive-brown, inside yellowish-green checkered dark red; pedicel 6–10cm. Outer
tepals narrowly elliptic, 2.5–4.7 × (0.8–)1.6–1.9cm; inner tepals broadly elliptic, (2.4–)5–5.2 × (1.4–)2–2.5cm. Filaments (0.4–)1.2–1.7cm; anthers
7.5–20mm, purplish; pollen cream. Ovary 8–12mm; style 1.4–2cm; stigma lobes
c.3mm. Capsule surrounded by persistent tepals, oblong-obovoid, c.2.4 ×
2.3cm, segments not winged. Seeds pear-shaped, c.8 × 6.5mm, reddish-brown,
wing c.1mm wide.
    Bhutan: C — Ha district (Kang La to Ha); **N** — Upper Mo Chu district
(Yale La, Thugphu, Nelli La); **Sikkim** (Chakalung La, Lachung, Tangka La,
Pata La); **Chumbi**. Screes and gravel, 4570–4880m. Fl. May–July; fr. October.

Bulbs eaten roasted in Sikkim; used medicinally in Bhutan.

2. F. cirrhosa D. Don. Bhutanese names: *tsika, cika.* Fig. 11h.
    Bulb c.1.2–1.8cm diameter, of 2 thick, whitish fleshy scales, covered with
whitish, papery tunic. Flowering stem (15–)18–51cm, leafy for at least upper
half. Upper leaves usually in 1 or 2 whorls of 3 (sometimes single or paired),
linear, apex filiform or cirrhose, 5–8cm × 3.5–5.5mm, veins 3 or more; lowest
leaves paired, broader, blunt-tipped. Flower subtended by whorl of usually 3
leaf-like bracts; pedicel to 3cm, recurved at flowering so flower drooping;
colour as in *F. delavayi*, scentless. Tepals oblong lanceolate to narrowly elliptic,
rounded to subacute, 2.7–5.1 × 1–1.8cm, inner wider than outer. Filaments
c.1.5cm; anthers 0.5–1.1cm. Ovary cylindric, 6–8mm; style 1–1.2cm; stigma
lobes stout, erect, c.7mm. Capsule oblong, truncate, c.1.4 × 1.5cm, segments
with 2 longitudinal wings on outer face. Seeds pear-shaped, c.5 × 4mm, very
narrowly winged.
    Bhutan: C — Ha (Kyu La), Thimphu (Kuma Thang, Paro Chu; Shodug
to Barshong La (F.E.H.2)), Punakha (Tang Chu, Ritang) and Bumthang
(Towli Phu) districts; **N** — Upper Mo Chu (Laya to Laum Thang, Goy to

Lingshi, Yale La (F.E.H.2)), Upper Bumthang Chu (Pangotang, Amlungnang) and Upper Kulong Chu (Shingbe) districts; **Darjeeling** (Sandakphu); **Sikkim** (Changu, Bikbari, Thangshing, Boktu, Dzongri, etc.). Grassy and rocky places in open, or among rhododendron scrub (occasionally in *Abies* forest), 3300–4880m. Fl. May–July.

5. LLOYDIA Salisbury ex Reichenbach

Bulbs small, membranous tunics forming an elongate collar. Leaves linear, basal. Infl. a few-flowered raceme or single flower, borne on scape with reduced, leaf-like bracts. Flowers campanulate, erect or drooping, tepals white or yellow in 2 whorls, outer usually narrower, basal nectary sometimes present. Anthers basifixed; filaments sometimes hairy. Stigma capitate or trifid. Fruit an erect capsule, splitting from apex to varying degrees.

1. Filaments glabrous; tepals white with purplish veins 2
+   Filaments hairy; tepals white or yellow with orange-brown basal spot . 5

2. Plant over 7cm; tepals over 1cm 3
+   Plant extremely dwarf, under 7cm (usually under 4cm); tepals under 1cm 4

3. Stigma obscurely 3-lobed; tepals under 1.8cm, apex rounded, outer narrowly elliptic .... **1. L. serotina** var. **serotina**
+   Stigma trifid; tepals over 1.8cm, contracted below apex, outer linear-oblong .... **3. L. yunnanensis**

4. Tepals rounded at apex, purplish midrib not reaching apex; nectary V-shaped at extreme base of tepal (very hard to see)
    1. L. serotina var. **parva**
+   Tepals acute at apex, purplish midrib reaching apex; nectary a round or transversely elongated swelling c.¼ way up tepal ..... **2. L. delicatula**

5. Tepals white with reddish, orange or brownish basal spot, hairy at base inside, apex subacute .... **4. L. longiscapa**
+   Tepals yellow with orange or brownish basal spot, glabrous at base inside, apex usually rounded .... **5. L. flavonutans**

1. L. serotina (L.) Reichenbach var. **serotina.**
Tunics pale, becoming fibrous. Scape 8–23cm. Basal leaf usually 1, apex blunt, half-length to equalling scape, 1–1.3mm wide. Leaf-like bracts 1–4, on

upper half of scape, linear to very narrowly lanceolate, blunt-tipped, margins membranous, clasping. Flowers 1–2, drooping, white with pinkish-purple veins and flushing from veins. Tepals 1.1–1.8cm; outer narrowly elliptic, rounded, 4.2–7.5mm wide; inner narrowly elliptic to narrowly obovate, rounded, 3.9–6.5mm wide. Nectary c.2mm from base of tepals, small (difficult to see in dry material), V-shaped. Filaments 4.2–5.5mm, glabrous; anthers oblong, 1.1–2mm. Ovary narrowly obovoid, 3–4.5 × 1.2–1.6mm; style 2.8–3.8mm; stigma capitate, lobes scarcely developed. Capsule broadly oblong-ovoid, apex emarginate, c.8 × 8mm.

Bhutan: N — Upper Mo Chu (foot of Pangte La, Laya), Upper Mangde Chu (Goktang La) and Upper Bumthang Chu (Marlung, Tolegang) districts; **Sikkim** (Tangu, Lachen; Phalut (F.E.H.1), Zemu and Lhonak valleys (Smith & Cave, 1911)); **Chumbi**. Open grassy hillsides, cliff-faces, stony plains and rock crevices, 3500–4490m. Fl. May–July.

var. **parva** (Marquand & Shaw) Hara.
Differs from var. *serotina* in its smaller size — scape 2–6cm; tepals 6.5–8mm, outer 2–2.5mm wide, inner 1.8–2.7mm wide; filaments 2.8–3.5mm, anthers ± circular, 0.7–1mm; ovary 2.2–3 × 1–1.4mm; style 2.5–2.6mm.

Bhutan: N — Upper Mo Chu district (Lingshi); **Sikkim** (Tangu, Muguthang). Alpine meadows; dry alluvial plain; peaty slopes, 3960–4570m. Fl. May–July.

2. L. delicatula Noltie.

Bulbs densely clumped, narrowly ovoid, c.4–5mm diameter; tunics 0.5–2.5cm. Scape 0.2–2cm. Leaf 1, basal, filiform, exceeding stem, c.0.5mm wide. Leaf-like bracts usually 3 (lower 2 subopposite) on upper part of scape. Flower single, erect. Tepals oblong, narrowly elliptic or narrowly rhombic, narrowed to subacute apex and base, 3.6–5.7 × 1.1–1.8mm, outer slightly wider than inner, white with prominent purplish midrib reaching apex and pair of lateral purplish veins arising from nectary. Nectary c.¼ from base of tepal, round or transversely elongate, yellowish, thickened. Filaments 2.1–2.9mm, glabrous; anthers ± circular, 0.4–0.5(–0.8)mm. Ovary narrowly ellipsoid to oblong-ovoid, 1.5–2.3 × 0.8–1.4mm; style 1.1–2.2mm; stigma capitate. Capsule c.2.8 × 1.9mm, lobes spathulate.

Bhutan: N — Upper Kulong Chu district (S side of Me La); **Sikkim** (Changu, Lachen, Nathui La, Kapoor, E of Bikbari, Chaunrikiang, Samiti, Thangshing). Mossy turf on boulders, grazed pasture, scree and rock-ledges, 3658–4450m. Fl. June–July.

3. L. yunnanensis Franchet; *L. himalensis* sensu Dasgupta & Deb (p.p.), non Royle. Fig. 11i.

Bulbs densely clumped, small; tunics purplish-brown. Scape to 10cm. Basal leaves to 4, tip blunt, darkened, often exceeding scape (to 11cm), c.0.5mm wide. Leaf-like bracts up to 6. Flower single, drooping, white with reddish-brown veins, reddish-brown at base outside. Tepals contracted below blunt or apiculate sometimes pleated tip, 1.8–2.2cm, outer linear-oblong, 2.5–4mm wide; inner narrowly oblanceolate, with pair of flange-like projections at base, 4.5–6.5mm wide. Filaments 4.2–6mm, glabrous; anthers linear-oblong, apiculate, 1.8–2.2mm. Ovary narrowly cylindric, blunt, 2.6–4 × 0.8–1.5mm; style 8.2–11.5mm; stigma trifid, lobes conspicuous, spreading or recurved, to 0.5mm.

Bhutan: C — Ha (Tare La) and Punakha (Ritang) districts; **N** — Upper Kulong Chu district (Me La); **Sikkim** (Changu, Namdee, Bikbari, Dzongri Pass to Dzongri, Onglakthang). Mossy rocks, moss on cliff-ledges and rock crevices, 3050–4420m. Fl. May–July.

4. L. longiscapa Hook.f.

Scape 10–25(–30)cm. Basal leaves to 6, apex blunt, sometimes darkened, almost equalling scape, somewhat flattened, 0.9–1.5mm wide. Leaf-like bracts to 5, lowest to 9cm. Flowers 1–2, drooping, white (sometimes flushed pinkish), with orange, brownish or reddish basal spot on lower half of tepals and veining of same colour in upper half. Tepals 1.5–2cm, deeply longitudinally ridged below, ridges clothed with short hairs; outer lanceolate to elliptic, slightly acuminate to subacute apex, sometimes laterally incised so trifid, 4.3–6.6(–8)mm wide; inner 4–6.2(–8)mm wide. Filaments with spreading hairs; anthers oblong, 1.4–2.5mm. Ovary (3–)4–5mm, cylindric, gradually tapered into style; style (3.5–)4–4.4mm; stigma capitate.

Bhutan: N — Upper Kulong Chu district (Shingbe); **Sikkim** (Yampung, Yumtso La). Among shrubs; grassy slopes, 3660–4270m. Fl. June–August.

5. L. flavonutans Hara; *L. delavayi* sensu F.E.H.1, non Franchet. Fig. 11j.

Tunics whitish. Scape 6.5–19(–26)cm. Basal leaves to 5(–9), apex blunt, sometimes darkened, filiform, shorter than to equalling scape, 1–2mm wide. Leaf-like bracts 2–3. Flowers 1–2(–3), drooping, yellow with orange or brownish spot at base (sometimes entirely yellow?). Tepals 1–1.7cm; outer narrowly oblong, rounded, 2–4mm wide; inner narrowly rhombic, sometimes slightly contracted below rounded apex, 3.8–7mm wide. Filaments 4.3–7mm, with spreading hairs; anthers oblong, 1.6–2.4mm. Ovary 3.5–5mm, cylindric, tapered into style; style 3.7–5.5mm; stigma lobes scarcely developed. Capsule c.1.8 × 0.4cm, narrowly oblong to clavate, splitting only at apex to form 3 subacute points. Seeds c.2.6 × 1mm, oblong, reddish brown with paler ?elaiosomes top and bottom.

Bhutan: C — Ha (Kyu La, Tare La), Thimphu (between Pajoding and the lakes), Punakha/Tongsa district (N side of Dunshinggang, Pobjeka) and Sakden (Orka La) districts; **N** — Upper Mo Chu (Jari La), Upper Pho Chu (Gyophu La), Upper Bumthang Chu (Pangotang) and Upper Kulong Chu (Shingbe) districts; **Sikkim** (Mome Samdong, Dzongri, Changu, Tangu, Jelep La, Muguthang, Bikbari, Thangshing to Lam Pokhri; Gamothang to Migothang (F.E.H.1)); **Chumbi** (Nathu La). Open grassy hillsides, sometimes among dwarf rhododendrons, 3050–4880m. Fl. May–July.

6. GAGEA Salisbury

Bulbs small, ovoid with membranous tunic. Leaf commonly single, basal, linear to lanceolate. Flowers 1 to several, umbellate, erect, subtended by pair of unequal leaf-like bracts with sheathing bases. Tepals similar, commonly yellowish, persistent, nectary absent. Anthers basifixed. Stigma scarcely expanded. Fruit a capsule splitting into 3 lobes.

1. G. lutea (L.) Ker Gawler; *G. elegans* Wall. ex D. Don.
Bulb c.0.8–1.1cm; tunic chocolate-brown. Basal leaf linear, to 20 × 0.5cm, with many parallel veins. Scape to 10cm. Infl. 1–4-flowered, lower bract lanceolate, spathe-like, overtopping infl., to 0.7cm wide; upper linear, to 1.6 × 0.1cm; pedicels slender, unequal. Tepals yellow, striped green on outside, oblong to narrowly obovate, gradually tapered to subacute apex, margins narrowly membranous, 12.6–14.6 × 2.4–3mm. Filaments 6.7–8.5mm; anthers oblong, minutely apiculate, 1.1–1.9 × 0.9–1mm. Ovary narrowly obovoid, blunt, 2.3–3.7 × 1.4–2mm; style slightly thickened upwards, 6.5–7mm. Capsule broadly oblong-trigonous, blunt, c.7 × 5.5mm.

Sikkim (Zemu Chu river, Thangu, Soonderdunga Glacier; Lachen (Dasgupta & Deb, 1986)). Moist, well-drained shade, 2740–3960m. Fl. March–May.

Family 208. IRIDACEAE

Perennial herbs with underground storage organs (rhizomes, bulbs, corms, etc.). Leaves usually narrow, with parallel veins and sheathing bases, arranged in two ranks. Flowers in terminal, cymose infls. which may be spike-like, or reduced to a single flower. Flowers usually subtended by 2 bracts (spathes). Flowers bisexual with 6 perianth segments in two whorls, free or united at base to form a tube. Segments all similar, or outer 3 different from inner. Stamens 3, opposite outer perianth segments. Styles with 3 branches, sometimes petaloid. Ovary inferior of 3 fused carpels; ovules numerous, axile. Fruit a 3-loculed capsule.

1. Inner and outer whorls of perianth segments strongly differentiated (Fig. 12d–h), outer usually reflexed, with expanded blades and narrower hafts, inner usually erect, narrower; styles petaloid; flowers solitary or in simply branched infls. .... **1. Iris**
+ Perianth not as above; styles not petaloid; flowers in more complex branched infls. or simple/branched spikes 2

2. Flowers zygomorphic; infl. a spike (sometimes branched), flowers borne singly 3
+ Flowers actinomorphic; infl. a compound cyme, flowers clustered 4

3. Perianth lobes over 1.5cm wide; flowers cream, yellow or scarlet; spike not branched .... **2. Gladiolus**
+ Perianth lobes under 1cm wide; flowers tawny-orange; spike usually branched .... **3. Crocosmia**

4. Bulbous (bulb red); leaves basal; flowers white .... **4. Eleutherine**
+ Rhizomatous; stem leafy; flowers reddish-orange .... **5. Belamcanda**

1. IRIS L.

Herbs, rhizomatous, or growing from cluster of fleshy storage roots. Leaves on lower part of stem and in fan-like non-flowering rosettes, linear to ensiform, usually in 2 ranks, bases overlapping. Flower stems branched or simple, flowers terminal, solitary or in few-flowered groups, subtended by two bracts (spathes). Two perianth whorls strongly differentiated, the outer (falls) with a narrow

---

FIG. 12. **Iridaceae.**
a, **Iris tectorum**: fall showing petaloid crest ($\times$ ½). b, **I. goniocarpa**: fall ($\times$ ½) and magnified enlargement of unicellular, clavate hairs. c, **I. decora** agg.: habit ($\times$ ⅙). d–j, **I. clarkei**: d–h, exploded ⅓ flower ($\times$ ½); d, ovary, e, fall, f, stamen, g, style arm showing stigmatic flap and terminal style lobes, h, standard; i, capsule ($\times$ ½); j, seed ($\times$ 2). k, **Crocosmia × crocosmiiflora**: habit ($\times$ ⅕). l, **Eleutherine bulbosa**: habit ($\times$ ⅓). m–o, **Belamcanda chinensis**: m, habit ($\times$ ¼); n, capsule ($\times$ ⅔); o, seed ($\times$ 3). Drawn by Glenn Rodrigues.

haft and expanded, usually reflexed, blade, the inner (standards) narrower and held vertically or at an angle; perianth tubular at base above ovary. Stamens attached to falls, filaments free, anthers basifixed. Styles petaloid, with two overlapping lobes projecting beyond stigmatic flap, arching over and closely appressed to stamens. Seeds arillate in some species.

The name *joen-shing-metog* (Sha) refers to a species of *Iris*.

1. Falls with central beard of single-celled or multicellular hairs or with feathery white petaloid crest 2
+ Falls without beard or white petaloid crest 6

2. Falls with white feathery petaloid crest (Fig. 12a) .... **1. I. tectorum**
+ Falls with beard of unicellular or multicellular club-shaped, yellow or orange-tipped hairs (Fig. 12b) 3

3. Robust plant; leaves 3–4cm wide; flowers to 12cm diameter; beard of multicellular hairs .... **2. I. germanica**
+ Plants more slender; leaves under 1.5cm wide; flowers less than 7cm diameter; beard of unicelluar hairs 4

4. Flower stem over 5cm; perianth tube c.1cm, flowers under 5cm diameter .... **3. I. goniocarpa**
+ Flower stem less than 5cm (flowers and fruit borne at ground level); perianth tube over 4cm, flowers over 5cm diameter 5

5. Flowers with dark mottling, mauve-violet (or paler); perianth tube to 8.5cm; standards erect; capsule ovoid, to 3cm .... **4. I. kemaonensis**
+ Flowers not mottled, uniformly dark purple; perianth tube to 11.5cm; standards spreading; capsule narrowly ellipsoid, to 5cm
5. I. dolichosiphon

6. Fall blades spreading horizontally, pale blue-mauve, with central yellow ridge on lower half; rootstock a bunch of fleshy storage roots
6. I. decora agg.
+ Fall blades deflexed, dark blue-purple with white or yellow markings at base; plants with spreading rhizomes .... **7. I. clarkei**

1. I. tectorum Maximowicz. Fig. 12a.
Rhizome stout, branched. Non-flowering shoots with fans of 3–4 curved, ensiform, very acute leaves, 20–30 × 1.5–2.5cm, prominently veined, thin-textured, yellowish-green. Flower stem 16–34cm, simple or with 1(–2) ascend-

ing branches, with few, narrow leaves at base. Spathes 2–3-flowered, 4.5–7cm, herbaceous. Flowers to 10cm diameter; tube 2.5–4.5cm; fall blades shallowly deflexed, blue-violet, darker-veined near base, oblong-orbicular, c.3.5 × 3 cm, abruptly contracted into haft, with central white, petaloid feathery crest; standards spreading horizontally, blade oblong-elliptic, c.3.5 × 2cm, claw dark brownish, very narrow, channelled. Style lobes toothed. Anthers white.

Bhutan: C — Thimphu (Taba, E of Simtokha Dzong) and Punakha (Menchunang) districts; **Sikkim** (Yoksum). Cultivated in gardens; sometimes naturalised near habitations or in disturbed oak/blue pine forest, 2370–2750m. Fl. April–May.

Native of C and SW China.

2. I. germanica L.; *I. nepalensis* Lindley.

Rhizome stout, branched. Leaves of non-flowering shoots ensiform, subacute, to 40 × 2.5–3.0(–4.0)cm, glaucous. Flower stem to 120cm, robust, with up to two branches. Spathes 1–2-flowered, herbaceous below, tinged purple, membranous above. Flowers to 12cm diameter; tube to 3cm; fall blades reflexed, dark crimson-purple, obovate, to 6.5 × 4cm, gradually narrowed into haft, with central beard of yellow, multicellular hairs; standards erect, paler than falls, obovate, to 8 × 3.5cm, haft short, channelled.

Bhutan: C — Thimphu district (Taba). Cultivated in garden. Fl. May.

Probably originated in the Mediterranean region but widely cultivated in numerous varieties; only one specimen seen from Bhutan, but other colour forms are probably grown.

3. I. goniocarpa Baker. Fig. 12b.

Plant slender, rhizome very compact, stems tufted, subtended by reticulate fibrous remains of old leaves. Leaves erect, grass-like to 40cm long, c.3mm wide. Flower stems 10–27cm. Spathes single-flowered, enclosing perianth tube, 3–3.5cm, herbaceous. Flowers to 4.5cm diameter; tube c.1cm; fall blades shallowly deflexed, mauve-purple, mottled darker, obovate, to 2.5 × 1.5cm, with central beard of club-shaped, orange-tipped, unicellular hairs; standards spreading, blades narrowly oblong, c.1.8 × 0.7cm, abruptly contracted into very narrow claw. Anthers orange. Capsule ellipsoid, tapering to pointed apex, 2.5–3.5cm, opening by lateral slits, enclosed by persistent papery spathes.

Bhutan: N — Upper Mo Chu (Lingshi) and Upper Kulong Chu (Me La) districts; **Sikkim** (Tangu, Yatung to Lachung La); **Chumbi** (Chumbithang). Dry, open hillsides, 3000–4000m. Fl. June–July.

4. I. kemaonensis D. Don; *I. kamaonensis* D.Don; *I. kumaonensis* auct.
Rhizomes very compact, stems tufted, subtended by fibrous remains of old
leaves. Leaves not fully developed at flowering, short, blunt-tipped and apiculate, pale green; mature leaves linear, tapering to acute apex, to 67 × 0.8cm.
Flower stem under 5cm, hidden by bract-like leaves. Spathe single-flowered,
shorter than tube. Flowers to c.6.5cm diameter; tube 4–8.5cm; fall blades
reflexed, usually mauve-violet (sometimes very pale) mottled and blotched
darker, oblong, blunt, c.3 × 2cm, with central beard of club-shaped, orange-
tipped unicellular hairs; standards erect, blade ovate-oblong, abruptly contracted into haft. Margins of style lobes crenate. Anthers mauve. Capsule borne
at ground level, ovoid, 2–3cm long, opening by 3 lateral slits. Seeds obovoid,
c.3mm diameter, dark brown with small, pale aril.

    Bhutan: C — Ha (Kung Karpo) and Thimphu (Paro Chu, Cheka,
Gunisawa, Kumathang) districts; **N** — Upper Mo Chu district (Lingshi Dzong
(Nakao & Nishioka, 1984)). Open grassy slopes, often south-facing,
3800–4300m. Fl. May.

5. I. dolichosiphon Noltie.
    Differs from *I. kemaonensis* in its uniformly dark purple, unblotched flow-
ers; flowers larger, to c.7.5cm diameter; tube wider at top, to 11.5cm long;
standards spreading horizontally; capsule larger (to 5cm), narrowly ellipsoid,
apex acute; seeds larger (to 3.5mm diameter) with a larger aril.

    Bhutan: C — Ha (Cheli La) and Thimphu (Barshong) districts;
N — Upper Mo Chu district (Laya). Open grassy slopes; juniper forest;
Berberis/Potentilla scrub, 3500–4130m. Fl. May–June.

6. I. decora Wall. agg. Fig. 12c.
    Not rhizomatous; stem and leaves arising from cluster of fleshy storage
roots, subtended by very few, short, stiff fibres. Leaves linear to narrowly
ensiform, (28–)30–80 × 0.6–1.7cm, conspicuously veined. Flower stems
(14–)23–67cm, rather slender, with 1 or 2 stiffly ascending branches. Spathes
1(–2)-flowered, linear-lanceolate, to 4.5 cm, very thin. Flowers 5–6cm diameter, lasting only a few hours, pale blue-lilac (?occasionally white); tube to
2cm, very slender; fall blades spreading horizontally, elliptic, c.3 × 2cm,
gradually tapered into haft, with yellow central ridge at base; standards spreading in same plane as falls. Style arms ± erect, lobes half-ovate, margins crenate.
Anthers cream. Capsule enclosed within persistent spathes, 2.7–3.8cm, opening
from top, valves straw-coloured, papery, narrowly elliptic, borne on pedicel
1.3–1.5cm. Seeds obovoid, 2–3mm diameter, dark brown, aril whitish, equalling seed.

1. IRIS

Bhutan: C — Thimphu (Chapcha), Punakha (Hinglai La, Wangdi Phodrang, Ritang, Samtengang), Tongsa (Punzor) and Mongar (Khinay Lhakang, Lhuntse) districts. Dry grassy slopes (sometimes under *Pinus roxburghii* or among *Artemisia* scrub); cliff-ledges; moist banks, 1200–2580m. Fl. April–July.

A depauperate specimen from Chungthang (Sikkim) may belong to this taxon.

Much work remains to be done on the taxonomy and typification of this aggregate: Bhutanese plants are different from Nepalese ones (whence the plant was described) and are almost certainly worth recognising at least at subspecific rank.

7. I. clarkei Hook.f. Bhutanese name: *cema*. Fig. 12d–j.

Rhizomes slender, spreading, upper parts clothed with brown, fibrous remains of old leaves. Leaves linear, (24–)35–75(–120) × 0.6–1.5cm, shining green on upper surface, greyish beneath. Flower stem solid, (25–)45–90cm, exceeding leaves, sometimes simple, usually with one branch. Spathes (1–)2-flowered; pedicels (2.5–)3.5–8cm carrying flowers almost free of spathes. Flowers to 9cm diameter, tube c.1cm, funnel-shaped; fall blades deflexed, dark blue to purple, with white or greenish-yellow tinged patch at base, oblong, c.3.5 × 2cm, not bearded; standards horizontal or recurved, lanceolate, acute, abruptly narrowed into very narrow, channelled haft. Style branches broad, forming flattened top to flower, lobes half-ovate. Anthers whitish-mauve or cream. Capsule oblong-trigonous, abruptly contracted at top and bottom (occasionally cuneate at base), (2.7–)3.6–5.5cm, opening from top. Seeds flattened, D-shaped (5–7mm long) or round (4–5mm diameter), flanged, dark brown.

Bhutan: S — Chukka district (above Lobnakha); **C** — Ha (Saga La, Damthang, near Ha), Thimphu (Hinglai La, Dochu La, Jyele La, Pomu La, Paro, Chelai La, Tsalimaphe), Punakha (Hinglai La), Bumthang (Kempe La, Ura) and Mongar (Sengor) districts; **N** — Upper Mo Chu district (Gasa); **Darjeeling** (Tonglu, Phalut, Sandakphu); **Sikkim** (Yakla, Reling, Changu, Laghep, Gnatong); **Chumbi**. Marshy meadows, swamps in clearings in conifer forest, 2280–4270m. Fl. June–August.

Dried leaves used as fodder for horses and yaks.

Specimens from Ju La and above Lambrang (Upper Mangde Chu and Upper Bumthang Chu districts) have similar flowers, but differ in having hollow flower stems and leaves glaucous on both surfaces — in these they resemble Chinese species such as *I. bulleyana* Dykes, but much work is required to resolve the taxonomy of subsection 'Sibiricae' in SW China.

An extremely interesting yellow-flowered specimen (*Ludlow, Sherriff & Hicks* 19496, BM, E) from Upper Bumthang Chu district (Dhur Chu, 3900m) cannot be placed with certainty. The flowers are described on the label as having: 'outer perianth segments white, inner yellow', but it is not clear as to which segments the colours refer as the inner segments (standards) are deflexed and so appear to be the outer ones. Falls narrowly oblanceolate, c.5.5 × 1.2cm, the blade tapering gradually into the haft; standards 3.0 × 0.4cm, linear-lanceolate, deflexed. It is possibly related to *I. wilsonii* C.H. Wright but more collections are needed.

2. GLADIOLUS L.

Cormose herbs. Infl. a terminal spike; flowers zygomorphic, often twisted to face one direction, each subtended by a pair of bracts. Perianth zygomorphic, of 6 petaloid lobes, usually tubular at base, tube widely funnel-shaped, lobes often unequal with 3 upper (laterals from outer whorl, central from inner) larger than 3 lower. Filaments free, inserted at top of perianth tube, anthers basifixed. Style linear; stigma with 3 linear lobes.

1. Perianth lobes white with conspicuous maroon spot near base

                                                                       3. G. callianthus

\+ Perianth lobes yellow or cream, lower ones sometimes with pink or reddish spot 2

2. Perianth yellow; tube stout, curved, less than 5cm ...... **1. G. natalensis**

\+ Perianth cream-white; tube slender, straight, over 5cm .. **2. G. undulatus**

1. G. natalensis (Ecklon) Hooker; *G. primulinus* Baker.

Leaves up to 5, sub-basal, to 130 × 2.4cm, prominently veined. Flower stem to 70cm with up to 3 bract-like reduced leaves in upper part. Spike to 7-flowered. Bracts lanceolate, apex twisted, outer mucronate, inner bifid, margins narrowly membranous, subequal, to 4cm. Flowers to 10cm; tube curved, to 4cm; upper 3 perianth lobes overlapping to form hood, elliptic, apiculate, c.4.3 × 2.5cm; lower lobes slightly reflexed, c.3 × 1.5cm, all uniformly yellow.

    Bhutan: C — Tashigang district (Tashigang). Cultivated in a garden. Fl. June.

Native of S and E Africa, but widely cultivated and used in hybridisation.

2. G. undulatus L.

Differs from *G. natalensis* as follows: leaves (up to 5) more evenly placed along stem, narrower, linear, to 54cm × 7–9mm; bracts unequal, outer longer, to 6.4cm long; flowers more regular, to 10.5cm long, tube straight and longer (to 7cm), lobes ovate to lanceolate gradually tapering to acute, wavy tips,

lower lobes slightly narrower than upper, colour cream-white or pinkish, lower lobes with pink or red markings.
Darjeeling (Darjeeling). Naturalised, 2100m. Fl. June.

Native of S Africa, but naturalised in some tropical countries.

3. G. callianthus Marais; *Acidanthera bicolor* Hochstetter.
Distinguished by its very characteristic, gracefully arching, fragrant white flowers: tube to 10cm, slender, curved, lower 5 lobes marked with maroon, rhombic spot at base.
Darjeeling (Darjeeling Town); **Sikkim** (below Rabong La). Cultivated as an ornamental.

Native of E Africa, widely cultivated.

The *Gladiolus* most commonly cultivated in gardens in Bhutan (e.g. Thimphu, Jakar, etc.) and less commonly in Sikkim (e.g. Yoksum) has scarlet flowers and is probably one of the 'Nanus' hybrids derived from *G.* × *insignis* Paxton. Large-flowered taxa in various pastel shades were observed in Bhutan (Thimphu) and Sikkim (above Tashiding) and can be referred to *G.* × *hortulanus* L.H. Bailey.

3. CROCOSMIA Planchon

Differs from *Gladiolus* as follows: corms small, spreading by creeping stolons so producing dense clumps; infl. usually branched; bracts small (just concealing ovary); at least lower part of perianth tube very slender.

1. C. × **crocosmiiflora** (Lemoine) N.E. Brown. Eng: *montbretia*. Fig. 12k.
Corm c.1.7cm diameter. Stem to 60cm. Leaves borne on lower part of stem, c.8, to 80 × 2cm, strongly veined and with conspicuous midrib. Infl. usually with 1 or more ascending basal branches; bracts small (to 8mm), subequal. Flowers tawny-orange, perianth tube c.1.5cm, lower part slender (c.1.2mm wide), abruptly expanded in upper third, curved; lobes to 2.5cm, narrowly lanceolate, lowermost of inner whorl wider (to 0.9cm) than rest. Stamens arching to top of flower; filaments attached at apex of slender part of tube, exceeding anthers. Style filiform, just exceeding anthers, stigma deeply trifid.
Bhutan: S — Chukka district (Gedu); **Darjeeling** (Darjeeling Town, etc.); **Sikkim** (Yoksum, etc.).

A hybrid of horticultural origin from S African parents; commonly cultivated in gardens in Darjeeling and Sikkim, less frequently in Bhutan and not becoming naturalised. The record of *Tritonia crocata* (L.) Ker Gawler for Darjeeling (Mukherjee, 1988) almost certainly refers to this taxon.

119

4. ELEUTHERINE Herbert

Bulbous herbs. Leaves basal, plicate. Infl. a one-sided, compound cyme, with ovoid partial infls. of 6–8 pedicellate, bracteate flowers, partial infls. pedunculate, subtended by spathe-like bracts, infl. subtended by leaf-like bract. Flowers actinomorphic, of 6 similar, obovate, spreading, white perianth lobes joined only at extreme base. Filaments free, attached to base of inner perianth lobes, anthers basifixed. Style with 3 equal, linear, erect branches.

1. E. bulbosa (Miller) Urban; *E. plicata* (Swartz) Klatt. Fig. 12l.
Bulb ovoid, c.5 × 3.5cm, with many layers of brittle, red-brown tunics. Leaves 1–4, narrowly lanceolate, 19–36 × 0.5–1.4(–2.3)cm, prominently 5(–7)-veined. Flower stem 22–30cm; infl. bract to 14cm. Floral bracts conduplicate, outer herbaceous, inner largely membranous. Flowers to 2.5cm diameter, perianth lobes obovate. Flowers opening in evening and lasting only a few hours.
Bhutan: S — Phuntsholing district (Phuntsholing). Cultivated in a park. Fl. April.

Native of S America, widely grown in tropical countries.

5. BELAMCANDA Adanson

Rhizomatous herbs. Stems stout, erect, bearing distichous, ensiform leaves. Infl. a much branched compound cyme, with stiffly ascending branches, partial infls. several-flowered, flowers pedicellate, bracteate, infl. bract small. Flowers actinomorphic, with 6 ± similar perianth lobes and extremely short tube, lobes spiralling after anthesis. Filaments free, attached to base of perianth lobes. Style linear, slender, stigma lobes 3, flattened, held vertically. Capsule of 3 valves, which reflex to reveal the persistent seed-bearing axis. Seeds round, with fragile black shiny coat.

1. B. chinensis (L.) DC. Nep: *tyang patare*. Fig. 12m–o.
Stems and small vegetative buds arising from swollen end of rhizomes. Stems to 2m, with up to 9 evenly spaced leaves. Leaves to 40 × 3(–3.7)cm. Infl. bract lanceolate, acute, to 12cm. Perianth lobes orange-red, spotted darker red, spreading, oblanceolate, to 3cm long, outer wider than inner. Capsule obovoid to 3.5cm long. Seeds c.6mm diameter.
Bhutan: S — Deothang district (Dantak Camp); **C** — Punakha district (Punakha Dzong, Wangdi Phodrang, Lobesa); **Terai** (Jalpaiguri); **Sikkim** (N

of Tashiding, Yoksum). Cultivated in gardens and occasionally naturalised. Fl. June–August.

Native of China and Japan.

Additional cultivated ornamental:

Schizostylis coccinea Backhouse & Harvey.
Rhizomatous perennial; infl. a 2-sided spike, flowers usually scarlet, unspotted, radially symmetric, tube straight.

Native of S Africa, widely cultivated as an ornamental; seen in Darjeeling.

Family 209. ORCHIDACEAE will form the concluding part (Volume 3 Part 3) of the *Flora of Bhutan*

Family 210. ARACEAE

Perennial herbs of diverse habit incl. climbers, floating aquatics, pachycaul shrubs and geophytes. Underground stems absent, rhizomatous or tuberous, aerial stems variously produced or not, often evergreen; bulbils for vegetative reproduction sometimes produced e.g. on leaf or on special shoots. Leaves alternate or apparently basal, usually petiolate with sheathing bases, often subtended by cataphylls, blades various e.g. linear, simple (base often cordate to sagittate), sometimes peltate or variously compound (e.g. pinnate, radiate, pedate). Infl. (sometimes precocious) subtended by membranous cataphylls, consisting of a spadix subtended by a spathe. Spathe commonly with tube-like base (margins fused or not) and deciduous blade. Spadix bearing bisexual or unisexual flowers, in latter case plants dioecious or monoecious (spadix female below and male above). Bisexual flowers: tepals 0, 4 or 6; stamens 4–6, filaments free, anthers bilocular; ovary usually 3-loculed. Unisexual flowers: male represented by single stamens or synandria of $2-\infty$ fused stamens, anthers often subsessile, usually dehiscing apically by pores or slits (straight or horseshoe-shaped); female flowers consisting of single ovaries (sometimes associated with a sterile staminode), commonly unilocular (sometimes with 3 or 4 locules), ovules 1–many per locule, commonly parietal, basal or apical. Neuter flowers derived from male or female flowers sometimes present at apex of female and/or male section. Spadix sometimes with a sterile, terminal appendix. Fruit usually a head of 1–several seeded berries, commonly red.

1. Floating aquatic (Fig. 13f) .... **17. Pistia**
+ Terrestrial herbs or climbers, if aquatic then rooted in substrate 2

2. Aquatic with narrow, membranous leaves, submerged for at least part of year; spathe with long, filiform tube and narrow blade (Fig. 13e)
   16. Cryptocoryne
+ Not as above 3

3. Climbers 4
+ Terrestrial or marsh herbs 6

4. Petiole with phyllodic wings; leaves entire .... **1. Pothos**
+ Petiole not winged; leaves entire or pinnately cut 5

5. Ovules many, small; leaves entire or cut .... **2. Rhaphidophora**
+ Ovule single, large; leaves entire .... **3. Scindapsus**

6. Petioles and peduncles prickly; marsh herbs .... **4. Lasia**
+ Petioles and peduncles not prickly; plants of dry land 7

7. Cultivated, stemless herb, apparently not flowering; leaf blade ± ovate, deeply cordate, veins of basal lobes exposed, forming base of sinus
   12. Xanthosoma
+ Not as above 8

8. Spadix lacking an appendix 9
+ Spadix terminating in a sterile appendix 12

9. Leaf peltate 10
+ Leaf not peltate 11

10. Spathe under 2cm, ± open; male part of spadix with conspicuous pores (Fig. 14b) .... **5. Ariopsis**

FIG. 13. Araceae I (habits), **Acoraceae**.
a, **Rhaphidophora grandis** (× ¹⁄₁₄). b, **Remusatia vivipara** (× ⅛). c, **Colocasia esculenta** (× ⅛). d, **Arisaema concinnum** (× ⅛). e, **Cryptocoryne retrospiralis** (× ½). f, **Pistia stratiotes** (× ⅓). g, **Acorus calamus** (× ⅓). Drawn by Glenn Rodrigues.

+   Spathe over 5cm, margins appressed (at least below) to form a tube; male part of spadix lacking pores .... **9. Remusatia**

11. Spathe ± closed at anthesis; spadix free; berries not red and fleshy, many-seeded .... **6. Homalomena**
+   Spathe open at anthesis; female part of spadix fused to spathe; berries red, fleshy, 1-seeded .... **7. Aglaonema**

12. Stalked, clavate neuters present immediately above female part of spadix and separated from male part by sterile zone (Fig. 14f) 13
+   Neuters absent or if present stalkless and like reduced male or female flowers 14

13. Margins at base of spathe overlapping but not fused; at least some leaves simple and hastate, commonly produced with infl. **13. Typhonium**
+   Margins at base of spathe fused to form a sac-like tube; leaf pedate, commonly not produced until after flowering .... **14. Sauromatum**

14. Leaves simple, ± oblong-ovate, peltate or deeply cordate; flowering with leaves 15
+   Leaves dissected (pedate, radiate or simply/compoundly trifoliate); sometimes flowering before leaves 16

15. Ovules many, small, parietal; spathe blade narrowly lanceolate, open only at base; leaves peltate, usually dull .... **10. Colocasia**
+   Ovules few, large, basal; spathe blade ovate to oblong, open; leaves peltate or cordate, often slightly glossy .... **11. Alocasia**

16. Leaf single with 3 ± pinnately (irregularly) dissected lobes; flowering before leaves; spadix appendix massive, not stipitate; monoecious
                                                     8. Amorphophallus
+   If leaf single then radiate or trifoliolate, if 2 or more on pseudostem then pedate (i.e. with 2 outer dissected lobes and single undivided central lobe); usually flowering with leaves; spadix appendix stipitate or minute, commonly filiform; monoecious or dioecious .. **15. Arisaema**

1. POTHOS L.

Perennial epiphytic and lithophytic climbers. Stems branched, rooting from lower nodes. Leaves distichous, blades simple, coriaceous, separated from winged, phyllodic petiole by a joint. Primary venation parallel. Infls. apparently axillary, shortly pedunculate, subtended at base by overlapping bracts; spadix ellipsoid to linear, sessile or stipitate; spathe persistent. Flowers bisexual; tepals 6. Stamens 6; filaments flattened, anthers attached basally, locules 2, not fused at apex, dehiscing laterally by slits. Ovary 3-loculed, each locule with a single ovule; stigma sessile, punctate. Fruit a 1–3-seeded, red berry.

1. P. cathcartii Schott. Nep (Darjeeling): *sanu kanchirna, chepari lahara*; Nep (Bhutan): *shutey lahara, kanchirna.* Fig. 14a.

Stem to 18m, 0.3–0.4cm diameter. Leaf blade narrowly elliptic, finely acuminate, base rounded to cuneate, 6.8–14.2 × (2–)2.3–5.8cm, longitudinal veins 2–3 each side of multiple midrib, secondary veins closely parallel, diverging from midrib, veinlets reticulate, coriaceous. Petioles narrowly triangular in outline, 2.3–9.2cm, widest (0.7–1.7cm) at truncate to shallowly emarginate apex. Bracts to 6 sheathing lower part of peduncle, elliptic, mucronate, coriaceous. Peduncle stiffly erect, 1.5–2.5cm. Spathe spreading, suborbicular, mucronate, strongly concave, 0.7–1.4cm, coriaceous, green. Spadix cream, subglobose to ellipsoid, 0.7–1 × 0.6–0.7cm, borne on short (0.5–0.7cm) stipe. Tepals square, apex triangular-hooded, c.1.5mm, keeled, rigid. Filaments elliptic, c.1.3 × 0.9mm; anthers c.0.7mm, cream. Ovary hexagonal-cylindric, truncate, c.1.5 × 1.8mm. Berry hexagonal-cylindric, apex truncate, 1.3–1.5 × 0.7–1cm.

Bhutan: S — Samchi (Tamangdhanra Forest, Chowrinal Hill), Phuntsholing (NE of Phuntsholing), Chukka (Marichong) and Gaylegphug (Karai Khola above Aie Bridge, Mon La, Betni, Hatipali) districts; **C** — Punakha district (Rinchu to Mishichen (F.E.H.2)); **Sikkim** (Kandam Chu); **Duars** (Jalpaiguri Duars, Buxa Reserve); **Darjeeling** (Tista, Darjeeling, Mongpu, Punkabari, Fogodhara, Glen Cathcart, etc.). On rocks and trees in subtropical forest, 200–1550m. Fl. December–April; fr. May–June.

Specimens from Sikkim and Darjeeling with leaves and petioles similar in shape but smaller, and smaller infls. (spadix 0.5–0.6 × 0.4–0.5cm) have been referred to *P. roxburghii* de Vriese (*P. vriesianus* sensu F.B.I.); they seem likely, however, to be starved forms of *P. cathcartii*. Further work is required on the distinction of taxa in SE Asia as a whole and on the elucidation of the considerable nomenclatural tangle.

P. scandens L. is recorded as being common in the Lower Hill Forests of Darjeeling (Cowan & Cowan, 1929) but only two poor and inadequately documented 'Sikkim' specimens and one from Gaylegphug have been seen. It differs from *P. cathcartii* in having much smaller infls. (spadix c.0.5 × 0.2cm) and smaller, acute (rather than acuminate) leaf blades often subequalling the petioles.

2. RHAPHIDOPHORA Hasskarl

Large lianes, epiphytic or scrambling over rocks. Stems thick, rooting. Petioles with pulvinus at junction with blade, margins sheathing. Blade symmetric or asymmetric, simple, or becoming pinnately dissected, commonly coriaceous. Spathe deciduous, coriaceous. Spadix sessile, cylindric. Flowers bisexual, lacking perianth. Ovaries hexagonal-cylindric, apex truncate or domed, 1-loculed, fibrous hairs embedded in tissue, ovules many, parietal; stigma sessile or raised. Stamens 4; filaments flat; locules 2, free at apex, attached basally, dehiscing laterally. Seeds many, small, oblong.

Difficult to collect, due to size and inaccessibility; specimens mostly poorly annotated. Measurements all given from herbarium specimens which are of small, flowering portions of stem which will fit onto a herbarium sheet and are not typical of whole plant.

NB Juvenile foliage of all species entire.

1. Mature leaves entire 2
+   Mature leaves pinnately cut 3

2. Leaf blades large (over 25cm), oblong-elliptic, cuspidate, membranous, one side shallowly cordate at base .... **1. R. hookeri**
+   Leaf blades smaller (under 20cm), lanceolate, abruptly acuminate, coriaceous, both sides rounded to cuneate at base ..... **2. R. calophylla**

3. Leaf blade under 37cm, sometimes glaucous beneath, outline ovate-acuminate; pinnae 2–5 per side, each with 3 subequal, parallel costae; spadix under 9cm .... **3. R. glauca**
+   Leaf blade over 40cm, not glaucous beneath, outline oblong, blunt; pinnae 6 or more per side, each with several weak costae at an angle to the strong midrib; spadix over 13cm 4

4. Pinnae broad (ones at mid-leaf over 4cm), truncate at apex, sinuses narrow, not reaching midrib; pinnae up to 10(–12) per side
    4. R. grandis
+   Pinnae narrow (under 3.5cm), apex gradually acuminate-curved, distant, sinuses broader, reaching midrib; pinnae more than 12 per side
    5. R. decursiva

1. R. hookeri Schott.

Stem 1.3–1.5cm diameter. Leaf blade entire, oblong-elliptic, slightly asymmetric, cuspidate (1–1.5cm), base asymmetric — one side rounded to oblique, other very shallowly cordate, 20–46.5 × 8–24.5cm, sides and base of midrib on underside shortly hairy, lateral veins all diverging at wide angle from midrib, of two distinct thicknesses. Petiole 14–25cm, sheath reaching pulvinus or stopping short by to 2cm, pulvinus distinct 1–1.5cm. Peduncle suberect, apex straight, c.8cm. Spathe obliquely deciduous, cylindric, margins almost appressed, 4–5.5 (± apiculus 1.5cm) × 2.5cm, very stiff, green outside, yellowish inside. Spadix stoutly cylindric, 4–5.5 × 1.8–2cm. Filaments (2 — immature)–6 × 1.2mm; anthers 2 × 0.8mm. Ovaries hexagonal-cylindric, apex granular, depressed, walls densely fibrous, 5–(8.5 in fr.)mm; stigma sunk, c.1mm diameter.

Bhutan: S — Samchi (Buduni), Sarbhang (Sarbhang to Chirang road) and Gaylegphug (Gaylegphug, Hatipali) districts; **C** — Mongar (Shongar Chu below Zimgaon) and Tashigang (Jiri Chu) districts; **Darjeeling** (below Kurseong); **Sikkim** (below Rumtek, Great Rangit; Singtam to Gangtok (F.E.H.1)). Subtropical forest, 300–1300m. Fl. March–July; fr. February.

2. R. calophylla Schott; *R. peepla* sensu F.B.I. non (Roxb.) Schott.

Differs from *R. hookeri* in having more slender stems (to 1cm diameter), leaf blades more strongly asymmetric, smaller (to 21 × 10cm), broadly lanceolate, abruptly shortly-acuminate, coriaceous, lateral veins all of same thickness, base asymmetrically rounded to oblique; petiole sheath reaching base of blade; spadix narrower (c.1.3cm wide in fl.); apex of peduncle curved so spadix held at an angle; apex of ovary smooth, truncate so stigma raised.

Bhutan: S — Phuntsholing district (Phuntsholing (F.E.H.2)); **C** — Tongsa district (6km SE of Shemgang); **Duars** (Buxa–Bhutan border); **Darjeeling** (Kurseong, Kotabari, Balasun); Sikkim records probably refer to Darjeeling. Subtropical forest, 300–2000m. Fl. August; fr. December–June.

3. R. glauca (Wall.) Schott. Dz: *tokchum*; Nep: *chuna champa, kanchirna.*

Stem 0.7–1cm diameter. Leaf blade ovate in outline, symmetric, acuminate, base truncate, oblique or shallowly cordate, 11.5–37 × 7.5–24cm, irregularly and asymmetrically pinnately cut; pinnae 2–5 per side, ascending obliquely, ones at mid-leaf 1.7–4(–5.8)cm wide, apex truncate with ascending falcate tip, usually with 3 equally strong, parallel costae, sinuses almost reaching midrib, rounded at base, sometimes glaucous beneath. Petiole 9–33cm, sheath reaching base of leaf blade, pulvinus indistinct. Peduncle spreading, curved at apex, 10–25cm, infl. ± erect. Spathe oblong-ovate, acuminate, widely open, 8–14 × 3–9cm, waxy, pale yellow. Spadix 4.5–8.5 × 0.8–1.3cm. Filaments flat, 2.7–3 × 0.8–1.5mm; anthers 0.9–1 × 1–1.4mm. Ovary 3–5.5mm, apex c.3–4mm, truncate, smooth, walls not fibrous; stigma flat, sessile, circular or elliptic, 0.6–1mm.

Bhutan: S — Phuntsholing (Ganglakha to Kamji), Chukka (Jumudag to Tala), Sarbhang (6km below Dara Chu on Chirang road) and Gaylegphug (Chabley Khola to Tama, Gaylegphug, Rani Camp) districts; **C** — Punakha (Rinchu to Mishichen (F.E.H.2)) and Tashigang (Shali) districts; **Duars** (Buxa); **Darjeeling** (Lloyd Botanic Garden, Rangirun, Lebong, Ging, Balasun, Latpanchor); **Sikkim** (Kulhait, Yoksum, below Rumtek, 33km W of Singtam, Kabi N of Gangtok). Subtropical and mixed broad-leaved (incl. oak) forest, (300–)1220–2130m. Fl./fr. September–May.

4. R. grandis Schott; *R. eximia* sensu F.B.I. non Schott. Dz: *tokim*; Sha: *brengla*; Nep: *kanchirna tula*. Fig. 13a.

More massive in all its parts than *R. glauca* and differing from it as follows: stem to 4cm diameter; petiole sheath stopping short of pulvinus; leaf blades larger (40–100 × 39–64cm), oblong, blunt, not glaucous below; pinnae 6–10(–12) per side, wider (ones at mid-leaf 4–6.4cm), lateral costae weak not parallel to the strong midrib, cross-veinlets obvious when dry; peduncle stouter; spadix larger (19–23 × 3–5cm); ovaries longer (to 13mm in fruit), fibrous, apex to 6mm diameter, domed so stigma raised.

Bhutan: S — Samchi (Dhamdhum), Phuntsholing (above Rinchending, Ganglakha to Kamji), Chukka (Marichong, Chukka to Taktichu) and Gaylegphug (Rani Camp) districts; **Darjeeling** (Rishap, Paik, Darjeeling, Lopchu to Jorebungalow); **Sikkim** (33km W of Singtam, above Yoksum, 11km N of Gangtok). Subtropical and mixed broad-leaved (incl. *Quercus/Castanopsis*) forest, 950–2070m. Fl./fr. May–February.

5. R. decursiva Schott; *R. eximia* Schott.

A large climber similar to *R. grandis* from which it differs in having leaves with usually over 12 pinnae per side; pinnae narrower (2–3.5cm), more widely spaced (sinuses reaching midrib), venation similar, apices finely acuminate, slightly curved.

Bhutan: S — Phuntsholing (above Phuntsholing) and Gaylegphug (Hatipali) districts; **C** — Tashigang district (Jiri Chu); **Darjeeling**; **Sikkim** (14km W of Singtam). Subtropical forest, 610–1220(–1520)m (at lower altitudes than *R. grandis*). Fl. August–November.

There has been much confusion over the last two species, which have commonly been regarded as synonymous. Of the two *R. grandis* is the more common. Because of confusion over identity and application of names literature records have not been included but *R. eximia* has been recorded for Deothang district (Narfong, M.F.B.) and *R. decursiva* for Punakha district (Rinchu to Mishichen, F.E.H.2).

3. SCINDAPSUS Schott

Similar to the entire-leaved species of *Rhaphidophora*, differing chiefly in its ovaries having a single, basal ovule which develops into a large seed.

1. S. officinalis (Roxb.) Schott. Lep: *tuffoor*.
Leaf blade elliptic to ovate, acute, base symmetric, rounded to oblique, 18–31 × 11.5–18cm, secondary veins ascending pinnately from midrib, closely parallel. Petiole 19–23cm, broadly sheathed. Spathe c.12cm. Ovaries at fruiting with broad (to 1.5cm diameter), flat apex, stigma raised with elongate slit.
Duars (Totekamia Duars); recorded for Sikkim in F.B.I. and Darjeeling (Mixed Plains and Lower Hill Forest (Cowan & Cowan, 1929)), but no specimens seen.

Description based on inadequate NE Indian material.

4. LASIA Loureiro

Prickly, glabrous, thicket-forming, rhizomatous marsh herbs. Stems commonly decumbent, with armed internodes. Leaves with long, prickly petiole, sheathing at base; blade sagittate, simple or divided, primary venation pinnate, prickly on underside, secondary venation reticulate. Infl. borne on long, spiny peduncle; spadix sessile; spathe narrow, long attenuate, convolute, usually caducous; flowers all bisexual; tepals 4(–6), free; stamens 4(–6), filaments free, flattened, thickened at apex, anthers attached basally, locules 2, cylindrical, dehiscing by circular, terminal pores; ovary with 1 apical ovule; stigma disc-like, sessile. Fruit single-seeded, faceted.

1. L. spinosa (L.) Thwaites; *L. heterophylla* (Roxb.) Schott.
Stem c.2.5cm diameter. Leaves to 2m; simple or with blades and lobes pinnatisect into narrow segments, apex and lobes cuspidate, blade lanceolate to broadly lanceolate, 28–39 × 11–34cm, basal lobes 22–27cm; petioles spongy. Peduncle slightly shorter than petioles. Spathe dark crimson on outside, slender, twisted above, tapering to very fine, slightly twisted apex, convolute above and below (shortly) spadix, 18–32cm. Spadix pale crimson, 2.5–5cm. Tepals oblong, apex triangular-hooded, c.2.5mm, keeled. Filaments c.1.5 × 0.8mm; anthers c.0.8 × 0.8mm. Ovary ovoid, c.1.5mm. Fruits c.1cm × 0.5–1.3cm; apex truncate, rugose-spiny.
Bhutan: S — Sarbhang district (Burborte Khola near Phipsoo); **Terai/ Duars** (Buxa Reserve, Dulkajhar); **Darjeeling** (Darjeeling); **'Sikkim'** (unlocalised Kurz specimen). Ravine in subtropical forest, 290m. Fl. March.

5. ARIOPSIS Nimmo

Small, perennial herb growing from tuber; flowering before or with leaves. Leaves asymmetrically peltate, petiole slender, subtended by membranous cataphylls; blade ovate, primary veins radiating from petiole, secondary veins reticulate. Infl. borne on slender peduncle; spathe persistent. Monoecious. Flowers unisexual, lacking perianth. Female part of spadix basal, fused on back to spathe; ovaries few, 1-loculed, ovules many, linear, parietal; stigmas 4–6, filiform, hooked, persistent. Male part of spadix terminal, swollen, stipitate, synandria of 3 fused stamens, embedded in tissue, dehiscing into chambers opening to outside by conspicuous, hexagonal pores. Berries angled, many-seeded.

1. A. peltata Nimmo. Fig. 14b.
Tuber 2–3cm diameter at flowering in precocious form, 1–1.5cm when leaves mature. Cataphylls oblong, apiculate, to 3.5–4.5cm, membranous, pale pinkish brown. Leaf blade ovate to suborbicular, cuspidate, base very shallowly cordate, 5.2–8.5(–15) × 3.8–6.8(–11)cm, thin-textured. Petiole 8.5–16.5(–20)cm. Peduncle very slender, 2–5cm. Spathe pinkish or mauve, narrowly obovate, strongly hooded, 0.8–1.7 × 0.6–1.2cm. Spadix almost equalling spathe; female part 2.5–6.5mm, ovules 6–8, ellipsoid, 1–2 × 0.6–1.6mm; male part ellipsoid, blunt, 4.5–8 × 2–3.8mm, yellow, shortly stipitate.
Darjeeling (Glen Cathcart, Ryang, Punkabari, Lebong, Kurseong, Rongtong); **Sikkim** (Gangtok (F.E.H.2)). On rocks, 460–1800m. Fl. May–June.

6. HOMALOMENA Schott

Evergreen, perennial, aromatic herbs, usually with short, erect stem. Leaves with long petioles, sheathing below. Leaf blades lanceolate to ovate, often cordate or sagittate. Spathe with closely appressed margins, opening slightly above, scarcely differentiated into tube below. Monoecious. Spadix with basal female flowers, male above, appendix absent. Flowers unisexual, lacking perianth. Ovaries partially 2–4-loculed, ovules many, parietal; stigma ± sessile; each ovary usually with a single, clavate staminode. Synandria of 2–4(–6) fused stamens; anthers attached basally, filaments short, stout, locules dehiscing extrorsely by slits.

1. H. rubescens (Roxb.) Kunth.
Stem to 10cm. Leaf blade triangular-ovate, acuminate, deeply cordate, 25–30 × c.21cm, basal lobes 5–7cm; primary veins radiating from petiole apex

and secondary veins closely parallel, diverging from compound midrib, arching towards marginal vein. Petiole 30–45cm, reddish. Peduncle 10–15cm, reddish. Spathe red, oblong in outline, cuspidate, 6–8 × 1.5–2cm. Spadix shortly stipitate; female part 2.5cm, reddish; male part 4 × 1cm, white. Ovary oblong-obovoid, 3-loculed, staminode 1, clavate, equalling ovary. Stamens mostly in groups of 4.

Terai (unlocalised Hooker specimen). 305m.

A single vegetative specimen seen; description of infl. taken from literature.

7. AGLAONEMA Schott

Stout, evergreen herbs; stems creeping or erect. Petiole with sheathing margins, encircling stem at base. Leaf blade simple, elliptic to oblong, primary venation pinnate, secondary veins parallel. Infls. in axillary groups, peduncu-late, subtended by cataphylls; spathes ovate, obliquely deciduous; spadix free or partly fused to spathe, lower part female, upper (longer) part male, decidu-ous. Flowers unisexual, perianth absent; synandria of 2 fused stamens, filaments short, anthers attached basally, locules dehiscing by an apical pore; female flowers consisting of a single ovary; ovary 1-celled, ovule single, basal. Fruit a fleshy berry with a single, large seed.

1. Leaves concolorous; fruits over 2cm long; female part of spadix fused to spathe on one side; native .... **1. A. hookerianum**
+   Leaves variegated, with pale bands along lateral veins; fruits to 1.5cm; spadix shortly stipitate completely free from spathe; cultivated
                                                           2. A. commutatum

1. A. hookerianum Schott.
Stem to 1cm diameter, apparently to 50cm. Leaf blade narrowly elliptic, slightly asymmetric, shortly acuminate, base oblique to rounded, 22–29 × 8.5–11.2cm, primary lateral veins arching, slightly thicker than the fine, parallel secondary veins, green. Petiole 16–21cm, sheathing for more than half its length, margins of sheath membranous. Peduncles c.11(–21)cm, borne in groups of 1–3. Spathe green, convolute at base, open above, c.4(–6) × 1cm. Spadix almost equalling spathe, lower part obliquely fused on one side to spathe, bearing few female flowers; male part free, narrowly cylindric c.3 × 0.5cm. Ovary ovoid, narrowed into short, stout style; stigma disc-like. Stamens: filaments very short, anthers c.0.7mm. Berry red, 2.5–3.2 × 1.1–1.7cm.

131

Darjeeling (unlocalised Cowan specimen). [Wet, shady forest to 1000m, fl. 4–7].

2. A. commutatum Schott.
Differs from *A. hookerianum* in having leaves with pale variegation along the lateral veins and spadix with a short (0.5–1cm) stipe-like base holding the female part free from the spathe. Immature (unfertilised?) ovaries yellow; berries red, oblong-cylindric, to 1.5cm long.
Bhutan: Cultivated in garden of Druk Hotel, Phuntsholing.

Native of the Philippines but widely cultivated (in a variety of forms) for its variegated foliage.

8. AMORPHOPHALLUS Blume ex Decaisne

Perennial herbs growing from a corm-like tuber, flowering before the usually single leaf appears. Petiole stout, often mottled, smooth or tuberculate; leaf divided into 3 segments, segments commonly once to several times pinnately divided into leaflets; secondary veins of leaflets diverging obliquely from midrib, connected by intramarginal vein; leaf sometimes bulbiliferous. Infl. borne on stout peduncle, subtended by membranous cataphylls; spathe persistent, open to campanulate, base funnel-shaped to convolute; spadix sessile to shortly stipitate, with stout, smooth or tuberculate, terminal appendix. Monoecious. Flowers unisexual, lacking perianth, female basal, male immediately above. Stamens single or 2–6 fused into synandria; anthers attached basally, often sessile, locules dehiscing by single or paired apical pores. Ovary 1–4-loculed; ovule 1 per locule; style short or long; stigma peltate, 1–4-lobed. Fruit a 1–few-seeded berry.

1. Spadix appendix smooth, whitish or pinkish; leaf bearing bulbils
    1. A. bulbifer
+   Spadix appendix tuberculate, yellow; leaf lacking bulbils
    2. A. napalensis

1. A. bulbifer (Schott) Blume. Sha: *olo-bantho.*
Tuber c.7 × 8cm, subglobose. Cataphylls brown, membranous. Leaf divided into 3 segments, segments shortly petiolate, almost immediately branched into large pinnatisect sections, leaflets narrowly oblong, finely acuminate, cuneate, veins sunk on upper surface; bulbils borne at primary and sometimes secondary divisions of leaf and leaflets. Petiole 0.3–1m, smooth, dark greyish-green, mottled pale green. Peduncle 13–16(–30)cm, stout, smooth, mottled.

Spathe ovate, subacute, margins strongly overlapping at base, 12–16cm, mottled greenish outside, pink to crimson on inside near base. Spadix almost equalling spathe, female section 2.5–3 × 1.3cm; male section 4–5 × 1.7–2cm; appendix swollen, subacute, 5–8 × 2.5cm, smooth, whitish or pink. Ovary c.2mm, 1(–2)-loculed, style very short (c.0.5mm); stigma capitate, 1.7–2.5mm diameter. Anthers c.2.5mm, prismatic, 4–6-angled.

Bhutan: S — Gaylegphug district (near Tori Bari); **Darjeeling** (Rangpo to Tista, Riang, Punkabari, Mongpu, below Tukvar); **Sikkim** (N of Jorethang; unlocalised Thomson & Treutler records). Among shrubs by roadside in subtropical forest, 150–1070m. Fl. May–June.

2. A. napalensis (Wall.) Bogner & Mayo; *Thomsonia napalensis* Wall. Dz: *gurba, dhow*; Sha: *ruginang asham*. Fig. 14c.

Differs from *A. bulbifer* as follows: leaf lacking bulbils, the 3 primary segments more elongate, divided into more numerous leaflets; leaflets more abruptly and finely acuminate; peduncle longer (to 70cm); spathe longer, narrower (18–29 × 7–11cm), not mottled outside; spadix much larger, male section 5–13.5cm; appendix 7.5–10cm, tuberculate, yellowish; style longer (1–3mm).

Bhutan: S — Phuntsholing (below Kamji) and Deothang (Deothang) districts; **C** — Tongsa district (near Nyakar, Shemgang (Nakao & Nishioka, 1984)); **Darjeeling** (Mongpu, Rishap; Kalimpong (F.E.H.1)); **Sikkim** (Legship, Rinchingpong to Pemagantsi; below Yoksum (F.E.H.1)). Dense, subtropical forest, 305–1500m. Fl. April–July.

A vegetative specimen from Kabi (Sikkim) probably belongs to this species.

Doubtfully recorded species:

A. paeoniifolius (Dennstedt) Nicolson (*A. campanulatus* Blume ex Decaisne) has been recorded from Phuntsholing (M.F.B.) but no specimens seen. Its leaf resembles that of *A. bulbifer*, but differs in having a rough petiole; spathe wider, with undulate margin, tubular at base, dark purple inside; appendix much shorter and wider (7–12 × c.5cm); style much longer (8–10mm), stigma deeply 4-lobed.

9. REMUSATIA Schott

Perennial herbs with cormose tuber. Bulbil bearing stolons arising from apex of corm; bulbils scaly, with hooked, filiform bristles. Leaf blade ± ovate, asymmetrically peltate; primary veins multiple, radiating from petiole and pinnately from midrib of main lobe; secondary venation arching-anastomosing, fine intramarginal vein present. Flowering with or before leaves. Spathe with persistent, closed, green, basal section enclosing female part of spadix and petaloid (yellowish to cream), deciduous blade. Monoecious. Flowers unisexual, lacking perianth. Spadix with terminal club-shaped male section; intermediate slender stipe composed of appressed male neuters; female section basal, sometimes with a single, apical whorl of female neuters. Synandria of 2–3 fused stamens, anthers attached basally to a short, common filament, locules 4–6, each opening by a single apical, circular pore. Ovaries subglobose, 1-loculed, ovules basal or parietal; stigma discoid, sessile.

1. Flowering before leaves; leaves glossy above; deciduous part of spathe lacking constriction near base 2
+ Flowering with leaves; leaves dull above; deciduous part of spathe constricted near base, forming a chamber enclosing male portion of spadix .... **3. R. pumila**

2. Bulbiliferous stolons stout, erect, simple, bulbils large (c.5mm); leaves usually oblong-ovate .... **1. R. vivipara**
+ Bulbiliferous stolons slender, spreading, much branched, bulbils small (c.1mm); leaves usually lanceolate, finely acuminate .. **2. R. hookeriana**

1. R. vivipara (Roxb.) Schott. Fig. 13b.
Corm 3.5–5cm diameter. Cataphylls 4 or more, broad, brownish, concealing peduncle, to 15 × 3cm. Stolons stout (c.5mm diameter), erect. Bulbils ellipsoid, stout, c.5 × 2.5mm, bristles to 1.5mm, stout. Leaf blade oblong-ovate to lanceolate, 11–33 × 7–19.5cm, sinus 1.5–3cm, intramarginal vein indistinct, glossy on both sides, yellowish-green, with pale, narrow (c.1mm) margin. Petiole 19–42cm, sheathing for lower ¼. Peduncle 6–12cm. Spathe: lower part 3–5 × 1.3–2cm, green outside; blade obovate, acute, apiculate, narrowed to base, 5.3–11.5 × 2.5–9cm. Spadix: female part 1.7–2 × 0.7–0.9cm, with 3–4 whorls of sterile ovaries at top and 1–2 whorls at base, stipe 1.1–2.5cm, tapering upwards; male part club-shaped, 0.9–1.9 × 0.4–1cm. Ovaries c.1.4mm diameter. Synandria c.1mm diameter usually 4-loculed.
Bhutan: S — Phuntsholing (Phuntsholing (F.E.H.2)), Chukka (Chukka Colony to Takhtichu), Gaylegphug (below Sham Khara, Mon La) and Deothang (near Chanari Village (M.F.B.)) districts; **C** — Punakha district (Punakha to Lometsawa, 15km E of Wangdi Phodrang, Mendengong; Rinchen to Khosa (F.E.H.2)); **Darjeeling** (Baobookhola, Sureil, Mongpu, above Sukna); **Sikkim** (Khyong, below Rumtek, S of Legship, S of Ranipul, Legship to

Tindong). Epiphytic and on boulders, banks and cliff-ledges in subtropical forest, 610–2000m. Fl. March–May.

2. R. hookeriana Schott; *Gonatanthus? ornatus* Schott.
Differs from *R. vivipara* as follows: leaves narrower, gradually finely acuminate, often purple beneath; stolons much branched, bulbils smaller as in *R. pumila*; infl. smaller, lower part of spathe 1–3 × 0.5–1cm, blade smaller (2–6 × 1–2cm), narrowly lanceolate; stipe broader, scarcely narrowing upwards (male and female neuters harder to distinguish); synandria usually 2-loculed.
    Bhutan: S — Phuntsholing (below Ganglakha) and Chukka (near Gedu, below Chimakothi, Chukka Colony to Taktichu) districts; **C** — Thimphu district (above Dotena, above Changri Monastery); **Darjeeling** (Rongbe, head of Rishap Ghora, Kurseong to Punkabari, Darjeeling, Rungirun, Mahalderam, W of Mongpu, Tiger Hill; Senchal to Takdah, Rimbick, Ramam (F.E.H.1)); **Sikkim** (Yoksum to Dzongri, Lachen, Lachung, 33km W of Singtam, Gangtok to Penlong La; Dentam to Pamianchi (F.E.H.1)). Mossy rocks and banks mainly in evergreen oak forest (i.e. higher than *R. pumila*), 1370–2660m. Fl. May–June.

3. R. pumila (D. Don) H. Li & A. Hay; *Gonatanthus pumilus* (D. Don) Engler & Krause; *G. sarmentosus* Klotzsch & Otto. Sha: *borang bozong*. Fig. 14d.
    Corm 1.5–2.5cm diameter. Cataphyll usually 1, slender, acute, 3–13cm. Stolons c.2mm diameter, spreading, covered by reddish-brown, papery scales, bearing few to many bulbils. Bulbils c.1 × 1mm, bristles filiform, to over 1cm. Leaf blade ovate or oblong-ovate, acute or slightly acuminate, base shallowly cordate (sinus 0.5–4.5cm), 8–23.5 × 5–14.5cm, dull on both surfaces, yellowish-green, areas between primary veins sometimes purple below and/or above. Petiole 9–33(–56)cm, sheathing for lower ¼. Peduncle 5–20cm. Spathe: basal section 0.9–2cm; blade in two parts, separated by a constriction, lower 1.5–2.5cm, opening to reveal male part of spadix, upper held at an angle, narrowly lanceolate, finely acuminate, 11.5–21 × 0.6–1.5cm. Spadix sessile: female part 5.5–7.5 × 3.5–4.5mm, with whorl of sterile ovaries at base and sometimes also at apex, stipe 5–8mm, purple; male part 9–12.5 × 4.5–5mm, purplish. Ovaries 1–1.5mm diameter, green, streaked white. Synandria 0.6–1mm diameter; filaments c.0.7mm.
    Bhutan: S — Phuntsholing (below Ganglakha), Chukka (Chukka Colony) and Deothang (between Morong and Narfong) districts; **C** — Punakha (between Lobesa and Tinlegang) and Tongsa (Changkha) districts; **Duars** (Buxa); **Darjeeling** (Lebong, Sureil, Kurseong, Punkabari, Kalimpong; Birch Hill, Happy Valley (F.E.H.1)); **Sikkim** (29km W of Singtam, Gangtok,

Namchi, Great Rangit, Yoksum). On damp mossy rocks, trees and banks in warm, broad-leaved forest, 550–2130m. Fl. May–July.

10. COLOCASIA Schott

Perennial, evergreen herbs with stout underground rhizome or short above-ground stem, sometimes stoloniferous. Leaves several, petiole sheathing below; blade asymmetrically peltate, base usually cordate, venation as in *Remusatia*. Flowering with leaves. Spathe with lower part persistent enclosing female part of spadix; upper part a deciduous blade. Monoecious. Flowers unisexual, lacking perianth. Spadix: basal part female, sometimes separated from male part by sterile flowers, appendix present. Synandria of 3–6 fused stamens, anthers subsessile, locules horseshoe-shaped in cross-section, dehiscing by semi-circular apical pores. Ovary 1-loculed; ovules minute, many, parietal; stigma sessile, sometimes extended upwards into a point.

The name *bozong* (Sha) refers to a species of *Colocasia*.

1. Spathe blade over 10cm; appendix shorter than or equalling male section of spadix; leaf blades usually large; plant cultivated or wild
   1. C. esculenta
+ Spathe blade under 6cm; appendix about equalling fertile (i.e. combined male and female) part of spadix; leaf blades under 20cm; plant wild 2

2. Narrowed zone of neuters between female and male section of spadix absent; leaves thick-textured .... **2. C. fallax**
+ Narrowed zone of neuters (to 0.5cm) present between female and male zones of spadix; leaves usually thin .... **3. C. affinis**

1. C. esculenta (L.) Schott; *C. antiquorum* Schott. Lep: *sankree*; Nep: *mane*; Eng: *taro*; Bengali: *kutchu*. Fig. 13c.
   Rhizome 3–5cm diameter, tuberous, vertical to horizontal. Leaf blade oblong-ovate to suborbicular, broadly and shortly cuspidate, base shallowly cordate (sinus 1–4cm), 13–45 × 10–35cm, dull, glaucous. Petiole 25–80cm, sheathing for ⅓ to ⅔ length, green. Peduncle 16–26cm. Spathe: tube 3–5 × 1.2–1.5cm, green; blade narrowly lanceolate, acuminate, lower part ± open, 10–19 × 2–5cm, cream to golden yellow. Spadix: female part 1.2–3.5cm; zone of sterile male flowers 1.2–2cm; male part 4.5–6.5cm; appendix 1.5–4.5 × 0.2cm (shorter than to equalling fertile male section), rough. Ovaries 1–3mm

diameter, not white-spotted; stigma broadly peltate (0.6mm diameter) not linearly extended. Synandria 0.7–1mm diameter.

Bhutan: S — Phuntsholing (below Kamji, 15km above Phuntsholing) and Chukka (Chongkha, Lobnakha) districts; **C** — Thimphu (Atsho Chhubar), Punakha (Punakha to Sinchu La, above Tinlegang) and Tongsa (viewpoint W of Tongsa) districts; **Terai** (Jalpaiguri); **Darjeeling** (Barnesbeg, Mongpu, Kalimpong, Lebong, Great Rangit Valley, Kumani, Punkabari); **Sikkim** (N of Jorethang, Legship to Tindong, Yoksum, 11km N of Gangtok). Naturalised or perhaps native in wet broad-leaved forest; cultivated near farmhouses, 90–2690m. Fl. July–September.

A very variable species cultivated pantropically. Rhizomes and petiole bases eaten in subtropical parts of Bhutan, Sikkim and Darjeeling.

A form with dark purple petioles and extremely glaucous leaves (sometimes narrower and with a deeper sinus than in the typically cultivated form) occurs in Bhutan (e.g. Chukka and Phuntsholing districts) and less commonly in Sikkim (below Rumtek) and is fed to pigs. Infl. not seen so cannot be assigned to any of the named varieties.

2. C. fallax Schott.

Rhizome c.1.5cm diameter, erect; stolons c.3mm diameter. Leaf blade narrowly oblong-ovate, apiculate, base shallowly cordate (sinus 0.2–1cm), 8–17.5 × 3.3–9.5cm, intramarginal veins several, slightly glaucous beneath. Petiole 12–27cm, sheathing for almost half length; sheaths reddish-brown. Peduncle slender, 6.5–10cm. Spathe: basal section 1.7–2 × 1.1–1.3cm, green; blade narrowly lanceolate, finely acuminate, margins appressed, 5–6cm, orange-yellow. Spadix: female part 1.2–1.4 × 0.5cm, with 4–6 rows of whitish sterile ovaries at base, no narrowed zone of neuters above; male section 1.1–1.4 × 0.4cm; appendix 2.5–2.7 × 0.1cm, acute, scaly-rough, with several rows of sterile male flowers at base, not narrowed at base. Ovaries subglobose, c.1.1mm diameter, green, speckled white; stigma discoid at base, tapering upwards into linear point to 0.5mm.

Bhutan: S — Chukka district (Raidak Valley); **C** — Tashigang district (Tashigang). Stream bed; moist mossy rock faces, 914m. Fl. May–July.

3. C. affinis Schott.

Differs from *C. fallax* as follows: rhizome more slender (c.1cm diameter); stolons filiform (c.1mm diameter); leaves usually more ovate (to 12.6cm wide), thinner-textured; petioles more slender; sheaths yellowish-green; infl. more delicate: spathe base c.1cm, blade to 4.5cm; spadix appendix smooth, shorter and stouter (1.5–2.5 × 0.3cm), blunt; narrowed zone of neuters (0.3–0.5mm)

present between female and male section; female section c.0.8 × 0.3cm, with
few ovaries; stigmas discoid, peltate.

Darjeeling (Pomong, Punkabari, Kurseong, above Sukna); **Sikkim** (Silak,
S of Legship, Legship to Tindong). Banks and on rocks in subtropical forest,
440–1070m. Fl. July–August.

As in *Remusatia pumila*, forms occur with attractive dark purple markings between the
major lateral veins. These have been called var. *jenningsii* (Veitch) Engler, but occur
mixed with unblotched plants and are not worth recognising taxonomically.

11. ALOCASIA (Schott) G. Don

Vegetatively very similar to *Colocasia* from which it differs in having basal
ovules. The Bhutanese species differ, however, in having at least a short, stout,
above-ground stem and leaves usually slightly glossy.

1. Leaf deeply cordate (not peltate); plant massive with stout stem to
    1m; cultivated .... **2. A. macrorrhizos**
+   Leaf peltate (sometimes very shortly); stem short (under 10cm); plants
    usually wild 2

2. Leaf hastate (basal lobes acute), distinctly peltate .... **1. A. fallax**
+   Leaf ovate (basal lobes rounded), shortly peltate .... **3. A. odora**

1. A. fallax Schott. Nep: *kala bako*. Fig. 14e.
Stem short (to 10cm), suberect. Cataphyll green. Leaf blade ovate, cuspidate, base cordate, sinus 8–14cm, 29–60 × 17–31.5cm, basal lobes strongly
angled, c.half length of main lobe, fused for ¼–⅓ length. Peduncle 15(–25)cm.
Spathe: tube green, swollen, 3 × 2cm, to 6cm in fruit; blade ovate, apiculate,
open, 9(–16) × 5cm, pale green or cream. Spadix: female part 1.2 × 0.7cm,
with c.2 rows of sterile ovaries at apex; middle section a slender stipe (1.3cm)
covered with appressed, sterile male flowers; male section cylindric, 1.8 ×
0.7cm, cream; appendix stout, conical, 1.8 × 0.7cm, rough. Ovaries ovoid, 2.5
× 2mm, green, not white-spotted; ovules c.5, large, basal, on erect funicles;
style c.0.3mm, brown, stout, with 3 spreading stigmatic lobes. Synandria
c.1.5mm diameter. Fruits c.6mm diameter, one-seeded, globose.
Bhutan: S — Samchi (Buduni) and Gaylegphug (Batase) districts;
Darjeeling (Great Rangit, Rungjo Ghora, Rangpo to Tista Bridge); **Sikkim** (N
of Jorethang). Subtropical forest, 305–1850m. Fl. May; fr. August–December.

Root poisonous.

2. A. macrorrhizos (L.) G. Don.
Stem massive. Leaves ovate, large (to 125 × 75cm), basal lobes rounded, deeply cordate, glossy, not peltate.
Bhutan: S — Phuntsholing district (Phuntsholing); **Terai**; **Darjeeling** (Rangpo to Tista); **Sikkim** (Gangtok).

Cultivated beside houses.

3. A. odora (Loddiges) Spach.
Like a small *A. macrorrhizos* — leaves similar in shape and texture, but shortly peltate.
Darjeeling (Peshok). Naturalised or perhaps native by house in subtropical forest, 400m.

12. XANTHOSOMA Schott

Vegetatively similar to *Alocasia* and *Colocasia* from which it can be distinguished as follows: spadix lacking appendix; ovary 2–4-loculed with numerous ovules; stylar discs coherent between adjacent female flowers.
The species cultivated in our area has not been seen to flower and may be distinguished vegetatively from any *Colocasia* or *Alocasia* occurring in Bhutan/Sikkim in having the main veins of the basal lobes exposed at the base of the sinus (i.e. leaf not peltate and lamina of basal lobes not reaching point of insertion of petiole).

1. X. brasiliense (Desfontaines) Engler.
Stemless; leaf blades hastate, to 40 × 15cm, veins of basal lobes exposed for c.1cm, intramarginal vein conspicuous; petiole to 40cm.
Darjeeling (Mongpu, Peshok); **Sikkim** (11km N of Gangtok). Cultivated on a small scale in subtropical areas to 1700m.

Native of tropical America but widely cultivated for its edible corms.

13. TYPHONIUM Schott

Perennial herbs from a cormose tuber. Leaves petiolate, blades hastate, 3-lobed or pedatifidly divided, present at flowering. Infl. pedunculate. Spathe with lower part persistent, swollen, margins strongly overlapping, upper part deciduous, narrow. Flowers unisexual, lacking perianth. Spadix with 4 sections — from base: female; sterile zone, basal part (contiguous with female) bearing filiform to clavate neuters, upper part with papillae (sterile male flowers); male; shortly stipitate, smooth, terminal appendix. Stamens single or fused in pairs;

anthers sessile, locules dehiscing by terminal pores. Ovaries 1-loculed with 1–2
basal ovules; stigma sessile, punctate. Berries 1(–2)-seeded.

1. T. diversifolium Wall. ex Schott var. **diversifolium.** Fig. 14f.

Corm c.1.5cm diameter. Cataphylls whitish, membranous, decaying early.
Leaf hastate in Bhutanese specimens (basal lobes divided into 2–3 linear
sections in some Nepalese specimens), upper part of blade triangular-
lanceolate, 3.2–14.5 × 1.3–4.5cm, secondary veins ascending from base or at
narrow angle from midrib, basal lobes 2–7cm. Petiole 6–16cm. Infl. usually
one per corm; peduncle 3–13cm. Spathe 6.5–16cm, blade oblong-lanceolate,
finely acuminate, margins inrolled, green with dark purplish-red lines, sometimes heavily marked with purplish inside. Spadix very shortly stipitate, shorter
than and enclosed by spathe; female zone 2.5–10 × 2.5–5mm; neuter zone
1.5–3mm, neuters with swollen bases, filiform stalks (to 2mm) and swollen
heads, upper sterile zone 12.5–25mm, with deflexed, appressed papillae; male
zone 4.5–13 × 1.5–3mm; appendix 20–47 × 3–5mm, dark purplish, smooth,
slightly swollen at base, distinct from short (1–3mm) stipe. Ovaries c.1mm
diameter, cylindric-faceted, apex truncate; ovule usually single. Anthers
0.3–1mm diameter, 2–4-loculed, laterally compressed.

Bhutan: C — Thimphu (Tashichho Dzong) and Sakden (Sakden) districts;
Sikkim (Lachen, Lang mang nang zo); **Chumbi** (La ree). Open grassy hillsides;
meadows; damp open ground, 2290–3660m. Fl. June–July.

var. **microspadix** Engler.

Vegetatively similar to var. *diversifolium* (basal lobes linearly divided in
Bhutan and some Chumbi specimens) from which it differs chiefly in its smaller
infl. Spathe 3–8cm, blade c.0.6cm wide; spadix more slender, appendix c.15 ×
1mm, gradually narrowed into relatively much longer (c.4mm) stipe, neuters
shorter-stalked with smaller heads; ovaries white-dotted.

Bhutan: C — Thimphu district (below Darkey Pang Tso); **Sikkim** (Dotha,
Lingtoo, Yeumtong); **Chumbi** (Galing). Dry rocky slopes, 3660–3930m. Fl/fr.
July–September.

FIG. 14. Araceae II (inflorescences).
a, **Pothos cathcartii**: single bisexual flower dissected from spadix (× 12). b, **Ariopsis
peltata**: spadix (× 3). c, **Amorphophallus napalensis**: spadix (× ½). d, **Remusatia pumila**:
spadix (× 4). e, **Alocasia fallax**: spadix (× 1). f, **Typhonium diversifolium**: spadix (× 2).
g, **Arisaema flavum**: spadix (× 2.5). h, **A. concinnum**: female spadix (× 1). i, **A. griffithii**:
male spadix (× 1). j, **Cryptocoryne retrospiralis**: section through pollination chamber
showing spadix (× 5). k, **Pistia stratiotes**: spadix (× 5). Drawn by Glenn Rodrigues.

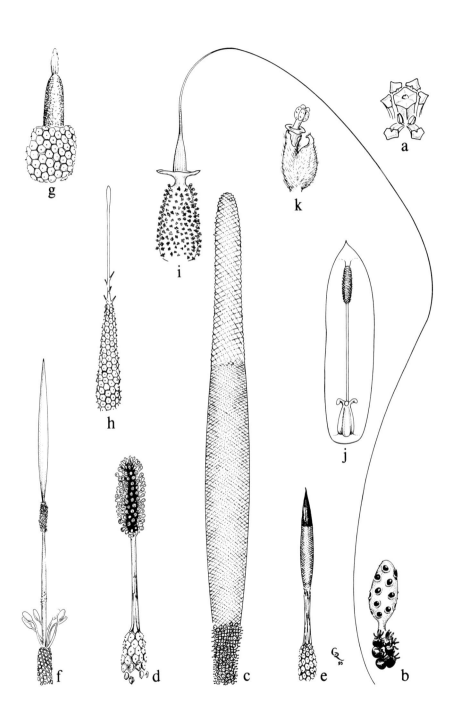

A specimen of what is almost certainly *T. trilobatum* (L.) Schott with an immature infl. has been seen from the **Terai** (Jalpaiguri); it differs from the above in having leaves with 3 subequal, erect lobes — central elliptic to lanceolate, laterals half-sagittate and the neuter flowers filiform, curled and deflexed.

14. SAUROMATUM Schott

Differs from *Typhonium* chiefly in having the margins of the basal part of the spathe fused to form a tube. Our species are further distinguished by having pedate leaves and flowering before leaves produced.

1. Leaf stout, to 50cm, central leaflet over 4cm wide; petiole pale green with rounded, dark blotches; infl. over 20cm, dark brown

                                                      1. S. venosum

+ Leaf slender, under 40cm, central leaflet under 3.5cm wide; petiole pinkish, unspotted; infl. under 15cm, pale greyish or flushed/spotted pinkish .... **2. S. brevipes**

1. S. venosum (Aiton) Kunth; *S. guttatum* Schott.
Corm to 6cm diameter. Central leaflet oblanceolate, acuminate, sessile, to 15.5–25 × 4–9.5cm, lateral leaflets 4–5 each side. Petiole 26 to 50cm, stout, pale green with round, dark blotches. Infl. single, sessile or peduncle to 11cm, subtended by several papery cataphylls. Spathe: tube cylindric, 4.5–10cm, dark brown; blade held at angle to spadix, oblong-triangular, acuminate, 16–39 × 1.5–4.5cm, heavily blotched dark brown on golden ground. Spadix: female zone 0.8–1.8cm; neuter zone 0.4–1.7cm, neuters to 0.5cm, clavate, stalked; sterile zone 3–6cm; male zone 0.8–1cm; appendix erect, 13–31cm, thickest ((0.2–)0.5–0.9cm) just above base, tapering slightly upwards.
Sikkim (Yoksum). Pathsides and field edges, 1850m. Fl. April–May; leaf mature July.

2. S. brevipes (Hook.f.) N.E. Brown; *Typhonium brevipes* Hook.f.
Differs from *S. venosum* in being much smaller and more slender in all its parts.
Corm 1–2cm diameter, forming several-flowered clumps. Leaflets 5–7(–9), narrowly lanceolate, finely acuminate to caudate, central 6.5–25 × 1–3.5cm. Spathe: tube swollen, 2–3 × 1–1.9cm, dark pink inside; blade very narrow, tapering from base (c.8mm wide) to very acute apex, to 12cm, pale greyish or flushed pink. Spadix: female zone c.3mm; neuter zone 4–6mm, neuters c.3mm; sterile zone c.5mm; male zone c.5mm; appendix to 7cm, very slender (0.6–1.2mm), even in thickness throughout length. Ovules 2 per ovary.

Darjeeling (Darjeeling, Rongbe, Jore Pokri, Sonada; near Ghum (F.E.H.2)). 1520–2740m. Fl. May–June; fr. July.

15. ARISAEMA Martius

Rootstock a subglobose corm or cylindric rhizome. Leaves few, basal or with overlapping petiole bases forming a pseudostem; blades trifoliolate, radiate, palmate or pedate; petioles stout, often mottled, smooth or verrucose, subtended by membranous cataphylls. Monoecious or dioecious (sex depending on nutrition and therefore variable from one year to another). Infl. borne with or before leaves, pedunculate; spathe tubular below, expanded into blade above, deciduous; spadix sessile, in monoecious plants female below, male above, neuters sometimes present on stipe of appendix; appendix usually produced sometimes ending in long filiform flagellum. Ovaries unilocular with several basal ovules; style sometimes distinct, stigma peltate, papillose. Synandria of 2–6 fused stamens, sessile or on a united filament, anthers dehiscing by 2 apical pores or a single horseshoe-shaped slit. Berries red, few-seeded.

Several species of *Arisaema* are of medicinal importance in SE Asia; for example, *A. jacquemontii* has recently been found to have anti-cancer properties (Habib-ur-Rehman *et al.*, 1992).

The following names refer to species of *Arisaema* in Bhutan: *roogi-nang-shoom* (Sha), *doaa, dowa* (Med), *gurbo, gurbu* (Nep).

1. Leaves pedate (with distinct central leaflet and outer sections divided into 3–7 leaflets decreasing in size outwards) or radiate (with (8–)11–20 similar, subequal leaflets) 2
+ Leaves trifoliolate or palmate (with 5–7 leaflets, central largest, decreasing in size outwards) 6

2. Leaves pedate 3
+ Leaves radiate 4

3. Plant small (under 33cm); spathe blade under 3.5cm, bright yellow (purplish at base); spadix monoecious .... **3. A. flavum**
+ Plant large (over 40cm); spathe blade over 4cm, yellowish-green; spadix monoecious or male .... **4. A. tortuosum**

4. Apex of spadix appendix spinulose .... **7. A. echinatum**
+ Apex of spadix appendix smooth or knobbly 5

5. Leaflets narrow (under 2cm wide), apex filiform-caudate (tip over 1cm); spathe yellowish-green, unstriped, blade distinctly broader than tube; apex of appendix smooth; fruiting peduncle recurved

5. A. consanguineum

\+ Leaflets broader (over 1.8cm wide), apex acuminate; spathe green or flushed purplish conspicuously white-striped, blade scarcely wider than tube; apex of appendix knobbly; fruiting peduncle erect

6. A. concinnum

6. Leaves palmate 7
\+ Leaves trifoliolate 8

7. Spathe blade with spreading basal auricles .... **1. A. nepenthoides**
\+ Spathe blade lacking basal auricles .... **2. A. jacquemontii**

8. Enclosed portion and usually lower part of exserted portion of spadix appendix conspicuously swollen 9
\+ Enclosed portion of appendix not conspicuously swollen or only at extreme base 12

9. Leaflets petiolulate, outer ones oblong, strongly asymmetric; tip of appendix long-filiform, pendent 10
\+ Leaflets sessile, outer ones elliptic-rhombic, slightly asymmetric; tip of appendix not filiform, sigmoidally ascending 11

10. Swollen part of appendix smooth .... **8. A. speciosum** var. **speciosum**
\+ Swollen part of appendix densely covered with whitish papillae

8. A. speciosum var. **mirabile**

11. Spathe dark purple, blade narrowly oblong, gradually narrowed to cuspidate apex .... **15. A. elephas**
\+ Spathe greenish with linear purple flecks, blade broadly oblong, abruptly contracted to cuspidate apex .... **11. A. dilatatum**

12. Spathe greenish; outer leaflets oblong, asymmetric 13
\+ Spathe dark purple; outer leaflets rhombic-elliptic 14

13 Spathe blade gradually attenuate, decurved .... **10. A. intermedium**
\+ Spathe blade small, cuspidate, rigidly deflexed .... **9. A. galeatum**

14. Appendix scarcely exceeding spathe blade; spathe blade under 8.5cm, lateral lobes not developed .... **14. A. propinquum**
\+ Appendix with long (over 10cm) filiform flagellum drooping beyond

15. ARISAEMA

apex of spathe blade; spathe blade usually over 10cm, lateral lobes
sometimes developed 15

15. Spathe blade dark blackish-purple with lateral lobes often greatly
developed (to 9cm wide) marked with pale reticulations; appendix with
disc-like swelling at base, included (thickened) part of appendix
under 4cm .... **12. A. griffithii**
+ Spathe blade reddish-brown, lateral lobes usually narrow, not reticulately veined; thickened part of appendix over 6cm, gradually tapering
from base .... **13. A. utile**

Measurements for leaves are given as at flowering time; they may grow considerably
afterwards. Plant height is to tip of tallest organ, whether leaf or infl. Many species
occur over a wide range of altitudes and are correspondingly variable in stature and
size of parts.

1. A. nepenthoides (Wall.) Martius ex Schott; ?*A. ochraceum* sensu F.B.I. Nep:
gurbu.
Dioecious. Corm 2–6cm diameter. Plant 29–100cm (to 170cm in fr.);
pseudostem, petioles and peduncle with large pinkish and brownish to blackish
chequered blotches; sometimes flowering before leaves produced. Cataphylls
brownish, with pinkish and dark blotches. Leaves usually 2, subopposite,
palmate, leaflets 5(–7), oblanceolate, sessile, abruptly acuminate to very acute,
base cuneate, dark green, glossy above; central leaflet 6–20.5 × 1.5–6.5cm (to
25 × 9 in fr.). Petiole 7.5–20cm, sheaths with conspicuous membranous apical
auricles. Peduncle usually exceeding leaves. Spathe: tube 3–8cm; blade arching
over spadix, oblong-ovate, acuminate, with two spreading, rounded, basal
lobes, 5–12.5 × 2.5–7.2cm, greenish, pinkish, brownish (occasionally blood-
purple) with broad, pale stripes, sometimes also with dark blotches. Appendix
cylindric, apex rounded, base truncate, 2.2–7.3 × 0.3–1.3cm, just exceeding
spathe tube, smooth, greenish or pinkish, stipe 0.2–1.3cm. Synandria widely
spaced, ± sessile, tinged dark purplish, usually of 2 anthers dehiscing by 4
apical pores. Fruiting peduncle erect.
Bhutan: S — Chukka (between Jumudag and Tala), Sarbhang (Dara Chu),
Gaylegphug (N of Chabley Khola) and Deothang (Keri Gompa) districts;
C — Ha (Shari), Thimphu (Dotena, above Taba, Dechenphu), Punakha
(Hinglai La, E side of Dochu La; Tsarza La (F.E.H.2)), Tongsa (between
Tongsa and Yuto La, 4km below Chendebi, Phobsikha, Tunle La, Tashiling)
and Bumthang (above Dhur) districts; **N** — Upper Mo Chu district (Tamji to
Pari La (F.E.H.2)); **Darjeeling** (Tonglu, Thosum La; Ramam (F.E.H.1));
Sikkim (Tumlong, Tendong, Lachen, Lachung, ridge between Tista and Rangit
rivers, Demthang, below Bakkim). Edges of and banks in forest (wet broad-

145

leaved, hemlock/rhododendron, oak/pine), sometimes beside streams, 1830–3440m. Fl. February–May.

Specimens from S Bhutan are larger and earlier flowering and probably worthy of further study.

2. A. jacquemontii Blume; *A. exile* Schott. Bhutanese name (Ha): *tou*. Dioecious. Corm 0.8–3cm diameter. Plant 13–54cm; pseudostem, petioles and peduncle pale green, unmarked. Cataphylls whitish, occasionally dark brown. Leaf 1(–2), palmate, leaflets 5(–7); central leaflet narrowly oblanceolate to elliptic, abruptly shortly-acuminate, base cuneate, sessile, 2–14 × 1.1–5cm, pale green; outer leaflets narrower, more gradually acuminate. Petiole 2–17.5cm. Peduncle exceeding leaves. Spathe pale green with narrow whitish stripes, margins and sometimes whole blade flushed dark purplish; tube 2.5–6.5cm; blade arching over spadix, oblong-triangular, 1.2–6 × 1–2.7cm, acuminate into ascending filiform tail 2.5–6cm. Appendix: upper part emergent from spathe tube, horizontal to decurved, tapering, base swollen, sometimes truncate, 3.5–6 × 0.1–0.2(–0.4)cm, smooth, greenish below, purplish above, stipe 0.1–0.5cm. Synandria very widely spaced, ± sessile, cream or tinged dark purplish, anthers 2–4 dehiscing by apical pores or slits. Fruiting peduncle erect.

Bhutan: C — Ha (Cchukata), Thimphu (Dochu La, E of Pajoding, hill above Hospital, Shodug), Tongsa (Chendebi, Yuto La), Bumthang (Dhur Chu), Bumthang/Mongar (Ura to Ghijamchu) districts; **N** — Upper Mo Chu (Gasa, Laya), Upper Bumthang Chu (Kantanang) and Upper Kulong Chu (Me La) districts; **Darjeeling** (Phalut, Sandakphu; Tonglu (F.E.H.2)); **Sikkim** (Dzongri, Thangshing, Potang La, Mejirdharn, Lating, Lachen, Lachung, Laghep, Chamnago, Tangu, Bikbari); **Chumbi** (Yatung). Open grassy places (sometimes among rhododendron and juniper); edges and clearings in forests (e.g. spruce), 2290–4270m. Fl. June–August.

3. A. flavum (Forsskål) Schott. Bhutanese name (Ha): *tou*; Dz: *dho*; Sha: *buchila-to*; Nep: *mane, boda sop, gurbo, birbanke, sarpa ko makai*. Fig. 14g.
Monoecious. Corm 1–2cm diameter. Plant 13–33cm; pseudostem, petioles and peduncle green, unmarked. Cataphylls pale, tinged brownish or purplish, unmarked. Leaves 1–2, pedate; central leaflet narrowly oblanceolate, finely acuminate, base cuneate, sessile, 3–7.5 × 0.8–2.2cm; outer sections petiolulate, each of 3–5 segments. Petiole 2–11cm. Peduncle usually exceeding leaves. Spathe: tube swollen, 0.5–1cm; blade suberect, very acute, 0.8–3.5 × 0.7–1.9cm, bright yellow, unstriped, purple at base and inside tube. Appendix ellipsoid, sessile, 1.9–3.2 × 1–2.2mm, just exceeding spathe tube, yellowish.

Male section about equalling female. Synandria ± sessile, dense, cream, anthers 2, each dehiscing by 2 apical pores. Fruiting peduncle erect.

Bhutan: C — Ha (Ha), Thimphu (Pyemitangka, Simu) and Punakha (Ritang) districts; **Sikkim** (F.B.I.); **Chumbi** (F.B.I.). Meadows; weedy cornfield, 2190–2800m. Fl. June–July.

Murata (1990) separated the E Himalayan and Chinese plant as subsp. *tibetica*.

Parker (1992) records this as a major weed of potato and other crops at about 2500m at Chapcha and Ha and also in Mongar (Tangmachu near Lhuntse), Punakha and Thimphu districts.

4. A. tortuosum (Wall.) Schott. Name at Tongsa: *jag*.
Monoecious (large specimens) or male. Corm 2.2–5cm diameter. Plant 42–200cm; cataphylls, pseudostem and petioles variously blotched with pinkish, grey, dark brownish or purplish. Leaves 2–3, well-spaced along pseudostem, pedate, dull above, dark green. Uppermost leaf: central leaflet narrowly to broadly elliptic to oblong-elliptic, shortly acuminate, base rounded to cuneate, petiolulate (0.5–7cm), 6.3–18(–25 in fr.) × 1.5–7.5(–8.5 in fr.)cm; outer sections petiolulate, each of 3–7 segments. Uppermost petiole 4.5–16cm, sheaths with conspicuous apical auricles. Peduncle usually exceeding leaves. Spathe: tube 1.5–4.5cm; blade spreading horizontally, oblong-lanceolate, acute or shortly acuminate, 4–11.5 × 1.5–4cm, green, yellowish-green or creamy yellow, not striped. Appendix sigmoidally ascending, gradually tapering from sessile base to very acute apex, 9–25 × 0.3–0.9 (at base) cm, greatly exceeding spathe, smooth, green, bluish-green or yellowish sometimes purplish at base. In monoecious plants male section of spadix usually exceeding female. Synandria widely spaced, stalked (1–3mm); cream to orange, composed of 2–3(–5) anthers each with 2 locules dehiscing by lateral slits. Fruiting peduncle erect.

Bhutan: S — Phuntsholing (Torsa River, below Ganglakah), Chukka (Gedu to Chukka Colony; Chimakothi (F.E.H.2)), Gaylegphug (Karai Khola) and Deothang (Diu Ri Valley) districts; **C** — Thimphu (Chapcha; Dochu La (F.E.H.2)), Punakha (Hinglai La, Lobesa to Lometsawa, Shenganga, Chachuphu; Rinchu to Mishichen (F.E.H.2)), Tongsa (Mangde Chu Bridge) and Tashigang (Gamri Chu) districts; **N** — Upper Mo Chu district (Khosa to Tamji (F.E.H.2)); **Darjeeling** (Rungbee, Tista, Punkabari, Kurseong, Tonglu, etc.); **Sikkim** (Nampok, Silake, Pemiongchi, Temi to Demthang, below Singhik, etc.). Broad-leaved and subtropical forest, especially at edges or disturbed areas e.g. pathsides, streamsides, 150–3050m. Fl. (March–)May–June.

Very variable in shape of leaflets; however, none of the specimens have leaflets narrow enough (i.e. under 1.2cm) to be referred to var. *curvatum* (Roxb.) Engler.

210. ARACEAE

5. A. consanguineum Schott.
Dioecious. Corm (1.2–)2–4.5cm diameter. Plant 17–69cm; pseudostem, cataphylls and stem with pinkish, purplish and/or brownish chequering. Leaf 1, radiate; leaflets 11–20, linear to narrowly oblanceolate, gradually acuminate to filiform-caudate tip (1–5cm), base cuneate, sessile, 5–21 × (0.35–)0.5–2cm. Petiole 9–21cm. Peduncle shorter than leaves. Spathe: tube (3–)3.5–6cm; blade arching over spadix, oblong-ovate, margins very narrowly auriculate at base, 3–6.5 × 1.3–4.5cm, greenish or greenish-yellow, sometimes striped purplish outside, sometimes tinged purplish towards apex, abruptly acuminate into drooping, filiform-caudate tip (5–18cm). Appendix narrowly club-shaped, apex rounded, gradually narrowed to base, (1.5–)2–3.5 × (0.16–)0.2–0.8cm, just exceeding spathe tube, smooth, greenish or yellowish, stipe short (0.2–0.5cm), usually bearing several, ascending, filiform neuters. Synandria dense, ± sessile, yellowish, 4-loculed, locules opening by apical pores. Fruiting peduncle recurved.

Bhutan: S — Phuntsholing (Phuntsholing Road) and Chukka (Chukka, Marichong, Chukka to Chimakothi) districts; **C** — Ha (Ha), Thimphu (common, for example between Cangnana and Paro, 6km N of Thimphu Dzong, Hinglai La), Punakha (Shenganga; Rinchu to Mishichen (F.E.H.2)), Tongsa (Mangde Chu Bridge), Bumthang (2km N of Jakar Dzong) and Tashigang (Tashi Yangtsi Dzong, Chorten Cora, Yangphula) districts; **N** — Upper Mo Chu district (Gasa); **Darjeeling** (Darjeeling, Tiger Hill, Batasia; Rimbick (F.E.H.1), Ghum to Peshok (F.E.H.2)); **Sikkim** (below Chiman). Open or disturbed forest (*Pinus wallichiana*, *Quercus lanata* and *Castanopsis*); meadows; under scrub; river banks, 1400–2900m. Fl. May–July.

The Tongsa specimen has curious flange-like projections at the base of the appendix.

6. A. concinnum Schott. Fig. 13d, Fig. 14h.
Similar to *A. consanguineum* from which it differs in its broader (1.8–5(–8)cm), abruptly acuminate leaflets; spathe blade oblong, scarcely wider than tube, not auriculate at base, green or often flushed dark purplish, with broad whitish or greenish stripes; appendix very slender, constricted below the knobbly apex; fruiting peduncle erect.

Bhutan: S — Phuntsholing (Phuntsholing Road, top of escarpment between Phuntsholing and Gedu), Chukka (Chimakothi (F.E.H.2)) and Gaylegphug (Chungsing) districts; **C** — Thimphu (Dotena (F.E.H.2)), Punakha (SW of Wangdi Phodrang, Lobesa to Lometsawa; near Punakha (Nakao & Nishioka, 1984), Bhotoka to Mishichen (F.E.H.2)) and Tongsa (Mangde Chu, 25.7km W of Yuto La, 1km S of Tongsa, 9km SW of Tongsa) districts; **N** — Upper Mo Chu district (Khosa to Tamji (F.E.H.2)); **Darjeeling**

(Senchal, Lebong, Algarah, Kurseong, Darjeeling, Garibans to Tonglu, Simonbong, Batasia to Palmajua; Takdah (F.E.H.1)); **Sikkim** (Lachen, below Chungthang, Rhikisum, Pemiongchi, above Dentam, Tsoka to Yoksum, Kabi; Gangtok (F.E.H.1)). Broad-leaved (*Castanopsis* and *Quercus*) and pine forest, often at edges or in disturbed areas, (305–)1370–3350m. Fl. (March–)April–July.

Small forms have been called var. *alienatum* (Schott) Engler, but do not seem worthy of recognition.

7. A. echinatum (Wall.) Schott.
Leaf similar to *A. concinnum*; infl. similar to *A. consanguineum* from which it differs in having the apex of the appendix spinulose and a relatively wider spathe blade, which is green marked with greyish.
Bhutan: C — Punakha (above Punakha (Griffith, 1848)), Bumthang (Pamprang — specimen incomplete; requires verification) and Sakden (Mera Valley) districts; **Sikkim** (Lachung). 2740–3050m. Fl. July.

8. A. speciosum (Wall.) Martius var. **speciosum**. Lep: *mongjing*; ?Nep: *sungure-to*.
Dioecious. Rootstock a cylindric rhizome to 15 × 2.5–4(–10)cm. Cataphylls reddish-brown. Leaf single, trifoliolate, leaflets petiolulate (0.5–4cm), with narrow red margin; central leaflet lanceolate to elliptic, abruptly shortly-acuminate, base cuneate, 11.5–33(–57 in fr.) × 5.5–14(–28 in fr.)cm, dull above, slightly bluish-green; outer leaflets oblong-lanceolate, abruptly shortly-acuminate, base strongly asymmetric (one side cuneate, other deeply cordate), 12–36.5 × 5–21cm. Petiole 19–55(–150 in fr.)cm, irregularly transversely striped with pale green and reddish-brown. Peduncle 2.5–21cm, much shorter than petiole. Spathe: tube 4–10cm, with broad white stripes; blade arching over spadix, oblong-lanceolate, gradually acuminate to very acute, drooping apex, margins spreading towards base, 8–21.5 (in total) × 4–9.5cm, shiny, bronze-coloured, purplish- or chocolate-brown. Appendix swollen, 3–7cm (excl. tail), widest (8–9mm) just above base, narrowed to base but scarcely stipitate, whitish, smooth, curving above and exserted from tube, gradually drawn into flagellum (to 50cm). Synandria shortly stalked (1–1.5mm), cream, composed of 4–5 anthers each opening by a horseshoe-shaped slit. Fruiting peduncle erect.
Bhutan: S — Chukka (Chukka Colony to Taktichu, W of Gedu; Chimakothi (F.E.H.2)) and Gaylegphug (Chungsing) districts; **C** — Thimphu (Dotena, Olaka), Punakha (Mishichen to Khosa (F.E.H.2)), and Tongsa (Tashiling to Charikachor (F.E.H.2)) districts; **N** — Upper Mo Chu district

210. ARACEAE

(Tamji: Nakao & Nishioka, 1984, plate 246, as *A. intermedium*); **Darjeeling** (Darjeeling, Rangirun, Lebong, Rungbee, Manibanjan to Batasi; Peshok, Rimbick, Palmajua (F.E.H.1)); **Sikkim** (Lachen, Yoksum, Tista Valley, Rate Chu N of Gangtok). Margins of broad-leaved (oak and subtropical) forest, 460–3050m. Fl. March–June(–July).

Vegetative parts used as pig fodder in Darjeeling/Sikkim (Pradhan, 1990).

var. **mirabile** (Schott) Engler.

Differs as follows: margins of spathe blade not spreading at base; swollen part of appendix densely covered with ridge-shaped papillae.

Bhutan: C — Thimphu (Dochu La, Tzatogang to Dotena (F.E.H.2)) and Mongar (Sengor, Reb La) districts; **N** — Upper Mo Chu district (Tamji to Gasa (F.E.H.2), (Nakao & Nishioka, 1984, plate 244)); **Darjeeling** (Tiger Hill, Senchal; Sandakphu to Garibans (F.E.H.1), Chitrey to Meghma, Tonglu (F.E.H.2)); **Sikkim** (Bakkim to Yoksum; Tendong (F.E.H.2)). 2000–3100m. Fl. (May–)June–July.

9. A. galeatum N.E. Brown.

Dioecious. Corm (not seen) 5–7cm diameter. Leaves as in *A. speciosum*, but petiole green, unmarked. Spathe tube 6–8cm, green, sometimes flushed purplish, striped white; basal part of blade strongly deflexed forming hooded apex to tube, margins reflexed, terminal lobe of blade pendent, oblong-ovate, cuspidate, 2.5–4 × 1–2.5cm, green or purplish, sometimes reticulately veined. Appendix 3.5–8cm (excl. flagellum), smooth, whitish, gradually tapering from wide (0.5–1.5cm), truncate base into pendent flagellum (to 25cm); stipe c.0.3mm. Synandria lax, shortly stalked (c.0.5mm), each of 3–4 anthers, anthers each dehiscing by a single, horseshoe-shaped slit.

Bhutan: S — Chukka (Putlibhir (F.E.H.2)), Gaylegphug (N of Chabley Khola) and Deothang (Samdrup Jongkhar Road) districts; **Darjeeling** (Kalimpong (Pradhan, 1990)); **Sikkim** (no locality). 2100m. Fl. March–June.

10. A. intermedium Blume f. **biflagellatum** (Hara) Hara; *A. biflagellatum* Hara.

Dioecious. Corm 2–3cm diameter. Cataphylls pale. Leaf single, trifoliolate, leaflets finely acuminate, sessile; central leaflet narrowly elliptic, base cuneate, (7–)9–11(–15) × (2.5–)3.5–4(–5)cm; outer leaflets lanceolate, base strongly asymmetric (one side cuneate, other deeply cordate), (6–)10.5–13(–20) × (2–)4–4.5(–8)cm. Petiole (10–)27–30cm, pale green. Peduncle (6–)18–25.5cm, unmarked. Spathe: tube 3–5cm; blade arching over spadix, lanceolate, 4–6 (excl. flagellum) × 1.5–2.5cm, yellowish green, sometimes with narrow white stipes, acuminate to pendent flagellum (8–16cm). Appendix swollen, 1.5(–3)cm (excl. flagellum), widest (3.5–5mm) just above base, shortly stipitate, whitish,

150

smooth, curving above and exserted from tube, gradually drawn into flagellum (to 25cm). Synandria shortly stalked (0.7–0.8mm), cream, composed of 3–4 anthers each opening by a horseshoe-shaped slit.

Darjeeling (Sandakphu to Garibans; Phalut to Sandakphu (F.E.H.1)). 3500m. Fl. June.

f. *intermedium* has been recorded from Darjeeling (Sandakphu, 3800m (F.E.H.2)), but no specimens seen. It is usually larger in all its parts and the spathe is often tinged purplish, but differs mainly in the spathe blade having a much shorter filiform apex. A specimen from Sakden district (Sakteng, *Bowes Lyon* 9109, E) might also be *A. intermedium* f. *intermedium* though it differs from W Himalayan plants in being larger in all its parts and having the appendix truncate (rather than attenuate) at the base.

11. A. dilatatum Buchet.
Dioecious. Corm c.7cm diameter. Cataphylls several, almost equalling petiole, very wide (to 4cm), apparently pale. Leaf single, trifoliolate, leaflets sessile, acute, subequal; central leaflet broadly elliptic(-rhombic), base cuneate, c.16 × 12.5cm; outer leaflets broadly elliptic, slightly asymmetric at base, to 19 × 13cm. Petiole c.50cm, green, unmarked. Peduncle c.15cm. Spathe: tube c.6cm, very wide (c.3.5cm); blade decurved, oblong-oblanceolate, shortly cuspidate, c.9 × 5cm, green with linear purple flecks. Appendix swollen, c.19cm, widest (c.1cm) at truncate base, smooth, flagelliform tip pendent; stipe c.1cm.

Bhutan: C — Mongar district (Pimi, Reb La). Broad-leaved forest, 2730–2990m. Fl. April–May.

The following three taxa appear to form an altitudinal cline and work is required in the field and laboratory before reaching a satisfactory taxonomic treatment. None of the characters traditionally used to define species seem to be reliable. Whether the individual has 1 or 2 leaves presumably depends on the size of the corm and growing conditions. The width of the lateral spathe blade lobes seems to vary continuously, despite the very dramatic differences between extremes of the range. The distinction between horseshoe and porose anther dehiscence (used by Engler to place some of the following in different Sections) simply cannot be upheld (see drawings in Hara (F.E.H.2) to see how this varies even within a species). Taxonomists such as Hara (who knew them in the field) have failed to resolve the tangle and have repeatedly changed their minds (e.g. Hara 1961, 1965, F.E.H.2). I have merely attempted to apply names only to plants matching the types representing 'nodes' of variation rather than attempting to judge appropriate ranks for the taxa based on herbarium specimens.

12. A. griffithii Schott; *A. verrucosum* Schott; *A. pradhanii* C.E.C. Fischer.
Fig. 14i.
Dioecious. Corm 2.5–7cm diameter. Cataphylls very wide (to 4.5cm), brownish, greenish or whitish mottled darker. Leaves 1–2, trifoliolate, leaflets subsessile, margins crisped, yellowish, brownish or reddish; central leaflet ovate-

or obovate-rhombic, acute or shortly-acuminate, base cuneate, 9.5–23.5 ×
7–28cm; outer leaflets asymmetrically ovate-rhombic, finely acuminate, base
cuneate, scarcely asymmetric, 10–44 × 7.5–35cm. Petiole 20–43cm, usually
dark-mottled, sometimes verrucose. Peduncle 6–20cm (much shorter than petiole). Spathe dark reddish-purple: tube 3.5–8cm, with broad white stripes; blade
6–20 × 8–18.5cm (in total), central part an oblong continuation of tube
forming deflexed hood, apex caudate (to 4.5cm), lateral lobes half-obovate (so
apex of blade retuse) to 9cm wide, cream- or green-reticulated. Appendix
1.5–4cm (excl. flagellum), base a swollen disc (0.7–1.9cm wide), abruptly
contracted into swollen middle section (0.2–0.7cm wide), gradually narrowed
into long (to 60cm), pendent flagellum, dark purplish, smooth, curving above
and exserted from tube; stipe 0.3–0.5cm. Synandria shortly stalked
(0.5–1.5mm), cream to orange, composed of 3–4 anthers each opening by a
pore or horseshoe-shaped slit. Fruiting peduncle erect.

Bhutan: C — Ha (Sele La), Thimphu (Dechencholing to Punakha;
Barshong to Tzatogang (F.E.H.2)), Thimphu/Punakha (Dochu La, Hinglai
La) and Tongsa (Phobsikha) districts; **N** — Upper Mo Chu district (Gasa to
Pari La); **Darjeeling** (Sandakphu, Tonglu, Senchal, Glen Cathcart, Phalut,
Kalipokri; Senchal to Ghum, Batasi to Palmajua (F.E.H.1)); **Sikkim** (Yakla,
Lachung Valley, Lachen, Chiabanjan, Dentam, Phedang to Bakkim, Tsoka,
Jalook). Coniferous/broad-leaved (with rhododendron and bamboo) forest,
often among rocks; open blue pine forest; occasionally in open, 2290–3700m.
Fl. April–May(–July).

The character of verrucose petioles seems trivial though as indicated by Hara (F.E.H.2),
who reduced the synonyms cited above to varietal rank, further work is necessary on
the genetic control and phenotypic plasticity of this character, which is in any case
apparently difficult to observe in herbarium specimens; some specimens labelled 'verrucosum' (which presumably were so in life) do not appear to be when dried. Verrucose
forms appear to be commoner at higher altitudes, but occur in mixed populations with
non-verrucose forms.

According to Pradhan (1990) 'sherpas living at high altitudes dig out the tubers of *A.
griffithii*, pound them, tie them in a cloth and keep them in running water for several
days. The powder is then dried and used for making bread in the winter months. In N
Sikkim the Forest Department allows collection of the tubers of this species for one
week before the onset of the winters for consumption and thereafter all collection
is banned'.

There has been much confusion between this and the following species and many
misidentifications. The species are very close and the following is probably only worth
varietal status though recent collections are needed.

15. ARISAEMA

13. A. utile Hook.f. ex Schott.

Differs from *A. griffithii* in having spathe blade paler (reddish-brown), lobes always narrow (to 2cm), never reticulately marked and especially in the appendix widest at the lobed, truncate base, gradually tapering upwards, the thickened part much longer (6–12cm). Petioles apparently unspotted.

Bhutan: C — Ha (Sele La) and Mongar (Sheridrang) districts; **Darjeeling** (Sandakphu); **Sikkim** (Lachen). Mixed forest; rhododendron jungle 2440–3960m. Fl. April–July.

Tubers fermented and eaten in Sikkim (Hooker, 1854, 2: 49).

Literature records not included as the name has been frequently applied in herbaria to what seem merely to be starved specimens of *A. griffithii* (see Clarke, 1885 for field observations on the species).

14. A. propinquum Schott; *A. wallichianum* Hook.f.; *A. sikkimense* Stapf ex Chatterjee; *A. ostiolatum* Hara.

Dioecious. Sometimes flowering before leaves. Corm 1.5–4.5cm diameter. Cataphylls sometimes wide (to 3cm), brownish mottled darker sometimes narrow and whitish. Leaves 1–2, trifoliolate, leaflets subsessile, margins crisped, sometimes blackish; central leaflet elliptic to obovate-rhombic, acuminate or apiculate, base cuneate, 5.5–11.8 × 4.8–8cm; outer leaflets asymmetrically ovate (occasionally suborbicular), finely acuminate, base cuneate, scarcely asymmetric, 6.5–10 × 4.5–8.7cm. Petiole 7.5–30cm, sometimes with small dark chequering or dark longitudinal stripes. Peduncle 7.5–23.5cm, often equalling or exceeding petiole at flowering, sometimes purplish. Spathe dark brownish-purple, glossy: tube 2.5–7cm, with broad yellowish or greenish stripes; blade oblong to narrowly obovate, acute to truncate, shortly caudate (0.2–2.5cm), 3.3–8.5 × 1.5–4.8cm, lateral lobes absent or spreading, reticulately veined. Appendix 6–13.5cm, widest (0.3–0.6cm) at truncate base, gradually tapered to filiform apex only just exceeding spathe blade, dark purplish, smooth; stipe 0.2–0.7cm. Synandria shortly stalked (0.5–1.2mm), cream to yellowish, composed of 3 anthers each opening by a horseshoe-shaped slit.

Bhutan: C — Ha (Damthang, Ha Chu) and Thimphu (below Darkey Pang Tso, above Naha; Tzatogang to Dotena, Barshong (F.E.H.2)) districts; **Darjeeling** (Sandakphu, Phalut, Ramam to Phalut); **Sikkim** (Phedang, Changu, Sakkargong, Jamlinghang, Prek Chu to Thangshing, Se-moo-do-ne, Dzongri). Conifer/rhododendron forest, often among rocks; yak pasture (in open or among shrubs), 3050–4270m. Fl. May–July.

Large forms come very close to small forms of *A. griffithii*.

The small forms described as *A. sikkimense* and occurring from E Nepal to W Bhutan will probably prove to be worth some level of taxonomic recognition if characters such as precocious flowering, narrow, acute spathe blades, narrow cataphylls and black-checkered petioles are constant. They seem, however (at least from herbarium specimens), to grade into *A. propinquum* (typically a larger, 2-leaved plant which extends to the W Himalaya).

A. ostiolatum Hara seems to be merely an abnormal form of the smaller form of *A. propinquum* (i.e. *A. sikkimense*) in which the sides of the spathe blade are inrolled.

15. A. elephas Buchet; *A. wilsonii* Engler. Plate 8.

Differs from *A. propinquum* in its very swollen appendix greatly exceeding the spathe, the exserted portion sigmoidally ascending. Leaflets red-margined, central one broadly obovate; petiole pale brown mottled, very faintly streaked pale greenish.

Bhutan: C — Thimphu district (below Darkey Pang Tso N of Paro); **N** — Upper Bumthang Chu district (Kantanang); **Chumbi** (Do-ree-chu). Among mossy rocks in fir/bamboo forest, 3530–3810m. Fl. June.

Doubtful and doubtfully recorded species:

A. costatum (Wall.) Martius ex Schott.

A photograph of a specimen taken at Kori La (Mongar district) appears close to *A. costatum* with its characteristic leaf venation (lateral veins parallel, very close set, conspicuously raised beneath and cross-veinlets not visible), its long peduncle (c.half length of petiole) and dark maroon, glossy spathe; it differs, however, in the mottling of the petiole which is like that of *A. speciosum*. Collections of this plant are highly desirable, and merit further investigation.

A. costatum is only known from Nepal. The record of this species from Sham Khara, 1500m (M.F.B.) should almost certainly be referred to *A. speciosum* from the description of its rootstock.

A. decipiens Schott.
Rhizomatous. Leaves pedate. Gomchu Hill, 2466m. (M.F.B.).

A. erubescens (Wall.) Schott.
Recorded for Sikkim in F.E.H.2, probably on the basis of old specimens (cited in F.B.I.) which have been re-determined as *A. concinnum*. It differs from that species in having narrower leaflets, a brownish spathe and smooth appendix apex.

A. petiolulatum Hook.f.
Rhizomatous. Trifoliolate, leaflets all petiolulate. Leaf single. Kamji, 1600m. (M.F.B.).

A. roxburghii Kunth.
Rhizomatous. Trifoliolate, outer leaflets sessile, median petiolulate. Leaves two. Sureylakha, 1500–1800m. (M.F.B.).

Pythonium ecaudatum Griffith.
This obscure name was dredged up from Griffith (1848) by Naithani (1990) and presumably refers to a species of *Arisaema*. There is no specimen at Kew, but the cryptic Latin diagnosis (which includes the phrase 'spatha triant') suggests *A. nepenthoides*.

16. CRYPTOCORYNE Fischer ex Wydler

Delicate rhizomatous aquatic herbs, submerged or emergent. Leaves in rosettes, linear with parallel venation or differentiated into petiole and simple blade with veins arching from base of blade and connected by transverse veinlets. Infl. borne singly, very shortly peduncled. Spathe slender, lower part swollen forming a chamber enclosing the spadix, middle part tubular with connate margins, upper part an open, usually narrow, blade, sometimes twisted. Monoecious. Flowers unisexual, perianth absent. Spadix filiform, fused at apex to flap closing pollination chamber, ovaries 4–7, connate, basal; central part of spadix naked; stamens borne at apex. Ovaries unilocular with many, basal ovules. Stamens borne singly, anthers sessile, dehiscing by 2 apical pores. Carpels connate, 2-valved, many-seeded.

1. C. retrospiralis (Roxb.) Kunth. Fig. 13e, Fig. 14j.
Rhizome cylindric, 0.2–0.4cm diameter. Leaves linear to narrowly oblong, apiculate, margins smooth (perhaps sometimes wavy), narrowly cartilaginous, sometimes minutely toothed near base, gradually or abruptly narrowed into sheathing, membranous, petiole-like base, 6–16(–27) × 0.44–1.2cm. Peduncle under 1cm. Spathe green on outside, white spotted purple inside; chamber c.9(–15) × 4.5mm, interior purple, coarsely reticulate above, vertically ridged below; tube 7–13 × 0.2–0.3cm; blade narrowly lanceolate, twisted, 3.5–4.5 × 0.5cm. Carpels 4(–6), tapering upwards, c.2.5mm, minutely white-granular, style stout, recurved, stigma discoid. Sterile section of spadix c.5mm. Male section c.1.4 × 0.8mm.
Terai/Duars (Tondu Forest, Buxa Reserve, Dulkajhar). Stony or sandy, seasonally flooded, water-courses, c.300m. Fl. December–February.

A very difficult genus, especially from herbarium specimens; however, Jacobsen (1980) was certainly wrong in identifying the Dulkajhar and Buxa specimens as *C. crispatula* Engler.

17. PISTIA L.

Floating stoloniferous aquatic; roots feathery. Leaves in rosettes, obovate, obcuneate or obcordate, densely hairy; primary veins parallel, rib-like, secondary veins reticulate. Infls. small, hidden in leaf bases. Peduncle short. Spathe densely hairy on outside, tubular below, blade short, suborbicular, whitish. Monoecious. Flowers unisexual, lacking perianth. Spadix almost equalling spathe; lower part consisting of a single ovary — fused on back to spathe, male section a single ring of stamens borne on short stipe subtended by green corona, appendix minute. Ovary subglobose, ovules many, parietal; style about equalling ovary; stigma not expanded. Stamens fused in pairs, dehiscing by 4 apical pores.

1. P. stratiotes L. Fig. 13f, Fig. 14k.
Description as for genus.
Terai (Booree Toorsa, Kurlanadi, Jalpaiguri, Mattighari). Tanks.

Additional cultivated ornamentals:
A number of exotic species are grown, especially in Darjeeling and Sikkim; these cannot be treated fully, but the following are among the more common:

Caladium bicolor (Aiton) Ventenat.
Tuberous herb easily recognised by its pink-spotted, ± elliptic, peltate, deeply cordate leaves.
Bhutan: S — Samchi district (Chengmari); **Sikkim** (common).

Native of northern S America widely grown for its ornamental foliage.

Epipremnum aureum (Linden & André) Bunting.
A trailing/climbing plant. Juvenile stage with ovate, acute, golden-variegated leaves; mature leaves (on climbing stage) large (to 80cm), oblong, irregularly pinnate; ovules 2–4, basal.
Sikkim (Legship); **Darjeeling** (Singla Bazaar).

Native of the Solomon Islands but widely cultivated in tropical gardens and as a pot plant for its ornamental foliage.

Monstera deliciosa Liebmann. Eng: *Swiss cheese plant.*
A massive climber with dark green, perforated leaves superficially similar to *Rhaphidophora* from which it differs in having 2-loculed ovaries with 2 ovules per locule.
Sikkim (Gangtok).

Native of central America but widely grown in the tropics and as a pot plant for its ornamental foliage and edible fruit.

Philodendron bipinnatifidum Endlicher.
Massive plant with initially erect stem to 2m; leaves dark green, glossy, ovate in outline pinnately cut into blunt, rather wavy lobes, sometimes again divided.
Sikkim (Legship, Gangtok).
Native of Brazil but widely cultivated for its ornamental foliage.

Syngonium podophyllum Schott.
Climber; leaves in the commonly cultivated form marked with large white patches: immature leaves hastate; adult leaves pedate with central lobe and 2 or more subequal lateral lobes.
Bhutan (Phuntsholing, cultivated in nursery).

Native of Central and northern S America widely cultivated for its ornamental foliage.

Zantedeschia aethiopica (L.) Sprengel. Eng: *calla lily.*
Stemless herb with tuberous rhizome; leaves ovate, cordate; spathe broadly ovate, apex recurved, white, waxy; female part of spadix with conspicuous yellow staminodes.
Sikkim (Yoksum); **Darjeeling** (Darjeeling town).

Native of S Africa widely cultivated for its decorative infls.

Doubtfully recorded species:

Plesmonium margaritiferum (Roxb.) Schott.
Similar to *Amorphophallus* vegetatively and at least superficially in its infl.; it differs from any of the Bhutanese species in having male and female sections of the spadix separated by a zone of large, inflated neuters. A Clarke sheet from Balasun (Darjeeling) labelled *Thomsonia napalensis* and bearing a leaf of that species also bears an infl. of *P. margaritiferum*. This latter species is not given for Darjeeling in F.B.I. or Engler's monograph, and it seems likely that there has been some confusion of labels; nevertheless it may occur in subtropical regions.

Steudnera colocasioides Hook.f.
Recorded for Sikkim in F.B.I. on the basis of a King drawing at Calcutta; no specimens have been seen and it seems unlikely. *Steudnera* is similar to

Colocasia but the infl. differs as follows: spathe open almost to base, lower part not tube-like; spadix lacking appendix, female part fused on back to spathe; ovaries surrounded by 2–5 sterile, clavate staminodes. It also possesses a stout, horizontal stem covered in reticulate, fibrous sheaths.

Family 211. ACORACEAE

Aromatic, glabrous, marsh or emergent aquatic herbs. Rhizome horizontal, bearing leaves at apex. Leaves linear, distichous, bases overlapping, venation parallel. Peduncle leaf-like. Infl. a dense, cylindric spadix-like spike, apparently lateral as subtended by erect, leaf-like bract. Flowers bisexual, tepals 6, free; stamens 6, filaments free, slender, flattened; anthers horseshoe-shaped, locules fused at apex, dehiscing introrsely by continuous slit parallel to outer edge; ovary 3-loculed, each locule with several apical ovules; stigma punctate.

Recent cladistic analyses of both morphological and molecular (Chase, 1993) data suggest that *Acorus* is basal to the monocots and should certainly be placed in its own family.

1. ACORUS L.

Description as for family.

1. Leaves with distinct midrib; rhizome stout (0.8–1.5cm diameter)
   1. A. calamus
+ Leaves lacking distinct midrib; rhizome slender (to 0.5cm diameter)
   2. A. gramineus

1. A. calamus L. Bhutanese names: *soka ta* (Ha), *ochu dha* (Ongdi), *chuta rechu* (Chapcha); Sha: *bar-tsi*; Lep: *ruk lop*; Nep: *bojho*. Fig. 13g.
Rhizome 0.8–1.5cm diameter, pinkish. Leaves asymmetric, apex very acute, 37–80 × 0.6–1.2cm, midrib conspicuous. Peduncle 9–19cm, trigonous. Infl. bract 13–48 × 0.6–1.6cm. Spadix straight or slightly curved, erect or oblique, 2.8–5.5 × 0.5–1cm, greenish. Tepals oblong, apex triangular-hooded, 1–2 × 0.5–0.8mm, keeled, membranous. Ovary 1.5–3.5 × 0.8–2.3mm, hexagonal-cylindric with conical, spongy apex. Filaments 1.2–2.4 × 0.3–0.5mm; anthers c.0.5mm, cream.
Bhutan: S — Samchi district (Dorokha (M.F.B.)); **C** — Ha (Puduna), Thimphu (Tsalimaphe, Paro, Thimphu, Chapcha; Dotena to Thimphu (F.E.H.2)), Punakha (below Pandagong; Dochu La to Wangdi Phodrang,

Samtengang to Kyebaka, Mishina (F.E.H.2)), Bumthang (Jakar to Kurje) and Tashigang (Tashigang) districts; **Darjeeling** (Sukhia Pokhri, Rishap); **Sikkim** (Dikchu, Yoksum, below Rumtek, Phodong to Kabi). Marshes, 610–2800m. Fl. April–July.

Used medicinally: in Bhutan fresh juice given for oedema and goitre or applied to dislocations and fractures (Bedi), and also used as a laxative (Cooper); in Darjeeling shoots crushed and made into paste to treat scabies, and shoots dried and made into pills as a brain tonic (Lama, 1989).

Forms with slender rhizomes, narrow leaves and a small, narrow spadix (e.g. Ha, Paro and Tashigang specimens) can be referred to var. *verus* L. (*A. griffithii* Schott) which is similar to the following species but differs in having slightly thicker rhizomes and leaves with a distinct midrib.

2. A. gramineus Aiton.
Differs from *A. calamus* in having a very slender rhizome (to 0.5cm diameter) and leaves lacking a prominent midrib; spadix commonly longer and more slender.
Sikkim (non-flowering, unlocalised, Hooker specimen).

Family 212. LEMNACEAE

Minute floating or submerged aquatic annuals. Fronds circular to lanceolate, flat or swollen beneath, reproducing from 1 or 2 budding pouches, daughter fronds becoming detached or remaining attached. Roots 0, 1 or several from lower side of frond. Flowers monoecious, reduced to a single stamen or single gynoecium; infls. of 1 stamen and one gynoecium naked in a cavity; or 2 stamens and 1 gynoecium enclosed by spathe in one of budding pouches. Gynoecium of 1-loculed ovary with 1–4 ovules and short, simple style. Fruit 1–4-seeded, winged or not, globose or compressed.

1. LEMNA L.

Fronds usually aggregated. Budding pouches 2, basal and lateral. Root single. Infl. of 2 stamens and 1 gynoecium enclosed by spathe in one of budding pouches. Anther bilocular.

No flowers observed on any Bhutanese material.

1. Root with buttressed sheath at point of attachment to frond (Fig. 15b);
    frond oblong, under 2mm wide .... **1. L. perpusilla**
+   Root not buttressed at base, arising apparently naked from frond;
    frond elliptic, over 2mm wide .... **2. L. minor**

1. L. perpusilla Torrey; *L. paucicostata* Hegelmaier; *L. minor* sensu F.B.I., p.p., non L. Dz: *liil.* Fig. 15a–b.

Mature frond oblong or narrowly oblong, apex rounded, 1.5–3.2(–4) × 0.9–2mm, daughter frond arising obliquely from base and remaining attached so fronds appearing unequally paired (or in 3s if further division occurs), flat, opaque, pale green, nerves not distinct, cells on undersurface very small (not visible with 10× lens). Root with acute root cap and buttressed translucent sheath at base.

Bhutan: C — Thimphu (Doteng, Paro), Punakha (10km E of Wangdu Phodrang) and Tashigang (above Tashigang, W bank of Dangme Chu, Cha Zam) districts (probably under-recorded). Rice paddies; roadside ditches and water-courses; muddy rocks by waterfall among mosses and liverworts, 1000–2400m.

2. L. minor L.

Differs from *L. perpusilla* in having relatively wider, more elliptic fronds (2.7–4 × 2.1–3mm) which are distinctly 3-nerved and with very large (visible with 10× lens) cells on undersurface; root lacking a buttressed basal sheath and root cap obtuse.

Bhutan: C — Tongsa district (Tongsa). Wet rock-face, 2350m.

Family 213. APONOGETONACEAE

S.J. Rae & H.J. Noltie

Perennial aquatic herbs, arising from tuberous rootstock. Leaves submerged and/or floating, commonly petiolate with lanceolate or elliptic blade.

FIG. 15. **Lemnaceae, Aponogetonaceae, Limnocharitaceae, Alismataceae, Hydrocharitaceae.**
a–b, **Lemna perpusilla**: a, habit (× 5); b, leaf undersurface showing buttressed root base and acute root cap (× 6). c, **Aponogeton undulatus**: habit (× ⅕). d, **Butomopsis latifolia**: habit (× ½). e, **Sagittaria sagittifolia** subsp. **leucopetala**: habit (× ⅓). f, **Hydrilla verticillata**: habit (× ½). g, **Otelia alismoides**: habit (× ¼). h, **Blyxa aubertii**: habit (× ½). Drawn by Glenn Rodrigues.

161

Usually monoecious. Infl. emergent, a terminal spike (sometimes forked) borne on a leafless scape, commonly 1–2 per plant, spike subtended by spathe-like bract, flowers sessile, ± spiral, sometimes secund. Perianth absent or of (1–)2(–6) often caducous tepals. Stamens 6, anthers dehiscing longitudinally outwards. Carpels 3, 4 (5), free, narrowed upwards into style, with stigmatic ridge on inside; ovules 2–12, commonly basal. Fruit of beaked, free follicles.

1. APONOGETON L.f.

Description as for family.

1. A. undulatus Roxb.; *A. microphyllum* Roxb. Fig. 15c.
Submerged leaves oblong, usually rounded, blunt, margins undulate, 10–25 × c.2cm, midrib wide with (2–)3(–4) parallel veins on either side; petioles 10–35cm. Floating leaves rare, to 20 × 3.5cm; petiole to 70cm. Scape single, slender, thickened above, 20–55cm. Infl. to 10cm, subtended by persistent spathe to 1.7cm, flowers facing in all directions. Tepals usually 2, caducous, spathulate, (2–)3.5–6(–12) × 1–2.5(–4)mm, white, pinkish (?or blue). Stamens 1.5–2.5(–4)mm, filaments widened towards base. Carpels 1.25–2 × 0.5–1mm, with 1–2 ovules. Follicles 5–7 × c.4mm, tapering into a short, curved beak. Sometimes reproducing vegetatively from plantlets borne on long, peduncle-like stalks.
Darjeeling (Darjeeling). [Ponds and ditches, 305m. Fl. July–November teste van Bruggen, 1985].

Only a single inadequate specimen seen; most information taken from van Bruggen (1985).

A. microphyllum Roxb. was described from 'damp places near the Bhotan mountains'. On the basis of the Roxburgh drawing, van Bruggen (1985) regards it as synonymous with *A. undulatus*, but clearly recent collections are needed. The drawing shows the tepals as white and larger than in *A. undulatus*, whereas the published description gives the tepals as blue.

Family 214. LIMNOCHARITACEAE

S.J. Rae

Annual or perennial, glabrous aquatic or marsh herbs with milky juice, stemless or with shortly creeping rhizomes. Leaves petiolate, bases expanded, blades with apical pore, main longitudinal veins arching, smaller cross-veins oblique. Flowers bracteate, in false umbels on leafless scapes, actinomorphic, bisexual; sepals 3, herbaceous, persistent; petals 3, thin, short-lived, white, pink or yellow. Stamens 6–9 or more, filaments free. Carpels 3–6 or more, free or coherent at base, styles short, stigmas slightly decurrent. Ovules numerous, parietal. Fruit a group of many-seeded follicles dehiscing by slits towards the centre. Seeds many, minute, U-shaped.

1. BUTOMOPSIS Kunth

Plant annual, stemless. Leaves all basal. Petals white. Stamens 8–12. Carpels 6–7.

1. B. latifolia (D. Don) Kunth; *B. lanceolata* (Roxb.) Kunth; *Tenagocharis latifolia* (D. Don) Buchenau. Fig. 15d.
Leaf blades narrowly elliptic-lanceolate, acute, base attenuate, 4–11 × 1–2(–3)cm, main veins 5–7; petioles 5–20cm. Scapes 15–30cm; umbels 3–15-flowered; pedicels 2–14cm, erect. Bracts forming an involucre, membranous with green midrib. Sepals broadly ovate, margins membranous, c.7mm (to 1cm in fruit). Petals ephemeral, obovate, 8–10mm. Stamens c.6mm. Carpels c.5.5mm. Follicles 10–12mm, tapering into short beak, papery. Seeds oblong with central channel, c.0.5mm, shining, brown.
Darjeeling (Katambari, Sukna). Marshes and paddy-fields, 300–900m. Fl. May–September.

Family 215. ALISMATACEAE

S.J. Rae & H.J. Noltie

Perennial, glabrous, stemless marsh or aquatic herbs, sometimes rhizomatous. Leaves basal; aerial leaves with lanceolate, elliptic or sagittate blades and petioles with expanded sheathing bases; submerged linear, bladeless phyllodic leaves sometimes present. Infls. terminal, on leafless scapes, paniculate (with flowers arranged in simple or compound whorls) or sometimes reduced to a single whorl or a single flower; whorls subtended by bracts forming an involucre. Flowers pedicellate, actinomorphic, bisexual or unisexual. Sepals 3, herbaceous, persistent. Petals 3, ephemeral, usually white. Stamens 3 or 6 to many, whorled. Carpels 3 or 6 to many, compressed, on subglobose receptacle. Ovule usually 1 per carpel. Fruit a head of achenes (occasionally follicles). Embryo strongly curved.

1. SAGITTARIA L.

Aerial leaves commonly sagittate. Infl. normally of simple, usually 3-flowered whorls each with 3 bracts. Flowers pedicellate, usually monoecious, upper male, lower female; bisexual flowers also sometimes present. Pedicels of female flowers sometimes thickened and recurved in fruit. Stamens numerous; filaments free, slender or flattened; anthers basifixed. Achenes usually winged, with persistent subapical or lateral stylar beak.

1. Leaf blades deeply sagittate .... **1. S. sagittifolia** subsp. **leucopetala**
+ Leaf blades linear-lanceolate .... **2. S. tengtsungensis**

1. S. sagittifolia L. subsp. **leucopetala** (Miquel) Hartog; *S. trifolia* sensu E.F.N. non L. Fig. 15e.

Leaves deeply sagittate, terminal lobe acute, 3.3–17 × 0.9–2.5cm, lateral lobes usually finely acuminate, to 14cm; petioles 11–45cm. Scapes 8.5–45cm, bearing 3–10 usually 3-flowered whorls; upper whorls with male flowers only, lower whorls with 2–3 female flowers (also occasionally producing secondary whorls), all flowers on divergent pedicels usually over 1cm long. Sepals boat-shaped, oblong, acute, to 4.5mm, herbaceous with membranous margins. Petals white, obovate to orbicular, to 8mm. Inner stamens c.2.5mm; filaments expanded towards base; anthers linear-oblong, equalling or slightly shorter than filaments. Achenes irregular, ± obovate, to 3.2 × 3.2mm, winged more or less equally all round, dorsal wing smooth or slightly waved; stylar beak usually curved, subapical, oblique.

Bhutan: C — Mongar district (Takila); **Terai** (Jaldaka, Tista Plains). Wet ground and ditches in paddy-fields, 610–1830m. Fl. August–February.

2. S. tengtsungensis H. Li.

Differs from *S. sagittifolia* subsp. *leucopetala* in its linear-lanceolate leaf blades, 7.2–10.5 × 0.4–1.2cm, gradually tapered to a blunt tip; pedicels of male flowers longer (to 4cm), more slender; female flowers not exceeding 2, restricted to lowest whorl; pedicels of female flower shorter (under 1cm); filaments lanceolate, flanged; anthers broadly ovate; achenes D-shaped.

Bhutan: C — Punakha (Punakha, Samtengang) and Bumthang (Byakar) districts. Marshes, 1830–2740m. Fl. May–October.

Family 216. HYDROCHARITACEAE

S.J. Rae & H.J. Noltie

Perennial or annual, usually submerged aquatic herbs. Leaves basal or spirally arranged or whorled along a stem. Flowers actinomorphic usually unisexual, rarely bisexual; plants sometimes dioecious. Flowers usually arising from two united bracts forming a spathe. Perianth segments free, with 3 sepals and usually also 3 petals. Stamens 2 to many, anthers basifixed. Ovary inferior, with 1 locule consisting of 2–15 fused carpels, apex often extended upwards into a beak. Ovules parietal. Fruits opening irregularly from top by decay.

1. Leaves all basal 2
+ Leaves in whorls along a stem .... **1. Hydrilla**

2. Leaves ± linear, tapering from base to acute apex .... **3. Blyxa**
+ Leaves with expanded lanceolate to suborbicular blades ...... **2. Ottelia**

1. HYDRILLA Richard

Dioecious (sometimes monoecious) aquatic herb with branched, leafy submerged stems, producing filiform roots from lower nodes. Leaves sessile in whorls of 3–8, variable in shape, linear to lanceolate (sometimes ovate), apiculate, margins serrulate, 5–22(–30) × (0.7–)1–3mm, midrib distinct, translucent, pale green with small brownish flecks, with 2 minute, fringed scales at base. Flowers unisexual borne singly in leaf axils. Male flowers very shortly pedicellate, enclosed in subglobose (c.1mm diameter) spathe with apical knob and corona of setae, opening to release flower which floats to surface; sepals 3, ovate, convex imbricate, petals 3, narrower. Stamens 3. Female flowers: spathe tubular, c.4mm long, apex bifid; sepals 3, narrowly oblanceolate; petals 3, narrower, transparent. Ovary cylindric, 1-celled, enclosed within spathe, prolonged upwards into filiform tube to 2.8(–10)cm long. Styles 3. Ovules few (to 6). Fruit cylindric, to 7mm.

1. H. verticillata (L.f.) Royle. Fig. 15f.
Description as for genus.
Bhutan: C — Punakha district (Punakha Dzong). Shallow backwater of river, 1300m.

2. OTTELIA Persoon

Monoecious or dioecious, glabrous aquatic herbs. Leaves basal, blade expanded, petiole with sheathing base. Flowers arising from tubular spathe borne on a peduncle. Spathe ribbed or with 2–10 wings. Female and bisexual flowers borne singly. Male flowers many per spathe. Sepals 3, persistent, borne on top of ovary beak. Petals 3, deciduous, conspicuous. Stamens 6–15. Ovary oblong, beaked, the single locule partially divided into 6 cells. Styles 6–15. Ovules numerous. Fruit oblong, 6-valved.

1. O. alismoides (L.) Persoon. Fig. 15g.

Leaves submerged or partly emergent; blade ovate to suborbicular, base usually cordate sometimes truncate or cuneate, to 15 × 18cm, with 7–11 arching longitudinal main veins and many oblique cross-veins, thin. Lanceolate phyllodic leaves sometimes also present. Petiole triangular in cross-section, to 27cm. Flowers arising from spathe, which encloses ovary. Spathe tubular, bifid, to 3.5cm, normally with 2 (sometimes more) wavy wings. Flowers sessile, bisexual. Sepals persistent, linear to narrowly oblong. Petals obovate, white. Stamens 6. Styles 6. Ovary oblong to narrowly ellipsoid, to 4cm, tapered above into short stout beak. Fruit oblong, with beak crowned by persistent calyx, opening irregularly at top.

Terai (Sukna, Jalpaiguri). Ditches, pools and streams, c.100m.

3. BLYXA Noronha ex Thouars

Submerged aquatics. Leaves basal, or inserted spirally along a stem, linear, venation parallel. Plants sometimes dioecious. Infl. basal or axillary, peduncle bearing a spathe enclosing a usually single (to 10 flowers in male infl. of dioecious species) flower. Flowers unisexual or bisexual, actinomorphic, sepals 3, persistent, petals 3. Stamens 3, 6 or 9; filaments free, filiform; anthers basifixed, dehiscing latrorsely. Ovary long-beaked above; ovules many; stigmatic lobes 3, fused below into long style. Fruit a linear capsule. Seeds variously ribbed and tubercled, sometimes with spiny appendages at end.

1. B. aubertii Richard. Fig. 15h.

Leaves all in basal rosette, widest at base tapering to very acute apex, 5–9(–18) × 0.35–0.55cm, translucent, midrib conspicuous, finer longitudinal veins connected by numerous transverse veinlets. Peduncle under 1cm; spathe bifid, 2.6–3cm. Flowers bisexual; sepals linear-lanceolate, hooded, 2.7–5.3 × 0.6–0.9mm, green; petals linear, c.3 × 0.4mm, white. Stamens 3; filaments c.2.3mm, inserted at base of sepals; anthers c.1.5 × 0.2mm. Ovary narrowly cylindric, 12–15 × 1.5–1.7mm, narrowed into filiform beak 1.5–4cm; styles c.7.5mm, relatively stout. Seeds narrowly oblong-ovoid, subacute, base truncate, c.1.5 × 0.9mm, c.6-ribbed, ribs obscurely tuberculate, pale green.

Bhutan: C — Thimphu district (near Simtokha, Atsho Chhubar). Rice paddies, 2300–2430m. Fl. July–August.

Family 217. JUNCAGINACEAE

S.J. Rae & D.G. Long

Usually perennial marsh or aquatic herbs with short, vertical, cormose rhizomes. Leaves mostly basal, linear, erect, entire, with sheathing bases. Flowers small, ebracteate, in spike-like racemes borne on simple leafless scapes, bisexual or unisexual, actinomorphic. Perianth of 3, 4, 6 or 8 free sepal-like segments in 2 whorls. Stamens attached to base of perianth segments, anthers ± sessile. Carpels 3, 4 or 6, free or partly united, each 1-celled, with one ovule per cell. Style usually very short, stigmas irregularly lobed. Fruit of free or connate achenes or follicles separating at maturity.

1. TRIGLOCHIN L.

Marsh herbs with short stout stems and linear basal leaves. Flowers in racemes on leafless scapes, bisexual with 6 perianth segments. Stamens 6. Carpels usually 3 fertile and 3 sterile, or all 6 fertile, connate at base; stigmas feathery. Achenes separating at maturity.

1. Leaf blades flat, 1–3mm wide, obtuse; fruit oblong-ovoid, 4–5mm, with 6 fertile achenes, rounded at base .... **1. T. maritima**
+ Leaf blades subulate, 0.4–1mm wide, acute; fruit narrowly clavate, 7–9mm, with 3 fertile achenes, tapering at base .... **2. T. palustris**

1. T. maritima L. Eng: *sea arrow-grass*. Fig. 16a–b.
Base of stems thickly clothed with old leaf sheaths. Leaves flat, obtuse, (2.5–)4–17(–30) × 0.1–0.3(–0.45)cm; sheaths whitish, papery. Scape stout, often curved, (3–)5–30cm; racemes 2–12cm. Flowers subsessile or on short pedicels c.1mm (to 4mm in fruit). Perianth segments green, often tinged red or pink, rounded, c.1mm. Fruits not appressed to scape, oblong-ovoid, 4–5mm; achenes 6, all fertile, separating when ripe.
Bhutan: N — Upper Mo Chu district (Shingche La, Tharizam Chu, Laum Thang, Phile La and Lingshi areas, locally common); **Sikkim** (Lhonak, Lachen, Tema La, Tangu, Lungnak, Naku La); **Chumbi**. Mountain marshes, 3660–5790m. Fl. May–July.

2. T. palustris L. Eng: *marsh arrow-grass*. Fig. 16c.
Similar to *T. maritima* but more slender throughout; leaves linear-subulate, acute, 0.4–1.0(–2.5)mm wide; scapes slender, elongating in fruit to 60cm; pedicels 1–4mm, scarcely elongating in fruit; fruit closely appressed to scape,

narrowly clavate, gradually tapered to base, 7–9 × 1mm, with 3 fertile and 3 sterile achenes.

Bhutan: N — Upper Bumthang Chu (below Lambrang) and Upper Kulong Chu (Shingbe) districts; **Sikkim** (above Lambi, Tang La, Lachen). Mountain marshes and streamsides, 3660–4570m. Fl. June–September.

Family 218. POTAMOGETONACEAE

Perennial aquatic herbs, usually with creeping rhizomes in substrate, and erect, flexible, leafy shoots in water; occasionally semi-terrestrial. Leaves alternate, the upper sometimes opposite, subtended at base by membranous stipules, bases sometimes sheathing, submerged leaves usually narrow, floating leaves with expanded blades sometimes also present. Flowers whorled, in cylindric spikes on axillary peduncles; 4-merous, bisexual, ebracteate, with 4 greenish claw-shaped tepals; stamens 4; carpels 1–4 developing into drupelets containing a single curved embryo.

1. POTAMOGETON L.

Description as for family.

1. Floating leaves with expanded blades present at time of flowering 2
+ Floating leaves not present, all leaves submerged, sessile, linear or linear-oblong 4

2. Floating leaves elliptic, over 3cm; linear, submerged leaves absent 3
+ Floating leaves narrowly oblong-elliptic, less than 2.5cm; linear, submerged leaves also present .... **3. P. octandrus**

3. Carpels 1–2 per flower; petioles slender .... **1. P. distinctus**
+ Carpels 4 per flower; petioles stout .... **2. P. nodosus**

4. Plant robust; leaves oblong, margins wavy, minutely toothed, more than 5mm wide .... **4. P. crispus**
+ Plant slender; leaves linear, usually less than 2mm wide 5

Fig. 16. **Juncaginaceae, Potamogetonaceae.**
a–b, **Triglochin maritima**: a, habit (× ⅔); b, fruit (× 7). c, **T. palustris**: fruit (× 10). d–e, **Potamogeton distinctus**: d, habit (× ½); e, drupelet (× 8). f, **P. octandrus**: habit (× 1). g–i, **P. berchtoldii**: g, habit (× 1); h, leaf base and stipule (× 6); i, section through base of stipule (open to base). j–k, **P. filiformis**: j, leaf base and stipule (× 6); k, section through base of stipule (tubular at base). Drawn by Glenn Rodrigues.

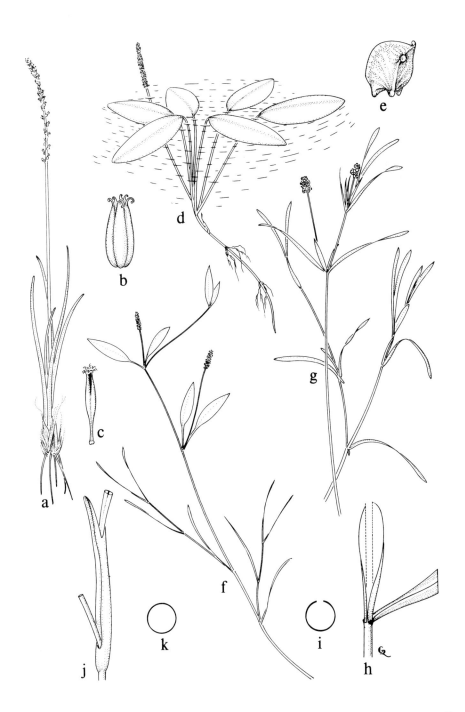

a

b

c

d

e

f

g

h

i

j

k

5. Stipules free from leaf base, forming an open tube around stem (Fig. 16h); leaf blade flat, not bitubular in section, apex mucronate
.... **5. P. berchtoldii**

+ Stipules fused to leaf base forming a closed tube around stem (Fig. 16j), prolonged upwards into free membranous ligules; leaf consisting of two tubes (seen in section), apex blunt .... **6. P. filiformis**

1. P. distinctus A. Bennett. Dz: *shochum, shoum*; Nep: *pani jhar*. Fig. 16d–e.

Overwintering as condensed sympodial rhizome system with slender, acute, lateral buds. Stem to 30cm, simple. Leaves all floating, upper usually opposite, variable in shape, elliptic or narrowly elliptic, obtuse to acute, base truncate (to cuneate), 3–5.7(–8.5) × (1–)1.5–3.1cm, shorter or longer than petiole, with 11–19 longitudinal veins and many fine cross-veins visible against light, coriaceous. Stipules mostly decayed at flowering, acute, to 3cm. Peduncle tapering very slightly upwards, 3.8–7cm. Spike dense. Carpels 1–2 per flower. Drupelets 3–4mm long, usually with 3 wavy keels, laterals prolonged into tubercles top and bottom, median with tubercle only at base, ventral face curved, dorsal face semicircular, beak less than 0.5mm.

Bhutan: C — Thimphu (Thimphu, Doteng, Paro Bridge, Drugye Dzong, Tashi Cho Dzong, Tsalimaphe), Punakha (Rinchu, Punakha Dzong, Wangdi Phodrang, Khuru, Khamayna), Bumthang (Gyetsa), Mongar (Lhuntse Dzong) and Tashigang (Tashi Yangtsi Dzong) districts. Rice paddies; ponds; forest pools; marshes; backwaters of rivers, 1200–3050m. Fr. July–August.

Plant occasionally semi-terrestrial, with more translucent leaves.

Parker (1992) records this as a dominant weed of rice in Central Bhutan and the most difficult of the rice weeds to control.

2. P. nodosus Poiret; *P. indicus* Roxb.; *P. oblongus* sensu F.B.I., p.p. non Viviani.

Differs from *P. distinctus* in having larger, more coriaceous leaves with stouter petioles; carpels 4 per flower.

Sikkim (Lachung). 2743m. Fr. August.

A recent paper by Wiegleb (1990) demonstrates the unsatisfactory state of knowledge of the *P. nodosus* group and he is probably correct in stating that *P. distinctus* may eventually be treated as a subspecies of *P. nodosus*.

3. P. octandrus Poiret; *P. javanicus* Hasskarl. Fig. 16f.

Stem to 60cm, filiform, branched. Floating leaves petiolate, blade narrowly oblong to elliptic, 10–25 × 5–8mm with 5–7 longitudinal veins, coriaceous; submerged leaves linear, gradually narrowed to acute apex, under 1mm wide,

3-veined, with air spaces (pale against light) both sides of midrib. Stipules of young leaves forming a tight tube around stem, but split to base, to 11mm, membranous. Peduncle to 2.5cm; spike slender. Drupelets c.1.7mm, dorsal face curved with a distinct shoulder, ventral face angled, 3-keeled, all keels tubercled at base, lateral keels with additional tubercles, beak conspicuous, c.0.4mm.
Terai (Jalpaiguri, Siliguri). Streams.

Immature specimens of a second heterophyllous species have been seen from Upper Bumthang Chu district (Kantanang, 3810m, BM, E) and Sikkim (Tangu, 4267m, LLOYD). They are probably *P. gramineus* L. and differ in having wider (over 2mm) submerged leaves and larger (to 4.5 × 1.1cm) floating leaves. Hooker's high-altitude records of *P. javanicus* (up to 2740m in Sikkim; F.B.I.) probably also refer to this.

4. P. crispus L.
Stem to 130cm, flattened, often branched. Leaves all submerged, linear-oblong, usually blunt, margins wavy, with many minute teeth, sessile, 3.5–7.5 × 0.5–0.7cm, usually 3-veined, with air channels adjacent to midrib. Stipules soon falling, less than 5mm. Peduncle to 5cm; spike few-flowered. Drupelets c.3mm, with a wavy keel, dorsal face semicircular, ventral curved, beak c.2mm, curved.
Bhutan: C — Punakha district (Punakha). In slow-flowing water, 1800m. Fl. April(–June).

5. P. berchtoldii Fieber. Fig. 16g–i.
Plant lacking rhizome; stems to 30cm, slender, simple. All leaves submerged, linear, mucronate, 1.5–2.0(–2.7)mm wide, veins 3(–5), with air spaces (pale against light) bordering midrib. Pair of small glands usually present at junction of leaf with stem. Stipules of young (upper) leaves forming tube around stem which is split to base, to 8.5mm, brownish, membranous. Peduncles short (less than 10mm); spikes short, few-flowered. Drupelets c.2mm, beak very short.
Bhutan: C — Ha (Ha Dzong) and Bumthang (Byakar) districts. In ponds and rivers, 2700–2740m. Fl. April–June.

6. P. filiformis Persoon. Fig. 16j–k.
Stem 20–40(–90)cm, slender, branched, especially at base. Leaves all submerged, linear, sessile, usually less than 1mm wide (up to 1.5mm), 3-veined but laterals not usually visible, with large air canals on both sides of midrib (leaf in section appearing to consist of 2 tubes). Stipules fused with leaf base, forming a sheath round the stem, tubular at base, to 2.5cm, prolonged above into free, acute, membranous ligules to 2cm. Peduncle to 10cm; flower whorls

distant. Drupelets c.2.5mm, dorsal face strongly curved, ventral slightly curved, beak c.0.2mm.

Bhutan: C — Tongsa district (Maruthang); **Sikkim** (Chamgong, Lhonak). Streams and springs, 3600–4600m. Fr. September.

Family 219. BROMELIACEAE

S.J. Rae

Short-stemmed herbs, roots poorly developed. Leaves in a dense basal rosette, thick rigid, margins often spiny, parallel-veined, bases often acting as reservoirs for water and humus. Infl. a terminal spike, flowers usually bisexual borne in axils of often brightly coloured bracts. Sepals 3, free. Petals 3, free. Stamens 6, attached to base of sepals and petals. Ovary inferior, 3-loculed; ovules numerous, axile; style simple bearing 3 stigmas. Fruit an aggregate of berries.

1. ANANAS Miller

Description as for family.

1. A. comosus (L.) Merrill. Eng: *pineapple*. Fig. 17a.
Leaves strap-shaped, sharply spinose-serrate, 24–60 × 1–4cm, leathery. Scape short, stout. Spike ovoid, c.6 × 4cm, many-flowered. Bracts reddish, triangular, leathery, serrate. Sepals broadly ovate, c.12 × 10mm, coriaceous, orange-tipped. Petals oblong-lanceolate, c.22 × 5mm, purple. Style 15mm. Ripe fruit (syncarp) orange, ovoid-ellipsoid, 15–20 × 9–12cm, with persistent bracts and apical tuft of leaves, rarely setting seed, sepals and base of bracts fusing to form rind.

Bhutan: S — Sarbhang district (Phipsoo); **Darjeeling** (below Tukvar). Cultivated, 300–1200m.

Native to tropical S America, cultivated for its edible fruits; leaf fibres sometimes used for weaving.

FIG. 17. **Bromeliaceae, Pontederiaceae, Sparganiaceae, Typhaceae.**
a, **Ananas comosus**: habit (× ⅕). b, **Monochoria vaginalis**: habit (× ⅓). c, **Eichhornia crassipes**: habit (× ⅕). d, **Sparganium fallax**: habit (× ⅕). e, **Typha elephantina**: habit (× ¹⁄₂₀). Drawn by Glenn Rodrigues.

173

Family 220. PONTEDERIACEAE

Annual or perennial, aquatic or marsh herbs. Stems short or thickly rhizomatous. Leaves commonly in basal rosette; petiole margins developed upwards into stipules. Infl. terminal on scape, commonly spike-like or condensed and sub-umbellate, subtended by 2 ± tubular spathes, lowermost often with leaf-like blade. Flowers short-lived, zygomorphic to actinomorphic, tepals 6, ± free or sometimes fused into tube below; bisexual, sometimes tristylous. Stamens usually 6, sometimes 1 or 3, anthers all similar or 1 large and 5 small. Ovary superior, (1–)3-loculed; ovules usually axile; style elongate; stigma capitate to trifid. Fruit a 3-loculed capsule, surrounded by remains of flower.

1. Flowers actinomorphic, blue; tepals free .... **1. Monochoria**
+ Flowers zygomorphic, pale mauve with yellow spot on upper tepal; tepals tubular below .... **2. Eichhornia**

1. MONOCHORIA C. Presl

Annual or perennial, commonly emergent aquatics. Stemless or rhizomatous. Leaves basal, usually with petiole and blade. Infl. racemose or sub-umbellate, rarely paniculate, subtended by sheathing spathes; lower spathe usually with leaf-like blade so infl. often appearing lateral. Flowers actinomorphic, tepals 6, free. Stamens 6; anthers all alike or one larger than others; filaments equal, one usually with tooth near apex.

1. Leaf blades truncate to shallowly cordate with rounded basal lobes; usually annual, usually under 20cm .... **1. M. vaginalis**
+ Leaf blade deeply cordate with sharply-angled basal lobes; rhizomatous perennial, commonly over 50cm .... **2. M. hastata**

1. M. vaginalis (Burman f.) Kunth. Dz: *damperu, olasam*; Nep: *piralay*; Man: *puktiwa*. Fig. 17b.
Usually annual. Leaves in basal rosette or on shortly decumbent stem, blades lanceolate (occasionally very narrowly) to ovate, acuminate, usually shallowly to deeply cordate, sometimes truncate, 1.5–5(–8) × 1–3.8(–7)cm. Petioles usually exceeding scape, margins of petioles and stipules membranous, sometimes purplish. Stipules 0.5–2cm. Scape 3–17(–41)cm. Infl. sub-umbellate to racemose, erect at first, strongly deflexed in fruit; upper spathe membranous, shorter than infl., lower spathe leaf-like, conspicuously exceeding infl. Flowers pedicellate; tepals 6, blue, persistent, 7–12mm. Anthers 5 small (1–1.2mm) on

simple filaments, 1 large (2.3–2.5mm) on filament with sharp, ascending tooth. Ovary ellipsoid, 2–2.7 × 1.5–1.7mm; style 3–3.5mm; stigma shortly 3-lobed. Capsule ellipsoid, 6–10 × 4–6mm. Seeds oblong, c.0.8 × 0.4mm, longitudinally ribbed, pale brown, tip darker.

Bhutan: C — Punakha (Wangdi Phodrang; Punakha (M.F.B.)) and Tashigang (Tashi Yangtsi Dzong) districts; **N** — Upper Mo Chu district (Gasa); **Terai** (Siliguri, Jalpaiguri); **Darjeeling** (Darjeeling). Rice fields, ditches, marshes and shallow ponds, 90–2400m. Fl./fr. August–October.

No doubt under-recorded; recorded by Parker (1992) as one of the most serious weeds of flooded rice, probably occurring in all districts [with cultivation].

2. M. hastata (L.) Solms.
Differs in being a usually taller (to 1m) rhizomatous perennial; leaf blades hastate.

Terai (between Phuntsholing and Siliguri, Salgara Terai). Flooded rice fields. Fl. July–August.

2. EICHHORNIA Kunth

Usually floating herbs. Stemless or with short, thick rhizome. Leaves basal, with spongy, often inflated petioles. Infl. a terminal spike, subtended by tubular spathes, lower with leaf-like blade. Flowers zygomorphic; tepals 6, tubular below. Tristylous, with stamens of two lengths; filaments curved upwards. Style elongate, of three lengths (differing from both stamen lengths so short, medium or long). Fruit a capsule.

1. E. crassipes (Martius) Solms. Eng: *water hyacinth*. Fig. 17c.
Floating perennial; roots feathery. Leaves in rosette; blades rhombic to suborbicular to widely, transversely elliptic, subacute to rounded, very shallowly cordate to cuneate, 3.5–10 × 4.5–11.5cm, firm-textured, shining. Petioles spongy, lower part becoming inflated. Stipules tightly appressed, to 7cm, membranous. Scape 4.5–24cm, usually exceeding leaves. Spike 6–20-flowered, peduncle largely hidden by 2 sheathing, membranous spathes, lower spathe bearing a small, leaf-like blade. Tepals pale mauve, uppermost with yellow spot near base surrounded by darker mauve ring, elliptic, to 3.5cm, dotted with sessile, brown glands; tube c.1cm, glandular-hairy outside. Stamens inserted near top of tube, filaments glandular-hairy, 3 short, 3 long, all upturned near apex; anthers all similar, c.3mm. Ovary narrowly ellipsoid, tapered upwards into style. Style intermediate, longer or shorter than two lengths of stamen, glandular-hairy, curved upwards near apex; stigma capitate.

Bhutan: S — Chukka (Parker, 1992) and Gaylegphug (Gaylegphug) districts; **Terai** (common); **Sikkim** (Phodong to Kabi). Marshy hollow by roadside; ditches; tanks, 300–1800m. Fl. May–July.

Native to S America, but widely introduced and now a troublesome, pantropical weed.

Family 221. SPARGANIACEAE

Emergent or floating aquatic perennials; sometimes rhizomatous or stoloniferous. Stems sometimes cormose below. Leaves sub-basal, with sheathing bases, linear, spongy below. Infl. monoecious, a simple or branched raceme of dense, globose heads, subtended by leaf-like bracts, lower female, upper male; female heads pedunculate or sessile, sometimes with peduncles fused to axis so appearing to arise above the bract (supra-axillary). Tepals 3–4(–6) chaffy, rather irregular, usually fused below. Ovary fusiform, usually of 1 locule; style persistent; stigmas 1(–3); ovule solitary. Male flowers in dense heads; stamens 1–8. Fruits drupe-like, crowded in dense head, stipitate below, mesocarp spongy, attenuate upwards into stylar beak.

1. SPARGANIUM L.

Description as for family.

1. S. fallax Graebner; *S. simplex* sensu F.B.I. non Hudson. Fig. 17d.
Stolons slender, spreading. Stem (to top of infl.) (15–)42–61cm. Leaves 1–3, exceeding stem, (0.3–)0.6–1.1cm wide; 1 intermediate bract-like leaf usually present. Infl. unbranched, lowest bract exceeding infl.; female part slightly zigzag, female heads 3–4, not overlapping, lowest pedunculate, upper sessile, some supra-axillary; male heads 6–13, at least upper convergent, distant from female, small (0.5–1cm excl. stamens); filaments long; anthers c.1.5mm, apiculate. Female tepals 3–5, rather irregular, at least some usually fused below, spathulate to obrhombic, rounded to subacute, margins minutely fimbriate, 2.8–4.2mm (at flowering), dark brown with pale margins. Style 1.3–2mm; stigma 1–2mm. Fruit (incl. style and stipe) 7.5–11 × 2–3mm, narrowly ellipsoid, tapered above into stylar beak and below into long, stipitate base.
Sikkim [presumably Chateng below Lachen; lakes, 2670m].

Above description based on Chinese and Indian material.

Originally described (in part) from Sikkim (Graebner, 1900) which is presumably the plant called *S. simplex* in F.B.I. The specimen on which this is based should be at Kew, but is presumed lost; the plant is almost certain to be that from Chateng referred to as '*S. ramosum*' (which in F.B.I. is only given for NW Himalaya) in Hooker (1854, 2: 96). There is a puzzling immature *Sparganium* specimen at CAL collected by Hooker and labelled 'Sikkim'. It has extremely narrow leaves (c.4.2mm wide) and has been determined as *S. simplex*, but it belongs neither to that species nor to the above.

Family 222. TYPHACEAE

Perennial, rhizomatous, marsh or emergent aquatic herbs. Leaves borne near base of flower stem, distichous, linear, flattened or keeled, sheathing below, spongy near base. Infl. terminal, densely cylindric, consisting of a usually brown, basal female section and a stipitate (in India) yellowish, upper male section. Male flowers subtended by usually several lobed or forked hairs; anthers 1–5(–8) borne on a united filament. Female infl. often with sterile, narrowly clavate scales, infertile flowers and fertile flowers consisting of a single stipitate ovary, stipe bearing whorls of hairs, ovary unilocular, bearing a single ovule.

The unmistakable genus *Typha* (Eng: *reedmace*), with its dense, cylindric infl., occurs in the Terai south of Bhutan and almost certainly in Bhutan itself, but modern collections are needed to check the identity of the species. Griffith (1847, p. 218) recorded *T. elephantina* Roxb. from Deothang/Tashigang district (between Sasi and Balfi), but apparently made no specimen. A specimen seen in the herbarium of the Lloyd Botanic Garden from Darjeeling district (Sivoke) was also identified as this species, but it was not possible to study it critically. *T. elephantina* (Fig. 17e) is a robust plant to 4m; male spike 20–30cm, distant from female, pollen 4-lobed; female spike 15–25 × 1–2.5cm, dark brown, female flowers bracteolate, stigma lanceolate (rather than linear), exceeding the hairs.

Family 223. MUSACEAE

Massive herbs, monocarpic or perennial and suckering producing long or short rhizomes from a large 'corm'. Leaves spirally arranged, sheaths forming a pseudostem. Petioles channelled. Leaf blade ± oblong, becoming tattered, pinnately veined. Infl. a terminal thyrse on a stout scape. Flowers unisexual, borne in commonly 2-rowed, lateral cincinni, subtended by a large bract; infl. developing acropetally, lower bracts bearing female flowers, apical ones male. Flowers with 3 outer and 2 inner tepals fused into compound tepal (toothed or divided), 1 tepal of inner whorl free (entire or lobed). Male flowers with 5(–6) stamens; anthers elongate; pistillode present. Female flowers with staminodes, ovary inferior, trilocular, ovules axile, 2 or more rows per locule. Fruit an elongate berry or fleshy capsule, with tough skin and seeds embedded in pulp.

1. MUSA L.

Plants not monocarpic, suckering from perennial corms. Stems not conspicuously swollen at base. Bracts usually deciduous. Compound tepal with 5 short teeth, outer two with filiform, dorsal appendages; free tepal conduplicate, truncate, mucronate.

No satisfactory account can presently be given of the bananas of Bhutan and Sikkim. Field study of flowering material is essential and recent collections and photographs are badly needed. The following account relies heavily on Simmonds (1957), Cheesman (1948) and a manuscript account of the wild bananas of Sikkim by King (copy at Kew) which was partly published by Baker (1893).

The Nepali name *kera* applies in Darjeeling to all 'plantains' according to Cowan & Cowan (1929); it is also used in Bhutan. According to Simmonds (1957) the Nepali name *bon kera* and Lepcha *kardung* apply to wild bananas generally. In Bhutan the following names have been recorded for *Musa* spp. (perhaps *M. balbisiana* though no vouchers seen): *nangla, gualha* (Dz) and *laishing* (Sha).

1. Apex of male bracts recurving 2
+ Apex of male bracts not recurving 4

2. Petiole margins spreading horizontally .... **3. M. griersonii**
+ Petiole margins erect or incurved 3

3. Male bracts purple streaked with yellow outside; young fruit forwardly
   directed .... **4. M. thomsonii**
+ Male bracts uniformly red-brown outside; fruit spreading even when
   young .... **5. M. flaviflora**

4. Leaves glaucous beneath; seeds irregularly globose, under 6mm diameter, minutely warty; occurring below 900m .... **1. M. balbisiana**

FIG. 18. **Musaceae**.
a–b, **Musa balbisiana**: a, fruit (× ½); b, seed (× 1.5). c–d, **M. sikkimensis**: c, male flower bud (× ¹⁄₁₀); d, seed (× 1.5). e–g, **M. griersonii**: e, habit (× ¹⁄₁₀); f, female flower (× ½); g, male flower (× ½). h, **M. flaviflora**: inflorescence (× ¹⁄₁₀). Drawn by Glenn Rodrigues.

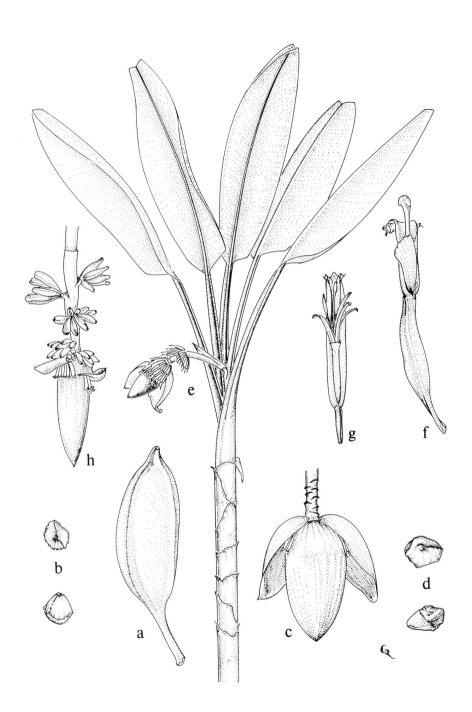

179

+   Leaves not glaucous beneath; seeds sharply angled, over 8.5mm diameter, smooth; occurring above 1000m .... **2. M. sikkimensis**

1. M. balbisiana Colla; *M. paradisiaca* subsp. *seminifera* var. *pruinosa* King ex Schumann; *M. sapientum* var. *pruinosa* (King ex Schumann) Cowan & Cowan. Lep: *reling* (*ralim*), *kargok* (*garkok*). Fig. 18a–b.
Pseudostem 3–7.5m, robust, (to 30cm diameter at base), green. Leaf sheaths often glaucous, upper parts often with black markings. Petiole 60–68cm, brown-blotched, narrowly channelled, edges almost meeting over channel, margins not becoming scarious, sometimes forming a black line in lower part against stem. Leaves spreading, blades oblong, base rounded or slightly cordate, 2.4–3 × 0.6–0.68m, shining above, very glaucous beneath, midribs green. Bracts sometimes persistent, broadly ovate, deep purple, glaucous outside, crimson inside. Male bud broadly ovoid to ellipsoid. Male flowers c.20 per bract, in two rows; compound tepal 4–5 × c.1.2cm, whitish variously flushed purplish below on inside, upper part yellowish, teeth c.5mm; free tepal c.half length of compound; stamens at first equalling, later exceeding perianth. Female flowers borne in up to 10 hands. Fruit bunch pendent; fruits crowded, standing out from axis, no callosity at junction with axis, 10–12.5 × 3.5–4cm, angled at maturity, abruptly narrowed into pedicel (1–2cm), at apex more gradually narrowed into short, broad acumen; pericarp pale yellow when mature, soon blackening. Seeds numerous, irregularly globose, minutely warty, 5–6 × 4–5mm, hard, black, surrounded by little, tough, whitish pulp.
    Darjeeling (Great Rangit, Mongpu; Rangpo, Coronation Bridge below Kalimpong (Simmonds, 1957)); **Sikkim** (by Tista River, Dikchu (Simmonds, 1957)). Lower Hill Forest, 305–914m. Fl./fr. December–May.

Description from Cheesman (1948) and King (Ms).

2. M. sikkimensis Kurz; *M. hookeri* (King ex Schumann) Cowan & Cowan; *M. paradisiaca* subsp. *seminifera* var. *hookeri* King ex Schumann. Lep: *tiangmoo-foo-goom.* Fig. 18c–d.
    Pseudostem to 4m, robust, (35–40cm diameter at base), tinged reddish. Leaf sheaths marked blackish-brown, waxy only when young. Petiole c.65cm, channelled, margins erect, narrowly blackish-scarious, forming a black line in lower part against pseudostem. Leaves spreading, blades oblong-lanceolate, base rounded or slightly cordate, 1.8–2.1 × 0.6m, yellowish-green, shining on both surfaces, usually purplish when young, sometimes beneath when older, midrib red or purplish beneath. Bracts broadly ovate, obtuse, not reflexing, deep purple to crimson, glaucous outside, inside concolorous, 1–2 male bracts opening at once. Male bud c.12 × 8cm, turbinate. Male flowers c.14 per bract;

compound tepal c.3.5cm, creamy-orange; free tepal c.1.6cm, translucent; stamens equalling compound tepal. Fruit bunch oblique, c.4 hands each of 7–9 fruits borne in two rows; fruits rather lax, arising from large, brown callosities on axis, 11–15 × 4cm, angled at maturity, abruptly narrowed into massive pedicel (c.2 × 1–2cm), apex bluntly rounded apiculate; pericarp green turning brown. Seeds numerous, sharply angled, smooth, 6–10.5 × 5–6mm, black, pulp scanty, dirty white to pale brownish-pink.

Bhutan: S — Phuntsholing district (below Ganglakha); **C** — Tongsa district (above Dakpai); **Darjeeling** (Sureil; waterfall below Darjeeling, Darjeeling to Peshok (Simmonds, 1957)); **Sikkim** (Gangtok (Simmonds, 1957)). Middle Hill Forest, 1320–1920m. Fl./Fr. October–April.

Description from Simmonds (1957) and King (Ms).

3. M. griersonii Noltie (?= *M. paradisiaca* subsp. *seminifera* var. *dubia* King ex Schumann; ?*M. sapientum* var. *dubia* (King ex Schumann) Cowan & Cowan. Lep: *luxom*). Fig. 18e–g.

Pseudostem c.3m, marked with purplish-brown. Petiole to 50cm, lower part glaucous, dark purple on abaxial side, margins developed into horizontally spreading, green wings. Leaves scarcely spreading, blades oblong, base rounded, 2–2.5m, glaucous beneath at least when young; midrib blackish-purple beneath. Male bracts deciduous, recurving at tip, dark reddish-purple outside, inside pinkish with cream margins. Male bud elongate, acute. Male flowers: compound tepal c.4.5 × 1cm, orange-yellow, teeth to 4mm; free tepal c.2cm; stamens shorter than compound tepal. Female flowers borne in 7 or more hands, each of 22 or more flowers borne in 2 rows on small callosities; compound tepal pinkish-cream, c.3.5cm, teeth c.4mm, free tepal c.2.5cm, staminodes 5, shorter than free tepal.

Bhutan: S — Sarbhang district (above Jhogi Dhanra, 11km above Sarbhang on Chirang road). Subtropical forest slopes, 740m. Fl./fr. March.

A very distinctive species known only from a single gathering; further specimens highly desirable.

4. M. thomsonii (King ex Schumann) Cowan and Cowan; *M. paradisiaca* subsp. *seminifera* var. *thomsonii* King ex Schumann. Lep: *kergel*.

Stem 3.6–4.5m, 17.5–23.5cm diameter at base, slender, green, speckled brown at base. Petiole c.90cm, compressed, narrowly channelled. Leaves scarcely spreading, blades c.2.4 × 0.6m, narrowly lanceolate, base asymmetrically cordate, thin-textured, glaucous when fresh, afterwards shining. Male bracts recurved at apex, deciduous, ovate, bright purplish-brown streaked paler outside, yellow inside. Male bud narrow, acute. Male flower unknown. Female bracts elongate ovate-lanceolate, acuminate. Female flowers borne in 3+? hands; each with c.18 flowers borne in 2 rows on callosities; free tepal ovate, acute, cordate. Fruits forwardly directed when immature, finally horizontal, small (6 × 1.5cm), pedicel slender, c.3.5cm. Seeds few, black, irregular in shape, 3.25mm diameter, surrounded by much soft, sweet pulp.

Darjeeling (Ryang Valley). Sheltered spots in Lower Hill Forest, up to 457m. Fl. November.

Description from King (Ms) and drawing by H. Baker at Kew.

5. M. flaviflora Simmonds. Fig. 18h.

Differs from *M. thomsonii* in having male bracts reddish-brown outside, not streaked; young fruits spreading, shortly pedicelled.

Male flowers: compound tepal orange-yellow, 3.6–4.6 × 1cm, teeth to 9mm; free tepal oblong, apiculate, conduplicate, 1.4–2cm; stamens exceeding compound tepal. Female flowers borne in 5–6 hands, each with 12–14 flowers borne in 2 rows on callosities.

Bhutan: S — Gaylegphug district (E bank of Thewar Khola near Gaylegphug). Damp gravelly slope on river bank, 300m. Fl. May.

According to R.C. Srivastava (pers. comm.), a member of Section *Rhodochlamys* (dwarf plants with erect infls.) occurs wild in W Sikkim and is taken into cultivation as a pot plant in Jorethang. Several species of this section are highly likely to occur in Bhutan.

Nothing is known of the cultivated bananas of Bhutan and Sikkim. No doubt various cultivars derived from hybrids between *M. balbisiana* and *M. acuminata* Colla are grown. Fruits of cultivars with more of the former in their parentage tend to be starchy and used for cooking, whereas those with more of the latter tend to be sweet, eating bananas. Male flower buds of *M. balbisiana* and its hybrids are eaten.

Ensete glaucum (Roxb.) Cheesman has so far not been reported from our area, though it occurs in E Nepal and is therefore likely to occur. It differs from any of the above species of *Musa* in being monocarpic with a swollen stem base; bracts green, persisting until fruit ripe; compound tepal divided almost to base into 3 lobes, free tepal 3-lobed at apex with central arista and 2 lateral teeth; seeds much larger.

Family 224. ZINGIBERACEAE

R.M. Smith

Rhizomatous herbs with aromatic oil cells. Leafy shoots few- to many-bladed, leaves distichously arranged, sheaths usually open. Infl. terminal on leafy shoot or borne directly on rhizome at base of leafy shoot or remote from it. Flowers solitary in axils of bracts or in cincinni, with or without secondary bracts (bracteoles). Calyx tubular, often unilaterally split. Corolla tubular with 3 petals, dorsal usually larger than laterals; lip (labellum = anterior staminode) adnate at base to corolla tube, generally the most conspicuous feature of the flower. Lateral staminodes petal-like or reduced to small subulate points. Fertile stamen single, anther ± sessile or with distinct filament; thecae parallel or divergent, connective sometimes prolonged into a crest. Style linear, held between thecae; stigma usually expanded. Ovary inferior, unilocular with parietal placentation or trilocular (sometimes incompletely so) with axile placentation. Epigynous glands (stylodes) forming outgrowths on top of ovary. Fruit a dry capsule or fleshy berry.

1. Style exserted well beyond anther-thecae and enfolded in a long anther-crest, giving a beaked appearance (Fig. 19e) .... **1. Zingiber**
+ Style not exserted much beyond anther-thecae; anther-crest, if present, not enfolding style 2

2. Bracts of infl. adnate to each other by their lower margins, forming pouches; anther versatile, spurred basally (Fig. 19j, 1) .... **3. Curcuma**
+ Bracts not forming pouches, usually free to base; anther versatile or not 3

3. Infl. terminal on a leafy shoot, shoot well-formed; rarely plant ± stemless and infl. arising from centre of a tuft of leaves 4
+ Infl. basal, borne directly on rhizome 10

4. Infl. surrounded by a bell-shaped involucre; plant few-leaved, ± stemless .... **6. Stahlianthus**
+ Infl. without a bell-shaped involucre; plants usually many-leaved and with distinct stems; flowers occasionally (some *Roscoea*) appearing before leaves 5

5. Filament strongly curved in upper part, style often becoming separated from it and forming a 'bow string' across curvature (Fig. 19h); ovary unilocular; flowers under 3cm long, sometimes replaced by bulbils
2. Globba

+   Filament not or slightly curved; ovary trilocular; flowers usually much
     larger; bulbils rarely formed (*Hedychium greenei*) 6

6.   Lateral staminodes reduced to small subulate points or swellings; plane
     of distichy of leaf blades transverse to rhizome .... **12. Alpinia**
+   Lateral staminodes well-formed, petal-like; plane of distichy parallel
     to rhizome 7

7.   Anther truly versatile with basal spurs (Fig. 19p) 8
+   Anther never truly versatile but sometimes (*Hedychium*) free from
     filament at base 9

8.   Flowers yellow; lateral petals connate to base of lip for about half
     their length .... **4. Cautleya**
+   Flowers purple and white; lateral petals free from base of lip
     5. Roscoea

9.   Plants delicate, under 40cm; leafy shoots often few-bladed; bracteoles
     open to base .... **7. Caulokaempferia**
+   Plants more robust, usually 1–2m; leafy shoots many-bladed; bracteoles
     tubular .... **11. Hedychium**

10. Infl. appearing before leaves 11
+   Infl. produced with leaves 12

11. Flowers white and lilac; lip c.5cm, deeply bilobed; ovary trilocular
     8. Kaempferia
+   Flowers yellow and pink; lip 1cm, suborbicular; ovary unilocular
     9. Hemiorchis

Fig. 19. **Zingiberaceae** I.
a–b, **Zingiber clarkei**: a, habit (× ⅙); b, flower (× ⅔). c–f, **Z. chrysanthum**: c, habit
(× ⅓); d, dissected flower (× ⅔); e, lateral view of stamen and corolla tube showing
anther crest (× ⅔); f, front view of stamen, corolla tube in l.s. to show style (× ⅔).
g–i, **Globba clarkei**: g, habit (× ⅙); h, side view of flower showing 'bow-string' style
(× 1); i, anther and stigma (× 4). j–l, **Curcuma aromatica**: j, habit (infl. precocious)
(× ⅓); k, flower opened out (× ⅔); l, anther (showing basal spurs) and stigma (× 4).
m–p, **Roscoea tibetica**: m, habit (× ½); n, flower (× 1); o, lip (× 1); p, anther
and stigma (× 2). Drawn by Mary Bates.

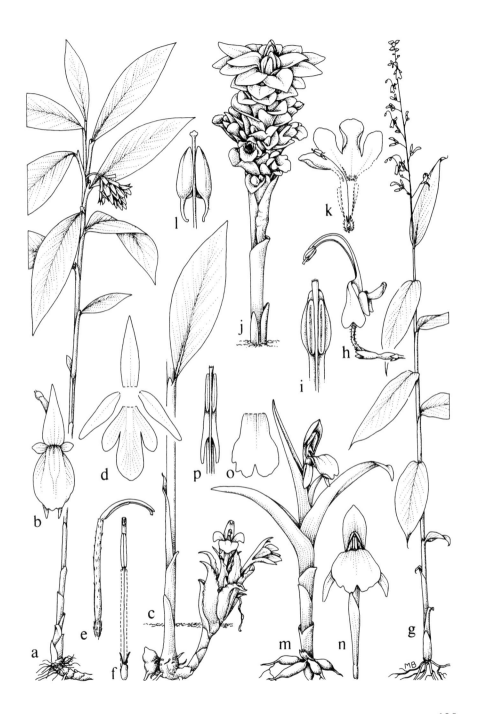

12. Leafy shoots 2–3-leaved, ± stemless; infl. narrowly fusiform, flowers well exserted .... **10. Curcumorpha**
+ Leafy shoots many-bladed, frond-like with well-developed stem; infl. usually broadly ovate, cone-like, only upper part of flower exserted .. 13

13. Lip with an elongate central lobe; base of lip and filament joined into conspicuous tube above insertion of petals .... **13. Etlingera**
+ Lip not elongate centrally; base of lip and filament not joined into tube above insertion of petals .... **14. Amomum**

1. ZINGIBER Boehmer

Leafy shoots usually frond-like. Infl. commonly basal, rarely terminal on leafy stem; bracts usually imbricate, each subtending a single flower, flowers rarely in cincinni (*Z. clarkei*); bracteoles open to base. Calyx tubular. Lateral corolla lobes usually joined together partly by their adjacent sides and to lip. Lateral staminodes adnate to lip thus forming a 3-lobed structure. Stamen subsessile or filament very short; anther connective prolonged into an elongated crest which is enfolded around upper part of style. Stigma scarcely wider than style, ciliate-margined.

1. Infl. terminal on leafy stem, sometimes appearing lateral; flowers in cincinni or borne singly 2
+ Infl. basal; flowers borne singly 3

2. Infl. breaking through uppermost leaf sheaths laterally; bracts not imbricating; flowers in cincinni .... **1. Z. clarkei**
+ Infl. terminal; bracts imbricating; flowers single .... **2. Z. capitatum**

3. Peduncle prostrate, often very short, infl. held at ground level 4
+ Peduncle elongate, infl. held erect 6

4. Bracts green, apex hairy, reflexed; lip bright yellow, unspotted, with prominent ovate lateral lobes (staminodes) .... **3. Z. chrysanthum**
+ Bracts red; lip white or whitish-yellow, spotted and streaked with red or reddish-purple, lateral lobes much less prominent, ± rounded 5

5. Lip yellowish-white, much spotted and streaked with reddish-purple; infl. appearing with leaves .... **4. Z. rubens**
+ Lip pinkish-white, with a few purple streaks and spots at base; infl. appearing before leaves .... **5. Z. sp.**

6.   Leaves oblong-lanceolate; lip yellowish-white, deeply bilobed
6. Z. purpureum
+   Leaves linear; lip purple with yellow spots, entire .... **7. Z. officinale**

1. Z. clarkei Baker. Fig. 19a–b.
Leafy stem to 2m. Leaves oblong-lanceolate, sessile, 20–40 × 5–9cm, very finely pubescent or ± glabrous beneath; ligule truncate, 0.4–0.6cm, pubescent. Infl. terminal, nodding, breaking through leaf sheaths below top of shoot, cylindric, to 10 × 3.5cm; peduncle tomentose; bracts each subtending 2–4 flowers, oblong, obtuse, 2.5cm, green at first, becoming red. Calyx 1cm. Corolla pale yellow, tube equalling bract. Lip yellow flushed with purplish-brown, oblong, 2.5cm. Stamen pale yellow. Capsule subglobose.
Bhutan: S — Phuntsholing district (Rinchending); **Darjeeling** (Sureil, Mongpu, Rishap, Kalimpong, Phubsering). Shady bank in subtropical forest, 610–1520m. Fl. July–September.

2. Z. capitatum Roxb.
Leafy stem 1–1.25m. Leaves linear-lanceolate, sessile, 20–30 × 1.5–2cm, usually shortly pubescent beneath (at least near midrib); ligule 1–2mm, puberulous. Infl. terminal, erect, sessile, narrowly elliptic, 5–10cm; bracts each subtending a single flower and bracteole, ovate, 3–4cm, coriaceous, green with narrow brown margin. Calyx 1cm. Corolla pale yellow, tube equalling bract. Lip pale yellow, c.1cm wide, 3-lobed, mid-lobe orbicular, shallowly emarginate, laterals (staminodes) oblong. Stamen pale yellow. Capsule bright red, elliptic, 2cm.
Terai (valleys); **Darjeeling** (Rangit); **Sikkim** (below Namchi). Fl. August.

3. Z. chrysanthum Roscoe. Fig. 19c–f.
Leafy stem 2–3m. Leaves oblong-lanceolate, sessile, 20–30 × 5–8cm, pubescent beneath; ligule bifid, 2cm. Infl. basal, globose-oblong, to 5cm diameter, sessile or peduncle very short; bracts each subtending a single flower and bracteole, ovate, apex acuminate, twisted, 6–7 × 3–5cm, hirsute, green. Calyx 2cm. Corolla tube just exserted from bracts; petals bright red. Lip yellow, 2.5 × 2.5cm, 3-lobed, mid-lobe orbicular, crenate, emarginate, laterals (staminodes) spathulate. Stamen yellow, crest tinged red. Capsule oblong, bright red.
Sikkim/Darjeeling (unlocalised Treutler specimen). 762m. Fl. May.

4. Z. rubens Roxb.
Leafy stem to 2.5m. Leaves oblong acuminate, sessile, 20–35 × 10–15cm, lightly pubescent or glabrous beneath; ligule bilobed, 0.5–1cm. Infl. half embedded in ground, globose, 3–4cm diameter, peduncle to 10cm, usually less; bracts bright red, each subtending a single flower and bracteole, 2.5–3.5cm, outer ones ovate, inner lanceolate. Calyx c.3cm. Corolla tube equalling bracts; petals

red, 2.5cm. Lip equalling petals, yellowish-white much streaked and spotted with reddish-purple, side lobes small, rounded. Stamen equalling lip, crest bright red.

Sikkim/Darjeeling (unlocalised *King* specimen). 610m. Fl. July.

Z. rubens was described from Rangpur (Rungpore) in N Bangladesh and is also known from Khasia. The Sikkim collection (*King* 1014) probably belongs here.

5. Z. sp. (*Grierson & Long* 2678).

Infl. precocious, leaves not seen. Peduncle very short at first, elongating to 6–[15]cm, prostrate [subterranean?]. Infl. c.6 × 2–3cm; bracts bright crimson, each subtending a single bracteole and flower, narrowly lanceolate, 5–5.5 × 1–1.3cm. Calyx c.1cm. Corolla tube 3.5–4cm; petals 3cm, pale red. Lip white, lightly spotted and streaked with reddish-purple, lateral lobes (staminodes) 1.2 × 0.8cm. Stamen 1.5cm, crest whitish purple.

Bhutan: C — Punakha district (between Lobesa and Tinlegang). Damp shady bank at foot of cliff in warm, broad-leaved forest, 1900m. Fl. July.

This seems to represent a distinct taxon, perhaps most closely allied to *Z. roseum* (Roxb.) Roscoe (Bangladesh), but differs in flower colour and precocious habit.

6. Z. purpureum Roscoe; *Z. cassumunar* Roxb.; *Z. montanum* auct. non (J. König) Dietrich.

Leafy stem to 1m. Leaves lanceolate, acute, subsessile, 20–30 × 7–8cm, pubescent beneath; ligule bilobed, very short, pubescent. Infl. basal, ovate, at least 5 × 8cm; peduncle 8–30cm, erect; bracts purplish-brown, each subtending a single flower and bracteole, broadly ovate, 3–3.5cm, pubescent with narrow membranous margin. Calyx c.1.5cm. Corolla pale yellow, tube shorter than bracts. Lip pale yellow, 2.5cm wide, mid-lobe suborbicular, emarginate, laterals (staminodes) oblong. Capsule ovoid.

Terai (Jalpaiguri (Sikdar & Samanta, 1984)).

Probably native from Bengal to Coromandel, doubtfully so in Bhutan and Sikkim. Cultivated throughout tropical Asia for its medicinal properties.

Z. zerumbet (L.) Smith is widely cultivated throughout the tropics, also mainly for its medicinal properties, and is often grown as a village plant. It is easily distinguished from *Z. purpureum* by the glabrous bracts and prominent (1.5–3cm) membranous ligules.

7. Z. officinale Roscoe. Sha: *saga*.

Leaves linear lanceolate. Lip dark purple with creamy yellow spots.

Terai (Jalpaiguri (Sikdar & Samanta, 1984)).

The root ginger of commerce, probably originating from China. The species has many races and is cultivated in tropical countries worldwide.

2. GLOBBA L.

Slender plants, usually under 1m. Infl. terminal (very rarely on a leafless shoot). Bracts remote. Flowers borne in cincinni, often replaced by bulbils. Bracteoles open to base. Corolla tube slender, well exserted, dorsal petal sometimes hooded. Lateral staminodes attached to corolla tube at same level as petals, petal-like. Lip connate to filament in a slender tube c.1cm above staminodes, auriculate at base, free part usually bilobed or emarginate. Filament long, slender, with inflexed edges, strongly curved in upper part, style often becoming separated and forming a 'bow-string' across curvature; anther small with parallel thecae each of which may bear 1 or 2 lateral appendages (not in Bhutanese species); crest not or hardly produced. Ovary unilocular with parietal placentation. Fruit a small, dehiscent capsule.

1. Bracteoles to 1cm, broadly ovate, ± persistent, folded round clustered flowers; capsule muricate .... **1. G. andersonii**
+ Bracteoles to 0.5cm, ± lanceolate soon deciduous; capsule smooth (but appearing verrucose when dry) 2

2. Cincinnus stalks at base of infl. rarely exceeding 1cm; flowers 4–6 per cincinnus, well spaced; bulbils produced at base of infl. **2. G. multiflora**
+ Cincinnus stalks at base of infl. 2–7cm; flowers 2–3(–4) per cincinnus, clustered; bulbils present or not 3

3. Bulbils formed as infl. matures; flowers yellow .... **3. G. clarkei**
+ Bulbils not formed; flowers orange 4

4. Lowermost cincinnus stalks 3–6(–7)cm; upper part of bud elliptic; flowers pedicellate (pedicels 1–3mm) .... **4. G. macroclada**
+ Lowermost cincinnus stalks to 2cm; upper part of bud ± globose; flowers sessile .... **5. G. racemosa**

1. G. andersonii Baker.
Leafy shoots to 60cm. Leaves lanceolate, long caudate, sessile, 15–30 × 3–6cm, sometimes slightly pubescent beneath; ligule entire, 2–3mm, pubescent. Infl. lax, 10–20cm, pubescent; bracts each subtending a clustered cincinnus of yellow flowers, to 1cm; cincinnus stalks 1–1.5cm; bracteoles persistent, broadly ovate, forming an involucre around flowers, 1cm. Calyx obscurely 3-lobed,

0.4–0.5cm; corolla tube 2–3 × length of calyx; petals 0.5cm, dorsal cucullate. Lateral staminodes equalling petals, ovate; lip longer, bilobed to half-way. Stamen 2–2.5cm. Capsule muricate.

Darjeeling (Rongsong, Tista, Punkabari, Ryang, Kurseong). 300–910m. Fl. July–August.

2. G. multiflora Baker.

Leafy shoots to 60cm. Leaves lanceolate, caudate, sessile, 20–30 × 3–6cm, pubescent beneath; ligule bilobed, 1–2mm, pubescent. Infl. lax, 20–30cm, pubescent; bracts small, soon deciduous, those at base of infl. producing bulbils, those from upper part subtending cincinni of 4–6 yellow flowers (flowers not clustered); cincinnus stalks 1cm or much less; bracteoles minute, soon deciduous. Calyx 3-lobed, 3–4mm. Corolla tube 3 × length of calyx, petals ovate, 3mm. Lateral staminodes 3–4mm; lip ± similar, shortly bilobed. Stamen c.1.5cm. Capsule smooth.

Bhutan: S — Gaylegphug district (Gaylegphug (M.F.B.)); unlocalised Griffith specimen; **Terai** (Jalpaiguri (Sikdar & Samanta, 1984)); **Sikkim** (Mamreng, Rishap, Rangit). 150–1520m. Fl. August–November.

3. G. clarkei Baker; *G. hookeri* Baker. Fig. 19g–i.

Leafy shoots to over 1m. Leaves lanceolate, long caudate, sessile, 15–30 × 3–8cm, ± glabrous or lightly pubescent beneath, upper surface usually sparsely hairy on either side of main veins; ligule bilobed, 2–3mm, pubescent. Infl. lax, 25–40cm, bulbils produced in upper part or ± throughout infl. as it ages; bracts to 1cm, soon deciduous, each subtending a cincinnus of 2–3 tightly clustered, bright yellow flowers; cincinnus stalks 2(–3)cm; bracteoles minute, soon deciduous. Calyx 3-lobed, 5–7mm, brown. Corolla tube rather more than twice length of calyx; petals ovate, 5–6mm. Lateral staminodes longer and narrower than petals; lip 1–1.5cm, bilobed, lobes ovate. Stamen 2–2.5cm. Capsule ± smooth (appearing somewhat verrucose when dry).

Bhutan: S — Chukka district (Marichong); **C** — Punakha (Rinchu, Ritang to Tsarza La, Hinglai La; Lometsawa (M.F.B.)) and Tashigang (Tashi Yangtsi) districts; **Darjeeling** (Rongbe, Ryang, Sureil, Rishap Ravine, Rungirun, below Observatory Hill); **Sikkim** (Lachen, Selak, Mamreng, Dhobjhua, between Mintok and Paha Kholas). Banks in dense, broad-leaved, evergreen forest, 610–2440m. Fl. May–September.

Following the suggestion in Hara *et al.* (E.F.N.), *G. hookeri* is here placed in synonymy. The type gathering (from Nepal) differs in the generally more pubescent habit, broader leaves and in more copious bulbils, but it is evident that intermediates exist, and, until a critical study can be made from living populations, the two taxa are best treated as a single species.

4. G. macroclada Gagnepain. Nep: *sanok*.

Leafy shoots under 1m. Leaves lanceolate, caudate, sessile, 10–30 × 4–5.5cm, pubescent to ± glabrous beneath; ligule bilobed, 2mm, ± glabrous. Infl. 10–15cm; bulbils not produced; bracts small, soon deciduous, each subtending a cincinnus of 2–3 yellow-orange, shortly pedicellate flowers; buds elliptic apiculate; cincinnus stalks 5–6(–7)cm at base of infl.; bracteoles ovate, 3–4mm. Calyx 3-lobed, 0.8–1cm, lobes shortly dentate. Corolla tube more than twice length of calyx, pubescent; petals 8–9mm, dorsal cucullate. Lateral staminodes equalling petals but narrower; lip 1.5cm, cuneate, shortly bilobed, lobes obtuse. Stamen 2–3cm; anther 6–7mm. Capsule smooth, appearing verrucose when dry.

Darjeeling (Riang, Sureil, Sukna, Great Rangit opposite Manjitar); **Sikkim** (Naya Bazaar to Legship; Tista (F.E.H.1)). Shady banks in subtropical forest; rocky hillside by river, 300–1520m. Fl. June–July.

5. G. racemosa Smith; *G. orixensis* auct. non Roxb. Nep: *alipat*.

Differs from *G. macroclada* in having shorter cincinnus stalks, sessile flowers and rounded buds.

Bhutan: S — Phuntsholing (Phuntsholing (M.F.B.)) and Gaylegphug (W bank of Thewar Khola near Gaylegphug) districts; unlocalised Griffith specimen; **Terai** (Siliguri; Jalpaiguri (Sikdar & Samanta, 1984)); **Darjeeling** (Rangit, Tista, Lebong). Stony ground in cleared jungle, 300–910m. Fl. May–August.

As with the *G. clarkei/hookeri* complex, *G. racemosa* highlights the need for field studies. Some collections resemble *G. macroclada* in the conspicuously dentate calyx lobes and cucullate dorsal petal and in these respects deviate from the type (described from Nepal).

3. CURCUMA L.

Infl. terminal on a leafy shoot (borne in centre of a leaf tuft), or on a separate shoot and sometimes (as in Bhutanese species) appearing before leaves. Infl. pedunculate, erect; bracts joined to each other for about half their length, forming pouches, free ends usually spreading, each subtending a cincinnus of 2–7 flowers; uppermost bracts often larger, differently coloured and sterile (forming a coma); bracteoles open to base. Calyx unequally lobed, unilaterally split; corolla tube ± funnel-shaped. Lateral staminodes petal-like, folded under dorsal petal. Lip with a thickened central portion and thinner side lobes which overlap staminodes. Filament short and broad, constricted at apex; anther versatile, usually spurred at base, sometimes with a small crest. Ovary trilocular with axile placentation. Fruit ellipsoid.

1. Leaves flushed red around midrib .... **3. C. zedoaria**
+ Leaves green 2

2. Coma pink; flowering before leaves .... **1. C. aromatica**
+ Coma white; flowering with leaves .... **2. C. longa**

1. C. aromatica Salisbury; *C. zedoaria* Roxb. non (Christmann) Roscoe. Dz: *düm*. Fig. 19j–l.

Rhizome yellow within, aromatic. Leaf tufts to over 1m, 5–7-leaved; lamina broadly lanceolate, acuminate, 40–70 × 10–14cm, shortly pubescent beneath; petiole sometimes equalling lamina. Infl. 9–20 × 6–10cm, commonly appearing before, and produced laterally to, leaves. Coma bracts pink. Fertile bracts recurved at tips, to 6cm, minutely pubescent at least in upper half; bracteoles c.2cm, lightly pubescent. Flowers whitish, pink tinged, not exserted, lip yellow, obscurely 3-lobed.

Bhutan: S — Samchi (Samchi), Sankosh (Balu Khola 7km W of Phipsoo) and Gaylegphug (Hatisar) districts; **C** — Punahka (1km E of Wangdi Phodrang, Bowdata Valley) and Tongsa (49km S of Tongsa on road to Shamgong) districts; **Darjeeling** (Ryang Valley, Berick; Kalimpong, Peshok to Lopchu (F.E.H.1)). Open slopes in hot, dry valleys; secondary scrub at margin of subtropical forest; shady forest floor, 150–1830m. Fl. March–July.

Root apparently not eaten.

2. C. longa L. Sha: *yongka*; Med: *gaser*; Eng: *turmeric*.

Differs from *C. aromatica* in flowering with the leaves and its white coma bracts.

Bhutan: S — Phuntsholing district (Suntlakha); **Darjeeling** (Barnesbeg, Mongpu). Fl. July.

Cultivated for its rhizome from which the spice turmeric is prepared.

3. C. zedoaria (Christmann) Roscoe.

Usually precocious in habit and distinguished from *C. aromatica* as follows: leaves glabrous, flushed red around midrib; coma bracts more intensely coloured; rhizome pale yellow, strongly camphor-scented.

Bhutan: C — Tongsa district (Berthi (M.F.B.)); **Terai** (Jalpaiguri (Sikdar & Samanta, 1984)); **Darjeeling** (Great Rangit, Punkabari, Labdah, Temi, Peshok); **Sikkim** (below Rumtek). Cultivated to 1100m. Fl. April–July.

Country of origin unknown, widely cultivated throughout the Indian subcontinent as a medicinal plant and also used in perfumery.

4. CAUTLEYA (Bentham) Hook.f.

Leafy shoots to 1m, few-leaved. Infl. terminal; bracts each subtending a single, yellow flower; bracteoles absent. Lateral petals connate to each other centrally and to lip in lower half, free at margins. Lateral staminodes petaloid, held erect in front of dorsal petal. Lip deeply bilobed. Anther truly versatile, spurred at base, ecristate. Ovary trilocular with axile placentation. Capsule splitting readily to base.

1. Slender plant; bracts much shorter than calyx; infl. with up to 12 flowers .... **1. C. gracilis**
+ Robust plant; bracts equal to or exceeding calyx; infl. many-flowered
2. C. spicata

1. C. gracilis (Smith) Dandy; *Roscoea gracilis* Smith; *R. lutea* Royle; *Cautleya lutea* (Royle) Hooker. Nep: *tembok*. Plate 9.
Leafy shoot to 40cm. Leaves 4–6, lanceolate, long caudate, sessile, 6–20 × 2.5–3cm, usually dark purple beneath; ligule entire, 3–5mm. Infl. 4–10cm (elongating with age); bracts usually red, remote, acute, 1–2.5cm. Calyx red, unilaterally split, 1–1.5cm. Corolla tube 1.5–2cm, petals oblong, rounded, 2cm. Lateral staminodes ± spathulate, 2cm. Lip bilobed to half-way, 2 × 1.25cm. Filament very short; anther 2cm. Capsule red, globose, to 1cm diameter.
Bhutan: S — Phuntsholing (30–40km N of Phuntsholing) and Chukka (Chasilakha (M.F.B.)) districts; **C** — Thimphu (near Changri Monastery, Dotena), Punakha (Tinlegang, Ritang to Pele La) and Tongsa (Sanka) districts; **N** — Upper Mo Chu (Gasa Dzong) and Upper Kulong Chu (Tobrang) districts; **Darjeeling** (above Sonada, Rongbe, Rissisum, Phalut to Ramam, Ghumpahar, Kurseong, Risom; Darjeeling (F.E.H.1)); **Sikkim** (Tsoka to Yoksum, Mamring, Lachen, above Chungthang, Phedang to Bakkim; Chia Bhanjang to Pamianchi, Gangtok (F.E.H.1)). Moist banks and rocks (sometimes epiphytic) in mixed and *Quercus semecarpifolia* forest, 1500–3000m. Fl. May–August.

C. cathcartii Baker is probably just a robust form of this species. It has been recorded from Darjeeling (Tonglu, Great Rangit) and Sikkim (Kulhait Chu above Dentam, Karponang).

2. C. spicata (Smith) Baker; *Roscoea spicata* Smith. Nep: *pahilo sana*.
Leafy shoots to 1m. Differs from *C. gracilis* in its more robust habit, petiolate leaves and in the densely flowered infl. which is larger in all its parts.
Bhutan: S — Phuntsholing (30–40km N of Phuntsholing; Kamji (M.F.B.)), Chukka (Gedu to Ganglakha; Chasilakha (M.F.B.)) and Deothang (between

Tshilingor and Riserboo) districts; **Darjeeling** (0.5km S of Tiger Hill, Lopchu to Jorebungalow, Labha, Batasi to Manibanjang, Rongbe, Balasun, Senchal); **Sikkim** (Prek Chu to Mintok Khola, Gangtok to Penlang La). Shady gullies and cliffs in mixed broad-leaved (incl. oak) and *Rhododendron grande* forest, 1500–2400m. Fl. June–September.

C. robusta Baker was described from inadequate, fruiting material from Darjeeling (Kurseong, Mongpu); it is characterised by its elongate spike, but is possibly synonymous with *C. spicata*.

5. ROSCOEA Smith

Small plants, 6–60cm. Infl. terminal on few-leaved stem, sometimes flowering before leaves develop. Lamina and sheath not clearly distinct; ligule present as curved line at junction between them. Bracts subtending a single purple, purple and white or sometimes white flower; bracteoles absent. Corolla tube slender, widening at throat, usually exserted from calyx. Lateral staminodes petal-like, held erect in front of dorsal petal. Lip entire to bilobed, clawed or not. Anther truly versatile, spurred at base, ecristate. Ovary trilocular with axile placentation. Capsule elliptic, cylindric, tardily dehiscent.

1. Bracts obtuse to truncate, very short (0.3–1cm); corolla tube well exserted from calyx; dorsal petal circular; lateral staminodes circular, symmetric .... **1. R. alpina**
+   Bracts acute, 2–13cm; corolla tube equal to or exserted from calyx; dorsal petal elliptic or obovate to broadly elliptic; lateral staminodes asymmetric 2

2. Leaves auriculate at base throughout plant .... **2. R. auriculata**
+   Leaves not auriculate at base or slightly so in lower part of plant 3

3. Plants generally 25–40cm; bracts 7–13cm; lateral staminodes spathulate with a long claw; lip 4.5–6.5cm .... **3. R. purpurea**
+   Plants generally 10–20cm; bracts 2–4cm; lateral staminodes elliptic without a long claw; lip 1.5–2.5cm .... **4. R. tibetica**

1. R. alpina Royle; *R. intermedia* Gagnepain.
    Leafy shoots 10–20cm. Leaves 2–3, usually not fully developed at flowering, linear to broadly elliptic, 2.5–15 × 1.5–3.5cm; ligule semicircular. Infl. to 5-flowered, flowers opening singly. Bracts narrow, obtuse or ± truncate, to 1cm. Calyx 2.5cm. Corolla tube long exserted; dorsal petal circular. Lateral

staminodes circular to elliptic, ± symmetrical. Lip obovate, bilobed, 1.5–2.5 × 1–2cm.

Bhutan: N — Upper Mo Chu district (Laya); **Darjeeling** (Sandakphu, Phalut; Sandakphu to Kala Pokhari (F.E.H.1)); **Sikkim** (Tari); **Chumbi**. Meadows; dry peaty soil, 3500–3960m. Fl. June–August.

2. R. auriculata K. Schumann; *R. purpurea* Smith var. *auriculata* (K. Schumann) Hara.

Leafy shoots 25–40cm. Leaves usually 5–7, linear to broadly elliptic, base auriculate, 5–25 × 1.5–6cm; ligule semicircular. Bracts acute, 7–8 × 0.5–2cm. Calyx 5–8cm. Corolla tube usually ± equalling calyx, sometimes well-exserted; dorsal lobe obovate to broadly elliptic. Lateral staminodes asymmetrically obovate. Lip shortly clawed, broadly obovate, entire or bilobed, 3.5–5 × 2.5–4cm.

Sikkim (Lachung, Lachen, Bakkim, Meiupup, Chungthang, Lamteng, Mon Lapcha, Lhonak). Open grassy banks; forest clearing, 2130–3960m. Fl. May–September.

3. R. purpurea Smith; *R. procera* Wall.; *R. purpurea* var. *procera* (Wall.) Baker. Nep: *rasgari*.

Leafy shoots 25–40cm. Leaves 4–8, lanceolate to elliptic, 10–20 × 1–5cm, lower leaves sometimes slightly auriculate at base; ligule semicircular. Bracts acute, 7–13 × 0.5–2cm. Calyx 5–9cm. Corolla tube hardly exserted from calyx; dorsal lobe narrowly elliptic. Lateral staminodes obliquely spathulate with long narrow claw. Lip obovate, entire or bilobed, 4.5–6.5 × 2–5cm incl. claw (c.1cm).

Bhutan: C — Mongar (Lhuntse Dzong, Saling) and Tashigang (Chorten Kora) districts; **N** — Upper Kulong Chu district (Tobrang). Banks in *Quercus/Lyonia* and mixed forest; open hillside, 1830–2650m. Fl. June–July.

4. R. tibetica Batalin. Fig. 19m–p.

Leafy shoots 10–20cm. Leaves 1–4, forming a rosette, lanceolate to oblong ovate, 2–20 × 1.5–5.5cm, first leaf auriculate at base; ligule ± semicircular. Bracts hidden in leaf bases, 2–4 × 0.4cm. Calyx 2.5–4cm. Corolla tube hardly to well-exserted from calyx; dorsal petal elliptic. Lateral staminodes elliptic or asymmetrically obovate. Lip obovate, usually deeply bilobed, 1.5–2.5 × 1–2cm incl. short claw.

Bhutan: C — Ha (Damthang), Thimphu (Chapcha, 6km N of Thimphu Dzong, Jato La, Phajoding, Dotena Chu, Tsalimape, Pumo La), Punakha (Wangdi Phodrang, Kotaka, Mara Chu Valley), Tongsa (Chendebi) and Bumthang (above Lami Gompa, Takhung) districts. Moist meadow among scrub; open pine forest; open grassy banks; clearing in forest; dry soil among bracken, 2130–3050m. Fl. May–August.

6. STAHLIANTHUS Kuntze

Small plants of *Kaempferia*-like habit, stemless. Infl. surrounded by a large bell-shaped involucre. Lateral staminodes petal-like; lip bifid. Epigynous glands absent (fide Kuntze).

1. S. involucratus (Baker) R.M. Smith; *Kaempferia involucrata* Baker.
Leafy shoots to 30cm. Leaf lamina lanceolate-oblong, 15–18cm; ligule unknown; petiole 12–15cm. Peduncle 4–10cm, involucre 4–5cm, enclosing several white flowers, each subtended by a lanceolate bract 1–1.5cm; bracteoles unknown. Calyx shorter than bract. Corolla tube c.3cm. Lateral staminodes oblong, to 0.75cm. Lip white with yellow blotch in centre, orbicular, bifid, 1cm. Anther with entire orbicular crest.
Darjeeling (Rungirun). Fl. April.

7. CAULOKAEMPFERIA K. Larsen

Infl. terminal; bracts each subtending a single flower or a cincinnus of flowers; bracteoles open to base. Lip usually entire. Lateral staminodes petal-like. Filament short; anther crested. Fruit ovoid.

1. C. sikkimensis (King ex Baker) K. Larsen; *Kaempferia sikkimensis* King ex Baker. Nep: *sannsana*. Fig. 20a–c.
Leafy shoots 10–30cm. Leaves to 10, narrowly elliptic, long caudate, sessile, 2–11 × 1–2cm; ligule entire, 2–3mm. Infl. sometimes reduced to a single bract (fide Larsen) and flower, sometimes with up to 3 bracts, each subtending 1 or 2 flowers; bracts lanceolate, 2–3 × 0.5cm; bracteoles linear, under 1cm. Calyx 1–1.5cm. Corolla tube 4–6cm, slender; petals narrowly oblong, 1–1.5cm, dark pink. Lateral staminodes oblong-cuneate, equalling petals, white; lip suborbicular, entire, white.

FIG. 20. **Zingiberaceae** II.
a–c, **Caulokaempferia sikkimensis**: a, habit (× ½); b, flower (× 1); c, anther showing crest and stigma (× 2). d, **Hedychium coronarium**: infl. (× ⅓). e, **H. spicatum**: flower (× ⅔). f–h, **H. gardnerianum**: f, habit (× ⅙); g, flower (× ⅔); h, anther and stigma (× 2). i–l, **Alpinia nigra**: i, infl. (× ¼); j, flower (× 2); k, lateral view of anther and stigma (× 2); l, fruit (× ⅔). m–n, **Amomum subulatum**: m, habit (× ⅒); n, flower (× ⅔). Drawn by Mary Bates.

Duars (Buxa); **Darjeeling** (Dalingkot, Rikisum, Rungbee); **Sikkim** (Phadonchen, S bank of Rate Chu N of Gangtok). Sandy, shady banks; on rocks and wet cliff-ledges, 760–1890m. Fl. June–August.

C. secunda (Wall.) K. Larsen is reported in error by Larsen (1964) for Sikkim. This is a Khasian species differing from *C. sikkimensis* in its broader, petiolate leaves and wider bracts.

8. KAEMPFERIA L.

Leaves few, tufted, pseudostem short or ± absent. Infl. terminal or basal and appearing before leaves; bracts each subtending a single flower; bracteoles open to base. Lip large and showy, deeply bilobed. Lateral staminodes petaloid. Anther sessile or with a short filament, connective prolonged into distinct, reflexed crest. Ovary trilocular with axile placentation.

1. K. rotunda L. Nep: *buin champa*.
Leaves oblong acuminate, c.20 × 6–10cm, purple beneath, mottled dark and light green above. Infl. appearing before leaves, 4–6-flowered, sessile; bracts imbricating, oblong, acute, to 3.5cm; bracteoles shorter, bidentate. Calyx 0.6cm. Corolla white, tube exceeding calyx, petals very narrow. Lateral staminodes white, oblong, held erect, c.5cm. Lip lilac, equalling staminodes, deeply divided into 2 suborbicular lobes. Anther crest exceeding thecae, divided into 2–4 segments.

Terai (near Sukna, Sivoke Terai); **Darjeeling** (Rongsong; Kalimpong (F.E.H.1)). Sal forest, 610m. Fl. April–May.

Widely cultivated for its ornamental qualities. Country of origin uncertain.

9. HEMIORCHIS Kurz

Infl. basal, appearing before leaves; peduncle erect clothed with lanceolate bracts. Flowers at first congested, borne singly, each subtended by a bract; bracteoles absent. Calyx 3-lobed. Petals ± equal, dorsal shortly cucullate. Lateral staminodes petal-like; lip concave, shortly bilobed with a thickened central ridge throughout its length. Filament short, curved; anther ± ecristate. Ovary unilocular with parietal placentation. Capsule ridged, globose or fusiform.

1. H. pantlingii King.
Leafy shoot to 80cm with well-developed pseudostem. Leaves oblong-lanceolate, acuminate, 15–40 × 3–6cm, margins and veins of upper surface

strigulose, glabrous beneath, narrowed at base into short, winged petiole. Peduncle 8–10cm; infl. at first congested, elongating to 3–5cm. Flowers yellow and pink, axis slightly tomentose; bracts linear-lanceolate, c.1cm. Calyx 0.6–0.8cm; corolla tube exceeding calyx; petals pinkish, 1.2cm. Lateral staminodes yellow with pinkish-brown spots, suborbicular, 0.5–0.6cm in diameter; lip yellow, central thickened part purplish, 1cm, suborbicular. Filament 0.5cm; anther 0.3–0.5cm. Capsule fusiform, 3cm, puberulous.

Darjeeling (Mongpu, Sivoke, Chunbati). Sand by sides of streams, 305–1220m. Fl. March–April.

The type collection of this species formed part of Pantling's Orchid herbarium and is, in general appearance, remarkably similar to some members of the Orchidaceae.

10. CURCUMORPHA A.S. Rao & D.M. Verma

Infl. basal on a short peduncle. Flowers opening from apex downwards, each subtended by a boat-shaped bract; bracteoles open to base. Calyx tridentate, deeply split unilaterally; corolla tube long exserted. Lateral staminodes petal-like, joined to base of filament to form a cup above insertion of petals; lip large. Anther subsessile with a minute crest. Ovary trilocular with axile placentation. Fruit unknown.

1. C. longiflora (Wall.) A.S. Rao & D.M. Verma; *Gastrochilus longiflorus* Wall.; *Boesenbergia longiflora* (Wall.) Kuntze.

Leafy shoot 30–50cm. Leaves 2–4, loosely clasping, blades oblong-lanceolate, caudate-acuminate, base slightly cordate, 10–45 × 4–13cm, ± glabrous; petioles 2–4cm. Flowers white and pink; calyx 2–3cm; corolla tube 7.5–12.5cm. Lateral staminodes white, obovate, 2–2.5cm; lip oblong-cuneate, entire with incurved margins, 3.5–5.5 × 2.5–5cm, white with a fleshy red median band, flushed with purple, base streaked with red. Anther 1–1.5cm.

Terai (Bamunpokri, Sukna; Jalpaiguri (Sikdar & Samanta, 1984)); **Darjeeling** (Punkabari, Tista, Rongsong). 150–760m. Fl. May–July.

11. HEDYCHIUM J. König

Leafy shoots many-leaved. Infl. terminal, many-flowered; bracts imbricating (main axis concealed) or remote (main axis visible), each subtending a single flower or a cincinnus of 2–6 flowers; bracteoles tubular. Corolla tube slender, petals linear, ± equal. Lip conspicuous, usually bifid. Lateral staminodes petal-like, narrow. Stamen often long exserted; anther thecae free at base (but not truly versatile), ecristate. Ovary trilocular with axile placentation. Fruit a capsule.

1. Bracts imbricate, concealing main axis 2
+ Bracts never imbricate; main axis visible 6

2. Flowers bright red .... **1. H. greenei**
+ Flowers white or yellow 3

3. Stamen not exserted .... **2. H. coronarium**
+ Stamen long exserted 4

4. Bracts convolute, becoming recurved; flowers white .. **3. H. thyrsiforme**
+ Bracts flat; flowers yellowish-white; stamen entirely orange-red or anther crimson 5

5. Flowers 9–11cm; lip 2.5–4cm; infl. flat-topped; stamen orange-red
   4. H. ellipticum
+ Flowers 7–8cm; lip 1cm; infl. elongate; stamen white, anther crimson
   5. H. griersonianum

6. Flowers to 4cm, orange or yellowish-orange, occasionally salmon; infl. dense, narrowly cylindric; flowers borne singly .... **6. H. densiflorum**
+ Flowers 8–14cm, variously coloured; infl. moderately dense, usually broadly cylindric; flowers single or in cincinni 7

7. Stamen shorter than lip .... **7. H. spicatum**
+ Stamen long exserted, much exceeding petals, lip and staminodes 8

8. Flowers single, white with a bright red stamen; stems slender
   8. H. glaucum
+ Flowers in cincinni (sometimes single in *H. gardnerianum*), not coloured as above; stems robust 9

9. Flowers lemon yellow, filament bright red .... **9. H. gardnerianum**
+ Flowers white, orange or red 10

10. Lip oblong-cuneate; flowers pure white .... **10. H. stenopetalum**
+ Lip clawed, limb suborbicular or broadly triangular; flowers red or orange, if white then red-tinged 11

11. Flowers red .... **11. H. coccineum**
+ Flowers orange or white tinged red 12

12. Flowers orange .... **12. H. aurantiacum**
+ Flowers white tinged red .... **13. H. elatum**

1. H. greenei W.W. Smith.
Leafy shoots 1–2m. Leaves broadly lanceolate, long acuminate, sessile, 10–30 × 2–9cm, dark purple beneath; ligule entire, pubescent; sheaths red. Infl. ellipsoid, bracts oblong, each subtending a cincinnus of 2–4 bright red flowers. Calyx 3cm. Corolla tube 5cm, not exserted from bracts, petals linear, c.3cm. Lateral staminodes slightly exceeding petals, narrowly spathulate; lip 4–5cm wide, 2-lobed, lobes rounded. Stamen barely exceeding lip. Viviparous bulbils often produced in infl.
Bhutan (described from 'low hills in W Bhutan'); **Sikkim** (Gangtok (Srivastava, 1985)). Fl. July.

2. H. coronarium J. König. Lep: *seldong dum*. Fig. 20d.
Leafy shoots 1–2m. Leaves lanceolate, acuminate, ± sessile, 10–60 × 5–11cm; ligule bilobed, 2–3cm. Infl. ± ellipsoid, 10–20cm. Bracts closely imbricating, oblong, rounded, 4–5 × 2–3cm, each subtending 3–6 fragrant, white flowers. Calyx 2.5–3cm. Corolla tube c.twice as long as calyx, petals linear, 3–4cm. Lateral staminodes oblong-lanceolate, 3–5cm; centre of lip often flushed yellow or yellowish-green, to 5cm wide and equalling staminodes, narrowed at base into short claw, limb shortly bifid. Stamen shorter than to slightly exceeding lip.
Bhutan: S — Deothang district (Sassi); **Terai** (unlocalised Hooker specimen); **Darjeeling** (Little Rangit Valley, Farsing, Labdah, Mongpu); **Sikkim** (Gangtok). Probably always cultivated, 610–1300m. Fl. August–September.

3. H. thyrsiforme Smith.
Leafy shoots 1–1.5m. Leaves lanceolate to narrowly elliptic, acuminate, subsessile, 10–30 × 5–10cm, sometimes sparsely pubescent beneath and on margins; ligule entire or emarginate, 1.5–2.5cm, pubescent. Infl. ellipsoid, 6–10cm; bracts convolute, becoming recurved, 2–3 × 1cm, each subtending one or two pure white flowers. Calyx shorter than bract. Corolla tube only just exceeding bract, petals linear, 2–2.5cm. Lateral staminodes linear-lanceolate; lip 3–4cm, clawed, bilobed to middle. Stamen 6–7cm.
Bhutan: S — Samchi (Changuna (M.F.B.)), Phuntsholing (Kamji (M.F.B.)) and Gaylegphug (Rani Camp to Tama (M.F.B.)) districts; **C** — Punakha district (Neptengka); **Terai** (Jalpaiguri (Sikdar & Samanta, 1984)); **Darjeeling** (Tista, Rongbee, Rangit below Badantam, Mongpu, Sureil, Tukvar, Farsing). Sometimes epiphytic, 305–2000m. Fl. August–November.

4. H. ellipticum Smith. Sha: *khui-see*; Lep: *tingbok rip.*
Leafy shoot 1–1.5m. Leaves elliptic, shortly acuminate, 7–15 × 25–30cm; petioles to 1cm; ligule bright red, entire or emarginate. Infl. ellipsoid, flat-topped; bracts imbricating, ovate, acute, to 2–2.5 × 1–1.5cm, each subtending a single white flower. Calyx equalling bract. Corolla tube 6–7cm, petals linear, 5–6cm. Lateral staminodes spathulate, 4–5cm; lip clawed, narrowly oblong, entire or emarginate, 2.5–4cm. Stamen orange-red, exserted, 8–9cm.

Bhutan: S — Chukka (Marichong) and Deothang (5km N of Deothang) districts; **C** — Mongar (near Lhuntse, Khoma) and Tongsa (Tongsa) districts; **N** — Upper Mo Chu district (Gasa); **Darjeeling** (Mongpu, Badamtam, Darjeeling, Great Rangit, Kurseong to Punkabari); **Sikkim** (Gangtok to Rangpo (F.E.H.1)). Dry hillside; rocky banks at margins of subtropical forest; sometimes epiphytic, 305–2440m. Fl. June–August.

5. H. griersonianum R.M. Smith.
Differs from *H. ellipticum* in its much more pubescent main axis, more elongate infl. and in its much smaller flowers. Calyx 2.5cm. Corolla tube 3.5–4cm; petals 3–3.5cm. Lateral staminodes 1.5cm; lip 1–1.3cm, clawed, shortly bilobed. Stamen 3.5cm, white with a crimson anther.

Bhutan: S — Sarbhang district (Sarbhang to Chirang road, 19km above Sarbhang). Steep rocky slope in hot forest, 1100m. Fl. June.

6. H. densiflorum Wall. Dz: *dyum.*
Leafy shoots to 1m. Leaves oblong-lanceolate, acuminate, subsessile, 12–30 × 3–10cm, glabrous; ligule entire, 1cm. Infl. dense, narrowly cylindric, 8–20cm; bracts oblong, inrolled, 1.5–2cm, each subtending a single usually yellowish-orange flower. Calyx ± equalling bract. Corolla tube usually only a little exserted from calyx; petals linear, 1.5–2cm. Lateral staminodes lanceolate, 1–1.5cm; lip cuneate, usually bifid, 1 × 0.5cm. Stamen 1.5–2cm.

Bhutan: S — Gaylegphug district (Rani Camp to Tama (M.F.B.)); **C** — Thimphu (Dotena), Punakha (Ritang, Zayla), Tongsa/Bumthang (Yuto La) and Tashigang (Tashi Yangtsi Dzong) districts; **Darjeeling** (Ghum); **Sikkim** (Lachen, between Mintok and Paha Kholas). Mixed broad-leaved (incl. oak) forest, wet grassy clearings, 1680–2590m. Fl. July–August.

H. densiflorum is variable in size and flower colour; particularly fine salmon-pink forms are known to occur.

Collections from Gasa and Tamji (Upper Mo Chu district), and Tinlegang (Punakha district), differ in having smaller, distinctly petiolate, leaves and shorter infls. In these respects they approach *H. aureum* Baker, described from the Khasia Hills where it is often epiphytic, but differ from this in their large, prominent ligule and glabrous ovary. A record of *H. aureum* from Deothang district (Narfong (M.F.B.)) is unconfirmed.

7. H. spicatum Smith. Nep: *ruksana.* Fig. 20e.

Leafy shoots 0.5–2m. Leaves lanceolate, acuminate, 10–45 × 3–10cm, pubescent beneath; ligule entire, 1–2cm, pubescent or not. Infl. narrowly cylindric, 15–30cm; bracts oblong, obtuse, convolute, 2–3cm × 1cm, each subtending a single flower; bracteole c.half length of bract and ± equalling calyx. Flower white tinged yellow or reddish at base; corolla tube 5–6cm; petals linear, 2–3cm. Lateral staminodes narrowly lanceolate, ± equalling petals; lip 3.5–4cm, shortly clawed with a ± orbicular, emarginate limb, or cuneate with limb more deeply divided into 2 ± acute lobes. Stamen shorter than lip, usually orange. Fruit yellow, orange inside, seeds red.

Bhutan: S — Phuntsholing (Phuntsholing), Chukka (Chimakothi to Chukka Colony), Sarbhang (2km below Dara Chu on Chirang Road) and Deothang (Narfong to Wamrung (M.F.B.)) districts; **C** — Thimphu (near Taktsang Monastery, Dotena), Punakha (1km above Wache; Lometsawa (M.F.B.)) and Tongsa (Tongsa bridge) districts; **Darjeeling** (S of Tiger Hill, Labha, Rungbee, Lopchu to Jorebungalow); **Sikkim** (Dhobijhora, Lachen, between Mintok and Paha Kholas). Broad-leaved (incl. *Quercus griffithii*) forest, sometimes epiphytic, 610–2630m. Fl. July–October.

Two varieties have been described: var. **acuminatum** (Roscoe) Wall. with a shorter, fewer-flowered infl. and consistently orange stigma, and var. **trilobum** Wall. with a small tooth between the lobes of the lip. Most Bhutanese collections belong to the former, but intermediates exist.

8. H. glaucum Roscoe; *H. gracile* auct. non Roxb.; *H. gracile* var. *glaucum* (Roscoe) Baker. Nep: *tembook.*

Leafy shoots c.1m. Leaves narrowly lanceolate, acuminate, base rounded, 10–25 × 4–10cm; petiole 1–2cm; ligule bilobed, 3–5mm, lobes rounded. Infl. narrowly cylindric, 10–20cm; bracts narrowly lanceolate, convolute, obtuse, 1.5–2cm, each subtending a single white flower; bracteole half length of bract. Calyx ± equalling bract or slightly shorter. Corolla tube only slightly exserted; petals linear, 2–3cm. Lateral staminodes narrowly lanceolate, acute, shorter than petals; lip narrowly oblong, bilobed to almost halfway, 1.5 × 0.5cm. Stamen red, 3cm.

Bhutan (unlocalised *Parkes* specimen); **Darjeeling** (Rungnoo Valley, Mongpu, Rongbee, Farseng, Samring, Tukdah, Kurseong); **Sikkim** (S bank of Rate Chu N of Gangtok). On rocks and wet cliff-ledges, 610–2130m. Fl. August.

9. H. gardnerianum Ker Gawler. Nep: *sarro*; Lep: *kongkoor.* Fig. 20f–h.

Leafy shoots 1–2m, robust. Leaves lanceolate, acuminate, 20–40 × 10–15cm; ligule 1.5–2.5cm, entire. Infl. broadly cylindric, 25–40cm; bracts narrowly cylindric, obtuse, 2–3cm, each subtending 1–2 very fragrant, lemon yellow flowers; bracteole half length of bract. Calyx ± equalling bract. Corolla tube 6–7cm; petals linear, 3.5–4cm, greenish. Lateral staminodes narrowly spathulate, 2.5–3cm; lip obovate, shortly bilobed or entire, 3 × 2cm. Stamen red, long exserted, 6–7cm.

Bhutan: S — Chukka district (2km N of Takhti Chu); **N** — Upper Mo Chu district (Tamji); **Darjeeling** (Rungbee, Sureil, Ghum); **Sikkim** (Yoksum to Bakkim, 6km W of Rabong La, Rate Chu N of Gangtok). Broad-leaved (incl. oak) forest, 910–2130m. Fl. July–September.

10. H. stenopetalum Loddiges.

Leafy shoots over 2m. Leaves linear lanceolate to oblong lanceolate, acuminate, sessile, 40–70 × 7–15cm, densely hairy beneath; ligule 1.5–2cm, pubescent; bracts markedly divaricate, oblong, convolute, 3.5–5cm, coriaceous, each subtending 3–5 white flowers. Calyx to 3.5cm, pubescent. Corolla tube 4–4.5cm; petals broadly linear, 3cm. Lateral staminodes linear lanceolate, 2cm; lip oblong-cuneate, bifid with acuminate lobes. Stamen 5–6cm.

Bhutan: S — Deothang district (6km N of Deothang on road to Narfong (M.F.B.) — requires confirmation); **Duars** (Muraghat Forest). Fl. September.

11. H. coccineum Smith; *H. angustifolium* Roxb.; *H. coccineum* var. *angustifolium* (Roxb.) Baker; *H. coccineum* var. *squarrosum* Wall. Lep: *tambok*.

Leafy shoots 1–1.5m. Leaves linear lanceolate, acuminate, sessile, 20–35 × 3–5cm, sometimes sparsely pubescent beneath, often glaucous; ligule entire, 1–2.5cm. Infl. broadly cylindric, 15–25 × 6–8cm; bracts convolute, 2.5–3.5cm, coriaceous, each subtending 2–3 brick-red to orange/salmon-coloured flowers. Calyx 2.5cm; corolla tube ± equalling calyx; petals linear, reflexed, 3cm. Lateral staminodes lanceolate, c.2.5cm; lip clawed, limb suborbicular, deeply bifid, 1.5–2cm wide. Stamen more than twice length of lip. Capsule globose, 2cm in diameter.

Bhutan: S — Gaylegphug (Rani Camp to Tama (M.F.B.)) and Deothang (Tsalari Chu, Keri Gompa) districts; **C** — Thimphu (Tsalimaphe), Punakha (Tinlegang), Tongsa (between Tongsa and Tashiling) and Mongar/Tashigang (Kolé La) districts; **Darjeeling** (Lopchu, Rungbee, Farseng, Darjeeling, Little Rangit Valley, Lebong, Rongsong, Pashok; Rayang (F.E.H.1)); **Sikkim** (Silak, Singtam, Yoksum). Warm broad-leaved forest; slopes in secondary scrub, 250–2130m. Fl. July–September.

A variable species, of which the following two species may be no more than colour forms.

12. H. aurantiacum Roscoe. Sha: *khui-see.*
Differs from *H. coccineum* in having bright orange-yellow flowers.
Bhutan: S — Chukka district (Marichong, Rydak valley); **C** — Punakha (Rinchu, Tinlegang) and Mongar (Mongar, Khoma) districts. Open hillsides; slope in deciduous oak forest, 1070–1980m. Fl. July–August.

13. H. elatum R. Brown ex Ker Gawler.
Differs from *H. coccineum* in having white, reddish-tinged flowers.
Darjeeling (Rungbee (Srivastava, 1985)); **Sikkim** (Srivastava, 1985). 1220–1880m.

Specimens from Punakha (Rinchu, 1524m) and Mongar (N of Lhuntse, 2850m) districts perhaps belong to this species, but the colour notes on the specimens are inadequate.

12. ALPINIA Roxb.

Leafy shoots many-leaved. Infl. terminal, usually a lax raceme or panicle. Flowers single or in cincinni, bracts sometimes absent, bracteoles present or not, sometimes tubular. Lateral staminodes non-petaloid, often tooth-like or reduced to swellings, sometimes absent. Lip large and showy or inconspicuous, variously lobed. Anther crested or not. Ovary trilocular with axile placentation. Capsule usually globose.

1. Bracteoles tubular, persistent (splitting with age but not usually to base); lip inconspicuous; fruit black .... **1. A. nigra**
+ Bracteoles open to base, often soon deciduous; lip large and showy; fruit red 2

2. Leaves to 2.5cm wide, glabrous beneath, margins with well-spaced bristles (sometimes reduced to bumps); filament at least twice length of anther .... **2. A. calcarata**
+ Leaves 7cm wide or more, pubescent beneath, margins hairy but lacking short bristles; filament ± equalling anther .... **3. A. malaccensis**

1. A. nigra (Gaertner) B.L. Burtt; *Zingiber nigrum* Gaertner; *Heritiera allughas* Retzius; *Alpinia allughas* (Retzius) Roscoe. Nep: *chulumpa*; Lep: *kongpala.* Fig. 20i–l.
Leafy shoots to 3m. Leaves narrowly lanceolate, acuminate, to 50 × 15cm, ± glabrous, sessile or very shortly petiolate. Infl. with some lateral branches in lower part, rather lax with remote cincinni; bracts 0.5–2cm, each subtending a cincinnus of 3–4 greenish-white flowers; bracteoles tubular. Lip pink, clawed,

to 2.5cm, limb cuneiform, 3-lobed, mid-lobe divided into two triangular lobes. Fruit black, 1–1.5cm in diameter.

Bhutan: S — Sankosh district (Balu Khola 7km W of Phipsoo); **Duars** (Soondree); **Terai** (Dulkajhar; Jalpaiguri (Sikdar & Samanta, 1984)); **Darjeeling** (Rhamti, Bamunpokri). Secondary scrub at margin of subtropical forest, 150–610m. Fl. July; fr. February–October.

The rhizome is used medicinally throughout Indo-Malaysia.

2. A. calcarata Roscoe.

Leafy shoots c.1m, slender. Leaves linear-lanceolate, to 35 × 2.5cm, glabrous except for bristle-like marginal hairs, sessile. Infl. to 10cm; bracts absent (or soon deciduous); bracteoles quickly deciduous, open, 1–1.5cm, usually each subtending a pair of flowers. Flowers greenish-white. Lip white, lined rose-purple, obovate, emarginate, 2–3 × 1.5–2cm. Fruit orange-red, 1–1.5cm in diameter.

Terai (Jalpaiguri (Sikdar & Samanta, 1984)).

Doubtfully native; widely cultivated throughout India.

3. A. malaccensis (Burman f.) Roscoe; *Maranta malaccensis* Burman f.

Leafy shoots to 3m, robust. Leaves lanceolate, acuminate, to 60 × 7cm, pubescent beneath; petioles 3–5cm. Infl. pubescent; bracts absent; bracteoles white, enfolding buds, soon deciduous as flowers open, 1.5–2cm, each subtending a pair of flowers. Flowers white and pink. Lip yellow orange, heavily lined with scarlet, emarginate, 3–5 × 3cm. Fruit red, 2.5cm diameter.

Bhutan: S — Gaylegphug district (Hatisar; Gaylegphug (M.F.B.)); **Terai** (Hattisuna; Jalpaiguri (Sikdar & Samanta, 1984)); **Darjeeling** (Katambari, Aloobari). 270–910m. Fl. April; fr. April–November.

Doubtfully native; cultivated for its large, showy flowers.

13. ETLINGERA Giseke

Leafy shoots many-leaved. Infl. borne separately from leaves, usually surrounded by at least some sterile bracts; fertile bracts each subtending a single flower and tubular bracteole. Corolla tube usually long and slender. Lip with an elongate central portion, lower part joined to filament and forming a conspicuous tube above insertion of petals. Lateral staminodes absent. Anther crested or not. Ovary trilocular with axile placentation. Fruit indehiscent.

1. E. linguiformis (Roxb.) R.M. Smith; *Alpinia linguiformis* Roxb.; *Amomum linguiforme* (Roxb.) Baker; *Hornstedtia linguiformis* (Roxb.) K. Schumann; *Achasma linguiforme* (Roxb.) Loesener. Lep: *kongola*.
    Leafy shoots to 2m. Leaves lanceolate-acuminate, 30–45 × 5–12cm, sub-sessile; ligule obtuse, 1cm. Infl. ellipsoid, lower half embedded in soil, c.7cm; peduncle to 14cm; bracts whitish-green, oblong, acute, to 5cm; bracteoles pink, tubular, ± equalling bracts. Calyx tridentate, unilaterally split, 8cm. Corolla tube red, shorter than calyx. Lip yellow-margined with some red in centre, 5–6cm with a bilobed, rhomboid limb. Filament to 1cm; anther slightly shorter, ecristate. Capsule ridged.
    Darjeeling (Rishap, Mongpu, Tista Valley). 150–910m. Fl. November.

14. AMOMUM Roxb.

Leafy shoots many-leaved. Infl. basal, ± cone-like with imbricating bracts. Flowers (in Bhutanese species) borne singly, each subtended by a bract and (always?) a bracteole. Corolla tube not exserted from calyx. Lateral staminodes forming small subulate teeth, occasionally absent. Lip usually showy, variously lobed. Anther usually with a well-developed crest. Ovary trilocular with axile placentation. Capsule round or ovate, sometimes winged or echinate.

The names *kakola* and *sukmel* (Med) refer to types of 'cardamom', but it is not known to which genus or species they refer.

1. Capsule winged or echinate 2
+ Capsule smooth (sometimes ± rugulose or ribbed) 3

2. Leaves glabrous beneath; bracts with prominent subulate tips; lip yellow; capsule echinate .... **1. A. subulatum**
+ Leaves pubescent beneath; bracts without prominent subulate tips; lip white, yellow in centre with radiating red veins; capsule 9-winged
2. A. dealbatum

3. Infl. globose; lip yellow tinged red in centre; capsule ovoid
3. A. aromaticum
+ Infl. oblong; lip white tinged yellow; capsule globose .... **4. A. kingii**

1. A. subulatum Roxb. Nep: *aleichi*; Eng: *large, greater, Indian or Nepal cardamom*. Fig. 20m–n.
    Leafy shoots to 2m. Leaves lanceolate, acuminate, sessile or very shortly petiolate, 30–60 × 6–10cm, glabrous; ligule emarginate, lanceolate acuminate.

Infl. subglobose to ovoid, 6–7 × 4–5cm; peduncle 5–7cm; bracts reddish-brown, ovate-obtuse with a 3–5mm subulate tip, 2.5–3 × 1.5–2cm, ± glabrous; bracteoles tubular, 3cm. Calyx pinkish-white, 4–4.5cm, lobes long subulate. Corolla tube white, shorter than calyx, petals yellow. Lateral staminodes absent? Lip yellow, oblong, acute, 3cm. Filament 0.5cm; anther 1cm, connective prolonged into an irregularly dentate crest. Capsule reddish, subglobose, 2–2.5cm, echinate.

Bhutan: S — Samchi (around Dorokha), Phuntsholing (Suntlakha), Sarbhang (Tschokana near Damphu) and Gaylegphug (Surey) districts; **Darjeeling** (Rithoo, Mongpu, below Roy Villa; Happy Valley (F.E.H.1)); **Sikkim** (common in W Sikkim e.g. below Yoksum). Commonly cultivated on steep slopes under wet forest (incl. *Alnus*), 700–2050m. Fl. April–November.

Much cultivated, but native to Bhutan, Nepal, Sikkim, Assam and W Bengal. The seeds are used for flavouring sweet dishes, cakes, pastries and curries, as a masticatory, and medicinally.

2. A. dealbatum Roxb. Nep: *chelumpa*; Lep: *tumbrup*.
Leafy shoots to 1.5m. Leaves lanceolate, acuminate, to 60 × 14cm, shortly and densely pubescent beneath, shortly petiolate or sessile; ligule 0.5–1cm. Infl. subglobose, c.6 × 5cm; peduncle to 5cm; bracts ovate to lanceolate, soon decaying; bracteoles not seen. Pedicels to 1.2cm. Calyx 2–2.5cm, 3-lobed. Corolla white, tube ± equalling calyx. Lateral staminodes 2mm. Lip white, yellow in centre with radiating red veins, oblong, entire or emarginate, 2.5cm. Filament 0.5mm; anther to 1.2cm, connective prolonged into an entire, truncate white crest. Fruit reddish, to 2.5cm, 9-winged.
Darjeeling (Punkabari, Sureil, Mongpu, Rongbe, Labdah). 910–1830m. Fl. May–August.

var. **sericeum** (Roxb.) Baker; *A. sericeum* Roxb.
Differs in having leaves densely silvery hairy beneath, long-petioled.
Terai (Dulkajhar).

3. A. aromaticum Roxb. Eng: *Jalpaiguri* or *Bengal cardamom*.
Leafy shoots to 1m or more. Leaves linear lanceolate, caudate-acuminate, 15–30 × 4–8cm, glabrous, subsessile; ligule entire, 0.3–0.5cm. Infl. globose, 3–4cm diameter, pubescent; peduncle subterranean, 3–6cm; bracts oblong, subapically mucronulate, 2–2.5cm; bracteoles tubular, 1.5cm. Calyx 1.5–2cm, 3-lobed. Corolla pale yellow, tube shorter than calyx. Lateral staminodes absent. Lip red tinged in centre, obovate-oblong, entire, 2cm. Filament c.4mm about equalling anther, connective prolonged into a 3-lobed crest. Fruit ovoid, c.2 × 1cm, somewhat rugose, pubescent.

Darjeeling (Rishap, Kurseong). 460–1220m. Fl. June.

Cultivated for the medicinal properties of the seeds; also used as a spice.

Records of *A. cornyostachyum* Wall. from Sikkim probably belong here.

4. A. kingii Baker.
Leaves lanceolate, 30 × 5–8cm. Infl. cylindric, 10–12cm; peduncle equalling spike; bracts ovate, 2.5–3cm; bracteoles? Calyx? Corolla white, tube 2.5cm. Lip tinged yellow, 1cm + wide. Stamen shorter than lip; anther crest obscurely 3-lobed. Fruit globose, 2.5cm diameter, smooth.

An imperfectly known species described from Sikkim (*King*). No material has been seen.

Family 225. COSTACEAE

R.M. Smith

Plants lacking aromatic oil cells. Leaves arranged spirally; sheaths tubular. Infl. dense, terminal on leafy stem or borne directly on rhizome. Bracts broad, imbricating, each subtending (in Asiatic species) a single large, showy flower and bracteole. Calyx 3-lobed. Lip conspicuous. Lateral staminodes absent. Filament broad and petal-like, anther thecae placed centrally, or at least well below apex. Ovary trilocular with axile placentation; external epigynous glands absent (septal nectaries present). Capsule dehiscent.

1. COSTUS L.

Description as for family.

1. Bracts soon disintegrating into fibres, not sharply pointed; calyx lobes obtuse .... **1. C. lacerus**
+ Bracts not disintegrating, tips sharply pointed; calyx lobes acute
2. C. speciosus

1. C. lacerus Gagnepain. Fig. 21a–c.
Leafy stems 1–3m. Leaf blade elliptic to obovate, acute or shortly acuminate, 20–45 × 6–15cm, sparsely to densely hairy beneath; petioles c.1cm, densely pubescent. Infl. terminal, ovoid, 4.5–8 × 4–7cm; bracts red, ovate, 2–5cm, chartaceous, glabrous to villous, apical portion soon disintegrating; bracteoles c.1–2cm, soon disintegrating. Calyx red, 2–3cm. Corolla tube white,

1.5–2cm. Lip white, tinged yellow in centre, broadly ovate when spread out, 5–7 × 3.5–9cm. Stamen white. Fruit ellipsoid, 1.5–3cm, woody.

Bhutan: S — Deothang district (6km N of Deothang); **C** — Mongar district (Shongar); **Darjeeling** (Labdah, Tista, Farseng, Kurseong, Rishap, above Peshok, Mongpu); **Sikkim** (Yoksum, 20km W of Singtam). Steep bank at margin of subtropical forest; edge of cardamom plantation; cultivated in gardens, 305–1400m. Fl. June–August.

Used medicinally in Sikkim.

2. C. speciosus (J. König) Smith; *Banksia speciosa* J. König; *C. speciosus* var. *nipalense* Roscoe. Nep: ?*sakupaani*. Fig. 21d.

Differs from *C. lacerus* in the non-disintegrating, sharp-tipped bracts, acute calyx lobes and pubescent corolla.

Bhutan: S — Phuntsholing district (Torsa River above Phuntsholing); **Terai** (Bamunpokri; Jalpaiguri (Sikdar & Samanta, 1984)); **Darjeeling** (Bakrikot, Silak, Jaldakar, Great Rangit Valley, Riang). In scrub, 150–610m. Fl. July–August.

The centre of distribution of *C. speciosus* probably lies in Malaysia. The species is widely cultivated throughout the tropics.

Both the above are very variable in the amount of indumentum.

Family 226. CANNACEAE

Perennial herbs, rhizomes sometimes tuberous. Stems usually unbranched. Leaves alternate, blades pinnately veined, petioles with sheathing bases. Infl. a spike or thyrse, sometimes branched. Flowers bisexual, irregular; sepals 3, small, persistent; petals 3, subequal, tubular below. Staminodes 4–6, petaloid in two whorls, fused below, outer whorl of 2–3, ± erect staminodes, inner whorl of 1 reflexed lip, 0–2 staminodes and one fertile stamen with a 1-celled

---

FIG. 21. **Costaceae, Cannaceae, Marantaceae.**
a–c, **Costus lacerus**: a, habit (× ¼); b, flower with half of lip removed and fertile stamen turned through 90° to show anther and style (× ½); c, bract (× ½). d, **C. speciosus**: bract (× ½). e–l, **Canna speciosa**: e, habit (× ⅛); f, rhizome (× ¼); g–l, flower (× 1): g, sepal; h, petal; i, outer staminodes; j, lip; k, anther; l, style. m–n, **Phrynium pubinerve**: m, habit (× ¼); n, prophyll (× ½). o, **P. placentarium**: prophyll (× ½). p, **Schumannianthus dichotomus**: habit (× ¼). Drawn by Glenn Rodrigues.

a

c

d

b

e

i-

i-

l

k

h-

-j

g-

f

m

n

o

p

G₂
94

anther borne on its edge. Style flattened, petaloid. Ovary inferior, papillose, of 3 locules with numerous axile ovules. Fruit a subglobose, 3-valved, bristly capsule.

Much more work is required on the taxonomy of the Bhutanese species and more collections needed. Additional species probably occur in cultivation.

1. CANNA L.

Description as for family.

1. Plant to 3m; leaves tinged purple; cultivated for its swollen, edible rhizome .... **3. C.** cf. **edulis**
+ Not as above 2

2. Outer staminodes obovate, over 3cm wide, brightly coloured (deep scarlet, yellow spotted orange, etc.) .... **4. C.** × **generalis**
+ Outer staminodes narrowly oblanceolate, under 1.5cm wide, dull orange/crimson 3

3. Outer staminodes 2; plant commonly over 1m .... **1. C. speciosa**
+ Outer staminodes 3; plant commonly under 1m .... **2. C. orientalis**

1. C. speciosa Sims. Plate 10. Fig. 21e–l.

Stem to 1.5m. Largest leaves broadly lanceolate, abruptly contracted to cuspidate apex, base cuneate, to 32 × 15cm, glaucous. Infl. c.30cm, with short, suberect branches; bracts erect, tightly appressed, to 12.5cm, whole infl. covered with conspicuous waxy bloom. Sepals narrowly oblong-lanceolate, subacute, to 1.5 × 0.3cm, membranous. Petals linear-lanceolate, very acute, to 3–4 × c.0.4cm; outer staminodes 2, erect, narrowly oblanceolate, bidentate, narrowed into claw-like base, 4.5–6 × 0.6–1cm, slightly unequal, dull crimson; lip recurved, narrowly oblong, 3.5–5 × 0.4cm, dull crimson streaked orange. Fertile stamen with recurved, petaloid apex; anther c.0.9cm. Free part of style narrowly oblong, c.1.7 × 0.2cm.

Bhutan: C — Punakha district (below Tinlegang); **Darjeeling** (Darjeeling); **Sikkim** (Jorethang). 500–1760m. Fl. June–August.

Cultivated occasionally in gardens for ornament.

2. C. orientalis Roscoe; *C. indica* s.l.

Stems commonly under 1m. Flowers differing from *C. speciosa* in having 3, slightly smaller (c.4 × 0.5cm) outer staminodes.

Terai (Kumergram). 90m.

Perhaps formerly grown as an ornamental but now seldom, if ever, cultivated in our area where it has been replaced by the more showy *C. × generalis*. Perhaps never grown in Bhutan, though there is a literature record for Deothang district (Chenari (M.F.B.)).

3. C. cf. edulis Ker Gawler. Nep: *rato phool*.
Rhizomes swollen, horizontal. Stems to 3m. Leaves oblong, to 60 × 30cm, tinged dark purple especially around margins and along veins.
   Bhutan: S — Chukka district (Surge Shaft Colony); **C** — Punakha (below Pandagong) and Mongar (Mongar) districts; **Darjeeling** (above Barnesbeg, Mongpu); **Sikkim** (S of Legship, Yoksum). 700–1970m.

Cultivated on a small scale in lower districts for its tuberous rhizomes which are eaten as a vegetable (though less frequently than formerly except by the poor) and to distil a chang-like liquor.

Identity not certain since no flowering material seen from Bhutan/Sikkim. The edible plant is normally assumed to be *C. edulis* (native of northern S America but widely cultivated) but there is some doubt over the diagnostic characters of that species, characterised by Kränzlin (1912) as having 3 outer staminodes. The type description and illustration show the third staminode to be reduced, whereas Roscoe (1828) and subsequent authors have stated it to have three equal staminodes.

Plants similar vegetatively to the Bhutan ones have been seen flowering in Nepal: of these most have only 2 staminodes, a single flower on one specimen having a reduced third staminode. Clearly work is needed on the constancy and diagnostic value of this character.

A green-leaved variety of the above was seen at Mongpu — perhaps the form with yellow flowers (Nep: *pialo phool*).

4. C. × generalis L.H. Bailey.
   Many varieties of this widely cultivated showy hybrid exist; they all have 3 widely (over 3cm) obovate, outer staminodes in various bright colours (dark scarlet, yellow with orange spots, etc.).
   Bhutan: Commonly cultivated in gardens to 2000m as in Chukka (Surge Shaft Colony), Thimphu (Thimphu) and Tongsa (Shemgang) districts; **Sikkim** (S of Legship, Yoksum).

Family 227. MARANTACEAE

Perennial, rhizomatous herbs or shrubs. Leaves usually distichous; petiole sheathing below, with jointed 'pulvinus' below attachment of leaf blade; blade with fine, pinnate venation. Infls. thyrsiform, with lateral, partial infls.

(cymules), sometimes very condensed, terminal or lateral, sometimes borne on petiole; flowers often bracteolate, often borne in pairs, subtended by a prophyll. Flowers zygomorphic, bisexual, sepals 3, free, corolla segments 3, tubular below. Staminal whorls 2, outer of 1–2, petaloid staminodes, fused below, inner of 2, petaloid staminodes (1 usually hooded) and 1 petaloid stamen bearing a single, lateral anther. Ovary inferior, of 3 locules, each (or 1 only) bearing a single, basal ovule; style fused with corolla tube, apical part hooked. Fruit 1–3-seeded, a loculicidal capsule, or indehiscent; seeds large, sometimes arillate.

1. Stemless herbs; infl. a condensed head borne laterally on petiole; leaf blade over 20cm .... **1. Phrynium**
+   Shrub; infl. terminal, a narrow, pedunculate raceme; leaf blade under 20cm .... **2. Schumannianthus**

1. PHRYNIUM Willdenow

Stemless herbs. Leaves large, long-petioled. Infl. a capitate head of condensed cymules borne laterally on petiole, or from rootstock. Flowers lacking bracteoles, in groups of 2–5 subtended by stiff prophylls. Fruit a 1–3-seeded capsule; seeds arillate.

1. Prophylls reddish-brown, broadly oblong, becoming fibrous-torn at apex; flowers pinkish or blue-veined .... **1. P. pubinerve**
+   Prophylls pale yellowish-brown, lanceolate, acute, not becoming torn; flowers pale orange .... **2. P. placentarium**

1. P. pubinerve Blume; *P. capitatum* sensu F.B.I.; *P. malaccense* Ridley. Nep: *kopat, kufyer, kawaipat.* Fig. 21m–n.

Rhizome thick bearing c.4-leaved rosettes. Leaf blade oblong-elliptic, very shortly cuspidate, 34–40 × 15–18cm, sides of midrib appressed-bristly beneath; infl.-bearing petiole c.130cm; pulvinus c.4cm. Infl. of c.3 narrow, condensed, spike-like cymules, longest 4–4.5cm. Prophylls all similar, broadly oblong, apex decaying becoming fibrous, reddish-brown, margins scarious, finely ribbed, very stiff, with scattered, appressed, whitish hairs on outside, lowest of each cymule 2.3–2.5 × 1.4–1.6cm. Flowers (not preserved in herbarium specimens — description from drawing on Clarke specimen): sepals linear, c.1cm, glabrous; corolla lobes oblong, blue-veined; petaloid staminodes ?3, outer 2 obovate (1 larger than other), 1 with retuse apex, pinkish-white; ovary glabrous. Capsule trigonous, oblong-obovoid, apex truncate, retuse, c.1.3 × 0.8cm.

Duars (Buxa); **Darjeeling** (Darjeeling, Mongpu, Rongbe, Great Rangit). Hot valleys, 150–760m. Fl. June.

Several taxa are possibly involved in the traditional usage of the illegitimate '*P. capitatum*' (e.g. F.B.I.); for instance, the illustration of *P. capitatum* in Wight (1853, t. 2016) and the description of Roxburgh differ from the Sikkim plant in having a hairy ovary. Characters such as hairiness of ovary and veins on the leaf underside are of unknown reliability and a revision of SE Asian species is needed.

According to Hooker (1854, 1: 131) *Phrynium* leaves are used in Lepcha 'umbrellas', enclosed between plaited bamboo. A specimen from Gaylegphug (300m) (*kouvepat*, *koupat* (Nep); used as elephant fodder) could belong to this or to the following.

2. P. placentarium (Loureiro) Merrill; *P. parviflorum* Roxb. Nep: *kufyer*, *kawaipat*. Fig. 21o.
Differs from *P. pubinerve* in having a denser, globose infl. of more than 4 cymules; cymule prophyll widely ovate, floral prophylls lanceolate, pale yellowish-brown, very acute, not becoming torn; flowers pale orange.
Bhutan: C — Mongar district (Saling); **Darjeeling** (Sukna, Rongsong, Churonti). Wet jungle, northern slope, 305–1070m. Fl. June–October.

2. SCHUMANNIANTHUS Gagnepain

Dichotomously branching shrubs. Infl. terminal, narrowly racemose; prophylls bearing 2 bracteolate flowers on a common peduncle. Sepals 3, small; corolla 3-lobed, tubular below; outer, petaloid staminodes 2. Fruit indehiscent, 2–3-seeded.

1. S. dichotomus (Roxb.) Gagnepain; *Clinogyne dichotoma* Salisbury. Bengali: *patipata*; Mog: *jao-ba*. Fig. 21p.
Shrub to 4.5m. Leaves narrowly elliptic, shortly acuminate, 8–15 × 3.5–7cm; petioles narrowly winged up to pulvinus; pulvinus c.0.5cm. Infl. to 13cm, with 3–7 prophylls; prophylls stiffly erect, narrowly lanceolate, acute, 4–4.5cm, pale brown. Flowers white; sepals narrowly triangular, 6 × 1.5mm; corolla lobes lanceolate, 3.5 × 0.6cm, tube short; outer, petaloid staminodes 2, oblanceolate, c.3.5cm. Ovary small, densely hairy.
Terai; **Darjeeling** (Darjeeling). Damp places.

Family 228. COMMELINACEAE

Commonly perennial, often semi-succulent herbs. Roots often fleshy, sometimes tuberous. Stems usually decumbent at base and rooting from lower nodes, upper parts erect or occasionally scandent, usually with line of hairs decurrent from fused margins of leaf sheaths. Leaves commonly lanceolate, narrowed or not into petiole-like base; sheaths tubular, fused margins usually shortly hairy, mouth often with long, jointed hairs. Infls. terminal and/or lateral, of cymose branches aggregated into thyrses, or simple when often subtended by spathe-like bracts. Flowers bisexual, or bisexual and male; sepals 3, persistent; petals usually 3, ephemeral, 2 similar and one smaller or absent (flowers zygomorphic) or 3 subequal (actinomorphic). Stamens (5–)6, all bearing fertile anthers, or variously modified, e.g. 3 with anthers (median sometimes distinct from laterals) and 2–3 'antherodes'; filaments glabrous or hairy, sometimes of 2 lengths. Ovary superior, 2–3-loculed, ovules axile 1, 2 or more per locule; style filiform, stigma often indistinct. Fruit a usually loculicidal capsule, 2- or 3-loculed (one locule sometimes indehiscent) sometimes whole capsule indehiscent. Seeds usually prismatic, often compressed dorsiventrally, bearing a circular boss-like embryotega apically or laterally; hilum linear or punctate; sometimes arillate.

As a family probably under-collected in Bhutan. Collections of pickled flowers (or photographs or field notes/drawings) are required to augment the following necessarily inadequate descriptions of the ephemeral flowers that have been based solely on herbarium material.

1. Climbing/scrambling plant; leaves deeply ovate-cordate on long, slender petioles .... **1. Streptolirion**
+   Erect to decumbent herbs; leaves never deeply ovate-cordate, petioles · if present short and stout 2

2. Infl. of dense, many-flowered clusters emerging through holes at base of leaf sheaths (Fig. 22d); leaf blades large, elliptic-acuminate
    4. Amischotolype

Fig. 22. **Commelinaceae** I.
a, **Streptolirion volubile**: portion of stem showing flowering shoot puncturing base of leaf sheath ($\times$ ½). b, **Floscopa scandens**: portion of flowering shoot ($\times$ ½). c, **Murdannia divergens**: portion of flowering shoot ($\times$ ½). d, **Commelina paludosa**: portion of flowering shoot ($\times$ ½). e–h, **C. sikkimensis**: e, spathe and flowers ($\times$ ½); f, schematic l.s. through dehiscent locules ($\times$ 5); g, schematic t.s. through capsule, indehiscent locule below ($\times$ 5); h, seed ($\times$ 7.5). i–j, **C. diffusa**: i, seed showing lateral embryotega ($\times$ 7.5); j, seed showing linear hilum ($\times$ 7.5). k, **C. appendiculata**: appendaged seed ($\times$ 6). Drawn by Glenn Rodrigues.

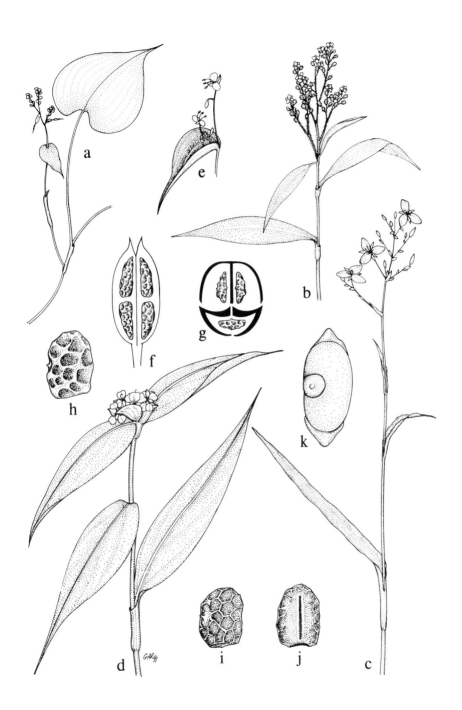

+ Infl. not as above, if borne within leaf sheaths then flowers emerging through sheath mouth and leaves linear-lanceolate 3

3. Some or all filaments hairy (except *Murdannia spirata*) 4
+ All filaments glabrous 7

4. Stamens 2–3; staminodes 3–4 bearing small, 3-lobed, bright yellow, sterile antherodes .... **7. Murdannia**
+ Stamens 6, all bearing fertile anthers 5

5. Flowers large (over 2.5cm diameter), usually violet; introduced
   5. Tradescantia
+ Flowers under 1cm diameter, blue or white; native 6

6. Flowers ± sessile, lower part of flowers (except for lobes) hidden by closely imbricating, sickle-shaped bracteoles or within leaf sheaths; petals tubular below .... **2. Cyanotis**
+ Flowers shortly pedicelled, held free from minute bracteoles; petals free .... **3. Belosynapsis**

7. Infls. of 1–2 short branches ± enclosed in spathes ...... **11. Commelina**
+ Infls. of many branches, not enclosed in spathes 8

8. Ovary and fruit densely covered by hooked hairs ...... **8. Rhopalephora**
+ Ovary and fruit glabrous 9

9. Infl. branches densely hairy; flowers pinkish-mauve, minute (under 5mm diameter) .... **6. Floscopa**
+ Infl. branches glabrous; flowers blue or white, larger 10

10. Fruits dehiscent, brown; seeds uniseriate; petals blue .. **9. Tricarpelema**
+ Fruits indehiscent, metallic-blue; seeds biseriate; petals usually white
   10. Pollia

1. STREPTOLIRION Edgeworth

Scrambling herb. Leaves long-petiolate; blade ovate, deeply cordate, caudate-acuminate. Infl. long-peduncled, a thyrse, arising from short branch puncturing base of leaf sheath. Most flowers bisexual, some male only; petals linear, free, all similar. Stamens 6; filaments with clumps of jointed hairs in upper part; anther connective thickened laterally, locules reniform, dehiscing longitudinally. Ovary 3-loculed, each locule with 2, superposed ovules; style filiform; stigma capitate. Fruit a 3-valved, loculicidal capsule. Seeds prismatic, coarsely rugose, hilum linear, boss lateral.

1. S. volubile Edgeworth. Fig. 22a.

Stems to 2m. Upper leaf blades 7–11.5 × 5.7–9cm, caudate apex to 2.5cm, margins densely ciliate, upper surface glabrous or with scattered, short bristles, undersurface glabrous; lower leaf blades to 18 × 15cm, sometimes with dark purplish markings; sheaths glabrous. Peduncle about equalling opposing leaf, subtended by tubular, membranous bract; panicle 2–5cm, narrowly pyramidal, subtended by leaf-like bract, thyrse branches shortly bristly to pilose, subtended by pinkish or whitish bracts. Sepals pinkish or greenish-white, lanceolate, acute, hooded, 6–7 × 1.5–2.3mm, sparsely pilose on outside; petals linear, 0.4–0.7mm wide. Filaments c.7mm, hairs yellow; anther locules 0.4–0.9mm. Ovary narrowly ellipsoid, gradually tapered into style, c.5 × 2mm, sparsely pilose; style c.3mm. Capsule ellipsoid-trigonous, acuminate, 6–11 × 2–5mm. Seeds c.3 × 2mm, reddish-brown.

Bhutan: S — Phuntsholing (S of escarpment W of Gedu), Chukka (W of Gedu; Chasilakha (M.F.B.)), Sarbhang (Loring Falls) and Deothang (Deothang to Narfong (M.F.B.)) districts; **C** — Thimphu (opposite Thimphu Dzong; Paro (M.F.B.)), Punakha (15km S of Wangdi Phodrang), Tongsa (Tongsa, between Tashiling and Chendebi), Mongar/Tashigang (Kori La) and Tashigang (Tashi Yangtsi) districts; **Darjeeling** (common around Darjeeling); **Sikkim** (Yoksum to Mintok Khola, Lachung). On rocks and bushes in damp, shady broad-leaved (especially oak) forest, 1000–2740m. Fl. June–October.

The Tashiling specimen can be referred to var. *khasiana* C.B. Clarke, having sheaths with dense, deflexed brownish hairs and hairy leaves. Hong (1974) raised this to subspecific rank, which may be justified if the cytological difference reported by Rao *et al.* (1968) is constant. This variety seems to have an eastern distribution (with specimens also seen from Yunnan and N Burma) but seems to be sympatric with the typical variety: further work is necessary on ecology and distribution.

2. CYANOTIS D. Don

Perennial or annual, succulent herbs, sometimes rooting and decumbent below, bases sometimes bulbous. Stems with line of hairs decurrent from fused margins of sheath, branches arising from within sheaths. Leaves linear to lanceolate, not narrowed at junction with sheath. Infl. usually of terminal and lateral scorpioid cymes (cincinni), with closely imbricating, sickle-shaped bracteoles inserted in two rows and subtended by leaf-like spathe bract; sometimes flowers all in axillary clusters. Flowers ± sessile, hidden (except for corolla lobes) by bracteoles, bisexual; sepals fused or not below; corolla tubular below, with 3 free lobes. Stamens 6, all similar; filaments bearing clump of jointed hairs near apex; anthers dehiscing by basal pores. Ovary 3-loculed, each locule usually with 2 superposed, axile ovules; style filiform widened below the acute apex, sometimes bearing hairs. Fruit a 3-valved, loculicidal capsule. Seeds with apical boss and punctate hilum.

The genus is clearly in need of revision in SE Asia and names at present must be regarded as provisional.

1. Flowers all in axillary clusters, corolla only exserted from leaf sheaths
    4. C. axillaris
+   Flowers in terminal and axillary scorpioid cymes, lateral ones usually pedunculate 2

2. Leaves oblong-lanceolate (over 1cm wide), subacute, leaf margins densely and shortly ciliate .... **3. C. cristata**
+   Stem leaves linear-lanceolate (under 1cm wide), very acute, leaf margins glabrous or with few, long cilia 3

3. Silky arachnoid pubescence absent; anthers c.1mm .... **1. C. vaga**
+   Silky arachnoid pubescence present on infl. and usually also on petiole sheaths and leaves; anthers c.0.6mm .... **2. C. cf. fasciculata**

1. C. vaga (Loureiro) J.A. & J.H. Schultes; *C. barbata* D. Don. Fig. 23a–b.
  Bulbous (?always) perennial. Stems decumbent, much branched, 6–65cm. Basal leaf in bulbous forms long (to 40cm); stem leaves lanceolate or linear-lanceolate, tapering gradually from base to acute apex, 2–8.5 × 0.4–0.7cm, decreasing in size upwards, with scattered, long hairs on both surfaces; sheath 0.5–1.3cm, sparsely long-hairy, mouth and fused margins densely long-whitish-ciliate. Cymes scorpioid, terminal (single or paired) and axillary (single or in groups), subtended by leaf-like spathes; bracteoles overlapping, sickle-shaped, margins with long, white hairs, 6.5–8.5 × 2.3–4mm, often purplish. Sepals fused at extreme base, oblanceolate, acute, 4.2–5 × 1–1.5mm, pale brown, membranous, with long white hairs near apex and on keel. Corolla pale blue, lower half tubular, 6.5–8mm, lobes oblong to obovate, c.2.5mm wide. Filaments long-exserted from corolla, hairs bluish (sometimes white); anthers oblong, emarginate, c.1 × 0.5mm, yellow. Ovary c.1.5 × 0.7mm, apex truncate with long, erect, white hairs; style exserted, filiform, 5.5–7.5mm. Capsule 3-lobed, truncate, 2.5–3 × 1.5–2mm; seeds c.1 × 1.2mm, rugose.

Bhutan: S — Chukka district (above Lobnakha, below Chimakothi); C — Thimphu, Punakha, Tongsa, Bumthang, Mongar and Tashigang districts; N — Upper Mo Chu (Gasa, Kencho) and Upper Kulong Chu (Tobrang) districts; **Darjeeling** (Darjeeling, Kurseong, Damsong Forest, Sureil); **Sikkim** (Chungtam, Lachen, Lachung, Yoksum to Bakkim); **Chumbi**. Very common on rocks, pasture, banks and slopes, in open or in scrub or pine forest, 910–3100m. Fl./fr. June–October.

Parker (1992) records this as a minor weed of dryland crops (e.g. potato) mainly over 2000m in Chukka, Thimphu, Punakha, Tongsa, Bumthang, Mongar and Tashigang districts.

Extremely variable, depending on habitat: robust forms (30–65cm, with leaves to 15cm) described as *C. nobilis* Hasskarl are said (F.B.I.) to be common in Khasia and a single specimen has been seen from 'Sikkim' (no locality). It appears very different but Clarke, who knew it in the field, 'had little doubt . . . that *C. nobilis* is only an extreme form of *C. barbata*' (1874, p. 57); it is perhaps a polyploid form. A form with stout, erect stems intermediate between var. *nobilis* and typical plants was seen in a marsh at Gyetsa (Bumthang).

Neotypification of Loureiro's and Don's names is needed. The names are probably not correctly applied at present. Our taxon to which the two names have been commonly applied and regarded as synonymous appears to be Himalayan and SW Chinese, whereas *C. vaga* was described from Canton (no type known). No type is known for *C. barbata*, which should be at BM. It was described from Nepalese material of Wallich, and the description could refer to the following taxon, which Wallich (Cat. No. 8990D) collected in Nepal in 1821.

2. C. cf. fasciculata (Heyne ex Roth) Roemer & Schultes.
Differs from *C. vaga* especially in its silky-arachnoid pubescence present at least on the infl. and usually on petiole sheaths and leaf undersurfaces; petiole sheaths ± inflated, papery when dry; leaves linear; flowers small (corollas c.5mm), sometimes pink; anthers usually small (c.0.6mm).
Bhutan: C — Thimphu (hill above Thimphu Hospital; Paro (M.F.B.)) and Tashigang (Dangma Chu Valley) districts; **Darjeeling** (Lingdam, Selim, Darjeeling, Sittong); **Sikkim** (Lower Burtuk Basti, Gangtok). Dry sand, usually at lower altitudes than *C. vaga*, 610–2500m. Fl./fr. August–October.

The smaller forms of this plant first collected by Wallich in Nepal (*Wallich* 8990D), which resemble the Darjeeling and Sikkim specimens, have never been identified satisfactorily: Clarke (1881) treated them as an alpine, arachnoid form of *C. barbata*; Rao (1964a) also placed them under *C. vaga* but identified some of the E specimens as *C. arachnoidea* C.B. Clarke. The well-developed Bhutan specimen clearly shows that a second taxon is involved, which appears to be closest to *C. fasciculata*. There seem very few diagnostic characters and Clarke seems to have separated them at least largely on geographical grounds.

3. C. cristata (L.) D. Don.

Occurring at lower altitudes than *C. vaga*, from which it differs in being a much more robust, fleshy annual; stem leaves wider; spathe bract oblong-lanceolate (4–7.5 × 1–1.6cm), wider, less acute, margins densely ciliate; cymes conspicuously elongating, finally prominently recurved and exserted from spathe bract; bracteoles glabrous.

Bhutan: S — Gaylegphug district (2km W of Bhur); **C** — Punakha district (Chuzomsa; Punakha (M.F.B.)); **Terai** (Bamunpokri); **Darjeeling** (Farsing, Mongpu, Ryang, Darjeeling, Kalimpong, Badamtam, Tista, Punkabari); **Sikkim** (1km W of Singtam). Damp, mossy rocks; bases of rocks in seasonally burnt scrub, 150–1830m. Fl./fr. May–September.

4. C. axillaris (L.) Sweet; *Amischophacelus axillaris* (L.) R.S. Rao & Kammathy. Fig. 23c.

Ascending part of stems 6–40cm. Leaves linear-lanceolate, narrowing from base to acute apex, narrow cartilaginous border sometimes with minute, forward-pointing bristles, 4–8.5 × 0.2–0.9cm, about equal in length throughout stem, glabrous; sheaths concealing sepals and capsules, 0.4–0.7cm, ± inflated, papery when dry, mouth long-ciliate, fused margins shortly hairy. Infls. all axillary, 1–6-flowered; pedicels very short. Sepals linear-lanceolate, acute, ciliate, c.7 × 1.4mm, membranous with green midrib. Corolla blue, tube narrow, to 1cm, lobes c.4mm. Filament hairs white; anthers elliptic, c.0.8 × 0.6mm. Capsule oblong-trigonous, apiculate, apiculus of each lobe bifid, ciliate, c.7.5 × 2.2mm, shortly stipitate. Seeds oblong-prismatic, c.2 × 1.5mm, coarsely pitted, dark brown, shining.

Sikkim/Darjeeling: ('Lower Hills, Sikkim'). Presumably in marshes, ditches, etc.

3. BELOSYNAPSIS Hasskarl

Differs from *Cyanotis* in having very lax, few-flowered lateral and terminal cymes, flowers distinctly pedicelled, held free of minute, lanceolate bracteoles, petals free and anthers dehiscing longitudinally.

No specimens from our area seen; the following description is based on SE Asian material.

1. B. ciliata (Blume) R.S. Rao.

Stems 7–17cm, much branched, with densely hairy line and scattered long white hairs. Stem leaves oblong-elliptic, subacute, margins densely appressed-hairy, not narrowed at base, 1.5–3.5 × 1–1.5cm, upper surface with scattered, long, white hairs, subglabrous beneath; sheaths funnel-shaped, to c.0.5cm, mouths long-ciliate. Basal leaves longer (to 4cm), narrower, less hairy. Bracteoles sepal-like, margins densely appressed-hairy, to 8 × 1.5mm, hairy near apex. Sepals linear-lanceolate, acute, c.5mm, long-white hairy. Petals all similar, pale blue, spathulate, c.6 × 3mm. Stamens 6; filaments long, coiled, with clump of white, jointed hairs below anther; anthers c.0.9mm. Ovary c.0.7 × 0.5mm; style filiform. Capsule oblong-trigonous, truncate, with apical tuft of hairs, each locule 2-seeded. Seeds ± cuboid, boss apical, c.1 × 1mm, shallowly reticulate-pitted, pale brown.

Bhutan: S — Phuntsholing district (Kamji (Rao & Deori, 1980)). [On rocks or trees in broad-leaved forest], 1000m.

This record was first reported (M.F.B.) as the S Indian *B. vivipara* (Dalzell) Sprague.

4. AMISCHOTOLYPE Hasskarl

Herbs; stems stout, rooting below. Leaves usually elliptic, acuminate, narrowed to petiole-like base; sheaths persistent, almost concealing stem. Flowers ± sessile, subtended by small, membranous bracteoles, borne in dense head-like clusters puncturing base of sheaths. Sepals accrescent in fruit; petals all similar, free. Stamens 6, filaments bearing a cluster of jointed hairs just below anther; anthers dehiscing longitudinally or by terminal pores. Ovary 3-loculed, each bearing 1 or 2 ovules; style filiform, long; stigma indistinct. Fruit a 3-valved, loculicidal capsule; seeds 4–6, covered by fleshy red aril (always?), hilum linear.

1. Capsule exceeding sepals, ellipsoid-trigonous, acute, scattered-hairy all over; sepals glabrous .... **1. A. hookeri**
+   Capsule shorter than sepals, globose, mucronate, densely long-hairy near apex; sepals hairy near apex .... **2. A. glabrata**

1. A. hookeri (Hasskarl) Hara; *Forrestia hookeri* Hasskarl. Fig. 23d–e.

Stem to 1.5m. Leaf blade elliptic, finely acuminate to caudate, margins with densely appressed, silky hairs, 14–31 (incl. tip) × 5.2–8mm, undersurface with short, appressed hairs, especially on veins, glabrous above, narrowed to petiole-like base to 3cm; sheaths widely cylindric, surface shortly hairy, veins (especially central one) with long, appressed hairs, mouth with long, slender cilia. Infl. to c.15-flowered, bracteoles small, ovate, membranous, ciliate. Sepals oblong, hooded, 4.5–7.5 × 2.5–5mm, keeled, glabrous; petals white or pink, oblong, 4.5–9 × 2–2.4mm. Filaments c.8mm; anthers c.2.5mm. Ovary narrowly ellipsoid, c.5 × 2mm, sparsely hairy; style c.10mm, slender. Capsule narrowly ellipsoid-trigonous, 10–13 × 5.5–6.7mm, apex acute, with long, scattered hairs; two locules normally 2-seeded, third locule 1- or 2-seeded. Paired seeds D-shaped, c.3.5mm, drying rugose, single seed c.5.5mm (colour and texture not known).

Bhutan: S — Gaylegphug district (Gaylegphug (M.F.B.)); **Darjeeling** (Balasun, Dulkajhar, Rishap, Garidoora, Mongpu, Selim, Tista, Darjeeling, Rungbee). Wet places in forest; apparently sometimes epiphytic, 120–610(–1220)m. Fl./fr. May–October.

2. A. glabrata Hasskarl; *Forrestia glabrata* (Hasskarl) Hasskarl. Fig. 23f.

Differs from *A. hookeri* in having veins of leaf undersurfaces glabrous; sepals hairy near apex; anthers smaller (c.0.9mm); capsules smaller than sepals, ± globose, mucronate with persistent style base, densely hairy near apex; seeds red arillate.

Darjeeling (Darjeeling, Pomong, Rishap, Rongbe, Rongsong); **Sikkim** (below Gesing). 610–910m. Fl./fr. July–October.

Records of *A. mollissima* (Blume) Hasskarl var. *hispida* (Lessing & A. Richard) Backer from Gaylegphug (Gaylegphug, 270m) and Deothang (Deothang to Narfong, 1400m) districts (M.F.B.) almost certainly refer to this taxon, but further work is required on the genus in SE Asia before deciding on the distinctness and appropriate ranks and names for *A. glabrata*, *A. mollissima* and *A. hispida*.

5. TRADESCANTIA L.

T. virginiana L. (or one of its many hybrid derivatives) has been recorded from Darjeeling (2000m). It is a tall plant (to 60cm) with long, linear leaves (15–35 × 0.5–2.5cm) and has large (2.5–3.5cm diameter), violet (occasionally white), radially symmetric flowers, and 6 similar stamens with hairy filaments, arranged in a condensed pair of terminal cymes subtended by 2 long, leaf-like bracts.

Widely cultivated in temperate gardens, native to N America.

Other species of *Tradescantia* are sometimes cultivated in gardens or pots for ornament:

T. pallida (Rose) Hunt '**Purpurea**'.

Shoots ascending from persistent bases; leaves narrowly oblong, dark purple. (Darjeeling, Yoksum, near Jorethang.)

T. zebrina Bosse.
Shoots trailing; leaves striped, elliptic. (Phuntsholing.)

6. FLOSCOPA Loureiro

Decumbent to erect herb, rooting from lower nodes. Stem leafy throughout. Leaves lanceolate, narrowed to short petiole-like base; sheaths tubular. Infl. a dense, terminal thyrse. Petals free, lowermost narrower than others. Stamens: upper 2–3 different from lower 3, filaments glabrous, fused below; 3 lower anthers, orbicular, locules curved, dehiscing laterally, filaments free. Ovary 2-loculed, each cell with a single, axile ovule; style filiform; stigma indistinct. Fruit a 2-valved, loculicidal capsule, 2-seeded. Seeds with linear hilum and lateral boss.

1. F. scandens Loureiro. Fig. 22b.
Stems 39–109cm, sparsely hairy near base, more densely so above. Leaves lanceolate, acuminate, margins shortly hispid-ciliate, gradually narrowed to base, 5.5–13 × 1.5–3.5cm, shortly (densely to sparsely) hispid above, glabrous beneath; sheaths widely cylindric, 1–2cm, mouth long-ciliate, fused margins ciliate. Infl. ± rhomboid, dense, epedunculate, 4–12.5 × 3–6cm, arising from leaf-like bract, sometimes with subsidiary racemes from upper axils, branches ascending, becoming flexuous; bracteoles of branches and flowers inconspicuous, membranous; pedicels short, erect; axis, branches and pedicels densely covered with jointed, wavy hairs. Flowers pinkish to purplish; sepals lanceolate, blunt, c.3 × 1.2–1.5mm, hairy on outside; petals lanceolate, blunt, c.3.6 × 0.9mm. Filaments c.3.5mm; anthers c.0.5 × 0.5mm. Ovary stipitate, stipe about equalling orbicular body (c.0.5 × 0.5mm); style 2.6–6mm. Capsule oblong, apiculate, ± plano-convex, with vertical groove on each face, 2.3–3.3 × 2.2–2.5mm, straw-coloured, shining. Seeds elliptic in outline, curved face with grooves radiating from central boss, c.1.7 × 1mm, glaucous.
Bhutan: S — Samchi (Changuna (M.F.B.)), Chukka (Marichong) and Gaylegphug (3km W of Kalikola; Gaylegphug (M.F.B.)) districts; **C** — Tongsa district (near Langtel); **Terai** (Kynanooka, Chalsa Duars); **Darjeeling** (Darjeeling, Jaldakar, Peshok Road, Tista, Kalimpong, Kalighora); **Sikkim** (Gangtok, Mintagong, Chakung). Wet places in broad-leaved (e.g. *Alnus*) forest; open *Carex*-rich marsh, 270–1520m. Fl./fr. August–February.

7. MURDANNIA Royle

Perennials or annuals. Roots tuberous or not. Leaves lanceolate with narrow cartilaginous borders, borne in a basal rosette and/or along stem, not narrowed into a petiole. Infl. a thyrse, a terminal, several-flowered cincinnus or fascicles of 1-flowered cymes in leaf or bract axils, usually borne on a leafy stem, occasionally on a leafless scape; bracteoles persistent or caducous, tubular or open. Petals free, all similar. Stamens 2–3, anthers dehiscing laterally, filaments hairy or glabrous; staminodes 3–4, antherodes commonly 3-lobed. Ovary with 3 locules, ovules axile, 1–many per cell. Fruit a 3-valved capsule. Seeds prismatic, variously rugose, hilum punctate or linear, boss lateral to dorsal.

1. Flower stems erect, stout, not branched near base; roots tuberous 2
+   Flower stems weak, decumbent at base, usually branched near base; roots not tuberous 4

2. Flowering stem bearing only bladeless sheaths .... **3. M. edulis**
+   Flowering stems leafy 3

3. Stem leaves lanceolate (over 3cm wide), bases usually clasping; lateral infls. usually present; petals under 5mm .... **1. M. japonica**
+   Stem leaves linear-lanceolate (under 1.6cm wide), bases not clasping; lateral infls. usually absent; petals over 9mm .... **2. M. divergens**

4. Bracteoles falling early, leaving scars; terminal infl. a simple (often capitate) cyme; stem leaves relatively long (usually over 3.5cm) and narrow; stamens 2 .... **4. M. nudiflora**
+   Bracteoles minute but persistent; terminal infl. with several, filiform cymose branches; stem leaves relatively short (usually under 3.5cm) and broad; stamens 3 .... **5. M. spirata**

1. M. japonica (Thunberg) Faden; *M. elata* (Vahl) Bruckner; *Aneilema lineolatum* (Blume) Kunth; *A. herbaceum* (Roxb.) Wall. ex Kunth. Nep (Darjeeling): *shinganijhar*; Nep (Bhutan): *garey malajhar*.

Perennial; some roots tuberous. Stem arising laterally from basal rosette, 48–90cm, glabrous. Stem leaves 3–6, oblong-lanceolate, tapered to acute apex

---

FIG. 23. **Commelinaceae** II.

a–b, **Cyanotis vaga**: a, habit (× ½); b, seed showing terminal embryotega and puncti-form hilum (× 15). c, **C. axillaris**: portion of middle section of stem showing axillary infl. (× 1). d–e, **Amischotolype hookeri**: d, habit (× ½); e, capsule with persistent calyx (× 2.5). f, **A. glabrata**: capsule with persistent calyx (× 2.5). g–h, **Tricarpelema gigantea**: g, capsule (× 2); h, seed (× 7.5). i–k, **Pollia hasskarlii**: i, fruit (× 3); j, seed from above (× 10); k, seed from side (× 10). Drawn by Glenn Rodrigues.

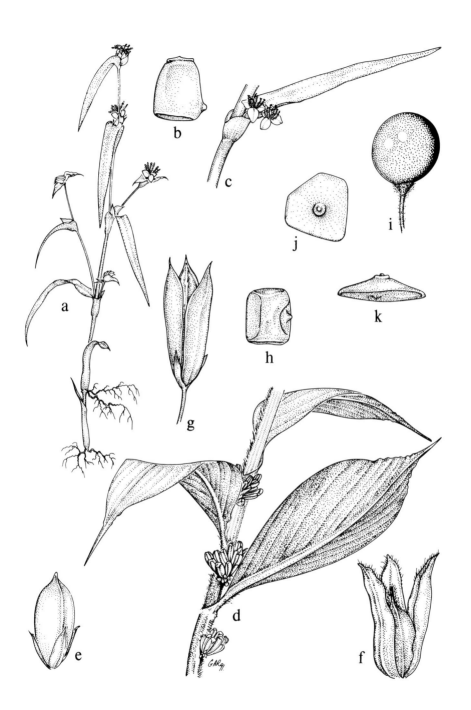

227

or shortly acuminate, base usually with clasping lobes, 3–4.5cm wide, decreasing in size upwards, glabrous; sheath narrowly cylindric, 1–1.5cm, glabrous or shortly hairy, mouth ciliate. Basal leaves narrower (2.3–2.7cm wide). Infl. a lax, terminal thyrse, usually with 1–5 subsidiary, lateral partial infls.; bracts of branches and flowers persistent, tubular, mouths widely oblique; branches ascending, bearing several, evenly spaced flowers on stiff pedicels 0.7–1cm. Sepals oblong-lanceolate, slightly hooded, 4.5–5.5 × 2.7–3mm, brownish-membranous; petals white, obovate, c.4.5 × 2.3mm. Fertile stamens: filaments c.3mm, densely white-hairy; anthers oblong-elliptic, 1.2–1.7 × 0.6–1mm. Antherodes bright yellow, 0.7 × 0.8mm, lobes sharply deflexed. Ovary ellipsoid, tapered upwards into style, 1.2–1.5 × 0.7–1.2mm; style persistent, hooked near apex, 2–4mm, stout; stigma papillose. Capsule trigonous, 4.5–6 × 4.5–6mm, segments ± circular, apiculate, pale greenish-brown, shining; seeds prismatic, boss small, on opposite face to punctate hilum, c.1.5mm, glaucous-papillose.

Bhutan: S — Gaylegphug district (between Gaylegphug and Tori Bari); unlocalised Griffith specimen; **Terai** (Mahanadi, Sivoke); **Darjeeling** (Darjeeling, Kurseong, Tista, Punkabari, Panchkilla, Bamunpokri, Selim Hill, Lish Block, Kalimpong); **Sikkim** (Namli, cultivated in garden). Among herbs and shrubs by roadside; wet places, 150–950m. Fl./fr. May–September.

2. M. divergens (C.B. Clarke) Bruckner; *Aneilema divergens* C.B. Clarke. Fig. 22c.

Differs from *M. japonica* in having all roots tuberous. Basal rosette lacking. Stem leaves narrower, linear-lanceolate, gradually tapered to subacute apex, base abruptly rounded, not clasping, longest 11–18.5 × 1–1.6cm about ⅓ way up stem, decreasing in size above and below, minutely red-spotted; sheaths long whitish-ciliate along mouth and fused margins. Infl. a single, terminal panicle (3–8 × 3–5cm); bracteoles clasping at base but not tubular, membranous, red-spotted. Sepals purplish-membranous; petals pinkish-mauve, larger (9.5–10 × 5.5–6.5mm). Filament hairs tinged purplish; anthers narrowly oblong (0.3–0.5mm wide). Style not hooked. Capsule segments narrowly oblong-elliptic (c.8 × 4mm); seeds with elongate hilum and larger, asymmetrically placed boss.

Bhutan: C — Punakha (Samtengang, Neptengka, Ritang, Wangdi Phodrang, Chuzomsa), Mongar (Ngasamp, Lhuntse) and Tashigang (Tashi Yangtsi) districts. Dry turf and among bushes in open, *Pinus roxburghii* forest; well-drained grassy slopes, 1200–2510m. Fl./fr. July–September.

3. M. edulis (Stokes) Faden; *Aneilema scapiflorum* (Roxb.) Kostelesky.

Leaves all in basal rosette, usually not present at flowering; coriaceous, reddish-brown remains of previous year's leaf bases persisting on corm-like base. Infl. similar (though narrower) to *M. divergens*, differing greatly in being borne on a leafless scape bearing only few, evenly spaced tubular, membranous, bladeless sheaths.

Bhutan (unlocalised record in F.B.I.); **Terai** (Singbi Ghora, Salgara). Sal forest. Fl. March–April.

4. M. nudiflora (L.) Brenan; *Aneilema nudiflorum* (L.) Sweet.

Slender annual, much branched from base, lacking basal rosette; roots not tuberous. Stems leafy throughout, 10–35(–47)cm, hairy on one side. Upper leaves linear-lanceolate (sometimes narrowly elliptic), tapered to acute or apiculate apex, base abruptly rounded, not clasping, 2–6 × 0.4–1cm, glabrous; lowermost leaves to 11.5cm. Sheaths 0.5–0.8cm, sometimes partly open, mouth and margins (fused and open part) white-ciliate. Infl. a simple, slender-peduncled, often capitate, 4–10(–20)-flowered terminal cyme (1–2.5 × 1.5cm), usually also with simple, lateral infls. from upper axils; pedicels 0.3–0.4cm; bracteoles falling early leaving scars. Sepals oblong-lanceolate (sometimes narrowly), slightly hooded, 3.3–4.3 × 1.7–2.5mm, glabrous, greenish-membranous. Petals lilac, mauve or pale blue, about equalling sepals. Fertile stamens 2: filaments 2.1–2.2mm, hairs whitish tinged purple; anthers elliptic, 0.9–1 × 0.5–0.8mm. Antherodes 4, 3-lobed, pale yellow. Style c.2mm, hooked near apex. Capsule segments widely oblong, apiculate, 3–4 × 2–3mm, pale brown, shining. Seeds 2 per locule, c.1.4 × 1mm, hilum punctate, surface deeply sculpted with grooves radiating from large, asymmetrically placed boss, grooves with small pits, surface covered with conspicuous grey papillae.

Bhutan: S — Gaylegphug district (Gaylegphug, Sham Khara and Maorey Forests (M.F.B.)); **C** — Punakha (½km N of Punakha Dzong, 10km N of Punakha, 15km S of Wangdi towards Chirang) and Tongsa (near Langtel, near Dakpai, Tintibi Bridge) districts; **Terai** (Balasun, Sukna, Siliguri); **Duars** (Khurul); **Darjeeling** (Darjeeling, Dikeeling, Tista, Rongsong, Sittong, Rishap, Ryang, Labdah, Mongpu, junction of Great and Little Rangit); **Sikkim** (Ratong Chu (F.E.H.1)). Roadsides; roadside ditches; paddy-fields; dry rocky banks; wall-tops around cultivated land; open *Carex*-rich marsh; marshes; sand by river, 150–2400m. Fl/.fr. March–November.

Specimens from our area determined as *A. nudiflorum* var. *terminalis* (Wight) C.B. Clarke seem to be large forms of the present species and not worthy of varietal recognition. They differ from the SW Indian plant (a stout ?perennial with basal rosette) that Rao recognises as *M. loriformis* (Hasskarl) R.S. Rao & Kammathy.

5. M. spirata (L.) Bruckner; *Aneilema spiratum* (L.) Sweet. Lep: *tuksyor*.

Differs from *M. nudiflora* in having leaves relatively shorter and broader (1.7–3.5 × 0.5–1.1cm). Terminal infl. paniculate with several filiform branches each bearing 4–10 flowers. Bracteoles persistent, base encircling axis but not tubular, minute, red-spotted. Filaments at least usually glabrous; fertile anthers 3, narrower (c.0.2mm). Capsules narrower (3.5–4.5 × 2–2.3mm). Seeds with smooth, flattened faces abutting adjacent seeds, other surfaces papillose, but not grooved and pitted.

Bhutan: S — Punakha (10km N of Punakha, Punakha) and Tongsa (near Langtel) districts; **Terai** (Jaldaka River, Siliguri, Sukna); **Darjeeling** (Tista, Rishap). Tops of banks around fields and paddies; rice paddies, open *Carex*-rich marsh, 150–2400m. Fl./fr. August–November.

M. vaginata (L.) Bruckner is recorded for Darjeeling/Sikkim on a distribution map on a herbarium sheet (*Kammathy* 81287, E). This species is distinct in its several 1-flowered cymes fascicled in the axils of bladeless sheaths on a leafless scape. Seed 1 per locule. Roots not tuberous.

8. RHOPALEPHORA Hasskarl

Erect herb. Stem leafy throughout. Leaves lanceolate, acuminate, narrowed to base. Infl. a lax, terminal thyrsoid panicle, sometimes with subsidiary panicles from upper axils. Petals free, subequal. Stamens 3, middle one on shorter filament and anther deformed, anthers dehiscing laterally; staminodes 2, antherodes dumb-bell-shaped; filaments glabrous, fused below. Ovary 3-loculed, ovules axile, 1 per locule; style filiform; stigma indistinct. Fruit a loculicidal capsule, 3-valved. Seeds 1 per locule, hemispheric, hilum linear, boss lateral.

1. R. scaberrima (Blume) Faden; *Dictyospermum scaberrimum* (Blume) Panigrahi; *Aneilema scaberrimum* (Blume) Kunth.

Stems 58–69(–90)cm, glabrous below, with short, hooked hairs above. Leaves 13–17 × 3–4cm, shortly scattered-hispid above, densely so on margins, glabrous beneath; sheaths 1.3–3.3cm, slightly inflated, conspicuously hairy, mouth long-ciliate. Infl. an obconic, terminal thyrse, branches arising from condensed axis (appearing to arise from single point); cyme branches hairy, ascending, becoming slightly decurved; pedicels long (1–1.5cm), glabrous, slender, ascending, bracteoles persistent, funnel-shaped with oblique mouths; ascending, slender-peduncled subsidiary infls. usually arising from upper axils. Sepals ovate, blunt, 2.2–2.5 × 2mm, glabrous, green with scarious margins. Petals very pale mauve. Fertile (outer) stamens: filaments c.6mm; anthers elliptic, 1.2–1.5 × 0.8–1.2mm. Antherodes bright yellow, c.2mm across. Ovary globose, 1–1.3mm, densely whitish-hairy, shortly stipitate; style persistent, accrescent, c.4mm. Capsule oblong-globose, retuse, c.3 × 3mm, densely hooked-hairy, greenish. Seeds oblong-elliptic in outline, 2.4–2.7 × 1.8–2.2mm, flat side with linear hilum, boss asymmetrical on curved face, coarsely rugose, minutely glaucous-papillose.

Bhutan: S — Deothang district (1km N of Deothang; Narfong to Deothang (M.F.B.)); **C** — Punakha (c.15km S of Wangdi Phodrang) and Tongsa (between Tama and Tintibi) districts; **Terai** (Selim); **Darjeeling** (Rongsong, Jinglam, Sittong, Labdah, Darjeeling, Lebong, Rungbee, Kurseong, Badamtam, Kumai to Rongu, Sandakphu to Rimbick); **Sikkim** (Dikchu, Chakung). Damp shady places; scrub on steep stony slopes; semi-shaded bank in wet forest, 460–1520(–3660)m. Fl./fr. August–October.

9. TRICARPELEMA J.K. Morton

Similar to *Rhopalephora* but differing as follows: infl. a lax, cylindric panicle with branches inserted on an elongate axis; bracteoles leaf-like (but much reduced) early deciduous. Staminodes 3. Filaments free. Ovary narrowly ellipsoid, glabrous, ovules c.6 per cell. Fruit a narrowly oblanceolate, sharply acuminate capsule, with c.18 seeds.

1. T. giganteum (Hasskarl) Hara; *Aneilema thomsonii* (C.B. Clarke) C.B. Clarke. Fig. 23g–h.

Stem to 85cm, ± glabrous. Leaves lanceolate, finely acuminate, margins hispid-ciliate, narrowed to very short, petiole-like base, 11–17(–24) × 1.5–5(–7)cm, upper surface papillose-hispid, hispid only on veins beneath; sheaths widely cylindric, 2.2–3cm, mouth and fused margins shortly ciliate, upper ones almost overlapping. Infl. axis 5–17.5cm, glabrous, bearing slender, evenly spaced, ascending branches; bracteoles leaving scars; flowers borne near ends of branches, pedicels short (under 1cm). Flowers pale mauve to bright blue; sepals oblong-elliptic, apex hooded, margins scarious, 5–6 × 4mm, glabrous; petals ovate, rounded, c.6.5 × 5.3mm, all equal. Anthers elliptic, 1.5–2.2 × 1.5mm; 2 outer filaments long (7–15mm), curved, middle filament short (3–5mm), straight. Antherodes c.1.2mm across, filaments c.0.5mm. Ovary narrowly ellipsoid, tapered upwards into style, c.3 × 1mm, glabrous; style c.9mm, filiform, curved at apex. Infl. branches becoming stouter in fruit, upward-curving, each bearing a single, terminal capsule. Capsule 1.5–2 × 0.4–0.5cm, segments narrowly oblanceolate, gradually narrowed to base, abruptly long-acuminate. Seeds prismatic (± cuboid), c.2 × 1.7mm, rugose, with radiating grooves from asymmetrically placed boss, glaucous.

Bhutan: S — Phuntsholing district (Gedu to Kharbandi); unlocalised Griffith record; **Darjeeling** (Darjeeling, Great Rangit, above Mongpu, Ghum, Senchal, Geille, Sonada, Kurseong, Rungbee, Ghumpahar, Mamring). Shady dripping scrubby cliff in broad-leaved forest; roadside scrub and grassy banks in open or disturbed oak forest, 610–2440m. Fl./fr. July–September.

The Burmese specimens I have seen bear glandular hairs on the infl. and so belong to *T. chinensis* Hong; whether this merits specific rank remains to be seen, but such glandular-hairy forms should be looked for in Bhutan.

10. POLLIA Thunberg

Perennial herbs. Leaves ± lanceolate, acuminate. Infl. a terminal thyrse. Petals free, similar. Stamens 6, anthers all equal and fertile or 3 larger and 3 smaller (sterile or fertile); filaments glabrous; anthers dehiscing laterally. Ovary 3-loculed; style filiform; ovules several per cell, axile. Fruit a bluish, crustaceous, subglobose, indehiscent capsule. Seeds vertically compressed, polygonal, boss central in face opposite hilum, hilum punctate to linear.

1. Infl. lacking peduncle, subumbellate, overtopped by leaves, infl. branches subglabrous, deflexed; stamens 3 fertile and 3 sterile
    3. P. subumbellata
+   Infl. pedunculate, finally pyramidal, overtopping leaves, infl. branches hairy, spreading to ascending; stamens all fertile or 3 sterile 2

2. Stamens all fertile; infl. very dense; peduncle short (to 5.5cm), stout; sepals hairy on outside .... **1. P. hasskarlii**
+   Stamens 3 fertile and 3 sterile; infl. rather lax; peduncle usually over 5 (to 10)cm, slender; sepals glabrous on outside .... **2. P. secundiflora**

1. P. hasskarlii R.S. Rao; *P. aclisia* Hasskarl *nom. illegit.* Fig. 23i–k.

Stems to 67(?–90)cm, grooved when dry, lower part with remains of old leaf sheaths. Leaves aggregated below infl., blades oblanceolate or narrowly elliptic, finely acuminate, gradually narrowed below into petiole-like base, 18–34 × 5–7cm, with scattered, short, hooked hairs above and beneath or subglabrous, veins tending to crinkle on drying; sheaths slightly overlapping, widely cylindric, 3.5–4.5cm. Infl. cylindric at first, becoming pyramidal, 5.5–11.5 × 5–9cm; peduncle 2.5–5.5cm, stout, subtended by small leaf-like bract; branches subtended by whitish bracts, lower spreading ± horizontally; flowers borne on short, erect pedicels; peduncle, infl. branches and pedicels densely clothed with short, hooked hairs. Sepals orbicular, concave, 4.3–5.5 × 3.5–5mm, thick-textured, hairy on outside. Petals white, suborbicular, concave, 3–4.5 × 3mm, thin-textured. Filaments c.2.5mm; anthers all fertile, c.1.5 × 1mm. Ovary globose, 1.3–2mm, glabrous; style 1–3.5mm. Capsule globose, c.6 × 6.5mm, shiny, bluish-black. Seeds 1.5–2mm, brown.

Bhutan: S — Chukka (Marichong) and Sarbhang (Loring Falls) districts; **Duars** (Numsonge); **Darjeeling** (Darjeeling, Rishap, Mongpu, Silak, Riang, Sittong, Pomong, Kurseong). Shaded bank in dense forest, 610–1220m. Fl./fr. June–August.

2. P. secundiflora (Blume) Bakhuizen f.; *P. sorzogonensis* (E. Meyer) Endlicher.
Differs from *P. hasskarlii* in its leaves which are more abruptly contracted into a short petiole; sheaths shorter (to 3cm); peduncle longer ((2–)5–10cm), more slender, sometimes with several intermediate leaf-like bracts; infl. branches slender, fewer-flowered, lowermost branches distant, sometimes again branched; sepals glabrous; anthers 3 fertile and 3 sterile.
Darjeeling (Darjeeling, Sonada, Tista, Rangit, Bakrikot). 150–2130m. Fl./fr. June–November.

3. P. subumbellata C.B. Clarke.
Similar to *P. secundiflora* in having 3 fertile and 3 sterile anthers and glabrous sepals; but differing from that and from *P. hasskarlii* in its smaller (to 14 × 4cm) leaves which are symmetrically narrowed at apex and base; infl. subumbellate, epedunculate, overtopped by leaves, branches subglabrous, deflexed; capsule to 5mm.
Terai (Bamunpokri); **Darjeeling** (Darjeeling, Mongpu, Rishap, Lebong, Tista, Badamtam, Kalighora, junction of Great and Little Rangit). Moist, rocky areas on forest margins, 305–1830m. Fl./fr. June–October.

This and the previous species are recorded for 'Bhotan' in F.B.I., but no Bhutanese specimens have been seen.

11. COMMELINA L.

Perennial or annual herbs. Roots sometimes thickened or tuberous. Stems usually decumbent at base, rooting from lower nodes; branches arising within and splitting leaf sheaths. Leaves commonly lanceolate; sheaths shortly hairy on fused margins often carried down as hairy line on stem. Infls. terminal and/or leaf-opposed, composed of conduplicate, or (when margins fused) funnel-shaped spathes, each enclosing 1 or 2 cymose branches; pedicels recurved in bud and fruit. Flowers bisexual, or bisexual and male, bilaterally symmetric, white or blue, petals, 2 large, clawed, third usually smaller, lanceolate or clawed or absent; sepals usually 2 wider than third; underground, cleistogamous flowers occasionally present. Stamens 3, 2 outer anthers similar, median usually larger with cordate base and divergent lobes; antherodes 3, cruciform, small; filaments glabrous, staminal ones longer than staminodal. Ovary 2–3-loculed, dorsal locule lacking or 1-seeded, dehiscent or indehiscent, ventral locules 1–2-seeded (rarely lacking), dehiscent. Seeds sometimes with terminal appendages, hilum linear, boss lateral.

1. Leaves oblong-elliptic, rounded to subacute, distinctly petioled; underground, cleistogamous flowers sometimes present .... **8. C. benghalensis**
+   Leaves lanceolate, acute to acuminate, narrowed petiole-like base extremely short (to 2mm) or absent; cleistogamous flowers never present 2

2. Spathes funnel-shaped (margins fused) 3
+   Spathes conduplicate (margins free) 4

3. Plants stout, glabrous; leaves over 13cm; spathes over 1.5cm; capsule over 5mm .... **1. C. paludosa**
+   Plants slender, usually hairy on leaves, spathes and upper stems; leaves to 7cm; spathes to 1cm; capsule to 4mm .... **2. C. maculata**

4. Spathes usually enclosing a single cyme 5
+   Spathes with 2 cymes, one with exserted, persistent peduncle 6

5. Stems stout, to 2m; spathes over 2cm; petals all similar, blue, blades c.1cm; capsule of 2 dehiscent, 2-seeded and 1 indehiscent, 1-seeded locules .... **9. C. coelestis**
+   Stems slender, to 0.5m; spathes usually under 2cm; petals white or blue, 2 clawed (blades c.3mm), 1 lanceolate; capsule of 2, single-seeded dehiscent locules .... **3. C. suffruticosa**

6. Spathes over 5cm; dehiscent locules single-seeded, seeds with whitish appendages at both ends (Fig. 21k) .... **7. C. appendiculata**
+   Spathe to 4.5cm; dehiscent locules 2-seeded; seeds without appendages 7

7. Seeds deeply reticulate-pitted (Fig. 21h), greyish, brain-like; leaves acuminate .... **4. C. sikkimensis**
+   Seeds smooth or with surface reticulation, blackish or brown; leaves acute 8

8. Seeds with surface reticulation (Fig. 21i) .... **5. C. diffusa**
+   Seeds smooth .... **6. C. caroliniana**

PLATES

PLATES

Plate 1
Paris polyphylla var. **wallichii** (× 0.7) with details of rhizome and fruit.

Plate 2
Polygonatum kansuense (× 0.6) with half flower (× 1.5).

Plate 3
Asparagus filicinus var. **giraldii** (× 0.6) with details of habit (× ¹⁄₁₀) and female flower (× 3).

Plate 4
Chlorophytum nepalense (× 0.5) with details of habit (× ¹⁄₁₀), capsule and flower (× 1).

Plate 5
Allium macranthum (× 0.6) with details of tuberous roots, capsule, seed and half flower (× 1.5).

Plate 6
Aletris glabra (× 0.7) with details of flower, half flower and capsule (× 1.6).

Plate 7
Disporum cantoniense (× 0.6) with details of young shoot and half flower (× 1.2).

Plate 8
Arisaema elephas (× 0.6) with details of female spadix, female flower and t.s. of ovary (× 3).

Plate 9
Cautleya gracilis (× 0.6) with details of half flower (× 1.2) and ripe and unripe capsules.

Plate 10
Canna speciosa (× 0.6) with details of fused fertile stamen and style (× 0.6), ripe and unripe seed and capsule.

Plate 11
Juncus concinnus (× 0.9) with details of leaf t.s. and capsule (× 2).

Plate 12
Carex duthiei (× 0.6) with details of (left to right) male glume, female glume, utricle (all × 3) and terminal gynaecandrous spike (× 1.6).

All painted by **Mary Benstead** from living plants grown at the Royal Botanic Garden Edinburgh.

Plate 1

Plate 2

Plate 3

Plate 4

Plate 5

Plate 6

Plate 7

Plate 8

Plate 9

Plate 10

Plate 11

Plate 12

11. COMMELINA

1. C. paludosa Blume; *C. obliqua* Buchanan-Hamilton ex D. Don, non Vahl. Fig. 22d.

Stem to 180cm, straggling, much branched. Upper stem leaves lanceolate, finely acuminate, base cuneate to extremely short, petiole-like base, 13.5–21.5 × 2.3–5cm, slightly rough-granular above, smooth, glabrous beneath; sheaths cylindric, 2–3cm, mouth with few long, brown cilia. Spathes terminal, shortly stalked, clustered, lanceolate, finely acuminate, base cuneate to extremely short, petiole-like base, 1.5–2.3 × 1.2–3cm; cyme single, c.7-flowered. Flowers white or pale blue, 2 petals with ± oblong blade c.4.5 × 3.5mm and claw 4–5mm; third petal lanceolate, c.5.5mm, or apparently absent. Filaments coiled; outer anthers ellipsoid, c.1.2 × 0.7mm; central anther 2–2.5mm; antherodes c.1.3mm. Ovary widely ellipsoid, c.0.7 × 0.5mm; style recurved at apex, c.6mm. Capsule widely oblong-trigonous, truncate, 5–5.5 × 4.5–6mm, pale greenish-brown, shining, 3-loculed, each locule 1-seeded. Seeds oblong in outline, 3.7–4.1 × 2–2.3mm, densely, minutely grey-granular, boss c.1mm.

Bhutan: S — Samchi (Changum (M.F.B.)), Phuntsholing (above Rinchending; Phuntsholing (M.F.B.)), Chukka (below Chimakothi, Gedu to Taktichu), Gaylegphug (Gaylegphug (M.F.B.)) and Deothang (near Chenari (M.F.B.)) districts; **C** — Punakha (Punakha to Sinchu La) and Tashigang (Damoitsa) districts; **Darjeeling** (Mongpu, near Tista below Temi, Rongbe Ghora, Rungbee, Lebong, Darjeeling); **Sikkim** (Chakung). Among scrub on shady bank in subtropical and oak forest; open hillside, 100–2180m. Fl./fr. May–November.

2. C. maculata Edgeworth; *C. obliqua* var. *viscida* C.B. Clarke and var. *mathewii* C.B. Clarke. Sha: *pishampi*; Sikkim name: *angulako*.

Perennating by means of narrow, vertical tubers. Similar to *C. paludosa* in its funnel-shaped spathes, but differs in being smaller (stems 12–46cm), more slender and usually hairy; upper stem leaves smaller (3.2–7 × 1.2–2.2cm). Stems with jointed hairs or glabrous. Leaves usually hairy, with jointed hairs especially on upper surface, sometimes also with short, conical bristles, occasionally glabrous; sheaths with many, long, brown (sometimes white) jointed cilia at mouth, fused margins and usually also surface hairy. Infl. smaller, occasionally reduced to a single spathe; spathe 1 × 1.2–1.6cm, usually hairy. Flowers smaller in all parts; claw of petals c.3mm. Outer anthers 0.9–1mm, median 1–1.5mm; antherodes c.0.8mm. Capsule 3–3.8 × 3–3.8mm; seeds elliptic in outline, 2.6–3.1 × 1.8–2mm, very sparsely papillose.

Bhutan: C — Thimphu (Dotena, Marichong), Punakha (Ritang), Bumthang (Dhur) and Tashigang (Damoitsi, near Khangma, Tashi Yangtsi) districts; **Darjeeling** (Darjeeling, Kurseong, Kalimpong, Lebong Woods, Rishap, Sukia Pokhri, Farsing, Labdah, Peshok, Sittong, Tista); **Sikkim**

(Domang, Lachen, 11km N of Gangtok, Yoksum). Sandy loam among moist herbs; weed of potato and maize fields; open rough grassland in forest; damp shady banks and field walls, (305–)914–3060m. Fl./fr. July–October.

Intermediates occur with *C. paludosa* of which it is perhaps merely a (higher altitude?) form. Rao (1966) was wrong in sinking *C. obliqua* var. *mathewii* under *C. undulata* R. Brown (which he equates with *C. kurzii* C.B. Clarke. and *C. striata* Edgeworth) and was correct in his earlier (1964a) treatment sinking var. *mathewii* under var. *viscida*.

Parker (1992) records this species as a common weed of dryland crops (especially maize and potato) and field borders (300–3000m), especially in Mongar and Tashigang districts, and also in Thimphu, Punakha and Bumthang districts.

3. C. suffruticosa Blume.

Stems 23–45(+?)cm, much branched above, sometimes shortly hairy above. Leaves lanceolate, acuminate, cuneate to short, petiole-like base, 8–16 × 1.8–4.8cm, minutely rough above, usually shortly hairy beneath; sheaths cylindric, 1.5–3(–4)cm, shortly hairy all over, mouth shortly ciliate. Spathes single, stalked, terminal on main and lateral branches, subtended by leaf-like bracts with persistent membranous sheaths and small, sometimes deciduous blades; spathes ovate, conduplicate, acute, 0.8–2.2cm, each half 0.6–1.2cm wide, densely hairy; cymes single, c.10-flowered. Petals usually white (sometimes blue), 2 clawed (claw c.2mm, blade c.3mm), one lanceolate. Anther filaments c.4.5mm; anthers subequal, c.1.5mm; antherodes minute (c.0.3mm). Ovary minute (c.0.5mm); style c.1.5mm, recurved. Capsule transversely oblong, retuse, compressed, 4.5–6.5 × 6–7.5mm, 2-loculed, sometimes stipitate (to 1.5mm). Seeds 1 per cell, D-shaped, 3.5–4 × 2.5–3.5mm, minutely papillose, sometimes coarsely rugose or pitted, brown.

Darjeeling (Tista, Rangit, Mongpu, Ryang, Selim, Bamunpokri, Labdah, Peshok). Natural habitat not recorded, but sometimes a weed of cultivation, 150–910m. Fl./fr. June–October.

4. C. sikkimensis C.B. Clarke. Fig. 22e–h.

Stems 18–85cm, much branched. Leaves narrowly lanceolate, finely acuminate, contracted into very short petiole-like base, 4–9 × 1–2.3cm, upper larger than lower, usually granular above, glabrous beneath; sheaths cylindric, 1.1–2cm, glabrous, red-spotted, mouth with jointed, white cilia. Spathes stalked, borne singly from sheath of an upper leaf on each branch; spathe lanceolate, acuminate, conduplicate, base sometimes cordate, 2–4.5 × 1.2–2.6mm; cymes in unequally peduncled pairs, longer peduncle exserted, persistent, bearing 2–4 deciduous male flowers; shorter peduncle included, bearing 5–6 bisexual flowers. Petals blue, claw c.3.5mm, blade orbicular 3.5–4.5mm. Anther filaments 6–7mm; lateral anthers 1.2–1.4mm, median anther c.1.8mm; antherodes 0.4–0.7mm. Ovary narrowly ellipsoid-trigonous, c.1.5 × 0.8mm, tapered into simple, filiform style 6–8mm. Capsule narrowly oblong-ellipsoid, 5–6 × 3mm, with apiculus 1–1.5mm, pale greenish-brown, 2 locules dehiscent, 2-seeded, 1 locule indehiscent, 1-seeded. Paired seeds 2–2.5 × 1.5–1.7mm, greyish, minutely papillose, coarsely reticulate with deep pits within the reticulations (brain-like), boss small; single seed 3.5 × 2.3mm.

Bhutan: S — Phuntsholing district (above Phuntsholing); **C** — Tongsa district (Tama); **Darjeeling** (Kurseong, Rungbee, Rishap, Mongpu, Darjeeling, Kalimpong); **Sikkim** (?Lachen). Rocky banks and streamsides in warm, wet, broad-leaved forest, 910–2130(?–3050)m. Fl. May–October.

5. C. diffusa Burman f.; *C. nudiflora* sensu F.B.I. non L. Fig. 22i–j.
Extremely similar to *C. sikkimensis* and not distinguishable with certainty from it without ripe seeds; it differs in having usually smaller (3.3–7 × 1.1–1.5cm), less finely acuminate leaves, spathes commonly smaller (1.5–2.2cm) and especially in the seeds which are blackish-brown and marked with a surface reticulation.

Bhutan: S — Chukka district (Kali Khola); **Darjeeling** (Tista, Rongsong); **Sikkim** (Gangtok). River banks; rice fields, 210–1220m (apparently at lower altitudes than *C. sikkimensis*). Fl./fr. April–October.

Immature specimens from Phuntsholing (Torsa River, 200m, *Grierson & Long* 731) and Sarbhang (above Sarbhang High School, 600m, *Wood* 6148) districts probably refer to this rather than to *C. sikkimensis*.

6. C. caroliniana Walter; *C. hasskarlii* C.B. Clarke.
Similar to *C. sikkimensis* but differs in having smaller (3–6.5 × 0.8–1.6cm), subacute leaves, shorter sheaths (to 1cm) and especially in its smooth seeds.

Bhutan: C — Punakha (Changhyo, Sirigang) and Mongar (Autsho 37km N of Mongar) districts; **Darjeeling** (on map on *Kammathy* 81293, K, E). Edge of rice-paddies, occasional in rice, commoner in non-flooded vegetable (e.g. chilli, maize) plots, 1000–1300m. Fr. June–July.

The two previous species can only be told apart with certainty in fruit and they have no doubt been confused in the field. Parker (1992) records the names *korum* (Dz), *humbatenang* (Sha) and *kaney jhar* (Nep) for the two species collectively and states that they are common or occasionally dominant weeds in dryland crops and rice occurring at lower altitudes (to 2000m), probably in all districts with cultivation.

7. C. appendiculata C.B. Clarke. Fig. 22k.
Resembles *C. sikkimensis* in its very acuminate, conduplicate spathes, but differs in having linear lanceolate leaves (10.5–13 × 1.2–1.7cm) gradually

228. COMMELINACEAE

narrowed into sheath; infls. several per branch; spathes longer (5–7 × 1.2–1.7cm), more gradually acuminate; peduncles of cyme branches subequal, the longer only just exserted from spathe; petals sometimes white, with a shorter (c.2mm) claw; median anther scarcely larger than laterals; capsule longer and broader (6–7 × 3.5mm), apiculus very small (under 0.7mm) and above all in the capsule and seeds. Dehiscent locules both 1-seeded, seeds large (4.7–5.2 × 2–2.5mm), flattened-ellipsoid, brown, smooth, with whitish appendages at each end; indehiscent locule (when present) with smaller, unappendaged seed.

Darjeeling Terai: (locality unknown). 150m. Fl./fr. June.

According to Kammathy (1983), recent field studies have failed to relocate this species.

8. C. benghalensis L. Sha: *humbatenang*; Nep: *kaney jhar*.

Stems 20–45cm, much branched, hispid especially above; base sometimes with weak, decumbent stolons bearing underground, cleistogamous flowers (capsules of which may look like bulbils). Leaves with distinct petiole-like bases (to 1cm), blades oblong-elliptic, rounded to subacute, margins densely, shortly ciliate, base rounded to truncate, 3.5–6 × 1.7–3.5cm, surfaces usually with many long, white hairs; sheaths funnel-shaped, 0.7–1.3cm, pale when dry, hairy all over, with conspicuous (often brown) jointed cilia at mouth. Spathes stalked, single or paired, terminal on main and lateral branches; spathes ovate, acute, margins fused so funnel-shaped, 1–1.6 × 0.8–1.2cm, densely hairy; cyme of c.3 bisexual flowers and a single male flower on long, exserted pedicel. Petals blue, 2 clawed (claw c.3.5mm, blade c.4 × 5mm), 1 lanceolate (c.4mm). Anther filaments c.6mm, lateral anthers c.1 × 0.3mm, median anther 1.2–1.5mm; antherodes c.0.5mm. Ovary c.2 × 1mm; style c.5.5mm, coiled at apex. Capsule oblong, retuse, 5–5.5 × 3–4mm, on stipe c.1mm, with 2 dehiscent, 2-seeded and 1 indehiscent, 1-seeded locules. Paired seeds plano-convex, truncate at one end, c.2 × 1.5mm, blackish-brown, pitted, boss large; single seed c.3.5 × 2mm.

Bhutan: C — Punakha (near Punakha) and Tashigang (Dangme Chu) districts; **Sikkim** (18km W of Rabong La, NW of Singtam); **Darjeeling** (Bamunpokri, below Tukvar, Ribong, Labdah). Roadsides and weed of cultivation; marshy places by stream, 305–1300m. Fl./fr. July–August.

Under-collected: Parker (1992) records this species as a common weed of lowland areas (to 2300m) in Thimphu, Punakha, Mongar, Tongsa and Tashigang districts.

9. C. coelestis Willdenow. Nep: *kaney*.

Roots tuberous. Stems 0.6–2m, stout (to 1cm wide), hairy above. Leaves lanceolate, acuminate, not constricted at base, 10.5–17.5 × 3–4cm, glabrous;

sheaths widely cylindric, 2.2–3.5cm, glabrous except fused margins, mouth minutely ciliate. Spathes long-stalked, terminal and from upper axils, widely ovate, acuminate, conduplicate, margins free, ciliate, 2.2–5cm, with strong longitudinal and transverse veins, densely hairy; cyme usually single, c.8-flowered, longer peduncled c.2-flowered cyme occasionally also present. Petals all similar, blue, blade obovate, c.1 × 1.3cm, narrowed into short claw (c.2mm). Anther filaments c.1cm; lateral anthers 1.8–3mm; median anther 2.3–3.5mm; antherodes c.1.5mm, golden-yellow. Style 1–1.4cm, curved, stigma minutely capitate. Capsule oblong, 6–7.5 × 4mm, with 2 dehiscent 2-seeded cells and 1 indehiscent 1-seeded cell. Paired seeds ± triangular, 2–3 × 2mm, pitted, brown or blackish, boss small; single seed c.3.4 × 2.3mm.

Darjeeling (Darjeeling — Lloyd Botanic Garden and Aloobari Road, Jalapahar, Katapahar). 2040–2440m. Fl./fr. August–October.

Native of Central and S America, cultivated for its showy flowers and becoming naturalised.

Family 229. XYRIDACEAE

Perennial or annual marsh herbs. Leaves linear, often compressed, in basal rosette and on lower part of scape, bases sheathing. Infl. a dense head borne on a scape. Flowers bisexual, subtended by stiff, glume-like bracts; petals 3, usually yellow, sometimes fused into tube below, blades free; sepals 3, lateral pair conduplicate, keeled, persisting in fruit, median sepal larger and deciduous after anthesis. Stamens 3, borne at base of blades, anthers basifixed, dehiscing laterally, sometimes alternating with 3 feathery staminodes. Ovary superior, of 3 fused carpels, unilocular, ovules parietal; style 3-branched above, each bearing a capitate stigma. Fruit a capsule.

1. XYRIS L.

Description as for family.

1. Leaves over 2.5mm wide; heads over 1cm wide; fertile bracts over 5mm wide, with small, whitish papillae all over upper ⅓ (Fig. 24e)
   3. X. indica
+ Leaves under 2mm wide; heads usually under 1cm wide; bracts under 5mm wide, papillae if present restricted to green, subapical keel 2

2. Subapical keel of fertile bracts not conspicuous, brown, not papillose; blades of scape leaves very reduced (under 1cm)
   1. X. capensis var. **schoenoides**

+ Subapical keel conspicuous, green, triangular, papillose (at least when young) (Fig. 24d); blades of scape leaves developed (similar to basal leaves) .... **2. X. pauciflora**

1. X. capensis Thunberg var. **schoenoides** (Martius) Nilsson. Fig. 24a–c.

Tufted ?annual. Leaves 1.5–2mm wide at middle, tapered to acute apex, laterally compressed, basal leaves less than half scape length, scape leaves 1–2, sub-basal, sheaths long, with hyaline margins tapering to oblique mouth, free blade very short (2–8mm). Scape 21–58cm, ridged. Infl. subglobose, flattish-topped, 7–9 × 6–11mm, (commonly wider than long); lower 2–3 bracts sterile, smaller than fertile ones; fertile bracts concave, widely obovate, rounded, margins narrow, paler at least when young, 5–6 × 4.5–5mm, 3–8-veined, dark brown, shining, coriaceous, slightly keeled below apex, keel not differing in colour or texture. Lateral sepals acute, conduplicate, 6–7mm, each half 1.5–1.7mm wide, pale brown, membranous, keel smooth; median sepal wider. Petals yellow, blades obovate, fringed-truncate, c.4 × 2.4mm, claws c.6mm. Filaments 1–1.3mm; anthers notched at apex, 1.5–1.7mm. Staminodes bifid, bearing clumps of viscid, yellow threads adherent to stigmas. Ovary narrowly ellipsoid, c.3.5 × 1.2mm; style c.2.2mm, arms about equalling style. Capsule narrowly ellipsoid-trigonous, apiculate, 4.5–5 × 1.8–2.5mm, dark brown. Seeds apiculate at ends, 0.7 × 0.3mm, surface reticulate with strong longitudinal ribs, pale brown.

Bhutan: S — Chukka district (above Lobnakha); **C** — Ha (Tare La), Thimphu (above Taba, Drukyel Dzong, Atsho Chhubar, near Motithang Hotel), Punakha (Wangdu Phodrang, Pubjikha Valley), Tongsa (Longte Chu, below Rukubji) and Bumthang (Bumthang, Dhur Valley, Byakar, Yuto La, Gyetsa) districts. Damp flushes, marshes and bogs in open or in *Pinus* forest 2130–3100m. Fl. June–September.

Fig. 24. **Xyridaceae, Eriocaulaceae, Juncaceae I.**
a–c, **Xyris capensis** var. **schoenoides**: a: habit (× ½); b, bract (× 4); c, petal showing stamen and staminodes (× 4). d, **X. pauciflora**: bract (× 4). e, **X. indica**: bract (× 4). f–h, **Eriocaulon viride**: f, habit (× ⅔); g, female flower (× 10); h, male flower (× 10). i–j, **E. cinereum**: i, female flower (× 12); j, male flower (× 12). k–l, **E. alpestre**: k, female flower with calyx removed (× 6); l, calyx of female flower opened out (× 6). m–n, **E. bhutanicum**: m, female flower (× 12); n, male flower (× 12). o, **Juncus bufonius**: habit (× ½). p–q, **J. inflexus**: p, habit (× ¼); q, l.s. of stem showing interrupted pith (× 6). r–t, **J. ochraceus**: r, habit (× ½); s, sterile partial infl. (× 3); t, half flower (× 3). u–v, **J. prismatocarpus**: u, habit (× ½); v, l.s. of pluritubular leaf with septa not crossing entire leaf (× 4). w–x, **J. sphacelatus**: w, habit (× ½); x, half flower (× 2). Drawn by Mary Bates.

2. X. pauciflora Willdenow. Fig. 24d.

Differs from *X. capensis* var. *schoenoides* in having basal leaves almost equalling scapes; scape leaves with long blades (about equalling basal leaves); infl. usually longer than wide, round-topped; bracts reddish-brown with broad yellowish margins, with fewer (3–5) veins and the distinct, triangular subapical keel greenish, papillose (papillae sometimes abraded); lateral sepals with keel bearing 1–2 teeth; capsule (to 4 × 2.4mm); seeds (0.4 × 0.2mm) smaller.

Bhutan (unlocalised Griffith specimen); **Darjeeling Terai** (Katanbari, Sukna, Jalpaiguri). Wet places, to 305m. Fl./fr. November–December.

3. X. indica L. Fig. 24e.

Differs from two previous species in having much broader leaves (over 2.5(–6)mm); larger heads (over 1cm wide), fertile bracts wider (6 × 5–6.5mm), scarcely keeled below apex and upper ⅓ covered with small, whitish papillae.

Darjeeling ('Darjeeling'). No details known, presumably from rice fields in the Terai.

Family 230. ERIOCAULACEAE

Marsh or aquatic, annual or perennial herbs. Bhutan species all tufted, stemless, rosette-forming, some other species with elongate stem bearing spirally arranged leaves. Leaves linear, often with fenestrate air-spaces. Flowers unisexual, infl. monoecious, a terminal capitate head on a long scape, subtended at base by tubular bract. Involucral bracts membranous. Male and female flowers intermixed, each subtended by a floral bract. Male flowers: calyx tubular below, blade spathe-like and undivided or variously divided into 2–3 lobes; corolla funnel-shaped, tubular below, 3-lobed, lobes equal or one larger than others; stamens (1–)6; anthers black or pale, 2-loculed, dehiscing latrorsely. Female flowers: sepals 0, 2 or 3, boat-shaped; petals 0, 2 or 3, reduced, sometimes with apical gland. Ovary sessile or stipitate, (2–)3-loculed, each with a single basal ovule; style slender, stigma lobes filiform (2–)3. Fruit a thin-walled capsule; seeds oblong, variously patterned.

1. ERIOCAULON L.

Description as for family.

1. Capitula black or with short white hairs on floral parts contrasting in colour with dark-grey to blackish background 2
+ Capitula silvery, glabrous or if hairs present on floral parts, not contrasting in colour 5

2. Female petals swollen, clawed, with narrowly elliptic blade (Fig. 24k)
    5. E. alpestre
+   Female petals membranous, minute, not differentiated into blade and
    claw 3

3. Involucral bracts reflexed .... **3. E. sollyanum**
+   Involucral bracts erect 4

4   Leaves broad (over 1.4mm), over ⅓ to equalling scape; scapes stout,
    erect; heads large (over 3mm diameter), dark grey .... **1. E. viride**
+   Leaves narrow (under 1mm), less than ⅓ scape length; scapes filiform,
    slightly flexuous; heads small (under 3mm), blackish .. **6. E. bhutanicum**

5. Anthers white; male calyx spathe-like, scarcely lobed; capitulum hemi-
    spheric, soft; receptacle sparsely villous; female flowers lacking petals
    and sepals .... **2. E. cinereum**
+   Anthers black; male calyx deeply divided into 2 equal sepals; capitulum
    elongate, hard; receptacle densely villous; female flowers with 2 sepals
    and 3 petals .... **4. E. hamiltonianum**

1. E. viride Körnicke. Fig. 24f–h.

Leaves oblong, blunt, ⅓ length to equalling scape, 1.4–5.5mm wide. Scape 3–14cm. Capitulum dark grey and white, hemispheric, 3–6mm diameter; receptacle convex, elongating in fruit, glabrous. Involucral bracts broadly oblong to obovate, blunt, 1.5–2 × 1.7–1.8mm, glabrous, yellowish-hyaline. Floral bracts oblanceolate, blunt to acuminate, 1.6–2 × 0.7–1.2mm, dark grey above with short white hairs. Female flowers: total length 1.5–1.8mm; sepals 3, oblong, acute, 1.1–1.5 × 0.3–0.5mm, hairy, grey; petals linear-oblanceolate, black-gland-tipped, 1.3–1.4mm, hairy, white-hyaline; ovary sessile; style c.1mm, lobed ⅓ to halfway; pedicel c.0.3mm. Male flowers: total length 1.3–1.8mm; calyx 1–1.5mm, grey, hairy, 3-lobed to at least halfway, lobes 0.5–0.7 × 0.2–0.4mm, tube 0.3–0.5mm; large petal hairy; anthers black. Seeds c.0.5 × 0.3mm, longitudinally grooved, with rows of tiny white papillae.

Bhutan: C — Thimphu (Paro, below Takstang), Punakha (Samtengang), Tongsa (Mangde Chu Valley S of Tongsa), Bumthang (above Dhur) and Tashigang (Tashi Yangtsi Dzong) districts; **Sikkim** (Dikeeling; Sabbia, Sureil (Fyson, 1923)). Marshes; edge of lake; shallow pond, 1370–2870m. Fl. July–October.

E. viride is sometimes treated as a synonym of the (earlier) *E. nepalense* Bongard; the description of that species is not very adequate, but it appears to be distinct in its

243

longer scapes and acuminate involucral bracts. *E. viride* is rather variable; the Bumthang specimen comes from a relatively high altitude and is therefore smaller and more slender.

E. luzulifolium Martius might also be expected to occur; it differs from *E. viride* in its densely villous receptacle, narrower (c.0.3mm) floral bracts and in having harder capitula.

2. E. cinereum R. Brown; *E. sieboldianum* Siebold & Zuccarini ex Steudel; *E. redactum* Ruhland. Fig. 24i–j.

Leaves gradually attenuate to piliferous apex, less than ½ length of longest scape, widest (0.7–1.5mm) at base. Scape (3–)9–12cm. Capitulum silvery-grey, hemispheric, c.4mm diameter; receptacle shortly elongate, with few hairs. Involucral bracts narrowly oblong-elliptic, rounded, 1.5 × (0.6–)0.9mm, glabrous, silvery-hyaline. Floral bracts linear lanceolate, subacute, c.2 × 0.5mm, glabrous, silver flushed grey. Female flowers: total length c.2mm; sepals 0; petals 0; style c.1.2mm, lobed in upper ⅓; pedicel c.0.3mm. Male flowers: total length 1.4mm; calyx spathe-like, scarcely lobed, dark grey, with few, sparse hairs on back, tube 0.4mm; large petal hairy, dark-tipped; anthers white. Seeds c.0.4 × 0.2mm, smooth, shining.

Bhutan: S — Samchi (Thoribhadi (M.F.B.)) and Gaylegphug (Sham Khara (M.F.B.)) districts; **C** — Punakha district (Centre for Agricultural Research and Development, Wangdi Phodrang); **Terai** (Siliguri (Fyson, 1923)). Flooded rice fields, 300–1350m. Fl. August.

Another variable species; the only specimen seen has female flowers lacking sepals and petals and is therefore the form described as *E. redactum*.

3. E. sollyanum Royle; *E. trilobum* Buchanan-Hamilton ex Körnicke.

Leaves gradually narrowed to subacute, callose-tipped apex, less than ⅓ length of longest scape, widest (2–2.5mm) at base. Scape 12.5–24cm. Capitulum dark grey and white, globose, 4.5–6mm diameter; receptacle elongate, sparsely pilose. Involucral bracts reflexed, narrowly oblong-elliptic, subacute, unequal, longest to 3.5 × 1.2mm, glabrous, silvery-hyaline. Floral bracts oblanceolate, acuminate, c.2 × 1.2mm, dark grey with short white hairs above. Female flowers: total length c.2mm; sepals 3, oblong, acute, 1.4–1.6 × 0.3–0.4mm, white-hairy; petals 3, linear-oblanceolate, dark gland-tipped, c.1.6 × 0.2mm, hairy, white-scarious; ovary shortly (c.0.2mm) stipitate; style c.1.3mm, lobed to halfway; pedicel c.0.2mm. Male flowers: total length 1.7–2.2mm; calyx variously 2–3-lobed, white-hairy, grey, tube c.1mm; large petal hairy; anthers black. Seeds c.0.5 × 0.3mm, with small, transversely elongate reticulations.

Terai (Dulkajhar, Kuntimari). Wet ground, 150m. Fl. October–January.

4. E. hamiltonianum Martius; *E. oryzetorum* sensu F.B.I. p.p. non Martius.
Nep: *kane*.
Leaves gradually narrowed to subacute, callose-tipped apex, less than ½ length of longest scape, widest (2.5–3.2mm) at base. Scape 4.5–13cm. Capitulum silvery-white, broadly ovoid, 4–5mm diameter; receptacle elongate, densely villous (hairs to 1.5mm). Involucral bracts obovate, blunt, c.2 × 1.5mm, silvery-hyaline. Floral bracts oblanceolate, acuminate, 2.3–2.5 × 0.8–1.1mm, glabrous above, pilose near base, silver flushed grey. Female flowers: total length 2–2.5mm; sepals 2, linear-oblanceolate, very acute, 1.2–1.5 × 0.2mm, glabrous, grey; petals 3, largest 1–1.4 × 0.1mm, darker, other 2 narrower, silvery, glabrous; ovary stipitate (to 0.2mm); style c.1mm, lobed to halfway; pedicel to 0.5mm. Male flowers: total length 1.5–2mm; calyx equally 2-lobed, lobes falcate-oblong, acute, 1–1.2 × 0.3–0.4mm, glabrous, grey, tube 0.6mm; large petal glabrous; anthers black. Seeds c.0.4 × 0.25mm, reticulate, shining.
Terai (Kutambaria, Siliguri). Wet ground, 150m. Fl./fr. November–December.

5. E. alpestre Hook.f. & Thomson ex Körnicke. Fig. 24k–l.
Leaves gradually narrowed to acute apex, about equalling scapes, widest (c.1(–4.5)mm) at base. Scape 3–8(–13)cm. Capitulum black, hemispheric, 2.5–4(–6)mm diameter; receptacle slightly convex, glabrous. Involucral bracts oblong to obovate, blunt, 1.5–1.8(–2.5) × 1.3(–2)mm, glabrous, greyish-hyaline. Floral bracts oblanceolate, blunt, c.1.5 × 0.6mm, black, shortly and sparsely white-hairy above. Female flowers sessile: total length c.1.8mm; calyx c.1.5mm, spathe-like, slightly 3-lobed, black, glabrous; petals 3, distinctly clawed, claw c.0.5mm, blade swollen, narrowly elliptic, dark gland-tipped, c.1 × 0.5mm, white, surface bullate, glistening; ovary sessile. Male flowers: total length 1.5mm; calyx spathe-like, slightly 3-lobed, black, glabrous; petals minute, subequal, glabrous; anthers black. Seeds c.0.8 × 0.6mm, fat, smooth, shining.
Sikkim (Lachung, Ching-goo, Yakla; Chamnago, Lachen (Fyson, 1923)); **Chumbi** (Kungboo). 2440–3960m. Fl. August–October.

The Sikkim specimens seen are all very small plants; measurements in parentheses are taken from larger specimens from Khasia and SE Tibet.

6. E. bhutanicum Noltie. Fig. 24m–n.
Leaves gradually tapered to acute apex, less than ⅓ length scape, c.0.8mm wide at base. Scape 2–6cm, filiform, weak, 3–5-grooved. Capitulum black and white, hemispheric, 1.5–3mm diameter; receptacle slightly convex, glabrous.

Involucral bracts broadly oblong to obovate, blunt, 1.3–1.5 × 0.8–1.2mm, glabrous, blackish-hyaline. Floral bracts oblanceolate, finely acuminate, c.1.2 × 0.4mm, black above with short white hairs. Female flowers sessile: total length c.1mm; sepals 3, oblong-elliptic, acute, c.1 × 0.3mm, white-hairy, grey; petals linear, black-gland-tipped, 0.8 × 0.1mm, hairy, membranous, whitish; ovary sessile; style c.1mm, lobed ⅓ to halfway. Male flowers: total length c.1.3mm, calyx 3-lobed in upper ⅔, lobes c.0.8 × 0.2mm, white-hairy, grey; petals subequal, minute, hairy; anthers 1–2(–?3), black. Seeds c.0.4 × 0.3mm, finely transversely reticulate, densely covered with rows of minute white papillae.

Bhutan: C — Thimphu district (above Serbitang). Open swamp with scattered, grazed *Arundinaria*, 2600m. Fl. September.

Doubtfully recorded species:

E. edwardii Fyson.
Differs from any of the above in having involucral bracts exceeding capitulum. Recorded for 'Sikkim' by Fyson (1923) on the basis of a specimen at CAL (*Gamble* 1875).

E. gracile Martius.
Recorded from the Terai by Fyson (1923) on the basis of a specimen at CAL (Dulkajhar, *Clarke* 36953). Fyson's record apparently refers to a species with hairy involucral bracts and hairy leaves, but there is some doubt over the identity and typification of *E. gracile*.

Family 231. JUNCACEAE

Usually perennial, rhizomatous herbs. Leaves linear with sheathing bases, glabrous or with long ciliate hairs; sometimes basal scale leaves only present. Infl. cymose, simple or compound, terminal or pseudolateral, sometimes very condensed or reduced to a single flower; usually subtended by a spathe-like bract. Flowers bisexual, wind-pollinated, tepals 6 in two whorls, usually brownish, greenish or whitish and partly membranous. Stamens free, 3 or 6. Ovary superior of 3 fused carpels, 1–3-locular; stigmas 3. Fruit a loculicidal capsule; seeds 3 or many, often with membranous outer integument, caruncle sometimes present.

1. Plants completely glabrous; capsule many-seeded .... **1. Juncus**
+ Plants with long ciliate hairs on leaf margins and also usually on infl.;
  capsule 3-seeded .... **2. Luzula**

1. JUNCUS L.

Glabrous herbs, usually rhizomatous perennials, sometimes annual (species 1), rhizomes bearing non-flowering basal rosettes and flowering stems, sometimes stoloniferous. Flower stems with basal scale leaves and usually also laminar leaves with sheathing bases. Lamina flat, channelled, uni- or pluritubular, sometimes transversely septate. Sheath often continued upwards into membranous auricles. Flowers often aggregated into 'capitula', or sometimes borne singly, infls. often compound with capitula arranged in anthelate cymes where peduncles of lower capitula overtop the upper. Tepals 6; stamens 3 or 6. Capsule many-seeded, the membranous testa often produced into tails at one or both ends.

Care is needed for correct identification, especially of the alpine species, many of which are extremely similar. The following characters are the most important and really require examination with a dissecting microscope.

1. Leaf anatomy: necessary to cut hand sections (with a razor blade), both l.s. to see if transversely septate, and t.s. to see whether unitubular, etc.

2. Presence or absence of upper stem leaf: traditionally used, but should be treated with caution (e.g. in *J. benghalensis*, populations are variable in this character and in *J. allioides* the upper leaf seems not to be developed in stunted plants).

3. Stigma lobes: a very important character; some can usually be found, even on fruiting material.

4. Anther length and degree of exsertion. NB Need mature material for full development of filaments.

When collecting care must be taken to collect basal parts.

The extent of hybridisation in our area is not known, but is possibly an added cause of confusion, as known to be common in the genus.

1. Densely tufted annual, with leafy flowering stems branched from near base; flowers sessile, borne singly (Fig. 24o) .... **1. J. bufonius**
+ Not as above 2

2. Infl. branched, of 2 or more cymosely arranged, normally capitate, partial infls., or unbranched and pseudolateral on leafless stem 3
+ Infl. unbranched, a single terminal 'capitulum' (1–25-flowered), or a group of 1–3 pseudolateral flowers on a stem bearing at least 1 leaf 22

3. Flower stems leafless, with only brown, chaffy scale leaves at base; infl. apparently lateral (Fig. 24p) 4
+   Flower stems bearing at least one true leaf in addition to basal scale leaves; infl. terminal 5

4. Pith of flower stem interrupted with air spaces (Fig. 24q); stamens 6; capsule exceeding tepals .... **2. J. inflexus**
+   Pith of flower stem solid; stamens 3; capsule shorter than tepals
    3. J. effusus

5. Much of infl. non-fertile, composed of short feather-like branchlets of golden-coloured 'bracts' (Fig. 24s) .... **4. J. ochraceus**
+   Infl. composed of fertile flowers only 6

6. Leaves septate, tubular (Fig. 25b) or pluritubular (Fig. 24v) 7
+   Leaves not septate, flat and grass-like or 1–3-channelled 15

7. Tepals cream or white 8
+   Tepals brown or greenish 9

8. Capitula usually more than 3; leaves stout (usually over 1mm wide); marshes .... **12. J. grisebachii**
+   Capitula 1–2; leaves filiform (under 1mm wide); on rocks and trees
    13. J. chrysocarpus

9. Lowest bract leafy, exceeding infl.; inner tepals over (4.5–)5mm; capitula usually over 10mm diameter; alpine plants, occurring over 2750m 10
+   Lowest bract shorter than infl.; inner tepals under 3.5(–4.4)mm; capitula usually under 8mm; usually occurring below 2500m 12

10. Capitula up to 12; tepals reddish-brown, acute; leaves gradually tapered to fine point; capsule exceeding tepals .... **9. J. himalensis**
+   Capitula up to 4; tepals chestnut or blackish-brown, very acute to finely acuminate; leaves wider, apex blunt; capsule shorter than tepals 11

11. Tepals 8.7–13mm, very finely acuminate; capitula usually 3 or 4
    10. J. sphacelatus
+   Tepals 5–9mm, acute; capitula 2 .... **11. J. sikkimensis**

12. Leaves flattened, pluritubular in cross-section, septa not crossing whole
width of leaf and not conspicuous (Fig. 24v) .... **5. J. prismatocarpus**
+ Leaves unitubular, septate (usually conspicuously so) 13

13. Tepals greenish; infl. lax, capitula not overlapping, branches spreading
at wide angles .... **6. J. wallichianus**
+ Tepals brownish; infl. more dense, capitula overlapping, branches
more erect 14

14. Stamens 3 .... **7. J. leptospermus**
+ Stamens 6 .... **8. J. articulatus**

15. Leaf blades flat, grass-like 16
+ Leaf blades channelled at least on upper surface (V-shaped in section),
sometimes with 3 or 4 grooves (Y- or X-shaped in section) 21

16. Rootstock knobbly, woody; tepals brown or if pale and flushed pinkish
then inner tepals under 3.5mm 17
+ Rootstock slender of short, spreading rhizomes; tepals white or cream,
inner ones over 4mm 19

17. Capitula more than 6; tepals whitish tinged pink; stigma lobes under
0.8mm; stamens exserted .... **20. J. spumosus**
+ Capitula 1–3; tepals dark brown; stigma lobes over 2mm; stamens not
exserted 18

18. Capitula 2–3; inner tepals usually over 5mm; anthers over 2mm; stigma
lobes over 3mm .... **19. J. amplifolius**
+ Capitula 1–2; inner tepals under 4mm; anthers c.1.3mm; stigma lobes
under 2.5mm .... **22. J. nepalicus**

19. Sheaths of stem leaves with short, membranous auricles at apex; terminal capitulum to 1cm diameter; inner tepals to 3.5mm
15. J. gracilicaulis
+ Sheaths of stem leaves lacking auricles; terminal capitulum over 1.5cm
diameter; inner tepals over 4mm 20

20. Lowest bract leaf-like, erect, exceeding pseudolateral infl.; flowers suberect so capitula hemispheric; anthers only partly exserted **17. J. clarkei**
+ Lowest bract bristle-like, shorter than terminal infl.; flowers spreading
so capitula subspherical; anthers completely exserted . **18. J. hydrophilus**

21. Leaves robust, channelled above, V-shaped (sometimes asymmetrically) in section; capitula usually 3 or more; capsule to 4.7mm; seeds to 1.2mm (incl. tails) .... **14. J. concinnus**

+ Leaves filiform, 3- or 4-channelled, Y- or X-shaped in section; capitula usually 2; capsule over 4.9mm; seeds over 1.7mm (incl. tails)
16. J. khasiensis

22. Tepals white (sometimes becoming darker in fruit) 23

+ Tepals brown 36

23. Two or more plantlet-bearing, bract-like leaves present on upper part of stem (Fig. 25f); infl. 1–3-flowered .... **27. J. trichophyllus**

+ Bract-like leaves on upper stem 0 or 1, not proliferating; infl. more than 4-flowered, usually hemispheric 24

24. Stigma capitate, lobes not developed (Fig. 25j) 25

+ Stigma obviously 3-lobed, lobes spreading to erect 26

25. Lowest bract chestnut, membranous, similar to other bracts and not exceeding infl.; stem not grooved; bract-like leaf present on upper part of stem .... **26. J. cephalostigma**

+ Lowest bract with leaf-like tip exceeding infl.; stem deeply grooved; upper stem leaf absent .... **32. J. brachystigma**

26. Leaves closely and strongly septate (septa visible externally when dry) 27

+ Leaves not or only weakly septate (septa not visible externally even when dry) 28

27. Slender plant to 26cm; tepals straw-coloured; lowest bract leaf-like, greatly exceeding infl.; leaves filiform, the uppermost overtopping infl.
13. J. chrysocarpus

+ Robust plant to 40cm; tepals white, contrasting with chestnut bracts which scarcely exceed infl.; upper stem leaf not reaching infl.
23. J. allioides

28. Lowest bract with leaf-like tip overtopping infl. 29

+ Bracts all similar, membranous (often brown), all shorter than infl. . 32

29. Leaf sheath gradually narrowed into blade .... **29. J. leucomelas**

+ Conspicuous oblong, membranous auricles (sometimes brown) present at apex of leaf sheath (Fig. 25h) 30

30. Rhizomes very short, plant forming dense clumps .... **31. J. perpusillus**

+ Stoloniferous, stems borne singly, distant 31

31. Stolons stout; auricles membranous; upper stem leaf absent
28. J. kingii
+ Stolons filiform (c.0.3mm wide); auricles brown; upper stem leaf present in some plants .... **30. J. benghalensis**

32. Bracts brown 33
+ Bracts whitish (occasionally pink) 35

33. Capitulum large, c.2cm diameter; leaves swollen, glaucous, transversely septate (when sectioned) .... **24. J. glaucoturgidus**
+ Capitulum usually under 1.5cm; leaves not swollen, not glaucous, not transversely septate 34

34. Basal sheaths chestnut-coloured, shining; upper stem leaf present
25. J. leucanthus
+ Basal sheaths reddish-brown, dull; upper stem leaf absent
33. J. thomsonii

35. Leaf channelled above, V-shaped (sometimes asymmetrically) in section .... **14. J. concinnus**
+ Leaf X- or Y-shaped in section .... **16. J. khasiensis**

36. Leaf blade flat 37
+ Leaf blade filiform or tubular 38

37. Capsule over 3mm wide, dark brown, apex truncate and apiculate; stems only just exceeding leaves, basal leaves over 3mm wide, upper stem leaf absent; infl. bract lanceolate (over 2.5mm wide), only just exceeding infl. .... **21. J. minimus**
+ Capsule c.2mm wide, light brown, gradually tapered into beak; stems c.twice length of leaves, basal leaves under 2.5mm wide, upper stem leaf usually present; infl. bract linear (c.1mm wide), exceeding infl.
22. J. nepalicus

38. Infl. a terminal 2–5-flowered capitulum .... **34. J. triglumis**
+ Infl. pseudolateral (infl. bract leaf-like, erect), if terminal then 1-flowered 39

39. Leaves septate, unitubular; flowers usually 2–3; tepals 4.5–7.7mm; stigma lobes 4.5–9mm; (if plant very dwarf and 1-flowered, then flower

sessile, appearing lateral with lower bract leaf-like and exceeding flower and anthers exceeding filaments) .... **35. J. duthiei**
+ Leaves filiform, non-septate, bitubular; flower always single; tepals under 5mm; stigma lobes under 4.6mm 40

40. Flower very shortly pedicellate, appearing terminal, lower bract equalling or just exceeding flower; anthers longer than filaments; stem with up to 2 green leaves .... **36. J. uniflorus**
+ Flower sessile, appearing lateral — lower bract to 1.5cm; anthers shorter than filaments; stem leafless .... **37. J. bryophilus**

1. J. bufonius L. Fig. 24o.

Slender, tufted annual. Flower stems to 20cm, branched from near base. Stem leaves several, sub-basal, filiform, solid, channelled above, to 5cm long, 0.8–2mm wide. Auricles absent. Flowers borne singly, ± sessile, enclosed by 2 transparent, ovate-acuminate bracteoles. Tepals unequal: outer finely acuminate, green, margins narrowly membranous, (3.8–)6–6.5mm; inner acute, membranous margins wider, (3–)4.5–5mm. Stamens 6, shorter than tepals; filaments (0.8–)1mm; anthers (0.3–)0.7mm. Ovary narrowly ellipsoid, 1.7–1.8mm; style very short; stigma lobes very short, deflexed. Capsule narrowly ellipsoid-trigonous, apex truncate, (3–)5 × (1.4–)2.3mm, shorter than inner tepals, straw-coloured. Seeds c.0.5mm, pale brown, without membranous testa.

Bhutan: C — Thimphu/Punakha (Thimphu to Wangdi Phodrang (F.E.H.2)) and Bumthang (Jakar, above Gortsam) districts; **Sikkim** (Giagong; above Thango (Smith, 1911)). Wet muddy patch in pasture; roadside ditch, 2740–4570m. Fl. April–August.

2. J. inflexus L.; *J. glaucus* Ehrhart. Name at Tongsa: *narhum*. Fig. 24p–q.

Flower stems bearing no true leaves, densely tufted, stiffly erect, cylindric, conspicuously longitudinally ridged (rough to finger-nail), 36–96cm, pith white, interrupted with air spaces. Scale leaves: lower blackish-brown, to 6cm; uppermost usually greenish, aristate, to 18cm. Infl. a lax irregular, compound head appearing lateral (stem continued into acute point), branches slender, flowers small, each subtended by several small, overlapping bracteoles. Tepals lanceolate, very acute, unequal, outer 2.1–3.4mm, inner 1.8–2.9mm, wider than outer, midribs brownish or greenish, margins membranous. Stamens 6, shorter than tepals; anthers 0.5–0.6mm, equalling filaments. Ovary oblong-ellipsoid, 0.6–1.4mm; style very short (0.1–0.3mm); stigma lobes short (0.2–0.6mm). Capsule oblong- to ellipsoid-trigonous, usually slightly exceeding outer tepals, 2.1–2.8mm; beak c.0.2mm. Seeds c.0.5mm, not tailed, with membranous ridge on one side.

Bhutan: S — Chukka district (Chimakothi, Chukka); **C** —Thimphu, Tongsa, Bumthang and Tashigang districts (very common); **Sikkim** (Yoksum, Lachen, Zemu Valley, Chungthang, etc.); **Chumbi**. Damp grassy slopes, swamps, beside ditches and streams, 1400–3200m. Fl. April–August.

3. J. effusus L.

Differs from *J. inflexus* in having orange-brown scale leaves; flower stem smoother (grooves shallower, more numerous), pith lacking air spaces; infl. fewer-flowered, more compact; stamens 3; capsule usually shorter than tepals, obovoid-trigonous, tip truncate to emarginate.

Darjeeling (Tiger Hill); **Sikkim** (Yakla). 2710–3050m. Fl. July–October.

Evidently much rarer than *J. inflexus*, with which it has often been confused.

4. J. ochraceus Buchenau; *J. tratangensis* Satake. Fig. 24r–t.

Flower stems tufted, 8–28cm. Leaves mostly basal, filiform, grooved, channelled above, bitubular, septate (septa not visible externally). Auricles blunt, membranous, pale brown, free parts c.0.3mm. Infl. branched, dense, to 12cm, majority of partial infls. non-fertile, composed of pedicellate feather-like fascicles of narrowly lanceolate, aristate, ultimately golden-brown sterile 'bracts'. Fertile flowers usually inconspicuous, situated at base of partial infls., occasionally predominating. Tepals lanceolate, acute or mucronate, equal, 3.3–4.2mm, straw-coloured. Stamens 6, shorter than tepals; filaments 1.2–1.5mm; anthers 1.3–2mm, often twisted. Ovary widely ellipsoid, 1.4–2.7mm, abruptly contracted into long style 1.8–2mm; stigmas erect, twisted, 1.4–2.7mm. Capsule widely ovoid-trigonous, c.2.4 × 1.5mm, abruptly contracted into beak c.1.8mm.

Bhutan: S — Chukka district (Chimakothi, Jumudag to Chasilakha, Putlibhir); **C** — Thimphu (Dotena, Taba, Thimphu to Dochu La, Mishina to Thimphu), Punakha (E side of Dochu La, W side of Pele La), Tongsa (S of Tongsa, Tongsa to Yuto La) and Tashigang (Kulong Chu) districts; **Darjeeling** (Senchal, Kurseong, Sureil, Tiger Hill, Sukia Pokhri to Manibhanjang, Tonglu, Peshok, Sittong; Ghum (F.E.H.1)); **Sikkim** (Keydangthang, Lachen, Karponang, Yoksum, Lachung, 11km N of Gangtok; Laghep (Smith, 1913)). Semi-shaded wet places: streamsides, rocks, cliffs, grassy places, roadsides, 1220–3660m. Fl. September–May.

The predominantly fertile form (described as *J. tratangensis*) has been recorded from Senchal, and Tongsa to Yuto La.

231. JUNCACEAE

5. J. prismatocarpus R. Brown; *J. leschenaultii* Gay ex Laharpe; *J. latior* sensu F.E.H.2, non Satake. Fig. 24u–v.

Untidy tufted perennial. Flower stems to 25cm. Leaves up to 3, evenly spaced along stem, blades laterally compressed, tapered to acute apex, 2–6cm, to 1.7mm wide, composed of 3 or more septate tubes (seen in cross-section), septa of adjacent tubes not aligned, usually visible when dry. Sheaths slightly inflated; auricles rounded, free parts c.0.2mm, transparent. Infl. much branched, subtended by long leaf-like bract to 4cm, flowers in groups of 3–9. Tepals narrowly lanceolate, very acute, equal or inner slightly shorter, 3–4.4mm, greenish or straw-coloured (often tinged red), margins membranous. Stamens 3, much shorter than tepals; filaments c.1.2mm; anthers 0.4–0.7mm. Ovary oblong to narrowly ovoid, 1.5–2.4mm, tapered suddenly to very short beak c.0.1mm; stigma lobes spreading, 0.9–1.2mm. Capsule trigonous, very shortly mucronate, segments narrowly lanceolate, exceeding tepals, 3.8–4.2mm, reddish-straw coloured, shining. Seeds c.0.5mm, pale brown with dark tip, not tailed.

Bhutan: S — Chukka district (W of Gedu); **C** — Thimphu (Taba, Bongde Farm, Atsho Chhubar, Chapcha), Punakha (Chuzomsa, Samtengang to Tashi Choling, Yuwak), Tongsa (Pertimi, Tongsa to Uto La Road), Mongar (Lhuntse Dzong) and Tashigang (Kanglung, Tashyi Yangtse) districts; **Darjeeling** (Rishap, Sittong, Mongpu, Tista, Sureil, Ghumpahar; Takdah, Rayang, Palmajua to Rimbick (F.E.H.1)); **Sikkim** (Mamring, Yoksum; Karponang, Phadonchen (Smith, 1913), Cheungtong (Smith, 1911)). Wet places (irrigation channels, marshy pasture, paddy fields, roadsides etc.), 250–2500m. Fl. April–August.

Parker (1992) records it as a minor weed of rice where flooding has not been continuous occurring over 1000m in Thimphu and Punakha districts and probably most other districts [with cultivation].

6. J. wallichianus Laharpe; *J. prismatocarpus* sensu F.B.I., p.p. non R. Brown.

Differs from *J. prismatocarpus* in its unitubular leaves which are visibly septate externally; auricles longer (free part c.1mm), more pointed; tepals slightly broader and shorter (2.2–3.3mm); capsule shorter (3–3.8(–4.2)mm), beak more pronounced.

Bhutan: S — Chukka district (Bunakha, below Chimakothi); **C** — Thimphu (Paro, above Thimphu Hospital, below Dotena, Bongde Farm), Punakha (Botakha), Tongsa (S of Tongsa, above Rukubji) and Bumthang (above Dhur) districts; **Darjeeling** (Rishi La, Kalimpong, Ghum to Kurseong, Tiger Hill, Mongpu, Senchal, Sukia Pokhri to Manibhanjang); **Sikkim**

254

(Lachen, Lachung, Gangtok, Phodong Gompa). Streamsides, marshes and wet slopes in open or in forest; rice paddies, 1100–3100m. Fl. May–August.

Flowers sometimes developing into plantlets.

7. J. leptospermus Buchenau.
Flower stems stiffly erect, 37–85cm. Leaves up to 5, arranged evenly along stem, blades cylindric, gradually tapered to acute apex, unitubular, septate, septa clearly visible externally, to 10–20cm. Sheaths slightly inflated; auricles acute, free parts c.2 × 1mm, pale brown, membranous. Infl. terminal, branched, with 9(–37) up to 19-flowered capitula, subtended by stiff, erect leaf-like bract up to ⅔ length of infl.; bracts of main infl. branches straw-coloured, ribbed, oblong, aristate or sometimes trifid. Tepals lanceolate, sub-equal or inner longer, 2.4–3mm, pale brown, margins membranous. Stamens 3, shorter than tepals; filaments flattened, c.1.6mm, whitish; anthers oblong, 0.6mm, cream. Ovary ovoid, tapered into beak, total length 2.4mm; style c.0.2mm; stigma lobes spreading, c.0.9mm. Capsule exceeding tepals, strongly trigonous, segments narrowly lanceolate, contracted above into short (c.0.7mm) beak, total length to 4.6mm, chestnut, shining. Seeds not tailed, 0.6mm, pale brown.
Bhutan: C — Thimphu (S of Dotena) and Tongsa (Tongsa) districts. Marshy hillside; open, wet grassy slope, 2350–2610m. Fl./fr. July.

8. J. articulatus L.; *J. lampocarpus* Ehrhart.
Differs from *J. leptospermus* in having flowers with 6 stamens; capsule only just exceeding tepals, c.2.5mm, very shortly (under 0.1mm) mucronate.
Bhutan: C — Thimphu district (Paro, Bongde Farm, Thimphu, Shaba). Shallow muddy pool beside river; roadside ditch; banks of rice paddies (occasionally in paddies), 2290–2500m. Fr. July–August.

9. J. himalensis Klotzsch.
Stoloniferous. Flower stems 19–59cm, tufted. Scale leaves brownish. Stem leaves up to 4, stiffly erect, arranged along stem, blade narrow, semi-cylindric, gradually tapered to fine point, channelled above, margins of channel very finely serrulate, septate (septa not visible externally), to 30cm, 0.9–1.9mm wide. Auricles conspicuous, free part c.1 × 2mm, sometimes tinged brown. Infl. terminal, branched, with up to 12 unequally peduncled, 3–10-flowered capitula, subtended by leaf-like bract exceeding to twice length of infl.; flowers subsessile. Tepals lanceolate, acute, equal, 4.5–6.2mm, reddish-brown, outer with greenish midribs, inner with membranous margins and tips. Stamens shorter than tepals; filaments 2.7–4.5mm; anthers 0.9–1.5mm. Ovary narrowly ellipsoid, 2.6–4.5mm, tapered into style 1–1.7mm; stigma lobes 1.9–3.6mm. Capsule

narrowly oblong- to ellipsoid-trigonous, conspicuously exceeding tepals, 6.1–8.3mm, beak 0.7–1.4mm, chestnut, shining. Seeds c.0.5mm, 2-tailed, total length c.3mm.

Bhutan: C — Thimphu (above Motithang, below Darkey Pang Tso) and Bumthang (above Gortsam) districts; **N** — Upper Mo Chu (Tharizam Chu) and Upper Bumthang Chu (below Ju La) districts; **Chumbi**; **Sikkim** (Prek Chu, Chamnago, Lachen, Lachung, Yumthang, Samding, Thangu, Gamothang, etc.). Hillsides and gravelly streamsides (in open or in *Abies* forest), 2750–4420m. Fl. June–August.

10. J. sphacelatus Decaisne. Fig. 24w–x.

Flower stems stout, 14–66cm, clothed at base with remains of old leaves. Stem leaves 2–3, blade semi-cylindric with large central hollow, tip blunt, often dark-coloured, channelled above, margins of channel not serrulate, weakly septate, septa not visible externally, 5–13cm, 1–2.3mm wide. Sheaths loose, usually pale; auricles conspicuous, pointed, free part to 3mm, often brownish. Infl. terminal, capitula (2–)3(–4), 3–6-flowered, finally distant; lowest bract leaf-like, overtopping infl., base inflated, spathe-like. Flowers long-pedicelled (to 4mm); tepals narrowly lanceolate, finely acuminate to almost filiform apex, subequal or outer longer (6.5–11(–13)mm) and wider than inner (6.3–9.7(–13)mm), dark reddish-brown or chestnut with paler tips and inner with paler margins, becoming blackish. Stamens shorter than tepals; anthers (0.9–)1.4–2.5mm shorter than filaments (2.5–4.4mm). Ovary ellipsoid, abruptly contracted into short style, 1–2.2mm; stigma lobes 2–3.7mm. Capsule narrowly oblong- to narrowly ellipsoid-trigonous, shorter (occasionally slightly longer) than tepals, 4.8–7.5(–8.5)mm, abruptly contracted into short beak. Seeds 0.6–0.7mm, 2-tailed, tails long, thin, total length 2.8–2.9mm.

Bhutan: C — Thimphu district (above Pajoding, below Laname Tso); **N** — Upper Mo Chu (Shingche La, Soe, Lingshi, Yale La) and Upper Bumthang Chu (Waitang, below Ju La) districts; **Sikkim** (Prek Chu, Samding, Lachung, Yakla, Dzongri, Changu, Llonakh, Phalut, Gurudongmar; Gamothang (F.E.H.1)); **Chumbi**. Among gravel and rocks by streams; rocky slopes, 3660–4880m. Fl. June–July.

Chumbi and Bhutan specimens have unusually long tepals (over 10mm) and may be worth recognising at subspecific rank. Specimens from Lachung and Samding (Sikkim) have unusually long capsules and small anthers and could possibly be hybrids with *J. himalensis* — similarly intermediate plants occur in Nepal and further west.

11. J. sikkimensis Hook.f.; *J. pseudocastaneus* (Lingelsheim) Samuelsson.

Stoloniferous, stolons creeping, stout (c.2.5mm diameter), shining reddish- or blackish-brown, clothed with oblong-lanceolate, striate scales. Flower stems

7–18(–22)cm, relatively stout, base with golden-brown (sometimes reddish-brown) remains of old leaf sheaths. Scale leaves: upper shortly aristate. Stem leaf 1, blade narrowed below apex to blunt, often darkened, tip, semi-cylindric with large central hollow, channelled above, septate (septa not always visible externally), 4–24cm, 1–2mm wide. Auricles short, free part c.0.3 × 0.7mm, often tinged brown. Infl. appearing lateral, with 2 unequally peduncled, 2–5-flowered capitula, lowest bract stout, erect, exceeding infl., base spathe-like, often darkened. Flowers shortly pedicelled; tepals lanceolate, somewhat irregular in length, outer usually longer, acute to mucronate, inner sometimes blunt and shorter, 5–8.5(–9)mm, dark brown to blackish (tips sometimes membranous and inner with membranous margins), somewhat shining. Stamens shorter than tepals; anthers often twisted (1.8–)2–2.5(–3.1)mm; filaments 0.5–1.5mm. Ovary 1.7–2.5mm, tapered into long ((2–)2.4–3mm) style; stigma lobes erect, twisted, 3.1–5(–5.5)mm, yellow-green. Capsule narrowly ellipsoid-trigonous, shortly beaked, shorter than or equalling tepals, 5.5–6.2 × 2.5–3mm. Seeds 0.5–0.9mm, 2-tailed, tails subequal, total length c.2.5mm.

Bhutan: C — Ha (Yakuna) and Thimphu (above Pajoding, below Darkey Pang Tso) districts; **N** — Upper Mo Chu (Shingche La, Chomo Lhari), Upper Bumthang Chu (below Ju La) and Upper Kulong Chu (Shingbe) districts; **Sikkim** (Gongchung, Lachen, Kunkola, Cheumsanthang, Yumcho La, Chakung Chu, Tosa, Zemu Valley, Kapoop, Bikbari to Chaunrikiang, near Kabur, Jemathang); **Chumbi**. Damp slopes; marshes and streamsides; peaty pools; sandy pasture, 3660–4880m. Fl. June–September.

12. J. grisebachii Buchenau.

Stoloniferous. Flower stems stiffly erect, 20–73cm. Stem leaves up to 3 near base and sometimes 1 on upper part, blade semi- to sub-cylindrical, channelled above, septate, septa nearly always visible externally, to 32cm, 0.8–2.2mm wide. Auricles conspicuous, blunt, free part c.3mm long, pale brown, membranous. Infl. terminal, branched, of 2–7(–13) hemispheric 7–12(–16)-flowered capitula, 1.2–1.8cm diameter; lowest bract erect, leaf-like, usually exceeding (to 2× length) infl. Tepals lanceolate, acute, outer 4.8–6.5mm, usually shorter than inner (5.3–6.7mm), whitish or pale straw-coloured. Stamens 6; filaments 4–6mm; anthers linear, 2.4–3.5mm, partly to completely exserted, cream. Ovary ellipsoid, 2.1–3.6mm, gradually tapered into long style (2.1–3.6mm); stigma lobes stout, twisted, 1–1.6mm. Capsule broadly ellipsoid-trigonous, sometimes longer and narrower and slightly curved, (4–)4.4–6.2mm, with short beak (0.3–1.8mm), orange-brown. Seeds 0.5–0.7mm, 2-tailed, tails subequal, total length 2–2.5mm.

Bhutan: C — Thimphu (Dochu La, hill above Thimphu Hospital, Paro, Chelai La, Pajoding), Tongsa (Rukubji, Pele La), Bumthang (Yuto La, Ura

La, Gyetsa, above Gortsam) and Tashigang (Sana) districts; **N** — Upper Mo
Chu (Pari La, N of Kohina) and Upper Bumthang Chu (above Lambrang)
districts; **Darjeeling** (Sandakphu, Phalut, Tonglu); **Sikkim** (Lachen, Chamnago,
Changu, Dzongri, etc.); **Chumbi**. Bogs; wet pasture; banks, often in conifer
forest; streamsides, 2300–4270m. Fl. July–October.

13. J. chrysocarpus Buchenau.

Stoloniferous, stolons slender, stems distant. Flower stems 10–26cm. Stem
leaves usually 2, inserted ⅓ and ⅔ way along stem, up to 2 basal leaves
sometimes also present; blades filiform, evenly tapered to acute apex, deeply
ridged, channelled above, unitubular, closely and regularly septate, septa usually visible at least when dry; upper leaf usually overtopping infl., to 14cm,
under 1mm wide. Auricles conspicuous, free part 0.5–1 × 0.5mm. Infl. a single
(occasionally 2 or 3) 3–15-flowered capitulum; lowest bract strongly ridged,
extended into leaf-like point equalling to twice length of capitulum. Tepals
narrowly lanceolate, acute, ridged, straw-coloured, outer usually shorter
(4.2–6.4mm), inner 4.7–7mm. Filaments 2.9–6mm; anthers partly exserted at
anthesis, 1.9–3.5mm. Ovary ellipsoid, 2.5–3.5mm, gradually tapered into style
1.8–3mm; stigma lobes erect 1–1.5mm. Capsule widely ovoid- to ellipsoid-
trigonous, 4.9–5.8 × 3–3.5mm, abruptly contracted into beak 0.5–2mm,
golden-brown, shining. Seeds c.0.8mm long, 2-tailed, total length c.2.2mm.

Bhutan: C — Thimphu (below Darkey Pang Tso), Tongsa (W side of
Yuto La), Bumthang (above Gortsam) and Tashigang (above Balfi) districts;
Sikkim (Kalapokri, Islumbo, Dzongri, Lachen, Lachung, Yakla, Sherabthang,
Yeumtong, Phullalong, Jelap La, Mon Lapcha to Phedang). On rocks and
trees in *Abies/Rhododendron* forest; wet cliff-ledges, 2440–4420m. Fl. August–
September.

Some of the Bhutan specimens are unusual in being 2- or 3-headed. Closely related to
J. grisebachii, of which it is perhaps an ecotype.

14. J. concinnus D. Don; *J. elegans* Samuelsson; *J. albescens* Satake, non
(Lange) Fernald; *J. yoskisukei* Goel; *J. luteocarpus* Satake. Plate 11.

Rhizomes shortly creeping, stems tufted. Flower stems 11–30cm. Scale
leaves 1 or more, pale, chaffy. Stem leaves up to 3, evenly spaced; blade
pluritubular, channelled above, V-shaped (usually asymmetrically) in cross-
section, not septate, tip blunt, 3.5–21cm, 0.8–2.5(–3.2)mm wide. Auricles
conspicuous, free parts 0.5–1mm, transparent or whitish-membranous. Infl.
with (1–)3(–9) unequally peduncled, 6–14(–17)-flowered capitula; lowest bract
leaf-like, equalling longest peduncle or sometimes exceeding infl. Tepals lanceolate, acute, unequal, outer keeled below, 2.1–4mm, inner longer (2.9–4.7mm),

shining, whitish or pale straw-coloured. Filaments 2.8–5.7mm; anthers 1–1.5mm, usually completely exserted at maturity. Ovary 1.1–2.6mm; style 1.1–1.8mm; stigma lobes 0.7–1.1mm. Capsule narrowly ovoid-trigonous, 2.6–3mm, contracted into short beak 0.6–0.9mm, whitish. Seeds 0.6–0.7mm, 2-tailed, usually one slender and longer than other, sometimes both short, total length 0.7–1.2mm.

Bhutan: C — Thimphu, Punakha/Tongsa, Tongsa, Bumthang and Mongar districts; **N** — Upper Mo Chu and Upper Kuru Chu districts; **Sikkim** (Chamnago, Talung, Rinchinping, Zemu, Rookah). Grassy hillsides; banks in spruce and pine forests; damp roadsides and streamsides, 2130–4300m. Fl. May–September.

Variable in flower size, infl. form and number of capitula and probably resolvable into more than one taxon, but further work is required. The following three are the most distinct forms.

1. Specimens from Sikkim (Lachen, *Hooker* s.n., K), Chumbi (*King's Coll.* 620, K), Upper Mo Chu district (Laya, *Sinclair & Long* 5120, E; Tharizam Chu, *Sinclair & Long* 5306, E) and Ha district (Kale La to Ha, *Bedi* 315, K) differ in being more robust with thicker flower stems (0.8–1mm), the lower bract always exceeding the infl. and above all in the large golden-brown capsules (3.5mm + beak 1.5mm) and seeds with a minute tail at one end only. It seems to be indistinguishable from the typical form at flowering.

2. A specimen from Chendebi (*Gould* 715, K) appears to be a form of this species (or of the form described in note 1), but merits further investigation. It is very tall (to 42cm), 3 out of 5 stems are single-headed and the flowers are very large (tepals to 5.2mm). It superficially resembles *J. membranaceus* Royle ex D. Don, but the leaves are of the *concinnus* type rather than unitubular and septate. Similar specimens have been seen from E Nepal.

3. A very attractive form with small flowers, very numerous capitula and conspicuous pinkish bracts was seen growing with the typical plant in Thimphu district (below Pajoding, below Ragyo) in *Picea* forest — also a Hooker specimen from Sikkim.

15. J. gracilicaulis A. Camus.

Differs from *J. concinnus* in its leaves which are flat (as in *J. clarkei*) and not channelled, 1.4–2.5mm wide, sometimes with conspicuous cross-veinlets.

Bhutan: C — Thimphu district (above Ragyo, Chelai La); **Sikkim** (Lachen, Nathost, Talung). Mossy rocks in fir forest, 2740–4300m. Fl. June–August.

Almost certainly better treated as a forma of *J. concinnus*, under which it was included by Hooker (F.B.I.).

16. J. khasiensis Buchenau.

Flower stems 6–12(–28)cm, slender. Scale leaves 1 or 2, pale, chaffy. Stem leaf 1, sub-basal, blade filiform with 3 or 4 deep channels, Y- or X-shaped in cross-section, pith solid, not septate, to 10cm, 0.4–0.7mm wide. Auricles very short 0.1(–0.6)mm. Infl. with (1–)2(–5) unequally peduncled 3(–6)-flowered capitula, secondary peduncle stiffly erect so capitula super-posed; lowest bract leaf-like, half length to equalling longer peduncle. Tepals narrowly lanceolate, acute, subequal or outer shorter (3.5–4.5mm) than inner (4–4.9mm), pale straw-coloured. Anthers (1.3–)1.7mm, exserted from tepals. Capsule narrowly ellipsoid, tapered into persistent style, exceeding tepals, total length 4.9–5.7mm, golden brown; stigma lobes 1–1.7mm. Seeds 0.8–0.9mm, 2-tailed, one wider than other, total length 1.7(–2.4mm).

Bhutan: C — Thimphu (above Ragyo), Tongsa (28km W of Tongsa) and Bumthang (above Gortsam) districts; **N** — Upper Mo Chu district (between Gasa and Pari La; Chamsa to Yabuthang (F.E.H.2)); **Sikkim** (Karponang). Wet cliffs and rocks in open or in juniper or fir forest, 2290–3510m. Fr. August–September.

17. J. clarkei Buchenau.

Rhizomatous, rhizomes slender. Flower stems 15–33cm. Stem leaves: blades flat, grass-like, gradually tapered to acute apex, upper usually over-topping infl., margins sometimes (var. *marginatus* A. Camus) narrowly mem-branous, minutely toothed at junction with sheath, to 23cm, 2–4.5mm wide, veins, especially midrib, conspicuous on undersurface. Sheaths often reddish, without auricles. Infl. apparently lateral of 1–4 unequally peduncled 4–12-flowered, hemispheric capitula; lowest bract erect, leaf-like greatly exceeding infl.; flowers ± erect, shortly pedicelled. Tepals lanceolate, outer acute, 4.1–7.5mm, midrib usually greenish, inner broader, usually longer, sometimes blunt, 4.9–7.5mm, pale straw-coloured. Filaments (3.4–)4–6mm; anthers 1.8–2.5mm, usually at least partly exserted. Ovary narrowly ovoid, inflated, gradually tapered into style, ovary + style 6.5–9.5mm, straw-coloured; stigma lobes 0.3–1.5mm. Capsule narrowly ovoid, tapered into persistent style, long-exserted from tepals, total length to 1.2cm, pale straw-coloured. Seeds not seen.

Bhutan: C — Thimphu (above Ragyo) and Bumthang (above Gortsam) districts; **N** — Upper Mo Chu district (a doubtful record (F.E.H.3)); **Darjeeling** (Tonglu); **Sikkim** (Gowsar, Cheungtong, Phadonchen, Chola, Kalapokri, Yoksum, etc.); **Chumbi**. Wet, mossy rocks, 1830–3760m. Fr. August–October.

18. J. hydrophilus Noltie.

Differs from *J. clarkei* in having stem leaves shorter than stem, narrower (under 3.3mm wide); infl. appearing terminal as lowest bract with blade shorter than infl., bristle-like; capitula ± spherical with flowers spreading in all directions, flowers more slender; anthers completely exserted.

Bhutan: C — Thimphu (Dotena) and Bumthang (above Gortsam) districts; **Sikkim** (Prek Chu below Bakkim). Very wet cliff-ledges in oak or oak/conifer forest, 2300–3510m. Fl./fr. July–August (earlier than *J. clarkei*).

19. J. amplifolius A. Camus.

Rootstock stout, knobbly, woody, bearing persistent, fibrous remains of old leaves, non-flowering rosettes and singly-inserted flower stems. Scale leaves 1 or more, reddish. Rosette leaves to 15 × 0.3cm. Flower stems 8–35cm. Stem leaves up to 3, evenly spaced, blades flat, gradually tapered to acute apex, smooth above, ridged beneath, margins with narrow transparent border, shorter than stems. Sheaths lacking auricles. Infl. terminal, with (1–)2–4 unequally peduncled 3–4(–6)-flowered capitula, capitula c.1cm diameter at flowering; lowest bract leaf-like, shorter than to just overtopping infl. Tepals lanceolate, acute to finely acuminate, subequal or outer shorter (4.4–5mm) than inner (4.5–6mm), c.1.5mm wide, dark reddish-brown, midrib sometimes greenish. Stamens shorter than tepals; anthers (1.5–)2–2.4mm, shorter than to equalling filaments. Ovary ellipsoid, 1.9–4.5mm, gradually narrowed into beak-like style (2–3.5mm); stigma lobes erect, twisted, 2–5mm. Capsule ellipsoid, c.5 × 2.5mm, abruptly contracted into slender, exserted beak c.3mm, golden-brown to chestnut. Seeds golden, c.0.5mm, 2-tailed, tails long, thin, total length to 3.7mm.

Bhutan: C — Tongsa district (Yuto La); **Sikkim** (Mon Lapcha, Gnathong). Wet pathside in *Abies/Rhododendron* forest; streamsides and bogs, 3270–4000m. Fl. June–July.

20. J. spumosus Noltie.

Similar to *J. amplifolius* vegetatively, but differing as follows: infl. decompound, with more (to 20) and smaller (to 7mm diameter at flowering) capitula; tepals smaller (outer 2.4–2.8mm, inner 3.2–3.5mm), whitish membranous, the outer with conspicuous greenish midrib, flushed pinkish, narrower (to 1mm) than inner; anthers smaller (under 1mm), exserted, much shorter than filaments; stigma lobes very short (c.0.6mm), ± spreading, not twisted.

Bhutan: C — Tongsa and Bumthang districts (W and E side of Yuto La). Gravelly slope (landslip) in wet *Abies* forest, 3350m. Fl./fr. August.

21. J. minimus Buchenau.

Rhizomes short. Flower stems 1–10cm, tufted. Scale leaves 1–2, pale brown. Stem leaves up to 4, sub-basal, blades flat, apex blunt, margins slightly wavy, to 4cm, to 4mm wide, prominently parallel-veined when dry. Margins of sheaths

membranous, tapered into blade, auricles lacking. Infl. a single terminal (sometimes appearing lateral) 2–6-flowered capitulum; lowest bract lanceolate, leaf-
like, blunt-tipped, only just exceeding capitulum at flowering, over 2.5mm wide;
other bracts hyaline, chestnut, acute. Tepals narrowly lanceolate, acuminate to
mucronate, outer shorter than inner, 3.8–5mm, brown with green midrib, becoming chestnut in fruit. Stamens usually shorter than tepals; anthers 0.8–1.3mm,
shorter than filaments (2.7–3.7mm). Ovary broadly ellipsoid, 2.5–3.4mm; style
short 0.5–1.5mm; stigma lobes erect, twisted, 1.5–2.5mm. Capsule oblong- to
ellipsoid-trigonous, exceeding tepals, 5–6mm, over 3mm wide, truncate with
short apiculate beak (c.0.4mm), dark chestnut, shining. Seeds pale brown,
c.0.5mm, 2-tailed, tails unequal, total length c.2.1mm.

Bhutan: N — Upper Kulong Chu district (N side of Me La); **Sikkim**
(Zemu Valley, Yumtso La, Kinchinjow Glacier). Open area in rhododendron
forest, 3900–5490m. Fl. July.

22. J. nepalicus Miyamoto & H. Ohba.

Differs from *J. minimus* as follows: basal leaves under 2.5mm wide; upper
stem leaf usually present; commonly with two capitula; lowest bract linear,
exceeding infl.; tepals smaller, inner 3.3–4mm; capsules narrower (under
2.5mm), paler brown, gradually narrowed into beak.

Sikkim (Lachen). 3050–3660m. Fl. June.

Intermediate between *J. minimus* and *J. amplifolius*.

Species 23–33 all have a terminal hemispheric capitulum-like infl. with white flowers,
the tepals are persistent and become suffused with purplish in fruit. They all grow in
damp places above the treeline, often many species growing together.

23. J. allioides Franchet; *J. concinnus auct.*, non D. Don. Fig. 25a–d.

Rhizomes short. Flower stems 8–40cm, densely tufted. Scale leaves not
tightly encircling stem, brown or reddish-brown, shining. Leaves of non-

---

FIG. 25. **Juncaceae** II.
a–d, **Juncus allioides**: a, habit (× ⅔); b, l.s. of leaf showing transverse septa (× 3);
c, capsule (× 4); d, seeds (× 8). e, **J. leucanthus**: stem base showing sheathing scale
leaves (× 1.5). f, **J. trichophyllus**: habit (× ⅔). g–h, **J. benghalensis**: g, habit showing
slender stolons (× ⅔); h, leaf-sheath auricles of stem leaf (× 3). i–j, **J. brachystigma**:
i, habit (× 1.5); j, capitate stigma (× 8). k, **J. duthiei**: habit (× ⅔). l, **J. uniflorus**:
habit (× 1.5). m–p, **Luzula multiflora**: m, habit (× ⅔); n, leaf margin (× 4); o, dehisced
capsule (× 4); p, seed showing terminal caruncle (× 8). q, **L. effusa**: habit (× ¼).
Drawn by Mary Bates.

flowering shoots slender, commonly shorter than flower stems. Stem leaves normally two, lower sub-basal, blade ± cylindric, shallowly grooved above (sometimes also beneath), unitubular, septate, septa normally visible externally when dry, tip blunt, 4–15cm, 0.4–1.6mm wide; upper leaf normally halfway along stem (sometimes absent in small plants) with conspicuous sheath and bristle-like blade to 4cm. Auricles brownish, blunt, large, the free part 0.7–1.4 × 1.7mm. Infl. dense, subglobose, 10–25-flowered; lowest bracts spathe-like in bud, equalling or slightly exceeding capitulum, broadly lanceolate, to 5mm wide, ribbed, shortly aristate, usually chestnut; flowers distinctly pedicellate. Tepals narrowly lanceolate, subacute, equal, 4.7–6(–7.6)mm, whitish. Filaments exceeding tepals at maturity; anthers normally exserted, linear 2–2.6(–3)mm. Ovary narrowly ellipsoid, 2.9–3.4mm, abruptly contracted into style 1.5–2.5mm; stigma lobes stout, 0.8–1.1mm. Capsule narrowly ellipsoid-trigonous, 3.9–5.3(–6.9)mm, beak partly exserted c.1mm. Seeds c.0.8mm, 2-tailed, total length 1.7–2.2mm.

Bhutan: C — Ha (between Kale La and Ha), Thimphu (Dotena, Pajoding, Ragyo to Darkey Pang Tso), Bumthang (above Gortsam) and Mongar (Ghijamchu) districts; **N** — Upper Mo Chu (Phile La, Tharizam Chu), ?Upper Mangde Chu (Jintang) and Upper Bumthang Chu (Waitang, above Lambrang) districts; **Sikkim** (Megu, Tangu, Lachung, Lachen, Rathong River, Tsomgo Lake, Chamnago, Bensar, Samiti, below Lam Pokhri); **Chumbi**. Bogs; river-banks; damp turf on hillsides; glacial sand by lake, 3050–4300m. Fl. June–August.

24. J. glaucoturgidus Noltie.

Differs from *J. allioides* as follows: leaves of non-flowering shoots equalling or exceeding stems; leaves swollen, glaucous, septa few, weak, not visible externally even when dry; bracts wider (7–8mm).

Bhutan: N — Upper Bumthang Chu district (Kantanang); **Sikkim** (Samiti). Open sandy slopes among dwarf shrubs. Fl. July.

25. J. leucanthus Royle ex D. Don. Fig. 25e.

Flower stems 8–32cm. Scale leaves up to 4, sometimes aristate, to 3cm, tightly encircling stem, chestnut, shining. Stem leaves usually 2, stiffly erect, usually inserted ½ and ⅔ way along stem; lower leaf blade cylindric, sometimes flattened, slightly grooved to deeply channelled above, not septate, reaching at least base of upper leaf, sometimes to infl., gradually tapered to subacute apex, to 10cm, to 1.4mm wide; upper leaf with long, slightly inflated sheath and short bristle-like blade to 2cm. Auricles conspicuous, oblong, blunt, free part to 0.9–1.3mm, often brown. Infl. hemispheric, 4–11-flowered, to 1.7cm diameter; lower 2 bracts subequal, equalling capitulum, lanceolate, dark chestnut, shining; pedicels of lower flowers to 3mm. Tepals lanceolate, subacute or inner blunt, equal 4.5–5.6mm. Filaments (2.2–)4.2–5.2mm; anthers 2–2.9mm, usually completely exserted. Ovary (1.5–)2.2–2.5mm, abruptly contracted into style 2–2.5mm; stigma lobes stout, short, erect, c.0.3mm. Capsule ellipsoid-trigonous, 3–3.3mm, beak 0.7–1.1mm, dark brown, shining. Seeds c.0.8mm, 2-tailed, tails small.

Bhutan: C — Thimphu district (above Pajoding, above Ragyo); **N** — Upper Mo Chu district (Timuzam; Chebesa to Lingshi (F.E.H.2)); **Sikkim** (Gamothang, Sherabthang, Phedang to Dzongri, Lachen, Chamnago, Jamlinghang to Bikbari). Damp ground and flushes in open or in rhododendron scrub; moraines; on mossy rocks, 3100–4280m. Fl. May–August.

In the field *J. leucanthus* is very difficult to tell from *J. allioides* (in which the septa are often not visible externally when fresh); the basal sheaths of the former are, however, characteristic.

A dwarf form from Dzongri (6.8cm high) was described as var. *alpinus* Buchenau.

26. J. cephalostigma Samuelsson.

Rhizomes short. Flower stems slender, 6–20cm, tufted. Scale leaves 1 or more, shortly aristate, tightly encircling stem, pale- to orange-brown. Stem leaves usually 2, lower sub-basal, upper ⅔ way along stem; lower leaf blade linear, cylindric, grooved above (sometimes also beneath), non-septate, apex ± acute, to 8cm, under 1mm wide; upper leaf with conspicuous, slightly inflated, often brownish sheath and short bristle-like blade to 0.8cm. Auricles blunt, free part c.0.2mm, transparent, membranous, sometimes not visible. Infl. hemispheric, 5–9-flowered, to 1.2cm diameter; lowest 2 bracts subequal, lanceolate, ribbed when dry, scarcely exceeding capitulum, usually pale brown; flowers shortly pedicellate. Tepals narrowly lanceolate, equal, 4.4–5mm. Filaments 4–5.4mm; anthers 1.8–2.5mm, at least partly exserted. Ovary ellipsoid, 2.3–2.8mm, abruptly contracted into style 2.5–2.7mm; stigma capitate, pin-like, with exceedingly short lobes. Capsule broadly ellipsoid-trigonous, 2.3–2.8mm, beak 0.9–1.2mm, straw-coloured to brown. Seeds c.0.8mm long, 2-tailed, tails short.

Bhutan: C — Thimphu (above Pajoding, above Talukah Gompa) and Mongar (Thrumse La) districts; **N** — Upper Kulong Chu district (Shingbe); **Darjeeling** (Sandakphu); **Sikkim** (Laghep, Changu, Sherabthang, Gnathong, Dzongri, Lachen, Tosar, Jelep La, Reshinangi, Jamlinghang to Bikbari, Laxmi Pokhri). Mossy rocks; gravelly roadsides by stream in *Abies* forest; shallow mossy pools and flushes, 3350–4540m. Fl. May–July.

27. J. trichophyllus W.W. Smith. Fig. 25f.
Rhizomes short, clump-forming. Flower stems slender, 5–14cm. Scale leaves straw-coloured, chaffy, becoming fibrous. Basal leaves filiform, bitubular, 0.2–0.5mm diameter. Stem leaves to 3, bract-like, blades short (to 1.5cm), filiform, sheaths reddish-brown, enclosing small, dark-coloured bulbils developing into plantlets. Auricles blunt, free part c.0.1mm, transparent or brownish. Infl. 1–3-flowered; bracts 2, boat-shaped, shorter than tepals, reddish-brown. Tepals narrowly lanceolate, acute, outer 2.6–3.2mm, inner longer (3.3–3.7 × c.1mm), narrower. Filaments exceeding tepals; anthers partly exserted, 1–1.5mm. Ovary ellipsoid, 2.5–3mm, abruptly contracted into style 0.8–1.8mm, white; stigma ± capitate, lobes very short (c.0.3mm), spreading. Capsule ellipsoid-trigonous, 2.6–3.1 × 1.5–1.8mm; beak c.0.5mm.
Bhutan: C — Thimphu district (above Pajoding, below Darkey Pang Tso); **N** — Upper Bumthang Chu district (below Ju La); **Sikkim** (Changu, Jamlinghang to Bikbari, Dzongri, Phedang, Laxmi Pokhri). Wet cliff-ledges; shallow boggy pools; streamsides; wet flushes, mossy boulders, 3800–4150m. Fl. July–August.

Perhaps hybridising with *J. cephalostigma* — intermediates seen at Dzongri.

28. J. kingii Rendle.
Stoloniferous, stolons stout. Flower stems single, distant, stiffly erect, 13.5–35cm. Scale leaves sheathing, upper with setaceous points, whitish-brown, dull. Stem leaf single, stiff, erect, linear, tip blunt, usually reaching midpoint of stem, semicircular to cylindric in section, channelled above, bitubular, with thin membranous longitudinal septum and weak transverse septa (septa not visible externally), to 1.7mm wide. Auricles rounded, free part 0.8 × 0.9mm, straw-coloured, membranous. Infl. 8–13(–24)-flowered, 1.4–2cm diameter; lowest bract extended into leafy point at least twice length of capitulum. Tepals narrowly lanceolate, acute, equal, 4.7–6.7mm. Anthers completely exserted at maturity, linear 1.9–2.5mm. Ovary ellipsoid, 2–4.2mm; style 1.3–2.7mm; stigma lobes 1–2.2mm. Capsule ellipsoid-trigonous, 3–4.1mm, abruptly contracted into partly exserted beak 1.5–2mm. Seeds orange-brown, c.0.8mm, 2-tailed, tails short.
Bhutan: C — Thimphu district (Pajoding); **N** — Upper Mo Chu district (Laya, Tharizam Chu); **Sikkim** (Lachen, Thangu, above Thangshing). Damp grassy turf on hillsides; among shrubs and boulders, 3750–4310m. Fl. July; fr. September.

29. J. leucomelas Royle ex D. Don; *J. bhutanensis* Satake.
Rhizomes very short. Flower stems 3–15(–19)cm, densely tufted. Scale leaves reddish-brown. Stem leaves usually 2, sub-basal, upper part of stem

leafless; blades linear, narrowed to subacute apex, which usually reaches ⅔ way up stem, unitubular, channelled above, to 5.5cm, to c.1mm wide. Sheaths with membranous margins gradually tapered into blade, auricles lacking. Infl. 5–10-flowered; lowest bract leafy, exceeding to twice length of capitulum. Tepals narrowly lanceolate, subacute, equal, (4.6–)5.2–6.4mm. Anthers completely exserted at maturity, linear (1.8–)2–2.9mm; filaments 3–5.5mm. Ovary ellipsoid-trigonous (2.3–)3–3.8mm; style 1.3–2.3mm; stigma lobes 0.5–1mm. Capsule ovoid to broadly ellipsoid, 2.6–3mm long, beak 0.8–1.5mm.

Bhutan: C — Thimphu district (Barshong to Nala); **Sikkim** (Chakalung La, Samding, Lachen, Zemu Valley; Migothang (F.E.H.1)); **Chumbi** (Yatung). Damp grassy places; on sand by rivers, 3050–4880m. Fl. May–July.

Many literature records are due to misidentifications.

30. J. benghalensis Kunth; *J. bracteatus* Buchenau; *J. sphenostemon* Buchenau; *J. membranaceus* sensu F.B.I., p.p. non Royle ex D. Don; *J. bhutanensis* sensu F.E.H.2, p.p. (Sikkim plants). Fig. 25g–h.

Stoloniferous, stolons filiform (under 0.5mm diameter). Flower stems slender, borne singly, 7–19cm. Scale leaves one or more, whitish-brown. Stem leaves usually 2(–3); lower leaf basal, filiform, acute, channelled above, bitubular in section, to 10cm, c.0.7mm wide; upper stem leaf usually present (sometimes absent), blade to 8cm, sometimes overtopping infl., sheath long, sometimes slightly inflated and brown-tinged. Auricles conspicuous, oblong, blunt, free part 0.7–1.2mm, usually brown. Infl. (2–)4–11-flowered; lowest bract developed into leaf-like point to 2.5cm. Tepals narrowly lanceolate, subacute, equal, 4–5.8(–6.5)mm. Anthers very narrow, (1.7–)2.2–3.3mm, exserted at maturity. Ovary ellipsoid, 2–3.5mm; style 1.4–2mm; stigma lobes erect, (0.3–)0.6–1.4mm. Capsule shortly stipitate, narrowly ellipsoid-trigonous, shorter than tepals, 2.6–4mm, tapered into beak 0.6–1mm. Seeds c.0.6mm, 2-tailed, total length to 1.2mm.

Bhutan: C — Thimphu (Pajoding, below Darkey Pang Tso, Chelai La), Bumthang (above Gortsam), Mongar (Thrumse La) and Tashigang (Chorten Kora) districts; **N** — Upper Mo Chu (Lingshi Dzong, Laya) and Upper Bumthang Chu (below Ju La) districts; **Darjeeling** (Sandakphu, Phalut); **Sikkim** (Gnatong, Chulong, Dzongri, Lachen, Jelep La, Tsomgo Lake, Yeumtong, Chamnago, Changu, Zemu Valley). River and stream banks; damp turf; damp gravelly banks and roadsides in *Abies/Rhododendron* forest; wet cliffs; sandy moraines, 3050–4570m. Fl. June–August.

Apparently the commonest alpine rush in Bhutan occurring over a wide altitudinal range.

31. J. perpusillus Samuelsson.

Like a dwarf form of *J. benghalensis* which it resembles in its leaf sheath auricles and elongate, leaf-like lowest bract; differing in its habit — rhizome highly branched and dense so stems forming dense clumps. Infl. sometimes reduced to a single flower, but cannot be confused with *J. duthiei* or *J. uniflorus* as tepals white.

Sikkim (Chaunrikiang). Moraines, 4550m. Fl. July.

32. J. brachystigma Samuelsson. Fig. 25i–j.

Rhizomes very short. Flower stems slender, tufted, 2–17cm. Scale leaves 1 or more, apiculate, brown. Stem leaf usually 1(–3), sub-basal, upper ⅔ of stem naked; blade blunt, channelled above and beneath, with up to 4 other shallower grooves, sometimes appearing bitubular in section, 0.2–5cm, to 0.8mm wide. Auricles very short, free part c.0.1mm, sometimes not produced. Infl. 4–13-flowered, 0.7–1(–1.4)cm diameter. Lowest bract equalling capitulum, or extended into aristate point to 3cm; flowers shortly pedicellate. Tepals narrowly lanceolate, equal, (3–)3.5–3.9(–4.5)mm. Anthers finally completely exserted, 0.8–1.6mm. Ovary ellipsoid, (0.9–)2.2–2.8mm, abruptly contracted into exserted style (0.9–)1.3–1.8(–2.8)mm; stigma lobes under 0.5mm, spreading. Capsule very shortly stipitate, narrowly ellipsoid, c.3mm, beak exserted, c.0.8mm, chestnut.

Bhutan: C — Thimphu (below Darkey Pang Tso) and Bumthang (above Gortsam) districts; **N** — Upper Mo Chu (Phile La, Lingshi Dzong, Chebesa) and Upper Kulong Chu (Shingbe) districts; **Sikkim** (Chulong, Lachen, Na-tut, Rookah, Kangpupehuthang, Dikchu, Chamnago, above Lambi, below Lam Pokhri); **Chumbi**. Damp silty ground above moraine; wet mossy cliffs; grassy streamsides; base-rich flush; dry sandy cliff, 3050–4880m. Fl. May–September.

33. J. thomsonii Buchenau; *J. leucomelas* sensu F.B.I., p.p. non Royle ex D. Don.

Rhizomes short. Flower stems 4.5–16(–26)cm, tufted. Scale leaves reddish-brown. Stem leaves usually 2, sub-basal, upper part of stem naked; blades linear, stiffly erect, tip blunt, usually reaching ⅓ length of stem, usually with weak transverse septa (not visible externally), pluritubular near base with up to 3 weak longitudinal septa, bitubular near apex, 0.8–5cm, 0.4–1.2mm wide. Sheaths with membranous margins often chestnut; auricles curved, acute, free part c.0.9mm. Infl. 3–10-flowered; lowest bracts subequal, chestnut, broadly lanceolate to ovate, boat-shaped, not exceeding capitulum. Tepals lanceolate, subequal, (3.2–)3.5–5mm. Anthers finally exserted, 1.2–2mm. Ovary ellipsoid, 2–4mm, tapered into style 0.5–1mm; stigma lobes 0.8–2mm. Capsule narrowly ellipsoid-trigonous equalling to just exceeding tepals, 3.8–5mm, beak 0.6–1mm. Seeds orange 0.5–0.7mm, 2-tailed, total length 2–2.4mm.

Bhutan: C — Thimphu (Shodu to Barshong) and Tongsa (Yuto La) districts; **N** — Upper Mo Chu (Shingche La, Lingshi, Soe, Yale La; Seanchu Passa to Chabecha (F.E.H.2)) and Upper Bumthang Chu (Waitang, above Lambrang) districts; **Sikkim** (Gurudongmar, Changu, Dzongri, Beeroom, Naku Chu, Llonak, Samiti to Jemathang; Sherabthang (Smith, 1913)); **Chumbi**. Sandy moraine; marshy meadows, streamsides, 3500–4760m. Fl. May–August.

Hooker specimens (K) probably from Lachen have long exserted fruits as in *J. triglumis*, but the long anthers of *J. thomsonii*; these and similar specimens from Nepal merit further investigation.

Usually only found at very high altitudes, but a single collection from much lower altitude (Bumthang district: Byakar, 2750m, *Grierson & Long* 1762) appears to belong to this species. It is exceptional only in its height (stems to 30cm), but further investigation is required.

34. J. triglumis L.

Very similar to *J. thomsonii* from which it differs as follows: tepals brown at flowering; anthers smaller (0.6–1.1mm); infl. 2–5-flowered; capsule much exceeding tepals (4.7–)5–6.1mm, lower part reddish-brown, upper part chestnut, tapered into persistent stylar beak 0.6–1.4mm; leaves narrower, filiform, under 0.5mm wide, channelled above, with up to 3 additional, shallower grooves, bitubular throughout.

Bhutan: C — Thimphu district (above Pajoding, Bimelang Tso to Dungtsho La, Ragyo to Darkey Pang Tso); **Darjeeling** (Sandakphu); **Sikkim** (Dzongri, Yeumtang, Chamnago, Chakung Chu, Phalut, Thangu, Chola, Bikbari, Phedang, Thangshing; Zemu (Smith & Cave, 1911)); **Chumbi** (To-koo-la). Shallow, boggy pools and flushes, 3350–4570m. Fr. July–October.

This almost certainly represents a distinct subspecies of *J. triglumis*, characterised by its long style which persists in the capsule as a beak (as observed by Buchenau, 1885), and perhaps also in its filiform leaves. It also occurs in Nepal (specimens seen in BM). Its relationship to subsp. *wakhaniensis* Snogerup requires investigation.

35. J. duthiei (C.B. Clarke) Noltie; *J. rohtangensis* Goel & Aswal; *J. sikkimensis* Hook.f. var. *monocephalus* Hook.f. Fig. 25k.

Rhizomes short, slender (c.1.5mm diameter). Flower stems 2–25cm. Scale leaves pale brown, dull, upper apiculate. Stem leaf usually 1, sub-basal, blade slightly contracted below blunt apex, unitubular, septate, 1.5–7.5cm (tip reaching halfway up stem or more), 0.3–1.4mm wide. Auricles blunt, oblong, apparently sometimes decaying, 0.1–1.1mm, transparent or brownish. Infl. appearing lateral, (1–)2–3-flowered, flowers shortly pedicellate (sessile if single); lowest bract usually erect, leaf-like, to twice length of infl., sheathing base brown. Tepals narrowly lanceolate, gradually tapering to acute apex, rather irregular in length even in a single specimen, (3–)5–7.7mm, reddish-brown to blackish, sometimes with paler tips and greenish midribs. Stamens shorter than tepals; filaments 0.4–1.2mm; anthers sometimes twisted, 1.2–3.5(–4)mm. Ovary oblong- to ellipsoid-trigonous, 1.3–2.2mm; style stout 1.5–4mm; stigma lobes very fine, long, erect, sometimes twisted, (2–)5–9mm, red. Capsule narrowly ellipsoid-trigonous, (1.4–)2.5–3.5mm, contracted into exserted beak (0.6–1mm). Seeds 0.6–0.7mm, 2-tailed, total length 0.7–0.8mm.

Bhutan: C — Thimphu (Pajoding, below Darkey Pang Tso) and Bumthang (Kamephu, above Gortsam) districts; **N** — Upper Mo Chu (Kohina, above Laya), Upper Pho Chu (Kesha La) and Upper Bumthang Chu (below Ju La) districts; **Sikkim** (Thangu, Lachen, Choktsu, Changu, Chakung Chu, Dzongri, Zemu Valley, Chola Valley, Tsoka to Jamlinghang, Prek Chu to Thangshing, Chemathang); **Chumbi**. Open grassy hillsides; damp turf; rock-ledges and streamsides, 3000–4570m. Fl. June–July.

36. J. uniflorus W.W. Smith. Fig. 25l.

Flower stems 2–3cm, densely tufted. Scale leaves long-aristate, shining yellowish-brown, striate. Stem leaves 1–2, exceedingly narrowly filiform, blunt, often twisted or recurved, bitubular, deeply channelled above, to 3cm, 0.2–0.4mm wide. Auricles acute, free parts c.0.2mm, transparent. Infl. a single, extremely shortly pedicelled flower, which appears terminal; lower bract aristate, equalling or only just exceeding flower, spreading horizontally at anthesis, sheathing base broad, boat-shaped, brown-membranous; upper bract ovate. Tepals reflexed and star-like at anthesis, narrowly lanceolate, acute to acuminate, margins narrowly membranous, subequal or inner slightly longer, 2.5–4(–4.6)mm, pale reddish-brown. Stamens shorter than tepals; filaments 0.4–0.5(–0.8)mm; anthers (0.7–)1.2–1.5(–2)mm. Ovary ellipsoid, (0.9–)1–1.6mm, narrowed upwards into style 1–1.5(–1.7)mm; stigma lobes erect, twisted, 1.8–3.2(–4.6)mm, red. Capsule narrowly ellipsoid-trigonous gradually tapered into beak, total length c.3mm, dark brown, shining.

Bhutan: C — Thimphu (Pajoding, above Talukah Gompa, below Darkey Pang Tso) and Bumthang (Penge La, above Gortsam) districts; **N** — Upper Kulong Chu district (Shingbe); **Sikkim** (Sherabthang, Jelep La, Changu, Laxmi Pokhri; Nathui La, Tosa, Chakung Chu (Smith, 1913)). Open, damp, peaty places (among moss, on rocks or bare ground, beside lake), 3660–4140m. Fl. June–July.

37. J. bryophilus Noltie.
Differs from *J. uniflorus* and dwarf forms of *J. duthiei* in its smaller anthers (c.0.7mm) which are shorter than the filaments (c.2.3mm); lowest bract erect, leaf-like, filiform, to 1.5cm; style short (0.7mm); stigma lobes short, (0.7mm), cream.
Bhutan: N — Upper Mo Chu district (Laya); **Sikkim** (Dzongri). Wet rock-ledge and mossy boulders, 4000–4450m. Fl. July; fr. September.

2. LUZULA DC.

Perennial tufted herbs, sometimes rhizomatous. Leaves mostly in basal rosettes, surfaces glabrous, margins usually long ciliate. Infl. terminal, cymose, many-flowered, on leafy stem, sheaths of stem leaves closed, without auricles. Tepals 6, stamens 6. Capsule unilocular, 3-seeded. Seeds with variously shaped white appendages (caruncles); membranous testa becoming mucilaginous and swelling in water.

1. Flowers sessile in small, dense capitula 2
+ Flowers borne singly on slender pedicels 3

2. Tepals brownish, narrowly lanceolate, finely acuminate, 2.5–3.6mm; capsule shorter than tepals; caruncle of seed 0.2–0.3mm
   1. L. multiflora
+ Tepals blackish, lanceolate, more suddenly contracted to mucronate tip, 1.6–2.2mm; capsule equalling tepals; caruncle minute, under 0.1mm .... **2. L. oligantha**

3. Infl. a long (to 36cm) raceme of cymes; plant large (to 80cm); leaves over 5mm wide .... **3. L. effusa**
+ Infl. a single terminal cyme with uneven, spreading to erect branches, mostly bearing a single flower, appearing as an irregular umbel; plant to 28cm; leaves under 5mm wide .... **4. L. plumosa**

1. L. multiflora (Retzius) Lejeune; *L. campestris* sensu F.B.I., p.p., non DC. Fig. 25m–p.
    Densely tufted. Flower stems 10–35cm. Stem leaves 3, evenly spaced, narrowly lanceolate, gradually tapered to blunt, callose-tipped apex, margins with long white hairs, to 14 × 0.4cm; basal leaves similar. Infl. hairy, with 3–10 unequal branches (sometimes again branched) each bearing a (3–)5–10-flowered head; flowers sessile, tightly enclosed by 2 bracteoles; branches sometimes not developed giving a dense head. Lowest bract leaf-like,

exceeding infl. in flower, finally shorter than fruiting pedicels. Bracteoles ovate, apiculate, irregularly toothed to ciliate, whitish-membranous. Tepals narrowly lanceolate, finely acuminate, subequal or outer slightly longer, (2–)2.5–3.6mm, reddish-brown to dark chestnut, margins paler. Stamens shorter than tepals; anthers 0.3–1mm, equalling or slightly shorter than filaments at maturity. Ovary ellipsoid-trigonous, 0.8–1.2mm; style 0.2–0.4mm; stigma lobes erect, 0.9–1.9mm. Capsule widely ellipsoid-trigonous, extremely shortly mucronate, shorter than tepals, 1.8–2.2 × 1.5–2mm, straw-coloured to chestnut. Seeds 0.9–1.3mm, straw-coloured; caruncle basal, blunt, 0.2–0.3(–0.5)mm.

Bhutan: S — Chukka district; **C** — Thimphu, Tongsa, Bumthang and Tashigang districts; **N** — Upper Mo Chu, Upper Bumthang Chu and Upper Kulong Chu districts; **Darjeeling**; **Sikkim**; **Chumbi**. Moist grassland in open or under scrub or forest (deciduous or coniferous, incl. *Abies*); peaty soil and on mossy rocks in rhododendron scrub, 2000–4570m. Fl. April–July.

The E Himalayan plant has been variously identified. Hooker (F.B.I.) called it *L. campestris*; Buchenau (1885) said that there was a spectrum present from that species to *L. multiflora*, citing Hooker's Sikkim specimens as nearer the former. These Hooker specimens are here placed under *L. oligantha*. The more common plant has more recently usually been identified as *L. multiflora*, which it resembles in its densely tufted, non-rhizomatous habit, but differs from typical material in its shorter caruncle.

2. L. oligantha Samuelsson.
Differs from *L. multiflora* in its smaller flowers with very dark, subequal tepals, 1.6–2.2mm, which are wider and more suddenly contracted to a mucronate apex; capsule ± equalling tepals; caruncle shorter, under 0.1mm.
Sikkim (Lachen, Zemu Valley, Yumthang). 3050–3660m. Fl. June–July.

Differs from Japanese and E Asian specimens in its longer, narrower stem leaves. Hara *et al.* (F.E.H.1, 2, 3) recorded this plant from Darjeeling, initially under the name *L. sudetica* DC., but all their specimens I have seen are *L. multiflora*.

3. L. effusa Buchenau. Fig. 25q.
Shortly rhizomatous. Flower stems to 80cm. Stem leaves up to 6, blades linear-lanceolate, gradually tapered to very acute apex, to 17 × 1.1cm, margins with few long white cilia; basal leaves shorter, narrower. Infl. laxly paniculate to 36cm, a raceme of cymes; primary branches widely spaced, subtended by leafy bracts; cymes corymbose, branches diverging at wide angles, subtended by linear, brownish, membranous, ciliate bracts; flowers borne singly, pedicellate, subtended by 2 bracteoles. Bracteoles ovate, margins sometimes ciliate near apex, chestnut. Tepals lanceolate, aristate, subequal, 2–2.5mm, chestnut, midrib greenish. Stamens shorter than tepals; anthers 0.5–1mm, about equal-

ling filaments. Ovary ellipsoid, 0.9–1.2 × 0.6–0.9mm; style 0.4–0.5mm; stigma lobes erect, 1.1–1.4mm. Capsule ovoid-trigonous, apiculate, about equalling tepals, c.2.3 × 1.5mm. Seeds pale brown, rounded on back, flat on ventral face, membranous testa produced into extremely short knobs at both ends — total size c.1.6 × 0.7mm.

Bhutan: C — Ha (near Ha), Thimphu (Thimphu to Dochu La, above Motithang, Dotena to Barshong), Tongsa (Pele La, W side of Yuto La), Bumthang (Lami Gompa, Byakar) and Mongar (Sengor) districts; **N** — Upper Mo Chu district (Kohina); **Sikkim** (Yakla, Rookah, Lachen, Laghep, Zemu Valley, Yeumtong, Tankra Mountain, Tsoka to Jamlinghang); **Chumbi**. Wet *Tsuga* and *Abies* forest; shady ravines, cliff-faces, 2250–3400m. Fl. May–August.

4. L. plumosa E. Meyer; *L. rostrata* Buchenau subsp. *darjeelingensis* Satake.

Tufted, sometimes shortly rhizomatous. Flower stems to 28cm, arising laterally. Stem leaves up to 4, blade narrowly lanceolate, gradually tapered from middle to blunt callose-tipped apex, margins hairy at least near base, to 6cm, to 4.5mm wide; basal leaves similar. Infl. a contracted cyme (appearing umbellate), with slender spreading to erect mainly single-flowered branches. Lowest bract leaf-like, much shorter than infl.; bracts of branches small, linear, membranous, ciliate; flowers tightly enclosed by 2 bracteoles. Bracteoles broadly ovate, irregularly ciliate, brown, edges pale. Tepals narrowly lanceolate, finely acuminate, equal or inner slightly longer, 2.4–3.5mm, chestnut, edges paler. Stamens shorter than tepals; filaments finally exceeding anthers; anthers (0.4–)0.7–1.1mm. Ovary widely ellipsoid, 0.8–1.4 × 0.5–1.1mm; style short 0.4–0.8mm; stigma lobes erect, 1.3–2.4mm. Capsule broadly ovoid-trigonous, very shortly apiculate, exceeding tepals, c.3 × 2.3mm, straw-coloured. Seeds widely ellipsoid, c.1.8 × 1.4mm, orange-brown; caruncle subterminal, oblique, horn-shaped 1.1–1.6mm.

Bhutan: S — Chukka district (Chimakothi); **C** — Ha (Ha Dzong), Thimphu (Drugye Dzong, Tzatogang to Dotanang, Motithang, etc.), Tongsa (Chendebi, Tunle La), Bumthang (Ura La) and Mongar (Donga La) districts; **Darjeeling** (Rimbick to Ramam; Garibans to Tonglu (F.E.H.1)); **Sikkim** (Chungthang, Rookah, Lachen, Tsoka to Jamlinghang); **Chumbi**. Moist conifer (incl. *Abies*) and oak forests, 1500–3500m. Fl. March–June.

Family 232. CYPERACEAE

Perennial or annual herbs, often rhizomatous. Stems commonly trigonous. Leaves basal and/or cauline, often 3-ranked, usually with grass-like blades

(blades sometimes absent); bases sheathing, sheaths open or closed, apex often ligulate, occasionally with ligule on side opposite blade. Infl. simple (e.g. spike) or compound (e.g. panicle or anthelodium) composed of 1–many spikelets, usually bracteate. Plants sometimes dioecious. Flowers bisexual or unisexual, perianth of scales or bristles or absent, subtended by glumes, arranged spirally or distichously in spikelets. Stamens 1–3, anthers basifixed. Ovary of 2 or 3 fused carpels, unilocular with a single ovule, stigmas 2 or 3, style often thickened at base, sometimes persistent in fruit. In *Carex* prophyll of female spikelets modified into a utricle; in *Kobresia* prophyll open to varying degrees and sometimes also enclosing 1 or more male flowers. Fruit usually a biconvex or trigonous nut.

1. Nuts large, hard, smooth or reticulate, white (or purplish), borne on usually 3-lobed, persistent disc, solitary and terminal in female or bisexual spikelets (Fig. 30n–s); if hidden by appressed glumes, then spikelets in sessile, axillary umbels evenly disposed along stem (Fig. 30t) .... **17. Scleria**
+ Nuts not as above, spikelets never in axillary umbels 2

2. Flowers bisexual, subtended by an open glume; spikelets all similar; perianth of scales or bristles often present 3
+ Flowers unisexual; prophyll of female spikelets modified as a perigynium which may be closed (utricle) or with appressed margins so nut hidden; spikelets commonly of 2 sorts (plants sometimes dioecious); perianth always absent 20

FIG. 26. **Cyperaceae** I.

a–c, **Hypolytrum nemorum**: a, habit (× ⅓); b, immature nut subtended by perianth scales (× 15); c, mature nut (× 7). d–e, **Scirpus ternatanus**: d, inflorescence (× ¼); e, biconvex nut and bifid stigma (× 16). f, **S. wichurai**: trigonous nut and trifid stigma, subtended by hypogynous bristles (× 13). g–h, **Erioscirpus comosus**: g, spikelet (× 3.5); h, nut subtended by hypogynous bristles (× 8). i–k, **Fuirena umbellata**: i, inflorescence (× ¼); j, glume (× 10); k, nut subtended by perianth scales (one hidden) (× 12). l–m, **Schoenoplectus juncoides**: l, habit (× ⅓); m, nut subtended by hypogynous bristles (× 8). n, **Eleocharis palustris** agg.: habit (× ½). o–p, **E. congesta**: o, apex of leaf sheath (× 3); p, nut with persistent style base and subtended by hypogynous bristles (× 25). q–s, **Bulbostylis densa**: q, habit (× ½); r, apex of leaf sheath showing hairs (× 18); s, nut and stigmas showing swollen style base (× 20). t, **Isolepis setacea**: habit (× ½). Drawn by Glenn Rodrigues.

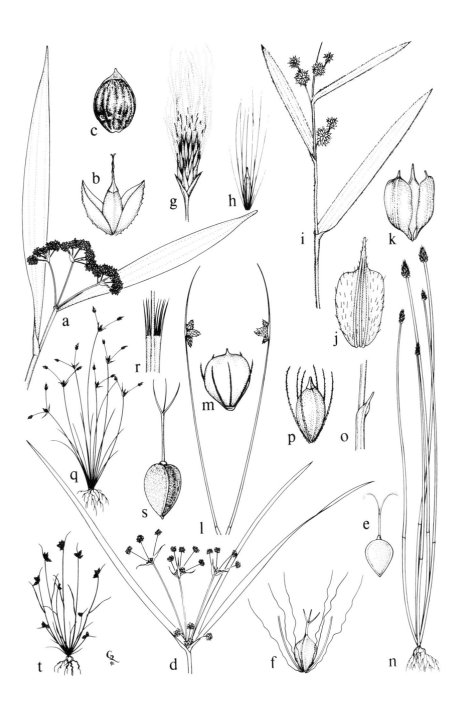

275

3. Perianth of hypogynous bristles and/or scales present, commonly persistent and subtending nut 4
+ Perianth of scales or bristles absent 13

4. Perianth of scales (bristles occasionally also present) 5
+ Perianth of bristles 7

5. Spikelets in a dense terminal head, usually whitish
                                                    14. Lipocarpha (part)
+ Infl. a corymbose panicle, or with heads in terminal and lateral partial infls., spikelets brownish or blackish 6

6. Infl. a corymbose panicle; glumes glabrous, apex rounded; leaves mainly basal .... **1. Hypolytrum**
+ Infl. with terminal and lateral partial infls. of heads; glumes hairy, aristate; stems leafy throughout .... **5. Fuirena**

7. Nut crowned with disc or beak, formed from persistent style base and distinct in texture from nut 8
+ Nut not crowned with disc or beak of distinct texture, though often apiculate 9

8. Infl. a single, narrow, terminal spike; leaves reduced to bladeless, tubular, basal sheaths .... **7. Eleocharis**
+ Infl. a loose panicle or dense hemispherical head; leaves with linear blades .... **16. Rhynchospora**

9. Leaves reduced to bladeless, tubular, basal sheaths; infl. apparently lateral .... **6. Schoenoplectus**
+ Leaves with blades; infl. terminal 10

10. Infl. dense, spike-like, of ± distichously arranged spikelets   **15. Blysmus**
+ Infl. anthelate 11

11. Hypogynous bristles 0–6, shorter than to slightly exceeding spikelet   12
+ Hypogynous bristles many, conspicuously exceeding spikelet in fruit so spikelets plumose .... **4. Erioscirpus**

12. Stems leafy, trigonous or rounded-trigonous .... **2. Scirpus**
+ Stems leafy only at extreme base, triquetrous, 3-winged
                                                    3. Actinoscirpus

13. Glumes spirally inserted (if distichous, then spikelet solitary); glumes always deciduous spikelet never falling entire 14
 + Glumes distichous, infl. never of a single spikelet; glumes deciduous or spikelets falling entire 18

14. Swollen style base persistent on nut; apex of leaf sheaths with long white hairs (Fig. 26q–s) .... **9. Bulbostylis**
 + Nut not crowned with persistent style base; apex of sheaths glabrous or with ligule of very short hairs across blade 15

15. Dwarf annuals; infl. of few, sessile, pseudolateral spikelets 16
 + Perennial (occasionally annual); infl. not pseudolateral (compound, partial infl. or spikelets stalked or sometimes reduced to a single, terminal spikelet) 17

16. Glumes not mucronate, blackish; spikelets not squarrose .. **10. Isolepis**
 + Glumes mucronate, base reddish, mucro green; spikelets squarrose
 14. Lipocarpha (part)

17. Stout erect perennial, stems leafy throughout; infl. compound; stigmas 2 .... **2.1. Scirpus ternatanus**
 + Stems leafy only at base; infl. simple or compound; stigmas 2 or 3
 8. Fimbristylis

18. Stigmas 3; nut trigonous .... **11. Cyperus**
 + Stigmas 2; nut biconvex 19

19. Glumes deciduous; spikelets usually over 6mm, axis persistent; infl. anthelate or sometimes condensed .... **12. Pycreus**
 + Glumes not deciduous; spikelets under 3.5mm, falling entire at maturity; infl. a dense head(s) .... **13. Kyllinga**

20. Prophyll of female spikelets with margins fused to varying degrees, enclosing a nut and either a racheola or one or more male flowers; stigmas 3; plants sometimes dioecious, usually densely tufted
 18. Kobresia
 + Prophyll of female spikelets completely closed to form a utricle, concealing nut, never enclosing male flowers, (if racheola present then

utricles few, large, deflexed); stigmas 2 or 3; Bhutanese plants never
dioecious, tufted or creeping .... **19. Carex**

1. HYPOLYTRUM Richard ex Persoon

Perennials, with stout rhizomes. Stems trigonous. Leaves commonly narrowly elliptic, mainly basal. Infl. a terminal panicle, often corymbose, sometimes contracted; bracts leaf-like. Spikelets with spirally inserted, persistent
glumes. Glumes each bearing a single, bisexual flower enclosed by 2 conduplicate (sometimes partly fused) scales. Stamens 2. Stigmas 2. Nut biconvex.

1. H. nemorum (Vahl) Sprengel; *H. latifolium* Richard. Fig. 26a–c.
Stem trigonous, 51–79cm, stout, 1.8–3mm diameter. Stem leaves 1–2.
Leaves mainly in basal rosette, margins serrate, about equalling stem,
1.1–1.8cm wide, strongly 3-veined, glabrous. Infl. corymbose or widely cylindric, 3.5–8 × 4–7cm, lowest 1–2 nodes each bearing 2–3 branches, subtended
by leaf-like bracts; primary branches stiff, 2–3cm, bearing short spikelet-bearing
branchlets. Spikelets ellipsoid (widely in fruit), blunt, 3.5–5 × 1.8–4mm, with
c.4 basal, sterile glumes and c.15 fertile glumes. Glumes oblong, rounded,
1.6–2 × 1.1–1.5mm, 1-veined, streaked dark reddish-brown to varying degrees.
Perianth scales lanceolate, acute, c.1.5mm, sharply keeled, keel minutely ciliate,
hyaline. Nut biconvex, elliptic, angles thickened to slightly winged, 2.5–2.7 ×
1.8mm, vertically rugose below, upper part forming a flattened, wrinkled
triangular apiculus, speckled dark purplish-brown to varying degrees.
Terai (Dulkajhar). 150m. Fr. October.

The nut described above is typical of virtually all Indian material seen, but one of the
Dulkajhar specimens has smooth nuts with a thick, corky testa produced upwards so
the apiculus filled with spongy tissue. More collections are required to know whether
this is of taxonomic significance, or merely represents a later stage of development than
that seen in most specimens.

2. SCIRPUS L.

Stout perennials; rhizomes short or sometimes stoloniferous. Stems stout,
often trigonous. Stem leaves usually present in addition to basal ones; bases
sheathing, blades well-developed. Infl. terminal, irregularly umbellate, several
times compound (anthelate), subtended by long, leaf-like bracts; infl. branches
and branchlets subtended by tubular prophylls. Spikelets inserted singly or in
groups, with many spirally inserted glumes on persistent axis. Flowers bisexual.

Perianth of 6 hypogynous bristles or 0. Stamens 2–3. Stigmas 2–3; style passing smoothly into nut, not swollen or articulated at base. Nut trigonous or biconvex.

1. Stigmas 2; nut compressed; hypogynous bristles 0 or short
                                                 1. S. ternatanus
+ Stigmas 3; nut trigonous; hypogynous bristles 6, conspicuous, especially in fruit .... **2. S. wichurai**

1. S. ternatanus Reinwardt ex Miquel; *S. chinensis* Munro; *S. rosthornii* Diels. Fig. 26d–e.

Leaves inserted at base of stem and with 4–7 on lower ⅔ of stem, blades overtopping stems, 0.5–1.7cm wide, midrib keeled beneath, coriaceous; bases sheathing, sheaths rather loose, with prominent transverse veinlets. Stems trigonous below infl., 28–105cm, stout. Infl. 3(–4) × compound, 9–30cm diameter; primary rays 5–8, rigid, divaricate, unequal (one usually subsessile, longest to 11cm), subglabrous; bracts leaf-like, exceeding rays. Spikelets in clusters of 4–13, narrowly ovoid, 2–4 × 1.3–1.5mm, reddish-brown. Glumes widely ovate, blunt, 1–1.5 × 1–1.2mm, hyaline, streaked orange-brown, midrib green. Stamens 2–3. Nut flattened, elliptic to obovate, minutely apiculate, 0.6–0.7 × 0.5–0.6mm, greenish-white; stigmas 2; bristles usually 0.

Bhutan: S — Chukka (Chimakothi to Thargyal Mathur bridge), Sankosh (5km E of Daga Dzong) and Deothang (Deothang) districts; **C** — Punakha (Mishichen to Khosa) and Tongsa (Tongsa, Tama) districts; **Darjeeling** (Kurseong, Riang, ?Kajell); **Sikkim** (Nampok, Lachen, Lachung). Wet cliffs; damp flushes; marsh in cleared broad-leaved forest, 457–2350m. Fr. April–June.

Infl. sometimes viviparous.

2. S. wichurai Boeckeler; *S. eriophorum* sensu F.B.I. non Michaux. Fig. 26f.

Differs from *S. ternatanus* in having leaves shorter than stems; primary infl. rays more slender, curved (usually all in same direction), shortly but densely hairy; glumes acute; stigmas 3; nut strongly triquetrous; hypogynous bristles 6, barbed, accrescent, becoming conspicuous in fruit.

Bhutan: S — Chukka (above Lobnakha) and Deothang (Rydang) districts; **C** — Thimphu (Pajoding, below Taba, Thimphu, Tsalimaphe to Pumo La, Paro Valley), Punakha (above Tinlegang), Tongsa (Niddapek) and Bumthang (Yuto La, Tang, Gyetsa) districts. Marshes; streamside in paddy field; among rocks in woods; swampy clearing in forest, 1520–3660m. Fl. July–August; fr. October.

Should perhaps be placed in the genus *Trichophorum*.

A specimen of **S. michelianus** L. has been seen from the **Terai** (Jalpaiguri). It is a dwarf, tufted plant with a dense capitate infl., superficially like a *Kyllinga* from which it differs in having spirally arranged glumes. It has biconvex nuts, with 2 stigmas and lacks hypogynous bristles. Its generic position is unclear and it is sometimes included under *Cyperus*.

3. ACTINOSCIRPUS (Ohwi) R.W. Haines & Lye

Stoloniferous perennials. Stems triquetrous. Leaves all basal, with well-developed blade. Infl. terminal, anthelate, subtended by leafy bracts. Glumes spirally inserted. Flowers bisexual. Hypogynous bristles 6, retrorsely barbed. Stamens 3. Stigmas 3. Nut trigonous.

1. A. grossus (L.f.) Goetghebeur & D.A. Simpson var. **kysoor** (Roxb.) Noltie; *Schoenoplectus grossus* (L.f.) Palla; *Scirpus grossus* L.f.
Roots bearing tubers. Leaves almost equalling stem, to 2cm wide, keeled beneath, bases sheathing, pale, with prominent transverse veinlets. Stem 3-winged, spongy, leafless, to 88cm. Infl. a narrow, several-times compound panicle; primary branches crowded near apex of stem, ascending at very acute angles, lowest 6.5–14cm, decreasing in length upwards; infl. bracts 2 or more, leaf-like, greatly exceeding infl.; prophylls sometimes with free filiform tips. Spikelets borne singly, narrowly ovoid, 3–4 × 2–2.5mm, mostly peduncled. Glumes ovate, oblong or obovate, truncate, emarginate or acute, margins narrowly hyaline, ciliate, c.3 (incl. mucro) × 2mm, minutely hairy, reddish-brown, midrib green, prolonged into recurved mucro (c.0.5mm).
Darjeeling (Darjeeling). No habitat details, but presumably in marshes, pools or ditches.

Recorded for Bhutan (near Phalang, Tashigang district) in Griffith (1847, p. 233), but no specimen seen so must remain uncertain.

4. ERIOSCIRPUS Palla

Densely tufted perennials. Stems solid, obscurely trigonous. Leaves basal, blades narrow, channelled. Infl. a terminal, several-times compound panicle or reduced to a few spikelets which may appear lateral. Infl. bracts leaf-like; bracts subtending infl. branchlets glume-like. Spikelets with numerous glumes spirally inserted on persistent axis. Flowers bisexual. Perianth of numerous hypogynous bristles, bristles papillose near apex, growing after anthesis so spikelets finally plumose. Stamens 1–2. Stigmas 3; style passing gradually into nut. Nut narrow, compressed-trigonous.

1.   Infl. a diffuse, compound panicle with numerous spikelets, surrounded by involucre of leaf-like bracts .... **1. E. comosus**
+   Infl. contracted, with 1–3 spikelets subtended by 1–2 filiform bracts
2. E. microstachyus

1. E. comosus (Wall.) Palla; *Eriophorum comosum* (Wall.) Wall. ex Nees. Sha: *zala bang*. Fig. 26g–h.
   Leaves greatly exceeding stem, 0.8–4.3mm wide, margins minutely serrate; sheaths persistent, chestnut to blackish, sometimes shining, margins fibrillose. Stems leafless, 27–128cm. Infl. a 3–4× compound panicle, 8–50cm, primary branches congested near stem apex; infl. bracts leaf-like, greatly exceeding infl. Spikelets borne singly or in pairs, narrowly ellipsoid (before bristles emerge), 4.5–10.2 (excl. bristles) × 1.4–3mm, sessile and peduncled. Glumes narrowly ovate to oblong, blunt to acute, 2.3–3 × 0.6–1.2mm, densely streaked orange-brown, midrib green, excurrent as short (c.0.4mm) mucro. Stamens 1–2, connective extended as red-brown, acute point. Stigmas 3, erect, reddish-brown papillose. Nut compressed-trigonous, oblong, slightly contracted below acute apex, c.2.8 × 0.3mm, dark brown, shining. Bristles many (c.40), fused at extreme base, white, not barbed, short at flowering, increasing to c.1cm, spikelets finally plumose.
   Bhutan: S — Samchi (Bhaintholi to Samchi, Dham Dhum), Phuntsholing (NE side of Phuntsholing, Torsa River, Sorchen) and Deothang (Samdrup Jongkhar to Deothang (M.F.B.)) districts; **C** — Thimphu (below Chapcha), Punakha (Punakha, Chusom to Mishina, Lobesa to Punakha), Tongsa (below Shemgang) and Tashigang (Cha Zam, Tashigang Dzong, Ghunkhara) districts; **Darjeeling** (Riang, Tista Valley, Raj Bhavan Darjeeling); **Sikkim** (Gangtok, Rangpo, S of Legship). Dry or damp rocks, cliffs and stream banks, 200–2240m. Fl./fr. April–December.

2. E. microstachyus (Boeckeler) Palla; *Eriophorum microstachyum* Boeckeler.
   Differs from *E. comosus* in being much smaller (stems (4–)6–23cm) and more slender; leaves filiform (c.0.2mm wide), channelled; infl. of 1–3 spikelets; bracts filiform; spikelets ovoid (3–4 × 2–3mm); glumes ovate, apiculate, c.1.5 × 1mm, dark brown above; nut wider (c.2 × 0.5mm); bristles shorter (to c.0.5cm).
   Bhutan: S — Deothang district (near Keri Gompa); **C** — Thimphu district (below Chapcha; Chimakhothi to Thimphu (F.E.H.2)). On rocks, 1980–2450m.

5. FUIRENA Rottboell

Annuals or rhizomatous perennials. Stems 3–5-angled, leafy throughout. Leaves with linear, often hairy blades, sheaths tubular with ligule at junction with blade. Infl. with terminal and usually several lateral partial infls. composed of clusters of heads of sessile spikelets, subtended by leaf-like bracts. Spikelets with spirally inserted glumes; glumes aristate, hairy. Flowers bisexual. Perianth of 3 clawed hypogynous scales (sometimes of 3 scales and 3 bristles or sometimes of 6 bristles). Stamens 3. Stigmas 3, style passing smoothly into ovary. Nut trigonous.

1. F. umbellata Rottboell. Fig. 26i–k.

Perennial. Rhizomes creeping. Stems 4–5-angled, 45–94cm, stout, hairy on angles, sometimes swollen at base. Leaves widest near base, tapering to acute apex, margins ciliate, longest ones (mid-stem) to 22cm, 0.5–1.6cm wide, sparsely hairy on upper surface, hairy on veins beneath (sometimes glabrous on both surfaces); sheaths hairy on veins, ligules blunt, brown. Infl. with terminal cluster of 4–6 subglobose heads and 2–3 shortly peduncled lateral partial infls. Spikelets narrowly ellipsoid, acute, 6–9 × 2–2.5mm, sessile. Glumes oblong, truncate, 3–3.5 (incl. mucro) × 1–1.6mm, dark brown, shortly hairy, with 3 prominent green ribs extended into stout, setose mucro (1–1.5mm). Stamens 3. Stigmas 3, slightly exceeding style; nut strongly trigonous, faces elliptic, narrowed below and above into acute point, 1.1–1.5 × 0.5–0.7mm; hypogynous scales 3, obovate, truncate or emarginate, minutely mucronate, margins minutely ciliate near apex, narrowed at extreme base into short (c.0.1mm) claw, slightly exceeding nut, 3-nerved, brown.

Terai (Dulkajhar, Meshbesli). Wet ground, 150m. Fr. October–December.

The few specimens seen from our area (with one exception) are all very hairy, whereas the species is typically more glabrous. *F. ciliaris* (L.) Roxb. probably also occurs, but no specimens have been seen. It differs in being annual with the square blade of the hypogynous scale about equalling the long claw.

6. SCHOENOPLECTUS (Reichenbach) Palla

Perennials or sometimes annuals; stems tufted or on short creeping rhizomes. Stems terete to triquetrous, nodeless, sometimes hollow and septate. Leaves reduced to basal bladeless sheaths. Infl. a dense, pseudolateral cluster of spikelets, sometimes with shortly stalked secondary infls., infl. bract appearing as continuation of stem. Spikelets with many spirally inserted glumes on persistent axis. Flowers bisexual. Perianth of usually 6 retrorsely barbed, hypogynous bristles, occasionally absent. Stamens 1–3. Stigmas 2–3, style not thickened or articulated at base. Nut compressed or trigonous.

1. Stems triquetrous 2
+ Stems terete 3

2. Spikelets all sessile forming a single head; stems clumped at apex of short rhizome .... **3. S. mucronatus**
+ Infl. compound with shortly stalked secondary infls.; stems inserted singly on slender spreading rhizome .... **4. S. triqueter**

3. Stems densely tufted; spikelets to 1.5cm; glumes to 4mm, golden brown above .... **1. S. juncoides**
+ Stems inserted in single row along creeping (sometimes short) rhizome; spikelets to 0.6cm; glumes to 3.2mm, at least the margins fuscous purple .... **2. S. fuscorubens**

1. S. juncoides (Roxb.) Palla; *Scirpus juncoides* Roxb.; *S. erectus* sensu F.B.I. non Poiret. Dz: *inchodum, chocksen, shesem, manitsan, gutsem, mani racha*; Mangdi: *dungtiwa*; Nep: *swirey, suire*. Fig. 26l–m.
?Annual. Stems densely tufted, ± terete, 9–73cm, 1–3mm diameter. Basal sheaths 1–3, apex usually apiculate, longest 2.5–6cm, mouth oblique. Infl. of 1–7 spikelets, infl. bract erect, (1.5–)4–12cm. Spikelets narrowly ovoid, to 1.5 × 0.5cm. Glumes widely ovate to suborbicular, strongly concave, blunt to subacute, apiculate, 2.5–4 × 2.3–3mm, shining, upper part golden brown, pale below, midrib green. Stamens 1–3. Stigmas 2(–3); nut unequally biconvex, widely obovate, truncate, apiculate, attenuate to base, 2–2.3 × 1.6–2mm, pale greenish-brown becoming dark brown; bristles 6, unequal, longest 1.9–2.3mm, just exceeding nut.
Bhutan: S — Samchi district (Samchi); **C** — Thimphu (c.1km above Paro, Taba; Dotena to Thimphu (F.E.H.2)), Punakha (Upper Gaseloo, near Rinchu, Toiberong Chu, Samtengang to Ritang) and Tongsa (Shemgang) districts; **Terai** (Jalpaiguri); **Darjeeling** (Darjeeling, Munsong, Mongpu, Riang); **Sikkim** (Lamteng, Chumthang, Phodong Gompa). Paddy fields; streams; damp roadside in broad-leaved forest; marshy meadows, 500–2500m. Fl./fr. February–October.

Parker (1992) records it as a major weed of flooded rice present in all districts [with cultivation].

2. S. fuscorubens (T. Koyama) T. Koyama; *Scirpus fuscorubens* T. Koyama.
Differs from *S. juncoides* in its single rank of stems inserted on a horizontal, creeping rhizome; much smaller infl. — spikelets to 6 × 3mm; infl. bract shorter (1.3–2.5cm), spreading at right-angles; glumes smaller (to 3.2mm), marked with dark fuscous-purple at least around margins.

Bhutan: C — Punakha (Gangtey Gompa to Phubjikah) and Bumthang (Byakar, Gyetsa, above Dhur) districts. Damp flushes, bogs and peaty meadows, 2700–2870m. Fl./fr. June–September.

3. S. mucronatus (L.) Palla; *Scirpus mucronatus* L.
Perennial, rhizome short. Basal sheaths rather wide, longest 14–24cm. Stems tufted, triquetrous, 85–200cm, 4–7mm diameter, stout. Infl. of 8–13 spikelets; bract triquetrous, (1.2–)2.7–8.5cm, stout. Spikelets narrowly ovoid, to 1.5 × 0.5cm. Glumes ovate to widely ovate, subacute to acute, sometimes apiculate, concave, 3–5 × 2–3mm, straw-coloured, streaked with pale reddish-brown, margins usually dark golden-brown, whole glume sometimes becoming dark, midrib wide, green. Stamens 3. Stigmas 3; nut unequally biconvex, obovate, truncate, minutely apiculate, attenuate at base, 1.9–2.2 × 1.3–1.7mm, olive-green becoming dark brown, smooth, shining; bristles 6, subequal (2.1–2.5mm), exceeding nut.

Bhutan: C — Thimphu (Paro) and Punakha (Punakha, Rinchu) districts; **Darjeeling** (Great Rangit, Darjeeling, Balasun, Gopaldora); **Sikkim** (Lachen, Yoksum, Phodong to Kabi, Gangtok); **Terai** (Balasun). Rice fields and other wet places incl. ponds, 305–2740m. Fl. May–June; fr. August–October.

4. S. triqueter (L.) Palla.
Differs from *S. mucronatus* in its compound infl. with shortly stalked secondary infls.; stems borne singly on a slender, spreading, reddish rhizome.

Bhutan: C — Thimphu district (Kitchu, Lango Block, Paro). Irrigation ditch, 2300m. Fr. July.

Not previously recorded for the E Himalaya and perhaps a recent introduction arising from Japanese agricultural activities in the Paro valley.

S. supinus (L.) Palla subsp. *lateriflorus* (J.F. Gmelin) T. Koyama is likely to occur and differs from all the above species in its slender habit (stems under 2mm wide) and nuts lacking hypogynous bristles.

A single specimen of **S. articulatus** (L.) Palla has been seen from the Terai (Jalpaiguri). It differs from the above species in having hollow, transversely septate stems.

7. ELEOCHARIS R. Brown

Perennials, commonly tufted with short rhizomes, sometimes stoloniferous, sometimes annual. Stems terete or 3–4-angled, leafless. Leaves reduced to basal, bladeless, tubular sheaths. Infl. a dense terminal spike of spirally (occasionally distichously) inserted glumes; spike subtended by 1–2 sterile glume-like 'bracts'. Flowers bisexual. Perianth of (0–)5–8 hypogynous bristles. Stamens 1–3. Stigmas 2–3; style swollen at base, with constriction between nut and swollen base. Nut biconvex or trigonous, crowned with persistent style base.

The genus appears to be under-collected in Bhutan and Sikkim; additional species are almost certain to occur.

1. Glumes distichous .... **5. E. retroflexa** subsp. **chaetaria**
+ Glumes spirally arranged 2

2. Stigmas 2 3
+ Stigmas 3 4

3. Small annual, stems filiform, under 10cm .... **6. E. atropurpurea**
+ Rhizomatous perennial, stems stout, over 10cm ...... **1. E. palustris** agg.

4. Stems strongly 3–4-angled; hypogynous bristles plumose with short, fat hairs .... **2. E. tetraquetra**
+ Stems terete or shallowly 5-grooved; hypogynous bristles with few, slender, deflexed hairs 5

5. Stems filiform (c.0.3mm diameter), basal sheaths transparent (scarcely visible); spikes small (to 6 × 2mm), slender, very acute; angles of nuts thickened; glumes c.1.3mm .... **4. E. thomsonii**
+ Stems stouter (c.1mm diameter), basal sheaths conspicuous; spikes usually over 8mm, fatter, apex rounded; angles of nuts not thickened; glumes usually over 2mm .... **3. E. congesta**

1. E. palustris agg. Fig. 26n.
    Rhizomes creeping. Stems loosely tufted, terete, 10–53cm, stout (usually over 2mm diameter). Sheaths membranous, pale- to dark-reddish-brown, upper ± truncate. Spike narrowly ellipsoid, acute, 0.5–1.4 × 0.2–0.4cm. Lowest two glumes sterile, widely ovate, blunt, green, with hyaline border; fertile glumes oblong-ovate, rounded, 3–4 × 0.8–1.8mm, brown with paler midrib, margins broadly hyaline. Stamens 2–3. Stigmas 2; expanded style base widely conical,

c.½ width of nut. Nut biconvex, oblong- to widely-ellipsoid, distinctly narrowed to style base, 1.2–1.5 × 1–1.2mm, yellow, smooth, shining. Bristles 5–7, exceeding nut, slender, reddish-brown, with few, slender, acute, deflexed hairs.

Bhutan: C — Thimphu (Paro Bridge, Drugye Dzong, Doteng) and Bumthang (Byakar, Gyetsa) districts; **Sikkim** (Lachen, Lhonak). Marshes; pools, 2400–3050(–4490)m. Fl./fr. June–August.

Work is needed on the Indian members of this aggregate. The Bhutanese plants possibly belong to *E. mitracarpa* Steudel and differ from *E. palustris* s.s. in the style base; in the latter the widest part is separated from the nut by a constriction whereas in the Bhutanese plants the base is wider than long and ± appressed to the apex of the nut.

Some immature specimens from **Chumbi** (Yatung, Gantsa, Lingmatang; 3050–3810m) differ from the above and are possibly referable to *E. uniglumis* (Link) Schultes. In these only the lowest glume is sterile, the glumes are dark red-brown, without a hyaline border and there are no hypogynous bristles.

2. E. tetraquetra Nees.

Differs from *E. palustris* in its more slender (under 1mm diameter), 3- or 4-angled stems; sheaths pale brown, upper minutely apiculate; stigmas 3; bristles plumose, densely covered with blunt, fat, deflexed hairs.

Bhutan: S — Chukka district (above Lobnakha); **C** — Thimphu (below Umsho, hill above Thimphu Hospital) and Bumthang (above Dhur) districts; **Sikkim** (Lachen, Lachung, Chungtam, N of Rangit). Marshes; damp flushes; rice paddies, (460–)2300–3660m. Fl./fr. May–October.

3. E. congesta D. Don. Fig. 26o–p.

?Annual. Stems densely tufted, terete, 12–23cm, c.1mm diameter. Sheaths purplish at base; upper truncate, apiculate. Spike oblong-lanceolate in outline, rounded, 7–10 × 3.5–4mm; sometimes viviparous. Basal sterile glumes oblong-ovate, c.3 × 2mm; fertile glumes oblong-elliptic, rounded, 2–2.5 × 1–1.9mm, purple with green midrib, margins narrowly hyaline. Stamens 2. Stigmas 3; faces of style base triangular, concave, c.½ width of nut. Nut trigonous, narrowly obovoid, angles not thickened, 1 × 0.6mm, yellowish-green, smooth. Bristles 6, just exceeding nut, slender, whitish, with minute, deflexed hairs.

Bhutan: C — Thimphu (Chapcha; Thimphu to Yuwak (F.E.H.2)) and Punakha (Toiberong Chu; Punakha to Botokha (F.E.H.2)) districts; **Darjeeling** (Jepi; Palmajua to Rimbick (F.E.H.1)); **Sikkim** (Gangtok, Penlong La, Yoksum, Phodong to Kabi). Paddy fields; marshes by streams, 1310–2300m. Fr. April–July.

4. E. thomsonii Boeckeler.
Differs from *E. congesta* in being much more slender (similar in habit to *E. atropurpurea*); stems 7–15cm, filiform (c.0.3mm diameter), 5-grooved; sheaths transparent, very thin; spike smaller (3–6 × 1–2mm), narrowly lanceolate, very acute; glumes smaller (c.1.3 × 0.6mm); stamen 1; nut smaller (0.8 × 0.6mm) with thickened angles.
Bhutan: C — Thimphu district (Atsho Chhubar N of Paro). Rice paddy, 2430m. Fr. August.

This species, previously known only from Khasia, was wrongly sunk under *E. afflata* Steudel in F.B.I. More work, however, is needed on the typification of the small species collected by Hooker & Thomson in Khasia under the number *Eleocharis* 3 and distributed under the name *E. gracilis*, elements of which were described by Boeckeler as *E. thomsonii* and *E. chlorocarpa* and others which he wrongly assigned to *E. ochrostachys* Steudel.

5. E. retroflexa (Poiret) Urban subsp. **chaetaria** (Roemer & Schultes) T. Koyama.
Small, slender ?annual. Stems densely tufted, 5-angled, to 9.5cm, filiform (c.0.3mm wide). Sheaths reddish-brown; upper oblique, subacute. Spike to 4 × 3mm, few-flowered. Glumes distichous, finally widely spreading; lowest sterile, oblong with broad hyaline border; fertile glumes to c.6, oblong-lanceolate, blunt, lowest to 2.4 × 1.4mm, deep reddish-brown, midrib greenish, margins hyaline. Stamens 3. Stigmas 3; expanded style base pyramidal. Nut trigonous, obovate, truncate, angles thickened, c.1.3 × 0.8mm, faces reticulately pitted. Bristles 6–7, unequal, longest equalling nut, slender, cream, retrorsely scabrid.
Bhutan: S — Gaylegphug district (Gaylegphug River towards Norboling); **Duars** (Chalsa Plains). Fine, moist sand by stream in river bed, 400m. Fr. February–May.

6. E. atropurpurea (Retzius) Presl.
Dwarf slender plant differing from *E. retroflexa* as follows: stems terete; spike smaller (to 3 × 2mm), narrowly ellipsoid, many-glumed; glumes spiral, remaining ± erect, smaller (lowest fertile one c.1.2 × 0.6mm); stamen 1; stigmas 2; style base disc-like; nut biconvex, narrowly obovate, smooth; bristles exceeding nut.
Bhutan: C — Thimphu (Doteng) and Punakha (Rinchengang) districts. Rice paddies, 1250–2400m. Fr. July–August.

8. FIMBRISTYLIS Vahl

Perennials usually with short rhizomes, occasionally stoloniferous, or annuals. Leaves basal and sub-basal, usually with linear blade and sheathing, membranous base; stem leaves sometimes reduced to sheaths. Ligule sometimes present as fringe of short hairs across base of blade. Stems usually tufted, often 4 (or more)-angled, sometimes compressed. Infl. irregularly umbellate, once- to several-times compound, subtended by involucre of unequal leaf-like bracts, primary rays unequal, surrounded at base by tubular cladoprophylls; sometimes reduced to a single or very few spikelets in which case bracts glume-like. Spikelets with spirally or distichously (sometimes placed in *Abildgaardia*) arranged glumes, axis persistent, fringed with persistent glume-bases; flowers

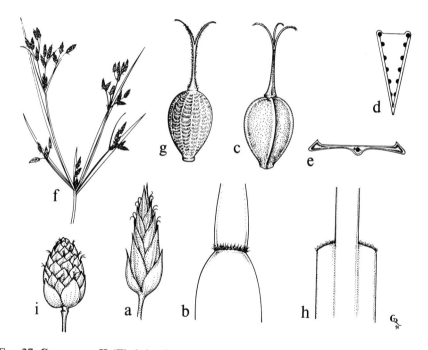

FIG. 27. **Cyperaceae** II (**Fimbristylis**).
a, **Fimbristylis ovata**: spikelet showing distichous glumes (× 3). b–c, **F. complanata**: b, ligulate leaf-sheath apex (× 4); c, trigonous nut with trifid stigma (× 25). d, **F. littoralis**: t.s. of leaf (× 15). e, **F. miliacea**: t.s. of leaf (× 25). f–g, **F. dichotoma**: f, inflorescence (× ½); g, biconvex nut and bifid stigma (× 25). h, **F. rigidula**: eligulate leaf-sheath apex (× 5). i, **F. schoenoides**: spikelet showing spirally arranged glumes (× 3). Drawn by Glenn Rodrigues.

bisexual. Stamens 1–3. Style flattened or triquetrous, fimbriate or glabrous, thickened at base, sharply demarcated from ovary; stigmas 2 or 3. Nut biconvex or trigonous, without terminal disc; smooth, tuberculate or reticulate.

The genus appears to be under-collected in Bhutan and Sikkim, no doubt due to its weedy appearance and preference for cultivated habitats; additional species are likely to occur.

1. Stigmas 3 2
+   Stigmas 2 9

2. Glumes distichously arranged (Fig. 27a) 3
+   Glumes spirally arranged (Fig. 27i) 4

3. Infl. usually a single spikelet (occasionally two); spikelet pale, usually over 8 × 4mm .... **1. F. ovata**
+   Infl. of usually 2–4 spikelets, reduced to 1 in depauperate plants; spikelets dark brown, under 6 × 3.5mm .... **2. F. fimbristyloides**

4. Flowering stem with bladeless sheaths at base; sheaths with oblique mouths, extended at most into filiform point to 2cm 5
+   Flowering stem bearing leaves with flat blades near base 7

5. Leaves laterally compressed, V-shaped in section, adaxial face a narrow groove (Fig. 27d); spikelets under 3mm, blunt .... **6. F. littoralis**
+   Leaves with flat blade (Fig. 27e); spikelets over 3.5mm, acute 6

6. Plant usually annual, without a rhizome; basal sheaths straw-coloured; glumes under 1.5mm; spikelets under 2mm wide .... **7. F. miliacea**
+   Plant perennial, rhizome short, stout; basal sheaths becoming fuscous; glumes c.2.5mm; spikelets over 2mm wide .... **8. F. filifolia**

7. Spikelets in tight clusters of 2–3; leaves eligulate (Fig. 27h), usually under half stem length .... **5. F. falcata**
+   Spikelets borne singly; leaves ligulate (Fig. 27b), commonly half or more stem length 8

8. Spikelets under 2mm wide; glumes without lateral veins; stems compressed, slender at base .... **3. F. complanata**
+   Spikelets over 2mm wide; glumes with 3 or more strong lateral veins either side of midrib; stems quadrangular woody and thickened at base
    4. F. thomsonii

9. Infl. a single (sometimes 2) spikelet; glumes thick-textured 10
+ Infl. of more than 3 spikelets; glumes membranous 11

10. Nut without transverse grooves; spikelet blunt; leaves with blades
16. F. schoenoides
+ Nut with conspicuous transverse grooves; spikelet acute; leaves reduced
to bladeless, membranous sheaths .... **17. F. acuminata**

11. Leaves eligulate (Fig. 27h) 12
+ Leaves ligulate (Fig. 27b) 14

12. Annual; glumes under 1.3mm; nut under 0.6mm .... **15. F. aestivalis**
+ Perennial with short, woody rhizome; glumes over 1.6mm; nut over
1mm 13

13. Glumes keeled, without lateral veins; infl. once to twice compound;
spikelets pale brown, angled; style glabrous .... **14. F. fuscinux**
+ Glumes not keeled, with 2–4 strong lateral veins each side of midrib;
infl. simple to once-compound; spikelets reddish-brown, rounded;
style ciliate .... **13. F. rigidula**

14. Plant spreading by slender stolons .... **12. F. stolonifera**
+ Plant tufted, lacking stolons 15

15. Nuts with more than 10 vertical rows of cells per face; spikelets ovoid
(3–4mm wide) .... **11. F. tomentosa**
+ Nuts with fewer than 10 rows of cells per face; spikelets narrowly
ovoid (under 3.5mm wide) 16

16. Glumes not keeled, over 1.8mm; spikelets rounded, usually over
1.5mm wide .... **9. F. dichotoma**
+ Glumes keeled, under 1.5mm; spikelets angled, under 1.7mm wide
10. F. bisumbellata

Plants perennial unless otherwise stated.

1. F. ovata (Burman f.) Kern; *F. monostachya* (L.) Hasskarl; *Abildgaardia monostachya* (L.) Vahl. Fig. 27a.
Rhizomes very short. Stems densely tufted, compressed, ± quadrangular, grooved, 3–18(–40)cm, 0.4–0.7mm wide. Leaves basal, acute, margins thickened, minutely serrate, c.half length to slightly exceeding stems, 0.6–1mm wide, inrolled, eligulate. Infl. a single, erect, terminal spikelet subtended by 2 glume-

like bracts. Bracts ovate, 2–3 (excl. awn) × 1.8–2.2mm, hyaline, with green midrib produced into awn 0.7–2(–5)mm. Spikelet compressed, narrowly ovate, acute, sometimes becoming twisted, 6.5–13 × 3.5–5mm. Glumes distichous, ovate, apiculate, 4.3–5 × 3.2–4mm, keeled, whitish-hyaline, sometimes reddish-streaked near midrib, midrib 3-veined, green; persistent glume bases streaked reddish. Stamens 3. Style triquetrous, angles fimbriate, thickened at base, 2.3–3.5mm; stigmas 3, c.0.7mm. Nut rounded-trigonous, faces broadly obovate, apiculate, narrowed to stipitate base, 2.3 × 1.5–1.7mm, tuberculate, straw-coloured.

Bhutan: C — Punakha district (1km N of Punakha Dzong); **N** — Upper Mo Chu district (Kencho). Sandy grassy ground by river, 1270–1830m. Fr. August–September.

2. F. fimbristyloides (F. von Mueller) Druce; *Abildgaardia fimbristyloides* F. von Mueller.

Dwarf annual. Stems 5-grooved, 2–8cm, 0.4mm wide. Leaves basal, curved, ± flat, apiculate, margins minutely serrate, much shorter than stems, to 1.3mm wide, eligulate. Infl. of 2–4 unequally rayed spikelets (reduced to a single terminal spikelet in depauperate plants), rays smooth, longest to 4mm. Bracts with glume-like bases and leaf-like points shorter than infl. Spikelets compressed, lanceolate, acute, sometimes becoming twisted, 5.5–6 × 1.7–3.5mm. Glumes distichous, ovate-triangular, acute, minutely mucronate, 2.8–3.2 × 1.6–1.8mm, minutely hispid near apex and on midrib, midrib 1-veined, dark reddish-brown, margins hyaline. Stamens 3. Style triquetrous, thickened at base, c.1.5mm, glabrous; stigmas 3. Nut rounded-trigonous, faces broadly obovate, apiculate, abruptly contracted to shortly (0.1mm) stipitate base, 1–1.1 × 0.8–0.9mm, tuberculate, straw-coloured, with shining whitish epidermis.

Bhutan: S — Chukka district (between Chukka and Chimakothi). Bare ground round rocks, where seasonally wet, in scrubby grassland, 1800m. Fr. October.

Further work is required on these small, seldom-collected, taxa related to *F. fusca* (Nees) C.B. Clarke in Sino-Himalaya; our single specimen differs from the description of *F. fimbristyloides* in Kern (1974) in having larger nuts and glabrous stems and rays.

3. F. complanata (Retzius) Link. Sikkim name: *jyakcha*. Fig. 27b–c.

Rhizome short. Stems densely tufted, upper part compressed, 13–32(–98)cm, 0.8–1.7mm wide. Leaves basal and sub-basal, rather abruptly contracted at apex, margins thickened, minutely serrate, less than half length (to equalling) stem, 0.8–4(–5.2)mm wide, stiff; sheaths pale brown, margins hyaline; ligulate. Infl. (once to) several times compound, 1–8(–15) × 1–9(–13)cm; primary rays flattened, longest 0.8–4.5(–9)cm; lowest bract leaf-like (others narrower), 1–5.4(–6)cm. Spikelets borne singly, linear-ellipsoid, angled, acute, 4.4–7 × 1.3–2mm, sessile and pedicelled. Glumes ovate, acute to apiculate, 2.1–2.6 × 1–1.8mm, midrib green, sometimes scabrid, strongly keeled, brown, upper part of margins narrowly hyaline; persistent glume bases dark reddish-brown. Stamens 3. Style swollen at base, c.0.9mm, glabrous; stigmas 3, erect, about equalling style; nut strongly trigonous, faces obovate, narrowed to base, 0.8–1 × 0.6–0.7mm, cream, shining, ± smooth, faces with c.10 vertical rows of transversely oblong cells.

Bhutan: S — Chukka district (below Chimakothi); **C** — Thimphu (near Simtokha, Olothang, below Sharadango Gompa), Punakha (Tinlegang) and Bumthang (Gyetsa, above Dhur) districts; **Terai** (Dulkajhar); **Darjeeling** (Mongpu, Punkabari, Manibhanjang, Ghum); **Sikkim** (Domang, Penlong La, Lachen, Phadonchen, Gangtok, above Yoksum, Phodong Gompa). Under-recorded, probably very common. Seasonally damp rough grassland; marshes; open, wet places in oak forest, 150–3210m. Fr. August–October.

The record of var. *microcarpa* [sic] (= *F. microcarya* F. von Mueller) for Sikkim in F.B.I. is an error, the specimen being typical *F. complanata.*

4. F. thomsonii Boeckeler.

Similar to *F. complanata* in its broad, ligulate leaves, but differing in its stouter, more woody stem bases; leaves more glaucous; stems quadrangular, not compressed; spikelets wider (over 2mm); glumes larger (c.3mm), more orange, with 3 or more lateral veins on either side of midrib.

Terai/Duars (Pulma, Buxa Reserve, Sivok). Apparently flowering after being burnt. Fl. February–April.

5. F. falcata (Vahl) Kunth; *F. junciformis* Kunth.

Rhizome stout, woody, shortly creeping. Stems borne singly, compressed, 14–36cm, 0.6–1mm wide. Leaves basal and sub-basal, tending to recurve, apex membranous, apiculate, margins minutely serrate, 3.5–7.5cm, much shorter than stem, 1–1.8mm wide; sheaths short, wide, margins hyaline; eligulate. Infl. twice compound, 3.3–4.5 × 2.5–4.5cm, longest primary ray 1.8–3cm; lowest bract linear, 1–2.2cm. Spikelets in tight groups of 2–3 (sometimes appearing to be single, fat spikelets), narrowly ovoid, acute, 3–7 × 1–2mm, sessile. Glumes ovate, acute to apiculate, 2–2.7 × 2–2.2mm, keeled, midrib 3-veined, dark reddish-brown, margins broadly hyaline. Stamens 3. Style gradually widened to base, c.1.5mm, glabrous; stigmas 3, c.1mm; nuts strongly trigonous, faces widely obovate, 0.9 × 0.8mm, cream, shiny, with small, widely spaced tubercles.

Bhutan: C — Punakha (1km N of Punakha Dzong) and Mongar (Lhuntse

Dzong) districts. Dry open hillsides; sandy grassy ground by river, 1220–1270m. Fl. April–September; fr. September.

6. F. littoralis Gaudichaud; *F. miliacea* Vahl. Fig. 27d.

Annual. Stems compressed, 3–4-angled, 26–53cm, 1.2–1.7mm wide, densely tufted, bases flattened. Basal leaves distichous, gradually tapered to very acute apex, c.half length to equalling (sometimes exceeding) stem, laterally compressed, V-shaped in section, inner face very narrow, channelled, 1.1–1.5mm deep; sheaths with overlapping, hyaline margins; eligulate. Stem leaves reduced to 1–2 sheaths, mouths long, oblique, apiculate tips to 3cm. Infl. 3–4× compound, 4.5–8.9 × 3–10cm; primary rays widely spreading, very unequal; bracts filiform, very acute, lowest 2.4–3cm. Spikelets borne singly, ovoid, blunt, 1.9–3 × 1.4–2mm, mostly pedicellate. Glumes ovate, rounded, concave, 0.7–1.2 × 0.5–0.8mm, midrib 3-veined, usually green, dark brown with narrow hyaline margins. Stamens 1–2. Style thickened at base, 0.4–0.5mm, fimbriate at base of stigma lobes; stigmas 3, 0.2–0.4mm. Nut rounded-trigonous, faces narrowly obovate, narrowed to base, 0.5–0.6 × 0.3–0.4mm, cream, faces with 3–4 rows very narrowly oblong transverse cells, conspicuously tuberculate.

Bhutan: C — Punakha district (Lobesa, Khuru, near Punakha Dzong); **Terai** (Balasun); **Darjeeling** (Tista, Mahanadi). Rice paddies, 150–1400m. Fr. August–October.

Parker (1992) records it as an important weed of flooded rice in Punakha district.

7. F. miliacea (L.) Vahl; *F. quinquangularis* (Vahl) Kunth. Fig. 27e.

Similar to *F. littoralis* in having stem leaves reduced to long sheaths with minute blades. Differs in being often larger and stouter (stems to 1m), stem bases not flattened, basal leaves not laterally compressed but with flat blades, with a prominent midrib and minutely serrate margins; spikelets longer (3.5–5.8 × 1.5–2mm), more acute and more strongly angled; glumes larger and wider (1.3–1.5 × 1mm), more acute, with midrib shortly excurrent, more strongly keeled.

Bhutan: C — Punakha district (Lobesa, Waecha Bridge, Wangdi Phodrang, Punakha); **N** — Upper Mo Chu district (Kencho); **Terai** (Sukna); **Darjeeling** (Rimbick, 1km above Mongpu, Tista); **Sikkim** (Phodong to Kabi). Marshes; paddy fields, 150–1860m. Fl./fr. July–October.

8. F. filifolia Boeckeler.

Similar to *F. miliacea* but much larger, with stout rhizomes, and densely tufted, stiffly erect stems 90–126cm, sharply 5-angled, c.2.7mm diameter; blade-less sheaths at base of stem to 27cm, yellow-brown with fuscous patches,

232. CYPERACEAE

shining, coriaceous; infl. larger (9.5–16.5 × 11–22cm), with widely spreading branches; spikelets wider (4.5–6 × 2–2.5mm); glumes larger (c.2.5 × 1.8mm); stamens 3.

Terai (Dulkajhar). 150m. Fr. October.

This plant is much larger than the type from Khasia and probably represents an undescribed species as indicated by E. Govindarajalu in a note on a specimen at K. It would be desirable to see more collections before describing it.

9. F. dichotoma (L.) Vahl; *F. diphylla* Vahl. Fig. 27f–g.

Rhizome very short. Stems densely tufted, compressed, grooved, 4–72cm, 0.5–1.7mm wide. Leaves basal and on lower ¼ stem, flat, acute, margins minutely serrate, to more than half length of stem, 1.5–2.6mm wide; inner face of sheaths brown-hyaline, shortly hairy; ligulate. Infl. 1–3 × compound, usually rather open, 1.8–12 × 1.5–14cm; primary rays usually unequal, longest 1–8cm, sometimes subequal when infl. very dense; lowest bract leaf-like, 5–11.2cm, shorter than to exceeding infl. Spikelets borne singly, narrowly ovoid, subacute, (2.5–)3.7–12 × (1.2–)1.5–3.5mm. Glumes ovate, (rounded–)acute to apiculate, concave, 1.8–2.7 × 1.4–2mm, midrib green, 3-veined, straw-coloured, flushed brown to reddish-brown in upper part to varying degrees, margins narrowly hyaline. Stamen 1. Style gradually widened towards base, 1.4–1.7mm, dark brown, margins white-fimbriate; stigmas 2, short, recurved. Nut biconvex, faces obovate, 0.8–1 × 0.7–0.9mm, cream, shining, faces reticulate, with 7–9 vertical rows of transversely oblong cells with thickened margins, borne on very short, dark gynophore.

Bhutan: S — Samchi (Thoribhadi), Chukka (Chimakothi to Phuntsholing) and Gaylegphug (Gaylegphug) districts; **C** — Thimphu (c.1km above Paro), Punakha (Lobesa, Wangdi Phodrang), Tongsa (W of Mangde Chu), Bumthang (above Dhur) and Mongar (below Mongar, Lingmethang) districts; **Terai/ Duars** (Balasun, Buxa to Santrabari); **Darjeeling** (Little Rangit, Kalimpong, Pashok, Labdah, Golondhora, Simulbari, Punkabari, Mongpu, Tista, Riang, Great Rangit opposite Manjitar); **Sikkim** (Gangtok, Dikchu, Yoksum, W of Singtam). Under-recorded; probably very common. Damp grassy ground; roadsides; marshes and shingle by stream; cultivated ground; paddy fields, 150–3210m. Fr. March–October.

10. F. bisumbellata (Forsskål) Bubani; *F. dichotoma* sensu F.B.I. non Vahl.

Differs from *F. dichotoma* in its smaller, narrower (3–5 × 1.2–1.7mm), more strongly angled spikelets; glumes smaller (1.2–1.5 × 0.7–1mm), often paler, with green midrib (sometimes hairy) strongly keeled and therefore more conspicuous.

Sikkim (Rayang (F.E.H.1)). 2500m.

294

11. F. tomentosa Vahl; *F. podocarpa* Nees; *F. dichotoma* subsp. *podocarpa* (Nees) T. Koyama.

Differs from *F. dichotoma* in its fatter (3–4mm) spikelets and wider nuts with over 12 vertical rows of cells per face, borne on more conspicuous gynophores.

Bhutan: C — Punakha district (Khuru); **Terai** (Dulkajhar). Rice paddies, 150–1250m. Fr. July–October.

12. F. stolonifera C.B. Clarke.

Similar to the forms of *F. dichotoma* with relatively simple infls., but differs in spreading by slender, creeping stolons covered with dull, fibrous scales; spikelets fatter (4–7.8 × 2.2–4.5mm); glumes (2.7–3.2 × 1.6–2.4mm) always dark reddish-brown (midrib green), shining; stamens 3; style fimbriate only in upper half and stigmas long, about equalling style. Nuts more like those of *F. tomentosa* having a conspicuous gynophore and c.12 vertical rows of cells per face, sometimes tuberculate.

Bhutan: S — Chukka district (Bunakha); **C** — Thimphu (above Thimphu Public School, near Motithang Hotel, above Yangchenphug School, hill above Thimphu Hospital, Paro), Bumthang (2km N of Byakar Dzong, Gyetsa, above Dhur) and Tashigang (Chorten Kora) districts; **Sikkim** (Phodong Gompa). Flushed grassy slopes; marshes; open turf; terraced grazing, 1680–2870m. Fl./fr. June–September.

13. F. rigidula Nees. Fig. 27h.

Rhizome thick, woody, shortly creeping. Stems borne singly, grooved, base swollen, clothed with fibrous remains of old leaf sheaths, (6–)15–21cm, c.0.7mm wide. Leaves basal and sub-basal, flat, rather abruptly contracted at apex, margins minutely serrate, to half length of stem, 1.7(–2.3)mm wide, rigid, slightly glaucous; eligulate. Infl. 1.8–3.3 × 0.8–2.8cm; longest ray 1–2cm; lowest bract leaf-like, shorter than infl. Spikelets ovoid, rather blunt, 5–6.5 × 2.5–4mm. Glumes broadly ovate, minutely apiculate, concave, c.2.8 × 2.5mm, midrib 3-veined, 2–4 prominent lateral veins each side of midrib, rich reddish-brown, margins narrowly hyaline. Stamens 2–3. Style thickened at base, c.1.5mm, sparsely ciliate below, densely ciliate on upper half and lower part of stigmas; stigmas 2, c.half length style, recurved. Nut biconvex, smooth, with many small, polygonal cells.

Jalpaiguri Duars (Kolabari). Grassland, c.150m. Fl. February.

14. F. fuscinux C.B. Clarke.

Rhizome short. Stems tufted, with fibrous remains of old leaf sheaths at base, compressed-quadrangular, 19–76cm, 1.1–1.8mm wide. Leaves basal and sub-basal, flat, tending to recurve, apex membranous, apiculate, to half length stem, 1.5–3mm wide; eligulate. Infl. rather lax, 3× compound, 5.2–18 × 2–9.5cm; longest ray 3–11cm; lowest bract leaf-like, shorter than infl. (2–2.7cm). Spikelets borne singly, narrowly ovoid, subacute, angled, 3.6–19 × 2.5–3.5mm, pedicellate and sessile. Glumes ovate, subacute, mucronate, 1.6–2 × 1.2–1.5mm, keeled, midrib green, 3-veined, pale brown, margins hyaline. Stamens 2–3. Style gradually thickened towards base, c.0.8mm, glabrous, dark brown; stigmas 2, about equalling style. Nut biconvex, rather fat, broadly ovate to suborbicular, narrowed to base, 1–1.3 × 1mm, cream or black, smooth, cells very small.

Terai (Balasun); Darjeeling (Great Rangit opposite Manjitar). Sandy shingle by river, 150–440m. Fr. August–October.

15. F. aestivalis (Retzius) Vahl.

Slender, tufted annual. Stems filiform, 1–9cm, 0.2–0.4mm wide. Leaves basal and sub-basal, acute, margins thickened, c.half length of stem, 0.2–0.7mm wide, with very few, scattered hairs, inrolled; sheaths pilose, margins pale brown, membranous; eligulate. Infl. umbellate, 1–2× compound, 0.3–2.5 × 0.2–2cm; longest ray 0.3–1.4cm; lowest bract to 2cm, shorter than to just exceeding infl., bases of bracts hairy. Spikelets linear-lanceolate, acute, angled, 3–5.4 × 0.9–1.6mm. Glumes oblong-elliptic, blunt to subacute, mucronate (to varying degrees), 1.1–1.3 (incl. mucro) × 0.9–1.6mm, keeled, lower ones hairy on midrib, midrib green, 1–3-veined, very pale brown, streaked reddish-brown. Stamen 1. Style 0.4–0.6mm, ± glabrous, sometimes minutely hairy on thickened base; stigmas 2, c.half length style, hairy at base. Nut biconvex, faces obovate, 0.4–0.5 × 0.3–0.4mm, pale yellow, translucent, smooth.

Bhutan: S — Samchi (2km W of Chengmari) and Gaylegphug (Gaylegphug River between Gaylegphug and Norboling) districts; **C** — Thimphu district (unlocalised record in Parker, 1992); **Duars** (Jaldaka River); **Darjeeling** (Singla; Rayang, Kalimpong (F.E.H.1)); **Sikkim** (Yoksum, Gangtok; Dentam to Pamianchi (F.E.H.1)). Fine sand on river banks; steep gravelly bank especially in hollows where water accumulates; lake margin; flooded rice-fields, 250–2000m. Fr. February–July.

16. F. schoenoides (Retzius) Vahl. Fig. 27i.

Rhizomes short. Stems densely tufted, grooved, 5–25cm, 0.4–0.8mm wide. Leaves basal and sub-basal, margins thickened, smooth or minutely serrate, inrolled, more than half length to equalling stem, 0.8–1.5mm wide; ligulate. Spikelet single (sometimes also with a secondary, peduncled spikelet), subtended by 1–2 sterile, glume-like bracts with midrib developed into awn to varying degrees (to 7mm, shorter than or exceeding spikelet). Spikelet ovoid, blunt, 6–8.3 × 3.5–4.5mm. Glumes very widely ovate, blunt, sometimes minutely apiculate, 2.5–3 × 3.2–4mm, not keeled, midrib 3-veined, not conspicuous, green above, with 3 or more strong, lateral veins each side of midrib, thick-textured, cream, upper part flushed or streaked pale reddish-brown. Stamens 3. Style flattened, thickened at base, margins fimbriate, 1.2–1.4mm; stigmas 2, to half length of style, recurved. Nut biconvex, faces obovate, c.1.4 × 1.3mm, smooth, cells polygonal, very small; gynophore distinct but minute.

Bhutan: C — Punakha district (0.5km N of Punakha Dzong); **Terai** (Balasun). Stream; grassy bank around paddy field, 150–1270m. Fr. September–October.

17. F. acuminata Vahl.

Rhizomes short. Stems densely tufted, grooved, 4–13cm, c.0.7mm wide. Leaves reduced to sheaths. Sheaths basal, c.3 per stem, longest to 2cm, membranous, brown, mouth oblique, apiculate. Infl. a single spikelet; bract indistinguishable from glumes, not aristate. Spikelets linear-lanceolate, very acute, 6.5–10.5 × 2–3mm, lowest 1–2 glumes empty. Glumes narrowly ovate, subacute, 3.8–4 × 1.8–2.4mm, midrib 3-veined, with 1 lateral vein each side of midrib, straw-coloured above, flushed reddish-brown near middle, margins narrowly hyaline. Stamens 2–3. Style gradually widened towards base, c.2.5mm, ciliate at extreme apex; stigmas 2, short. Nut biconvex, suborbicular, 1.5 × 1.5mm, pale brown with c.5 very prominent transverse grooves.

Terai (Balasun, Siliguri). Presumably wet places (e.g. paddy fields), 150m. Fr. October.

Doubtful species:

F. tenera Roemer & Schultes has been recorded from Deothang district (9km from Wamrong (M.F.B.)) but no specimen seen and not very likely.

9. BULBOSTYLIS Kunth

Differs from *Fimbristylis* in being much more slender, with filiform stems and leaves, sheaths with long white hairs at apex and above all in the thickened style base which persists as a disc on the nut (rather than whole style deciduous).

1. Infl. usually umbellate (occasionally reduced to a single spikelet); glumes acute; nuts densely but finely papillose .... **1. B. densa**
+   Spikelets sessile forming a dense head; glumes with midrib developed into a short, recurved apiculus; nuts smooth .... **2. B. barbata**

232. CYPERACEAE

1. B. densa (Wall.) Handel-Mazzetti; *B. capillaris* var. *trifida* (Nees) C.B. Clarke. Fig. 26q–s.
Densely tufted annual. Stems 4–20cm, filiform. Leaves basal and sub-basal, erect, very acute, half or more stem length, filiform (c.0.2mm wide); sheaths pale brown, membranous. Infl. umbellate (occasionally reduced to a single spikelet), spikelets sessile and rayed, rays 1–3, 0.4–2.5cm each bearing a single spikelet or again branched; lowest bract with filiform tip shorter than infl. Spikelets narrowly ovoid, acute, angled, 2.5–5.8 × 1.2–2.5mm. Glumes ovate, acute, margins narrowly hyaline, minutely fimbriate, 1.5–1.9 × 1–1.8mm, keeled, midrib green, 3-veined, reddish-brown, darker brown near midrib. Stamens 2, anthers minute (c.0.3mm). Style 0.6–0.8mm, glabrous; stigmas 3, c.0.5mm. Nut crowned with small brown disc, strongly trigonous, faces obovate to shallowly obcordate, 0.6–0.7 × 0.6mm, cream initially, finally greyish, surface densely but finely papillose.
Bhutan: S — Chukka district (Chukka to Chimakothi, above Lobnakha); **C** — Thimphu (Dechhenphu, hill above Thimphu Hospital, above Yangchenphug School), Punakha (Lometsawa), Tongsa (Mangde Chu) and Bumthang (Gyetsa, Bumthang, above Dhur) districts; **Darjeeling** (Ghumpahar, Kolbong, Raysing, Darjeeling, Manibhanjang; Kalimpong (F.E.H.1)); **Sikkim** (Lachen, Gangtok, Chungthang, Zemu Valley, Thanka La, Yoksum, Phodong; Dzongri (F.E.H.1)). Damp gravelly roadsides and terraces; open places in oak/pine forest; open marshy meadows; arable weed, 1520–4100m. Fr. May–October.

2. B. barbata (Rottboell) C.B. Clarke.
Differs from *B. densa* in its densely congested terminal head of spikelets 0.2–1.5cm diameter; glumes uniformly orange-brown, 1.3–1.4 × 1–1.4mm, with midrib continued into short, slightly recurved apiculus; nut smooth.
Bhutan: S — Chukka district (Kalikhola); **Terai** (Mahanadi, Siliguri, Balasun); **Darjeeling** (Great Rangit opposite Manjitar, Tista; Rangpo to Tista (F.E.H.1)). Shingle and sand beside river, 150–610m. Fr. May–October.

10. ISOLEPIS R. Brown

Dwarf, tufted annuals. Stem filiform, bearing 1 sub-basal leaf with filiform blade and basal bladeless sheath. Infl. of 1–2 spikelets, usually appearing lateral, subtended by bract appearing to continue stem. Spikelets with spirally inserted glumes. Flowers bisexual. Perianth absent. Stamens 2–3. Stigmas 3; style passing gradually into nut. Nut obscurely trigonous.

298

1. I. setacea (L.) R. Brown; *Scirpus setaceus* L. Fig. 26t.
Stems often curved, 1.5–11cm, under 0.7mm wide. Blade of stem leaf scarcely developed to almost equalling stem, filiform (under 0.5mm wide), sheaths often reddish. Infl. bract with glume-like base and long-excurrent, green midrib 3–8mm (if shorter than spikelet then infl. appearing terminal). Spikelet ovoid, 2–4 × 1.5–2mm. Glumes oblong, rounded or emarginate, strongly concave, c.1 × 0.7mm, purple to blackish, midrib green, margins narrowly hyaline. Nut oblong to obovoid, apiculate, 0.7–0.8 × 0.4–0.7mm, brownish-pink, longitudinally ribbed and transversely very finely rugulose.

Bhutan: C — Thimphu (above Bongde Farm, 6km N of Thimphu Dzong, Doteng, Chapcha, Hongtso, etc.), Punakha (Gon Chungang to Punakha (F.E.H.2)), Tongsa (Pele La) and Bumthang (Byakar, above Gortsam) districts; **N** — Upper Mo Chu district (Gasa Dzong); **Sikkim** (Lachen, Tungu, Zemu Valley, Tookit). Wet, bare open ground (e.g. shingle, mud, by streams and ditches, roadsides), 1400–3660m. Fl./fr. June–October.

11. CYPERUS L.

Annuals or perennials. Rhizomes short (stems tufted) or long, sometimes stoloniferous, stolons sometimes bearing tubers. Stems trigonous, with sheathing leaves at or near base. Infl. terminal, commonly irregularly umbellate (simple or compound to various orders), bearing spikelets or partial infls. on unequal peduncles ('rays'), rays sometimes not developed so infl. capitate; rays surrounded at base by tubular bracts ('cladoprophylls'); infl. subtended by an involucre of leaf-like bracts usually greatly exceeding infl. Spikelets arranged in spikes along a rachis, or in heads or umbels (rachis 0). Spikelets compressed, with many distichously arranged glumes on persistent axis ('rachilla'), glume margins sometimes developed downwards forming a wing to the rachilla; glumes deciduous or (in subgenus *Mariscus*, spp. 4–9) entire spikelets deciduous. Glumes concave, midrib sometimes keeled. Perianth absent. Stamens 1–3. Stigmas 3. Nut trigonous.

I have followed Goetghebeur (1989) in his suggested generic delimitation of *Cyperus*, i.e. incl. *Mariscus*, but not the genera with 2 stigmas and biconvex nuts (*Pycreus* and *Kyllinga*).

No doubt under-collected, many of the species from the area have only been seen from a few, old collections. More species have been recorded for Nepal.

The letters following the species names in the keys to *Cyperus* and *Pycreus* refer to the representative infl. types shown in Figs 28–29.

1. Stems to 1.5m, leafless; bracts leaf-like forming a slightly interrupted
   involucre; cultivated ornamental .... **25. C. involucratus** (D)
+ Not as above 2

2. Glumes distinctly mucronate, mucro over 0.2mm — erect or recurved   3
+ Glumes not mucronate 7

3. Perennial; leaves (and bracts) over 6mm wide; spikelets in very lax
   compound umbels .... **16. C. laxus** (D)
+ Annual; leaves (and bracts) under 3mm wide; spikelets in dense heads
   or spikes 4

4. Mucros stout, ± erect, developed from wide (c.1mm) midrib; spikelets
   c.3mm wide; stout annual (stems usually over 15cm)
             11. C. compressus (A or C*)
+ Mucros slender, recurved, midrib narrower; spikelets to 2mm wide
   (excl. mucros); slender annuals (stems usually under 10cm) 5

5. Plant smelling of curry when dry; spikelets arranged in blunt, cylindric
   spikes (rachis over 3mm) .... **1. C. squarrosus** (B or rayless version)
+ Plant not smelling of curry; spikelets arranged in heads (i.e. rachis not
   developed) 6

6. Nut with obovate faces (to 0.7 × 0.4mm); stamens 2–3; glumes with
   curved margins, sides golden- to reddish-brown
         2. C. cuspidatus (A or 1 × compound version of D)
+ Nut with linear faces (to 0.8 × 0.3mm); stamen 1; glumes with straight
   margins, sides reddish-purple
         3. C. castaneus (A or 1 × compound version of D)

FIG. 28. **Cyperaceae** III (representative inflorescence types of **Cyperus**).
a, Type A, **C. niveus** (× 1): simple infl. with spikelets sessile in dense head. b, Type B,
C. cyperoides (× ½): 1 × compound infl. with indehiscent spikelets spreading, arranged
along elongate rachis, in narrowly cylindric spikes (rayless versions also occur). c, Type
C, **C. rotundus** (× ⅔): 1 × compound infl. with spikelets suberect, arranged along
shortly elongate rachis, in broadly cylindric spikes. d, Type C*, **C. compressus** (× ⅔):
as for C but spikelets spreading. e, Type D, **C. laxus** (× ⅔): 2 × compound umbellate
infl., with spikelets sessile in umbels with no rachis developed (1 × and 3 × compound
versions also occur). Drawn by Glenn Rodrigues.

a

b

c

d

e

GARAI

301

302

7. Infl. a head of sessile or rayed, densely cylindric spikes; spikelets falling entire at maturity, small, linear, with 4 or fewer nuts 8
+   Infl. not as above; glumes falling at maturity and rachilla persistent (if spikelets not disarticulating, spikelets orange-brown and leaves and sheaths with conspicuous transverse veinlets or midrib of glumes spongy-keeled); spikelets usually elliptic to oblong, with more than 4 nuts 11

8. Spikes long-rayed (1–9cm); spikelets very narrow (to 0.8mm), spreading horizontally at maturity; nut usually 0.5mm wide
                                                          5. C. cyperoides (B)
+   Spikes sessile (occasionally with rays to 2.5cm); spikelets usually over 1mm wide (if less then plant stoloniferous and spikelets with 1 nut), obliquely erect or slightly recurved; nut 0.6–0.8mm wide 9

9. Lowest fertile glume under 2.7mm
                                  4. C. sikkimensis (rayless version of B)
+   Lowest fertile glume over 3.2mm 10

10. Stoloniferous; stems with reddish-brown fibres at base; spikelets narrow (c.0.5mm), recurved; spikes short and narrow (to 1.5 × 0.7cm)
                                      7. C. paniceus (B or rayless version)
+   Not stoloniferous; stems not fibrous at base; spikelets wider (c.1mm), straight; spikes longer and wider (to 1.8 × 1.2cm)
                                      6. C. cyperinus (rayless version of B)

11. Underside of leaves and leaf sheaths with prominent transverse veinlets; spikelets orange-red, with 3–7 fertile glumes, falling entire at maturity
                                                          8. C. compactus (F)

FIG. 29. **Cyperaceae** IV (representative inflorescence types of **Cyperus**). a, Type E, **C. difformis** (× ½): 2× compound infl. with spikelets in very dense, compound heads (versions with rayed secondary heads also occur). b, Type F, **C. pangorei** (× ½): 2× compound infl. with spikelets suberect, arranged along rachis, in broadly cylindric spikes. c, Type G, **C. pilosus** var. **obliquus** (× ⅔): 2× compound infl. with spikelets stiffly spreading, arranged along rachis, in cylindric spikes. d, Type H, **C. iria** (× ⅔): 2× compound infl. with spikelets erect, arranged along rachis, in narrowly cylindric spikes (3× compound versions also occur). Drawn by Glenn Rodrigues.

+ Leaves and sheaths lacking prominent transverse veinlets; if spikelet orange red then with more than 10 fertile glumes and disarticulating at maturity 12

12. Infl. a dense head of sessile, whitish, elliptic spikelets   **10. C. niveus** (A)
+ Infl. not as above 13

13. Spikelets in dense, subspherical heads 14
+ Infl. laxer, spikelets either in umbels or in wide or narrow spikes (i.e. inserted along a rachis) 16

14. Spikelet about equalling length of glumes, falling entire at maturity; fertile glumes 3 or fewer, midribs keeled, spongy-winged
                    9. C. pseudokyllingioides (E or 2 × compound version)
+ Spikelets much longer than glumes, disarticulating at maturity; fertile glumes 10 +, midribs not spongy-keeled 15

15. Glumes obovate, blunt, sides chestnut-coloured ...   **12. C. difformis** (E)
+ Glumes narrowly ovate, subacute, sides hyaline to pale brown
                                                        13. C. silletensis (E)

16. Spikelets suberect in narrowly cylindric spikes 17
+ Spikelets spreading in very wide spikes or umbellate heads 18

17. Stout perennial; spikelets large (over 7mm); glumes subacute
                                            14. C. nutans var. **nutans** (H)
+ Slender annual; spikelets small (under 8mm); glumes blunt
                                                        15. C. iria (H)

18. Spikelets in umbellate heads (no rachis developed); glumes under 1.2mm 19
+ Spikelets inserted along rachis of a spike; glumes over 1.5mm 20

19. Spikelets over 1mm wide; glumes erect at maturity concealing nuts; nuts cream, ± smooth, trigonous
                            17. C. haspan (D or 1 × compound version)
+ Spikelets under 1mm wide; glumes spreading at maturity to reveal nuts; nuts white, tuberculate, swollen-trigonous
                            18. C. tenuispica (D or 1 × compound version)

20. Adjacent glumes on one side of spikelet not overlapping; spikelets under 1mm wide .... **19. C. distans** (F)
+   Adjacent glumes strongly overlapping; spikelets over 1.5mm wide ... 21

21. Spikelets over 2.2mm wide; spikes narrow, with many stiffly spreading spikelets, rachis usually over 1cm 22
+   Spikelets under 2mm wide; spikes very wide, with few, suberect spikelets, rachis usually under 0.5cm 23

22. Rachis of spikes shortly hairy; glumes under 2mm ... **21. C. pilosus** (G)
+   Rachis of spikes glabrous; glumes over 3.5mm .... **22. C. thomsonii** (G)

23. Infl. 2× compound; glumes narrow (c.1mm), rather lax so rachilla visible .... **20. C. pangorei** (F)
+   Infl. nearly always 1× compound; glumes over 1.5mm wide, densely overlapping so rachilla hidden 24

24. Glumes under 3.3mm, sides dark reddish-brown; slender, tuber-bearing stolons present .... **23. C. rotundus** (C)
+   Glumes over 3.6mm, sides golden-brown; lacking tuber-bearing stolons .... **24. C. tenuiculmis** (C)

1. C. squarrosus L.; *C. aristatus* Rottboell; *Mariscus aristatus* (Rottboell) Chermezon.

    Dwarf, tufted annual smelling of curry when dry. Stems trigonous to triquetrous, 3.5–15cm, 0.9–1.7mm wide. Leaf single, sub-basal, blade flat, about equalling stem, 1–3.4mm wide; sheaths reddish-purple. Infl. 1–3.5 × 1–3cm, a single head or 1× compound with 2–4 rays, longest ray 0.5–2cm. Partial infls. spicate, spikes elongate-hemispheric, 0.6–1.5 × 0.7–1.1cm, with 8–25 spreading, spikelets, rachis 3–8mm. Spikelets oblong-fimbriate, 3.5–5.5 × 1.4–2mm (excl. mucros). Glumes 7–13, suberect, scarcely overlapping, linear to narrowly elliptic, mucronate, 1.4–2 (excl. mucro) × 0.8–1mm, midrib green, produced as long (0.7–0.8mm) recurved mucro, sides pale hyaline-brown to reddish-brown, strongly 3–4-veined each side of midrib, gradually tapered into mucro, margins not carried down as wings. Stamen 1. Style c.0.4mm; stigmas c.1mm. Nut shouldered above, apiculate, 0.8–0.9 × 0.3–0.5mm, faces linear to narrowly obovate, pale brown, minutely pitted.
    Bhutan: S — Chukka district (road to Surge Shaft Colony); **C** — Thimphu (around Thimphu), Tongsa (Tongsa) and Bumthang (Bumthang) districts; **N** — Upper Mo Chu district (above Gasa Dzong); **Sikkim** (Lachen, Lachung, Keadom). On mossy boulder; damp roadsides; bare earth banks in scrub; fields; arable weed, 1960–2900m. Fr. August–September.

2. C. cuspidatus Kunth.

Small, densely tufted annual. Stems trigonous, 2.5–8.5cm, slender (c.0.5mm wide). Leaves sub-basal, blade slightly exceeding stem, 0.7–1.3mm wide; sheaths reddish. Infl. a head of 5–31 sessile spikelets 1.2–3cm diameter, sometimes 1 × compound with 1–4 rays, longest ray 0.8–1.7cm. Partial infls. umbellate. Spikelets oblong-fimbriate, 5–15 × 1.2–1.5mm (excl. mucros). Glumes 14–34, erect at first, finally spreading obliquely, scarcely overlapping (nuts visible), oblong-elliptic, apex sharply notched, mucronate, 0.7–1.4 (excl. mucro) × 0.4–1mm, midrib green, 3-veined, keeled, excurrent as long (0.5–0.7mm), recurved mucro, sides golden to reddish-brown, margins carried down into very narrow (c.0.1mm) wings. Stamens 2–3. Style c.0.2mm; stigmas c.0.6mm. Nut shouldered above, apiculate, 0.5–0.7 × 0.4mm, faces obovate, narrowed to base, dark brown, minutely pitted.

Bhutan: C — Punakha (1km N of Punakha Dzong) and Tashigang (Changpu) districts; **Terai** (Balasun); **Darjeeling** (Punkabari, Tista, Great Rangit opposite Manjitar); **Sikkim** (unlocalised Hooker specimen, probably from Darjeeling). Sandy grassy ground by river; weed of cultivation, 150–2130m. Fr. August–November.

3. C. castaneus Willdenow.

Differs from *C. cuspidatus* in its reddish-purple, more closely overlapping, parallel-sided glumes; stamen 1; nut longer (0.8 × 0.3mm), narrowly oblong with parallel sides.

Darjeeling (Punkabari). 366m. Fr. August.

In species 4–9 the spikelets do not disarticulate at maturity; for this reason sometimes separated into the genus *Mariscus*.

4. C. sikkimensis Kükenthal; *Mariscus hookerianus* C.B. Clarke.

Tufted perennial, rhizome very short. Stem trigonous, 62–73cm, 1.8–2.2mm wide. Leaves 1–2, sub-basal, blade shorter than stem, 3.2–5.2mm wide; sheath long, reddish-brown at base. Infl. 1 × compound, a dense head of c.15(+) spikes, 4–5cm diameter; involucral bracts 4–5, long, leaf-like. Spikes sessile, cylindric, to 2.5 × 0.8cm, with many obliquely spreading spikelets. Spikelets 3–6 × 0.8–1.9mm, with 2 small, sterile, basal glumes and 1–4 suberect, loosely overlapping, fertile glumes. Glumes 2.1–2.7 × 1.2–1.4mm, narrowly ovate, subacute, midrib 3-veined, sides straw-coloured, strongly 2–3-veined each side of midrib, margins carried down into broad (c.0.3mm), hyaline, wings. Stamens 3. Style 0.3mm; stigmas 0.9mm. Nut curved, shortly apiculate, 1.7 × 0.6–0.7mm, faces oblong-elliptic, 2 flat, 1 concave, brown at first, turning black, minutely papillose.

Bhutan: S — Chukka district (Chukka Colony to Taktichu); **Sikkim** ('hot valleys' — unlocalised Hooker record, possibly from Darjeeling). Open, wet, gravelly slope in broad-leaved forest, 1940m. Fr. July.

5. C. cyperoides (Retzius) Kuntze; *Mariscus sieberianus* Nees ex Steudel; *M. sumatrensis* (Retzius) T. Koyama. Fig. 28b.

Perennial, rhizome short. Stem trigonous 18–85cm, 0.8–3.8mm wide. Leaves sub-basal, shorter than to equalling stem, 2.6–9mm wide; sheaths purplish-red at base. Infl. 1× compound, 2.5–13 × 2.5–16cm, rays 6–14, longest 1–9.5cm. Partial infls. spicate, spikes densely cylindric, blunt, 1–4 × 0.7–1.3cm, with 36 to many, finally spreading spikelets, (occasionally with 2 short basal branches). Spikelets linear, acute, 3–7.6 × 0.5–0.8mm, glumes appressed, 2 basal, small, sterile, upper 1–4 fertile. Fertile glumes linear-lanceolate, acute, lowest 3–3.8 × 1–1.4mm, midrib green, 3-veined, sides straw-coloured, sometimes with minute reddish flecks, usually strongly 3–4-ribbed each side of midrib, margins carried down as narrow (c.0.2mm) wings. Stamens 3. Style 0.2–0.9mm; stigmas 1.5–3mm. Nut slightly curved, 1.7–2.1 × 0.4–0.5(–0.7)mm, faces linear, narrowed above, apiculate, dark brown, minutely papillose.

Bhutan: S — Samchi (Sibsoo Bazaar), Phuntsholing (Phuntsholing, Chimakothi to Phuntsholing), Chukka (Surge Shaft Colony, Marichong) and Gaylegphug (Bhur to Toribari, W bank of Thewar Khola, Gaylegphug) districts; **C** — Ha (Parker, 1992), Thimphu (Damgi, hill above Thimphu Hospital, below Lobnakha, etc.), Punakha (Shenganga), Tongsa (W of Mangde Chu) and Mongar (Saling) districts; **N** — Upper Mo Chu district (Tamji to Goen Gaza); **Terai** (Buxa–Bhutan road); **Darjeeling** (Rangit, Tista, Darjeeling to Kalimpong, Rimbick, Labdah, etc.); **Sikkim** (Gangtok, Selim, Yoksum, Phodong Gompa). Under-recorded, very common. Roadsides and field-borders; sometimes a weed of cultivation (incl. rice paddies); stony ground in cleared jungle; wet jungle; moist hollows in seasonally burnt, rough, stony bushland, 270–2950m. Fl./fr. April–September.

Very variable in degree of development of rays (compact forms therefore grading into the next species) and number of nuts per spikelet (2 in the commonest form). The variation is continuous, probably environmentally induced and scarcely worthy of taxonomic recognition. Small forms with 1 nut per spikelet approach var. *khasianus* (C.B. Clarke) Kükenthal, but differ from that variety in having rayed spikes; forms with basal branches to the spikes have been called var. *subcompositus* (C.B. Clarke) Kükenthal; forms with 3–4 nuts per spikelet have been called var. *evolutior* (C.B. Clarke) Kükenthal.

232. CYPERACEAE

6. C. cyperinus (Retzius) J.V. Suringar; *Mariscus cyperinus* (Retzius) Vahl.
Differs from *C. cyperoides* in its very dense heads of sessile spikes (var.
bengalensis (C.B. Clarke) Kükenthal), though occasionally with rays to 2.5cm,
spikes coarser, shorter and fatter (1.2–1.8 × 0.7–1.2cm) with the wider spike-
lets (5–7 × 1–1.1mm) oblique at maturity; glumes wider ((3.3–)3.5–4.2 ×
(1.4–)1.6–1.8mm), sometimes flushed brown; nut wider, (1.9–)2–2.2 ×
(0.6–)0.7–0.8mm.
Bhutan: S — Samchi (below and to N of Chengmari, Sibsoo Bazaar),
Phuntsholing (Phuntsholing, Chimakothi to Phuntsholing) and Chukka
(Marichong) districts; **Terai** (Balasun); **Darjeeling** (Panchkilla, bed of Little
Rangit, Sivoke, Mongpu, Barnesbeg, Balukup, Kalijhora); **Sikkim** (NW of
Singtam, Gangtok). Dry, open forest; rough grassland in partial shade; bank
at edge of field, 150–1830m. Fl./fr. April–October.

C. cyperoides is so variable and the differences between it and *C. cyperinus* so slight
that there seems much wisdom in treating them as varieties of the same species (as
done by Clarke (1884) who used the name '*C. umbellatus* Benth.').

7. C. paniceus (Rottboell) Boeckeler; *Mariscus paniceus* (Rottboell) Vahl.
Nep: *harkotey*.
Perennial, spreading by slender, creeping stolons. Stems 26–53cm, clothed
at base with reddish fibres. Infl. of 4(–12) sessile to shortly rayed spikes. Rays
curved, to 2(–2.5)cm. Spikes small (1–1.5 × 0.5(–0.7)cm). Spikelets c.3 ×
0.5mm, slightly recurved, with one fertile flower. Fertile glume 2.8–3.2 ×
1.2–1.8mm. Nut c.1.8 × 0.8mm.
Bhutan: S — Phuntsholing district (Phuntsholing Industrial Estate);
Darjeeling (Sukna, above Peshok). Perhaps under-recorded at low altitudes.
Roadside ditch; roadside in open *Tectona/Shorea* plantation, 300–500m. Fr.
May–July.

The specimens seen may be referred to var. *roxburghianus* (C.B. Clarke) Kükenthal,
being slightly larger and stouter than typical *C. paniceus*, thus forming a link with the
previous species.

8. C. compactus Retzius; *Mariscus compactus* (Retzius) Boldingh;
M. microcephalus Presl.
Perennial, rhizome short. Stems tufted, rounded-trigonous, 46–59cm, stout
(2.7–3mm wide). Leaves 1–2 on lower half of stem, blades greatly exceeding
stem, c.6mm wide, transverse veinlets prominent on underside; sheaths very
long, pale reddish-brown, prominently transverse-veined. Infl. 2(–3)× compound, 6.5–11 × 9–15cm, primary rays 5–8, longest 6–7.5cm; secondary rays
to 1.5cm. Partial infls. rather dense. Spikes broadly hemispheric, 1.2–1.5 ×

1.2–2.5cm, with many spreading spikelets, rachis under 1cm. Spikelets linear, acute, 6.5–12 × 0.6–1.2mm, glumes appressed, overlapping, 2 or more basal, sterile, upper 3–7 fertile. Fertile glumes linear-oblong, acute, 3.8–4.2 × 1–1.4mm, midrib narrow, green, 3-veined, sides red-brown, 2–3-veined each side of midrib, margins carried down into broad (c.0.3mm), hyaline wings. Stamens 3. Style 0.7–1.2mm; stigmas 1.5–2mm. Nut curved, linear, long-apiculate, 2.1–2.3 (incl. apiculus) × 0.5–0.6mm, minutely pitted, pale brown.

Terai (Falakata, Katanbari, Simulbarie, near Siliguri). Wet places, 90–305m. Fr. November.

9. C. pseudokyllingioides Kükenthal; *Courtoisia cyperoides* (Roxb.) Nees; *Mariscus cyperoides* (Roxb.) A. Dietrich.

Annual. Stems tufted, trigonous, 18–57cm, 1–2mm wide. Leaves 2–3 on lower half of stem, blades shorter than to exceeding stem, 2.5–4.5mm wide; sheaths very long, pale. Infl. 2×-compound, 2–12 × 3–9.5cm; primary rays 3–6, longest 1–9cm; secondary rays 0–1cm. Partial infls. 0.6–1cm diameter, composed of dense heads of sessile spikelets. Spikelets flat, elliptic, acute, 3–4 × 1.3–2.4mm, glumes 2–3 basal small, sterile, upper 2(–3) enclosing 1(–2) nuts, rachilla 0. Glumes conduplicate, each face semi-elliptic to semi-ovate, 2.8–3.8 × 0.6–1.5mm, midrib acutely keeled, 0.2–0.4mm wide on each face, shortly apiculate, green, spongy, sides 0.5–1mm wide membranous, pale to dark reddish-brown, apex acute, emarginate or truncate. Stamens 3. Stigmas c.0.7mm, sessile. Nut sharply trigonous, 2.5–3 × 0.4–0.7mm, faces linear-elliptic, contracted above into long (c.0.3mm) apiculus, pale brown.

Bhutan: S — Samchi district (Dwarapani); **Sikkim** (no specimens seen, but one cited by Kükenthal, 1936). Wet places (presumably rice paddies), 150–305m. Fr. December.

Sometimes placed in a distinct genus, when its correct name is *Courtoisina cyperoides* (Roxb.) Soják, the generic name *Courtoisia* Nees being an illegitimate homonym.

10. C. niveus Retzius. Fig. 28a.

Perennial; rhizomes woody, very shortly creeping. Stems trigonous, bases swollen, covered with dark scales, 10–46cm, 0.6–1.1mm wide. Leaves sub-basal, shorter than stem, very acute, 0.6–2.6mm wide. Infl. a dense head 1.2–3.5cm diameter of 4–13 sessile spikelets; involucral bracts 2, filiform, spreading. Spikelets whitish, oblong-elliptic, acute, 9–16.5 × 3.5–5.5mm. Glumes 12–31, oblique, very closely overlapping, ovate, subacute, 3.3–4 × 2.4–2.8mm, midrib green, 1-veined, sides creamy-white, finally minutely reddish-flecked, 4–6-veined each side of midrib, margins whitish-hyaline, margins not carried down into wings. Stamens 3. Style 1.4–2.3mm; stigmas c.1mm.

Nut strongly trigonous, 1.3–1.4 × 0.8–1.1mm, faces elliptic to obovate, apiculate, finally dark brown.

Bhutan: C — Punakha district (Punakha, Chusom to Mishina); **Terai/ Duars** (Balasun, Sivoke, Siliguri, Jalpaiguri Duars); **Darjeeling** (Darjeeling). Burnt savannahs (presumably also other habitats), 120–1830m. Fl./fr. February–August.

11. C. compressus L. Fig. 28d.
Annual. Stems tufted, trigonous, 7.5–43cm, 1.1–1.5mm wide, stiff. Leaves sub-basal, blades c.half to equalling stems, 1.6–2.6mm wide; bases of sheaths reddish. Infl. a head of sessile spikelets 2.5–4.5cm diameter, or 1 × compound with 1–4 stiff rays, longest 2–6cm. Partial infls. spicate, spikes of 3–4 widely spreading spikelets, rachis under 3mm. Spikelets strongly compressed, quadrangular in section, oblong-elliptic, acute, 1–2.6 × 0.3cm. Glumes 11–30, erect, closely overlapping, ovate, 3.5–4 (incl. mucro) × 2.4–2.8mm, midrib very wide (c.1mm), sharply keeled, 6–9-veined, excurrent as sharp mucro (0.5–0.7mm), sides hyaline or flushed golden or dark red near midrib, 3–5-veined each side of midrib, acute or emarginate, not carried down into wings. Stamens 3. Style 1.3–1.7mm; stigmas very slender, 0.5–1mm. Nut 1.2–1.5 × 1–1.4mm, shouldered above, faces concave, obovate (sometimes elliptic), strongly narrowed to base, brown, shiny.
Bhutan: S — Phuntsholing (Phuntsholing) and Gaylegphug (Bhur to Toribari, Gaylegphug, by Sarbhang Bridge) districts; **C** — Punakha district (1km N of Punakha Dzong, Wangdi Phodrang); **Terai** (Dulkajhar); **Darjeeling** (Lebong, Darjeeling, Mongpu, Great Rangit opposite Manjitar); **Sikkim** (W of Rabong La, NW of Singtam, N of Jorethang). Common in low areas at roadsides, field-edges, beside rivers etc., 150–2130m. Fr. May–October.

12. C. difformis L. Dz: *guchem, ochumani, chow*; Nep: *mothey*. Fig. 29a.
Annual. Stems tufted, triquetrous, 13–29(–43)cm, 1.9–3.1mm wide. Leaves on lower ¼ of stem, blades, acute, about equalling stem, 2.3–4.5mm wide, rather narrow. Infl. (1–)2× compound, 1.5–3 × 2–5.5cm; primary rays 4–6(–8), longest 1.5–3cm; secondary rays not developed or very short. Partial infls. 0.6–0.8cm diameter, usually compound, very congested heads. Spikelets oblong, blunt, 2.2–2.5 × 0.9mm. Glumes 10–11, oblique, just overlapping, rachilla visible, obovate, rounded, emarginate, 0.6–0.7 × 0.6–0.8mm, midrib conspicuous, green, sides chestnut, margins widely hyaline near apex, not carried down into wings. Stamens 1–2. Style c.0.2mm; stigmas c.0.1mm. Nut c.0.6 × 0.4mm, faces narrowly elliptic, acute, cream.
Bhutan: S — Gaylegphug district (Gaylegphug River towards Norboling); **C** — Thimphu (Chapcha, Doteng; Paro (M.F.B.)) and Punakha (1km S of

Rinchu, Upper Gaseloo) districts. Rice paddies; damp roadside in forest, 400–2500m. Fr. May–September.

Probably under-recorded; Parker (1992) states it to be a major weed of flooded rice probably occurring in all districts [with cultivation].

13. C. silletensis Nees.
Perennial, rhizome short. Stems densely tufted, trigonous, 19–33cm, c.1.1mm wide. Leaves sub-basal, blades shorter than stems, 1.5–2.5mm wide; bases of old sheaths persistent. Infl. 2 × compound, 1.5–3 × 4–6cm; primary rays 5–9, longest 1.5–2.5cm; secondary rays very short. Partial infls. compound, with many congested heads. Spikelets oblong, blunt, 4.5–5 × 1.4–1.6mm, pale brown. Glumes 20–28, spreading, rather closely overlapping, narrowly ovate, subacute, c.1 × 0.8mm, midrib 3-veined, keeled, ending in blunt tip, sides hyaline, sometimes with narrow dark red line next to midrib, margins not carried down into wings. Stamen 1. Style 0.2–0.3mm; stigmas 0.3mm. Nut c.0.6 × 0.3mm, faces narrowly elliptic, apiculate, pale brown, minutely pitted.
Terai (Siliguri, Bhutanghat); **Sikkim** (Rangpo to Tista Bazar (F.E.H.1)). 150–350m. Fr. June.

14. C. nutans Vahl var. **nutans**. Name in Punakha: *agaseep*.
Perennial, rhizomes short. Stems 97–125cm, stout (3.8–5mm wide). Leaves sub-basal, blades c.6mm wide; sheaths papery, reddish-brown. Infl. 3 × compound, 30–37 × 16–23cm; primary rays 6–8, stout, longest 10–30cm; secondary rays to 5cm; tertiary rays extremely short. Partial infls. subtended by leafy bracts. Spikes narrowly cylindric, with c.25–30 erect spikelets, rachis to 3cm. Spikelets oblong, 7–9 × 1.5–2mm. Glumes 8–10, oblique, overlapping but rachilla visible, oblong-ovate, apex hyaline, subacute becoming emarginate, 1.8–2.3 × 1mm, midrib keeled, green, 3-veined, minutely apiculate from below apex, sides straw-coloured, streaked dark red near midrib, strongly 2–3-veined each side of midrib, margins hyaline, carried down into narrow (c.0.2mm) hyaline wings. Stamens 3. Style c.0.2mm; stigmas c.1.5mm. Nut concave on 2 sides, 1.4–1.7 × 0.6mm, faces oblong, apiculate, golden- to reddish-brown, minutely pitted.
Bhutan: C — Punakha district (Waecha Bridge, c.8km S of Wangdi Phodrang); **Darjeeling** (Rungnoo Valley, Rangit, Kolbong, Dulkajhar, S of Rangpo, Lodoma, Sivoke); **Sikkim** (NW of Singtam). Steep marshy bank in open; marshy streamside, 305–1620m. Fl./fr. May–October.

Kükenthal (1936) gives a Clarke record of var. *eleusinoides* (Kunth) Haines (*C. eleusinoides* Kunth) for Sikkim without citing a specimen. The record is not included in

F.B.I. or Clarke (1884) and I have seen no specimen. This variety differs in its blunt, much more compact partial infls. and smaller, more strongly apiculate glumes.

15. C. iria L. Dz: *guchem, ochumani, chow*; Nep: *mothey*. Fig. 29d.
Usually annual. Stems tufted, 24–57cm, 1–2.8mm wide. Leaves on lower half of stem, blades c.half to equalling stem, 2.4–4mm wide; sheath bases reddish-brown. Infl. 2 × compound, 6.5–9.5 × 5–10cm; primary rays 3–6, longest 2–6cm; secondary rays c.2mm. Partial infls. to 4 × 6cm, with 3–5 ± sessile spikes, lower spikes spreading widely. Spikes narrowly oblong, with 10–16 erect spikelets, rachis 1.2–2.6cm. Spikelets oblong, blunt, 3–8 × 1.5–1.7mm. Glumes 5–13, oblique, overlapping but rachilla visible, widely obovate to suborbicular, blunt, emarginate, 1.2–1.5 × 1.2–1.6mm, midrib green, 3-veined, keeled, minutely apiculate from below apex, sides golden, translucent, with 1 rib close to each side of midrib, margins hyaline near apex, not carried down into wings. Stamens 3. Stigmas c.0.2mm, sessile. Nut 1.1–1.3 × 0.6–0.8mm, faces concave, narrowly elliptic to obovate, apiculate, pale brown, finally blackish.

Bhutan: S — Samchi (Samchi), Phuntsholing (Chimakothi to Phuntsholing, Phuntsholing) and Gaylegphug (Bhur to Toribari, Gaylegphug) districts; **C** — Punakha (N of Punakha Dzong, Wangdi Phodrang, Upper Gaseloo) and Mongar (Lingmethang) districts; **Duars** (Buxa); **Darjeeling** (Rangpo to Tista, Kalimpong, Mongpu, Sonada, Birick, Riang, etc.); **Sikkim** (Gangtok, Singtam, Selim, below Rumtek). Common in low areas as weed of cultivated ground (e.g. paddy fields), roadsides, beside rivers etc., 250–2500m. Fl./fr. May–September.

Parker (1992) records it as a common weed of partially flooded rice paddies probably occurring in all districts [with cultivation].

A curious form from Samchi Town Park (*Grierson & Long* 3435, E) differs in having spikelets with over 20 glumes and the glumes very closely overlapping (rachilla not visible).

16. C. laxus Lamarck; *C. diffusus* Vahl. Fig. 28e.
Perennial, rhizomes short. Stems tufted, sharply trigonous, 24–58cm, 1.8–3.7mm wide. Leaves sub-basal, blades equalling stem, 6.5–14.5mm wide, many-veined; sheaths papery, dark reddish-brown. Infl. 2(–3)× compound, lax, widely and irregularly umbellate, 5–18 × 7–34cm; primary rays 6–9, longest 3–15cm; secondary rays to 2cm; tertiary rays to 1cm; involucral bracts very wide. Partial infls. usually 1–2× compound. Spikelets in umbels of 1–4, oblong, 3.5–9.3 × 2–2.5mm. Glumes 7–18, spreading (rachilla visible), scarcely overlapping, oblong-ovate, 1.5–1.8 (incl. mucro) × 1.2–1.6mm, midrib c.0.6mm wide, strongly keeled, 5-veined, excurrent into hooked mucro

0.2–0.4mm, sides 2-veined each side of midrib, brown, acute, margins narrowly hyaline near apex, carried down into narrow wings (c.0.2mm wide). Stamens 3. Stigmas c.1mm, sessile. Nut c.1.4 × 0.7mm, faces concave, narrowly elliptic, acute, dark brown to blackish, shining.

Bhutan: S — Phuntsholing (Phuntsholing) and Gaylegphug (Hatipali, c.5km N of Kali Khola) districts; **Terai** (Jalpaiguri Duars); **Darjeeling** (Great Rangit, Mendong, Tista, Slake, Barnesbeg, Sivoke, Selim, etc.); **Sikkim** (NW of Singtam). Well-drained, stony slopes and roadsides in subtropical forest, 240–1520m. Fl./fr. March–October.

C. laxus var. *macrostachyus* (Boeckeler) Karthikeyan has been recorded from Phuntsholing (M.F.B.), but I have seen no specimens. It differs in its longer (to 12mm), narrower (c.2mm) spikelets with up to 30 closely overlapping, erect glumes (rachilla hidden).

17. C. haspan L. subsp. **haspan.**
Annual or perennial, roots fibrous or with short rhizome. Stems tufted, trigonous to triquetrous, 6.5–28cm, 0.8–1.3mm wide, many bladeless sheaths near base. Leaf single, sub-basal, blade very acute, much shorter than stem, 1.6–2mm wide. Infl. 1–2(–3)× compound, irregularly umbellate, 2–5.5 × 2.5–7cm; primary rays (2–)5–9, longest 1–4.5cm; secondary rays to 1cm; involucral bracts 2–3, shorter than to just exceeding infl. Partial infls. umbels of sessile spikelets or compound with shortly rayed umbels. Spikelets in umbels of 3–4, linear, acute, 6–11.5 × 1.2–1.7mm. Glumes 14–50, erect, overlapping (concealing nuts), oblong, subacute, truncate or minutely emarginate, 1–1.2 × 0.8mm, midrib green, 3-veined, slightly keeled, sometimes minutely apiculate from below apex, sides hyaline, pale brown or reddish-brown, margins hyaline, not carried down as wings. Stamens 1–2. Style c.0.3mm; stigmas c.0.3mm. Nut stipitate, rounded-trigonous, 0.4–0.5 × 0.3mm, faces narrowly obovate, apiculate, cream, minutely granular at first, becoming smooth.

Bhutan: S — Samchi district (Samchi); **C** — Tongsa district (Tama); **Darjeeling** (Simulbarie, Labda, Manabari, Rayang, Tista; Kalimpong (F.E.H.1)); **Sikkim** (Yoksum, NW of Singtam; Gangtok (F.E.H.1)). Paddy fields; lake margin; marsh by stream, 250–1820m. Fl./fr. February–July.

subsp. **juncoides** (Lamarck) Kükenthal.
Differs in its slender, creeping stolons covered with reddish scales; stems taller (76–81cm), wider (2–2.8mm), without leaves but with long sheaths with short, acute, erect blades (to 5 × 0.6cm) with crimped margins; infl. larger (to 17 × 15cm, longest primary ray to 13cm), several-times compound; stamens 3.
Terai (Dulkajahr). 150m. Fr. October.

18. C. tenuispica Steudel; *C. flavidus* sensu F.B.I. non Retzius.

Differs from small, annual forms of *C. haspan* in its narrower (under 1mm) spikelets, with spreading, scarcely overlapping glumes which do not conceal the white, prominently tubercled, rather swollen nuts.

Bhutan: C — Mongar district (Yayung Research site near Mongar). Rice field, 900m. Fr. June.

Recorded for Darjeeling (Kalimpong and Rangpo to Tista Bazar (F.E.H.1)), but a voucher specimen of the other record cited in the same work is *C. haspan*.

19. C. distans L.f.

Perennial; rhizomes short, ascending, covered with purplish-brown scales. Stems 22–89cm, 1.6–3.2mm wide. Leaves on lower ⅓ stem, blades about equalling stem, 3.8–4.8mm wide. Infl. 2(or more)× compound, rather open, 4.5–23 × 5–18cm; primary rays 5–7, longest 3–15cm; secondary rays 0.5–4.5cm. Primary partial infls. 1.3–6 × 2–6cm, with leafy bracts, bearing rayed and unrayed spikes. Spikes broad, with 7–20 widely spreading spikelets, rachis 0.7–1.3cm. Spikelets linear, 5–17 × 0.4–0.8mm, rachilla wavy. Glumes 7–14, not overlapping, appressed, oblong, truncate, 1.5–2 × 1mm, midrib green, 3-veined, sides flushed reddish-purple to varying degrees, 1–2-veined each side of midrib, margins whitish, carried down into narrow (c.0.1mm), hyaline wings. Stamens 3. Style c.0.3mm, stigmas very slender, c.1.5mm. Nut 1.4–1.8 × 0.4–0.5mm, faces narrowly oblong, apiculate, yellowish-brown, minutely pitted.

Bhutan: S — Phuntsholing (Druk Motel, Phuntsholing, Chimakothi to Phuntsholing), Gaylegphug (Gaylegphug to Bhur) and Deothang (Samdrup Jongkhar to Deothang) districts; **C** — Mongar district (Saling); **Terai** (Bagdogra, Bhutanghat to Jalpaiguri); **Darjeeling** (Rangpo to Tista, below Barnesbeg); **Sikkim** (N of Ranipul, NW of Singtam; Gangtok (F.E.H.1)). Common in low areas on roadsides, field-edges, etc., 300–1650m. Fl./fr. May–August.

20. C. pangorei Rottboell; *C. tegetum* Roxb. Fig. 29b.

[Rhizomes short, not creeping. Stems triquetrous, 50–90cm, 2.4–3.9mm wide; long basal sheaths apparently sometimes developing laminar blades]. Infl. 2(or more)× compound, widely spreading, (7–)15–19 × (7–)15cm; primary rays (3–)7–10, longest (4–)10–12.5cm; secondary rays to 3cm; involucral bracts with prominent cross-veinlets. Partial infls. bearing rayed and unrayed spikes. Spikes broadly cylindric, with up to 13 spreading to suberect spikelets, rachis to 1cm. Spikelets linear, 11–22(–28) × 1.5mm. Glumes 14–22, erect, rather lax so rachilla visible at maturity, oblong (sides parallel), subacute, folded but not keeled, 2.6–2.8(–3.2) × 1mm, 2–3 veins on each side of green midrib, sides usually reddish-brown, margins narrowly hyaline; basal wings 0.1(–0.2)mm wide, pale at first, finally chestnut, deciduous. Stamens 3. Style c.0.5mm; stigmas c.2mm; nut 1.2(–1.5) × 0.4(–0.6)mm, faces oblong, triangular at apex, inner face concave, very dark brown, minutely pitted.

Terai (unlocalised Hooker specimen); **Darjeeling** (Sittong, Mongpu). 910–1220m. Fr. September–December.

The Mongpu plants are said to have been introduced.

Only incomplete specimens seen from our area so description of vegetative parts taken from other Indian material.

There is confusion over the identification and nomenclature of this species and *C. corymbosus* Rottboell, which probably also occurs. I have followed the usage of Kükenthal (1936). Much of the confusion arises from the lack of basal parts in the majority of specimens and reliance on rather dubious characters such as degree of overlap of glumes. *C. corymbosus* supposedly differs in having long, creeping stolons and more densely overlapping glumes; other characters of *C. corymbosus* appear to be a usually more contracted infl., shorter and narrower spikelets which are often paler due to the wider hyaline glume-margins, glumes elliptic, with more numerous veins. Intermediates, however, seem to occur.

21. C. pilosus Vahl var. **pilosus**.
Apparently with long, slender stolons. Stems sharply triquetrous, (20–)60–67(–75)cm, 2.6–5.5mm wide. Leaves sub-basal, blades (10–)25–38 × 0.2–0.65mm; sheaths pale reddish-brown, with prominent cross-veinlets. Infl. 2× compound, (3.5–)5–12.5 × 4–15cm; primary rays 4–7, longest (1.5–)3–17cm (occasionally all short so infl. congested); secondary rays extremely short. Partial infls. c.2.5 × 2.5cm, with 3(–5) subequal spikes, basal ones spreading. Spikes broadly ovoid, with up to 30 horizonatally spreading spikelets, rachis hairy, 1.5–2cm. Spikelets linear, 8–13 × 2.2–3mm, with 14–23 glumes; glumes oblique, lax, rachilla showing at maturity. Glumes 1.7–2 × 1.4–1.8mm, ovate, truncate to subacute, straw-coloured with narrow, hyaline margins, becoming flushed fuscous-red to varying degrees, 2 strong veins either side of green, 3-veined midrib, bases not decurrent. Stamens 3. Style c.0.2mm; stigmas c.0.7mm. Nut 1.2 × 0.7–0.9mm, strongly trigonous, faces rhombic-elliptic, apiculate, very dark brown, minutely papillose.

Bhutan: S — Gaylegphug district (towards Norboling); **C** — Punakha district (Samtengang); **Terai** (Falakata); **Sikkim** (Rinchinpong). Edges of swamp; damp borders of paddy field, 90–1980m. Fr. July–November.

var. **obliquus** (Nees) C.B. Clarke. Fig. 29c.
Differs in its cylindrical spikes (2–2.5 × 1cm); shorter (3–5 × 1.5–2mm), 5–9-flowered spikelets; smaller (1.6–1.7 × 1.2–1.4mm), minutely apiculate glumes and smaller (0.9–1 × 0.7mm) nuts.

Bhutan: S — Gaylegphug district (Gaylegphug); **C** — Punakha district (Changhyo); **Terai** (Siliguri); **Sikkim** (Rumman, NW of Singtam, Phodong Gompa). Marshy roadsides, meadows and streamsides, 90–460m.

Although the two extreme varieties look very different, intermediates occur — specimens seen from **Sikkim** (Chungtam) and **Terai** (Dulkajhar). Variety not known of the following records from Darjeeling: Little Rangit, Kalimpong, Tista.

22. C. thomsonii Boeckeler.
Rhizome woody, ?short. Stem faces ridged, base clothed with dark reddish-brown fibres, 53–75cm, stout (2.5–4.5mm diameter). Leaves rather few, sub-basal, blades ?almost equalling stem, 2–6.5mm wide; sheaths long, membranous, pale. Infl. 2× compound, rather dense, 8.5–14 × 9–16cm; primary rays c.6, longest to 7cm; secondary rays extremely short. Partial infls. of many, very dense spikes. Spikes widely cylindric, c.2 × 2cm, spikelets spreading, rachis 0.6–1cm. Spikelets, linear, acute, 11–12 × 2.5mm. Glumes 7–12, oblique, overlapping, narrowly ovate, subacute, 3.8–4.5 × 2mm, midrib green, strongly 3-veined, stopping below apex, sometimes minutely apiculate; sides narrow, reddish-brown 1–2-veined each side of midrib; margins broad, thick-textured, straw-coloured with minute reddish flecks, continued downwards as narrow (to 0.3mm) wings. Stamens 3. Style c.0.5mm; stigmas c.1.5mm. Nut 1.7 × 0.6mm, faces oblong, apiculate, inner concave others ± flat, blackish, shining.
Terai (Siliguri). c.150m. Fr. May.

23. C. rotundus L. Dz: *guchem*; Nep: *mothey*; Eng: *nut sedge*. Fig. 28c.
Stolons slender, spreading, bearing tubers with black, fibrous coats. Stems 14–38cm, 0.8–1.3mm wide. Leaves rather closely crowded near stem base, blades c.half length to equalling stem, 1.6–3.8mm wide; sheaths membranous, pale. Infl. usually 1× compound, (2–)3–10 × (1.5–)3–6cm; primary rays (0–)1–5, stiffly erect, longest 1.5–7cm; involucral bracts shorter than to just exceeding infl. Partial infls. consisting of broadly cylindric spikes with 3–7 suberect spikelets (largest spike sometimes with 1 branch bearing 2–3 spikelets), rachis 2–7mm. Spikelets linear, acute, 9–21.5 × 1.5–2mm. Glumes (8–)10–22, strongly overlapping, narrowly ovate, blunt, 2.7–3.3 × 1.6–2mm, midrib green, strongly 3-veined, stopping below apex, sides reddish-brown, with 1–3 weak ribs close to midrib, margins hyaline, carried down into wide (0.3–0.5mm) wings. Stamens 3. Style c.1mm; stigmas long (c.4mm). Nut (not always

developing) c.1.6 × 0.8mm, faces concave, oblong-elliptic, scarcely apiculate, pale brown, minutely papillose.

Bhutan: S — Phuntsholing district (Toribar, Druk Motel, Phuntsholing); **C** — Thimphu (Sisina) and Punakha (Punakha) districts; **Terai** (Bagdogra Airport); **Darjeeling** (Rayang, Barnesbeg, S of Rangpo; Tista Bazar, Kalimpong to Darjeeling (F.E.H.1)); **Sikkim** (Gangtok, near Jorethang, Legship, N of Ranipul). Not uncommon in low areas in damp habitats such as roadsides, waste-ground, rice paddies, etc., 250–2300m. Fl./fr. April–August.

Parker (1992) records it as an important weed, especially in perennial crops, probably occurring in all districts [with cultivation].

24. C. tenuiculmis Boeckeler; *C. zollingeri* sensu F.B.I., non Steudel.

Very similar to *C. rotundus* in its infl. and above-ground vegetative parts, but lacking slender, tuber-bearing stolons. Differing also in its larger glumes (3.6–3.7 × 2.4–2.8mm), with wider (c.0.8mm), acutely keeled midribs, minutely apiculate from below apex, sides golden brown and wider (1.7–1.8 × 1mm) nuts with elliptic faces.

Bhutan: C — Punakha district (Tinlegang, Changhyo, Lometsawa); **N** — Upper Mo Chu district (Tamji); **Terai** (Dulkajhar); **Darjeeling** (junction of Great and Little Rangit); **Sikkim** (Rishee, Yoksum, below Rumtek). Damp borders of rice paddy; open slopes; roadside; sandy riverside, 150–2440m. Fr. July–October.

25. C. involucratus Rottboell.

Perennial, rhizomes short. Stems leafless, tufted, obscurely trigonous, spongy, to 1.5m, to 0.8cm wide. Basal sheaths papery, to 19cm. Infl. to 3 × compound, umbellate; primary rays to 10(+), often recurved, longest to 9cm; secondary rays to 2cm; tertiary rays short; bracts leaf-like, numerous, forming a very shortly interrupted involucre, greatly exceeding primary rays, longest to 30 × 1.6cm, margins hispid. Spikelets in umbels of 5–9, strongly compressed, narrowly elliptic, 4–5 × 1.6mm. Glumes appressed, strongly overlapping, lanceolate, acute, 1–1.5 × c.1mm, midrib narrow, 3-veined, strongly keeled, sides veinless, pale straw-coloured, becoming flushed chestnut. Stamens 2–3. Style 0.3–0.5mm; stigmas 0.8–1.3mm. Nut c.0.7 × 0.5mm, faces oblong-elliptic, acute, pale yellowish, minutely papillose.

Darjeeling (Singla Bazar); **Sikkim** (Gangtok). Cultivated in gardens.

Native to Africa, but widely cultivated as an ornamental.

Doubtfully recorded species:

C. digitatus Roxb. has been recorded from Takvar, Darjeeling (Mukherjee, 1988). It is a very distinctive large perennial with spreading spikelets arranged in long cylindric spikes (infl. type G), with rather small glumes.

12. PYCREUS P. Beauvois

Differs from *Cyperus* in having biconvex nuts and 2 stigmas.

Letters following specific names in key refer to representative infl. types — see under *Cyperus*.

1. Glumes with midrib developed into distinct mucro tending to recurve
                                                    1. P. pumilus (A or C*)
   + Glumes at most minutely apiculate 2

2. Nuts smooth or minutely dotted 3
   + Nuts coarsely reticulate or with transverse, wavy wrinkles 7

3. Nuts obovate to suborbicular (over 0.9mm wide); glumes over 1.4mm wide; spikelets oblong, rather blunt, not strongly compressed 4
   + Nuts linear to narrowly oblong (under 0.6mm wide); glumes usually under 1.2mm wide; spikelets linear, rather acute, strongly compressed   5

4. Glumes with purple sides and broad green midrib
                                              2. P. sanguinolentus (A or C*)
   + Glumes golden .... **4. P. unioloides** (C*)

5. Spikelets erect, rays often not developed so infl. densely capitate
                                                5. P. polystachyos (A or C)
   + Spikelets spreading, rays developed 6

6. Nut with grooved depression on each face; spikelets very narrow (c.1.5mm wide); glumes golden .... **6. P. sulcinux** (C*)
   + Faces of nut not grooved; spikelets often broader (c.2mm); glumes commonly dark purple, sometimes golden .... **8. P. flavidus** (C*)

7. Spikelets erect, narrow (to 2mm wide)
                              7. P. stramineus (simple version of C)
+   Spikelets spreading, (c.3mm wide) .... **3. P. diaphanus** (C*)

1. P. pumilus (L.) Nees; *P. nitens* Nees.
Slender annual. Stems tufted, 3.5–9(–19)cm, 0.5–0.7mm wide. Leaves on lower ⅓ stem, blades exceeding stems, 1.5–1.7mm wide; sheaths pale brown. Infl. a congested head of sessile spikelets, or 1× compound with 1–4 rays, longest ray 0.9–1.8(–4)cm; involucral bracts 2(–3), greatly exceeding infl. Partial infls. spicate, spike with 5–15 spikelets, rachis to 3mm. Spikelets narrowly oblong, 3.5–12 × 1.5–1.8mm. Glumes elliptic, oblong or obovate, blunt to acute, 1.1–1.4 (incl. mucro) × 0.8–1mm, midrib green, 3-veined, developed into a prominent (to 0.2mm) mucro tending to recurve, sides dirty cream. Stamen 1(–2). Style 0.1–0.2mm; stigmas 0.5–0.7mm. Nut oblong-obovate, truncate, apiculate, c.0.6 × 0.4mm, reddish-brown, minutely granular.
  Bhutan: S — Phuntsholing (Phuntsholing) and Gaylegphug (Gaylegphug River towards Norboling) districts; **Terai** (Sukna, Sivoke); **Darjeeling** (Ryang); **Sikkim** (NW of Singtam). Silt by damp streambed; marsh by stream, 150–610m. Fr. May–October.

2. P. sanguinolentus (Vahl) Nees.
Two distinct forms of this polymorphic species occur in our area, but their nomenclature and taxonomic status is uncertain; they seem not to conform with the treatments of either Kükenthal (1936) or Kern (1974). From the inadequate annotations on the specimens it is impossible to tell if they are associated with different ecologies. Their apparent differences in distribution could merely reflect accidental patterns of collection.

form 1:
  Slender, tufted ?annuals. Stems not decumbent at base, not rooting at basal nodes, 4.5–45cm, slender (0.6–1mm wide). Leaves on lower ⅓ stem, blades half length to equalling stems, (0.7–)1.1–2mm wide; sheaths reddish-brown. Infl. a dense head of (1–)3–23 spikelets; involucral bracts 2–3, longest 4.5–20cm. Spikelets ± oblong, subacute, 6–19 × 2.4–3.5mm. Glumes ovate, blunt, 1.7–2.1 × 1.4–2mm, keeled, midrib green, 3-veined, sides dark reddish-brown 0.6–0.7mm wide, margins narrowly hyaline. Stamens 2. Style short (c.0.3mm); stigmas c.0.8mm. Nut ellipsoid, apiculate, 1.1–1.2 × 0.9–1.1mm, black, ± smooth.
  Bhutan: C — Thimphu (above Bongde Farm) and Bumthang (above Dhur) districts; **Darjeeling** (Kurseong, Darjeeling, Punkabari); **Sikkim** (Kopup,

232. CYPERACEAE

Yoksum, Phodong Gompa). Wet places; shingle by stream, marshy meadow, 150–3960m.

form 2:
Stouter perennial. Stems decumbent at base and rooting from lower nodes, taller (13–91cm), wider (1.1–3.5mm). Leaves wider (1.5–6mm). Infl. 1 × compound with 1–5 rays, rays 1.5–11.5cm. Partial infls. spicate, spike with rachis to 0.5cm. Glumes with narrower (0.3–0.5mm) red-brown sides and sunken green band between these and outer vein of midrib. Stamens 3. Style longer (0.7–1.6mm); stigmas about equalling style. Nut broader — suborbicular (1.1–1.4 × (0.8–)1–1.2mm).

Bhutan: C — Punakha (Samtengang, below Chusom, Gon Chungnang to Punakha) and Tongsa (Mangde Chu S of Tongsa) districts; **Terai** (Balasun); **Darjeeling** (Rongbi Ghora, below Kalimpong); **Sikkim** (Lachen). Marshy edge of lake; ditch; forest slopes, 150–3050m.

Under-recorded, common. Records from Chukka (Surge Shaft Colony), Thimphu (Paro, Simtokha to Dochu La, below Dotena) and Tongsa (W of Mangde Chu) districts; Darjeeling (Kalimpong, Rishap, Sureil, Balasun, Punakabari) and Sikkim (Gangtok, Phadonchen) probably mainly refer to form 1.

I have seen only two specimens (from Samchi district and Darjeeling) intermediate between the two forms, both agreeing in having 3 stamens and glumes as in form 2, but with the growth form, habit and narrow nuts of form 1.

3. P. diaphanus (Schrader ex Roemer & Schultes) S.S. Hooper; *P. latespicatus* (Boeckeler) C.B. Clarke.
Annual. Stems tufted, 8–45cm. Leaves on lower ¼ stem, c.half stem length, blades 1.6–2.6mm wide; sheaths pale- to reddish-brown. Infl. sometimes capitate, usually 1 × compound; rays 3–4, longest 2–9.5cm; involucral bracts 2–3, about equalling infl. Partial infls. spicate, spike with 3–9 finally widely spreading spikelets, rachis 3–5mm. Spikelets oblong-elliptic, acute, 8–27.5 × 3–4mm. Glumes narrowly ovate, truncate, 2.5–2.8 × 1.8–2mm, keeled, midrib green, 3-veined, sides straw-coloured, or reddish-brown. Stamens 2. Style c.1mm; stigmas about equalling style. Nut broadly obovate to suborbicular, 1–1.1 × 0.9–1mm, black, with paler, wavy, transverse wrinkles.
Bhutan: C — Bumthang district (above Dhur); **Darjeeling** (Rongbi Ghora). Marshy pasture in clearing in blue pine forest, 2870m. Fr. August–September.

4. P. unioloides (R. Brown) Urban.
Very similar in overall appearance to *P. diaphanus* from which it differs in its nuts being ± smooth (black covered with minute whitish papillae), lacking transverse wrinkles and its larger (c.4 × 2.6mm), more rigid glumes; stamens 3.

320

Bhutan: C — Thimphu district (near Motithang Hotel); **Darjeeling** (Tista). Flushed grassy slope, 305–2660m. Fr. August.

5. P. polystachyos (Rotboell) P. Beauvois.
Annual or perennial. Stems tufted, (5–)42(–60)cm, 1.2–2.3mm wide. Leaves on lower ¼ of stem, blades flat, much shorter than stem, 2.5(–4)mm wide; sheaths pale- to reddish-brown. Infl. usually a dense 1 × compound head 2–3.5cm diameter, rays very short (under 0.5cm), occasionally more open with rays to 7cm. Partial infls. spicate, spikes with up to 16 ± erect spikelets. Spikelets linear, acute, 9–14.5 × 1.5mm. Glumes lanceolate, acute or minutely apiculate, 1.7–2 × 1–1.2mm, keeled, midrib 3-veined, green, sides straw-coloured to reddish-brown, margins hyaline. Stamens 2. Style c.0.9mm; stigmas about equalling style. Nut narrowly oblong-elliptic, truncate, minutely apiculate, c.1 × 0.5mm, finally brownish-black, minutely pitted.
Sikkim/Darjeeling (unlocalised Beddome specimen).

6. P. sulcinux C.B. Clarke.
Slender annual. Stems tufted, 7–45cm, 0.7–1.3mm wide. Leaves sub-basal, blades very acute, almost equalling stem, 1.2–1.5mm wide; sheaths pale- to reddish-brown. Infl. 1× compound, rays 3–7, longest 1.5–8cm; involucral bracts 3–4, longest exceeding infl. Partial infls. spicate, spikes with 4–11 spikelets, rachis 3.5–5mm. Spikelets linear, acute, 13–38 × c.1.5mm. Glumes oblong, rounded, 1.8–2 × 1–1.2mm, keeled, midrib green, 3-veined, sides straw-coloured to golden brown. Stamen 1. Style 0.3–0.5mm; stigmas about equalling style. Nut oblong, abruptly contracted to apiculate (c.0.1mm) apex, slightly narrowed to base, 1.3–1.4 × 0.5–0.6mm, with shallow depression on each face, brownish-black, cells very small, surface minutely granular.
Bhutan: S — Phuntsholing (Phuntsholing) and Gaylegphug (2km W of Bhur) districts; **Terai** (Balasun, Siliguri); **Darjeeling** (Little Rangit, Badamthan, Soodoong, Great Rangit opposite Manjitar). Shallow, seasonally burnt soils around rocks in rough, bushy grassland; sandy riverside, 150–1220m. Fr. May–October.

7. P. stramineus C.B. Clarke; *C. substramineus* Kükenthal.
Similar to *P. sulcinux* in its slender habit and slender spikelets but differing as follows: infl. a simple, ± sessile, terminal spike of 3–8, erect spikelets; glumes ovate, 2 × 1.6mm; stamens 2; nuts broader (oblong-obovate, c.0.9 × 0.7mm), faces not grooved, coarsely reticulately rugose.
Bhutan: S — Samchi (Dorokha) and Gaylegphug (2km E of Lodrai) districts; **Duars** (Buxa (Sikdar, 1981)). Grassy roadside in subtropical forest, 300–1000m. Fr. October–November.

8. P. flavidus (Retzius) T. Koyama; *P. capillaris* (J. König ex Roxb.) Nees. ?Annual. Stems tufted, 6.5–88cm, 0.7–1.9mm wide. Leaves on lower ⅓ stem, blades channelled, to almost equalling stems, 1–1.5mm wide. Infl. sometimes a congested head, usually 1 × compound, rays 2–6, longest 2–9cm; involucral bracts 2–3, greatly exceeding infl. Partial infls. spicate, spikes with 6–20 spikelets, rachis 5–8mm. Spikelets linear-lanceolate, acute, 6.5–25.5 × 1.5–3mm. Glumes oblong, rounded to subacute, 1.6–2.9 × 0.8–1.4mm, keeled, midrib green, sometimes streaked red, 3-veined, sides golden, brown or dark reddish-black, with conspicuous hyaline margins. Stamens 2. Style 0.4–1mm; stigmas about equalling style. Nut narrowly obovate, conspicuously apiculate, 0.8–1.1 × 0.5–0.6mm, pale to dark reddish-brown, minutely pitted.

    Bhutan: S — Chukka district (below Chimakothi — glumes golden); **C** — Thimphu (above Yangchenphug School, Paro, Chapcha, hill above Thimphu Hospital), Punakha (Tikijampha near Chusutsa — glumes golden), Tongsa (S of Tongsa), Bumthang (Jakar) and Mongar (E side of Kuru Chu below Mongar) districts; **Terai** (Dulkajhar, Siliguri); **Darjeeling** (below Kalimpong, Great Rangit opposite Manjitar); **Sikkim** (Rinkinpung). Probably under-recorded, common. Wet, usually open, places (banks of rice paddies, roadsides, by streams, marshes), 150–2800m. Fr. June–December.

The form with golden glumes appears to predominate in Sikkim/Darjeeling whereas the dark purple-glumed form is the prevalent one in Bhutan. The reliability and significance of this apparent distinction is hard to assess, but the dark forms also occur in Khasia and were described under *Cyperus globosus* as f. *khasianus* C.B. Clarke.

13. KYLLINGA Rottboell

Perennials with short (stems tufted) or extensively creeping rhizomes, sometimes annual. Stems trigonous to triquetrous. Leaves basal and sub-basal,

---

Fɪɢ. 30. **Cyperaceae V.**
a–b, **Kyllinga brevifolia**: a, habit (× ½); b, spikelet showing wingless glume (× 8). c, **K. nemoralis**: glume showing winged keel (× 8). d–g, **Liphocarpa chinensis**: d, inflorescence (× ⅔); e, single flower as in life (× 13); f, flower with hypogynous scales pulled apart to show nut (× 13); g, glume (× 15). h–i, **Blysmus compressus**: h, habit (× ⅔); i, flower showing bifid stigma, biconvex nut and hypogynous bristles (× 6). j–k, **Rhynchospora rugosa** var. **griffithii**: j, inflorescence (× 1); k, flower showing hypogynous bristles and swollen style base (× 30). l–m, **R. rubra**: l, inflorescence (× 2); m, nut (× 10). n–q, **Scleria terrestris**: n, inflorescence (× ½); o, contraligule (× 1.25); p, male spikelet (× 3); q, female spikelet (× 5). r, **S. lithosperma**: nut (× 7). s, **S. parvula**: nut (× 9). t, **S. caricina**: habit (× 1). Drawn by Glenn Rodrigues.

with linear blades. Infl. terminal, dense of 1–3 heads of sessile spikelets, subtended by spreading, leaf-like bracts. Spikelets small, compressed, falling entire at maturity; glumes 3–5, distichous, conduplicate, lower 1–2 sterile, middle bisexual, upper male or sterile. Perianth 0. Stamens 1–3. Stigmas 2. Nut biconvex.

1. Midrib of upper glumes bearing conspicuous teeth or a membranous, ciliate wing 2
+ Midrib of upper glumes green, forming outer edge of glume sometimes sparsely ciliate 3

2. Rhizomes spreading, stems distant; midrib of glumes bearing a flat-tened, curved, ciliate keel (Fig. 30c) .... **3. K. nemoralis**
+ Plant tufted; midrib of glumes bearing long, flattened, whitish teeth
4. K. squamulata

3. Infl. drying white, cylindrical (over 1cm), sometimes with small subsidi-ary heads at base; spikelets over 1.3mm wide
5. K. odorata subsp. **cylindrica**
+ Infl. drying green, a single subspherical head (less than 0.9cm) or several subequal ones; spikelets under 1mm wide 4

4. Infl. a single head; rhizomes extensively spreading so stems ± distant
1. K. brevifolia
+ Infl. of 3 or more subequal heads; densely tufted .... **2. K. tenuifolia**

1. K. brevifolia Rottboell. Fig. 30a–b.
Perennial; rhizomes extensively creeping. Stems distant, triquetrous, 2–42cm. Leaves sub-basal, blades shorter than stem, 2–3.7mm wide; leaf sheaths and basal, bladeless sheaths reddish-brown. Infl. hemispheric or slightly elongate, sometimes with 1–2 small subsidiary heads, 0.5–0.9 × 0.5–0.7cm, greenish in life; involucral bracts 2–4, unequal, longest 2–18cm. Spikelets lanceolate, sessile, 2.3–3.5 × 0.9–1mm, usually with 3 glumes. Lowest glume sterile, ovate, 0.6–1.3 × 0.6–1mm. Middle glume sterile, conduplicate, oblong-ovate, mucronate, mucro recurved, 2.1–2.5 × 1.2–1.4mm, hyaline, sometimes streaked with red, with 2–3 strong ribs either side of midrib; midrib keeled, green, commonly ciliate. Upper glume fertile, 2.3–3.1 × 1.6–2mm, similar to middle. Stamens 1–3. Style 0.3–0.7mm; stigmas 1–1.5mm. Nut narrowly oblong-obovate to oblong-elliptic, truncate, apiculate, 1–1.4 (excl. apiculus) × 0.7–0.9mm, pale yellowish- to reddish-brown, ± smooth.
Bhutan: S — Samchi (Daina Khola), Phuntsholing (Druk Motel,

Phuntsholing, Chimakothi to Phuntsholing), Chukka (below Chimakothi, Surge Shaft Colony) and Gaylegphug (Gaylegphug) districts; **C** — Thimphu (Gidakom, below Dotena, Chapcha, Paro), Punakha (Talo, Lobesa to Mendegong, Lometsawa), Tongsa (Berthi), Bumthang (above Dhur) and Mongar (E side of Kuru Chu below Mongar) districts; **Darjeeling** (Darjeeling, Great Rangit, Tista Bridge, below Peshok, Mongpu, Rongchong, Riyang Ghora); **Sikkim** (Gangtok, Lachen, Yoksum, N of Ranipul). Under-recorded, very common. Wet, grassy places in shade, by rivers, in tea gardens; shingle by stream; gravelly wasteground, 200–2870m. Fl./fr. March–August.

2. K. tenuifolia Steudel; *K. triceps* Rottboell *nom illegit.*
Differs from *K. brevifolia* in being tufted, often with persistent fibrous leaf bases; infl. of 3 or more subequal heads, central one less than 0.7cm, glumes as in *K. brevifolia*.
   Bhutan: C — Mongar district (Autsho 37km N of Mongar). Weed in maize, 1000m. Fr. June.

3. K. nemoralis (J.R. & G. Forster) Dandy ex Hutchinson & Dalziel; *K. monocephala* Rottboell (*nom. illegit.*); *Cyperus kyllingia* Endlicher. Fig. 30c.
Differs from *K. brevifolia* chiefly in its upper glumes which have a prominent, curved, ciliate membranous wing developed along the centre (between the lateral ribs), rather than a green, keeled midrib. Infl. conspicuously whitish in life, central head largest, several subsidiary ones usually present, glumes often drying pale reddish-brown.
   Bhutan: S — Phuntsholing district (Chimakothi to Phuntsholing); **Darjeeling** (Barnesbeg, Riang); **Sikkim** (Great Rangit, N of Jorethang, above Singtam; Rangpo to Tista Bazar (F.E.H.1)). Common on roadsides at low altitudes in Sikkim and Darjeeling — under-recorded; 300–1100m. Fl./fr. April–July.

4. K. squamulata Thonning ex Vahl.
Differs from *K. brevifolia* in being tufted, stems shorter and stouter (usually shorter than leaves); midrib of glumes bearing long (to 0.5mm), flattened, whitish teeth, sides of glumes white. Fragrant when crushed.
   Bhutan: C — Thimphu (Yosepang), Bumthang (Bumthang) and Tongsa ((Parker, 1992)) districts. Arable weed among potatoes, soyabean etc., 2600–2700m. Fr. August.

5. K. odorata Vahl subsp. **cylindrica** (Nees) T. Koyama; *K. cylindrica* Nees; *Cyperus sesquiflorus* (Torrey) Mattfeld & Kükenthal subsp. *cylindricus* (Nees) T. Koyama.

Rhizomes short. Stems ± tufted. Infl. cylindric 1–1.1 × 0.6cm, drying white, sometimes with one or more subsidiary heads. Spikelets ovate, 2.3–2.7 × 1.3–1.7mm. Glumes 4, lower 3 sterile, uppermost fertile: lowest linear, c.1.2 × 0.2mm, sub-basal one ovate, c.1.1 × 0.8mm; upper 2 ovate, acute (not mucronate), midrib green, keeled, smooth, not winged — lower 2.1–2.3 × 1.8–2.2mm, 4–6-veined each side of midrib, fertile glume 2.3–2.5 × 1.8–2mm, 3-veined either side of midrib. Stamens 2. Style 0.1–0.2mm; stigmas 0.6–0.8mm. Nut obovate, truncate, apiculate, 1.3–1.4 (excl. apiculus) × 1–1.3mm, ± smooth, yellow.

Bhutan: C — Punakha district (Talo); **N** — Upper Mo Chu district (Tamji); **Sikkim** (Rangit; Rangpo to Tista (F.E.H.1)). Grassy hillside, 300–2130m. Fr. June–August.

14. LIPOCARPHA R. Brown

Tufted perennials or annuals. Leaves sub-basal, with flat, linear blades. Infl. a dense, terminal cluster of sessile spikelets, involucral bracts leaf-like and spreading or erect and stem-like so spikelets pseudolateral. Spikelets with glumes spirally inserted on persistent axis. Flowers bisexual. Perianth of 2 persistent, linear, conduplicate, hyaline, hypogynous scales, or absent. Stamens 1–2. Stigmas 2–3, style continuous with ovary. Nut trigonous or compressed, sometimes tightly enclosed within perianth scales.

1. Spikelets whitish; glumes acute; bracts spreading, so infl. appearing terminal .... **1. L. chinensis**
+ Spikelets brownish; glumes mucronate; bract erect, so infl. appearing lateral .... **2. L. squarrosa**

1. L. chinensis (Osbeck) Kern; *L. argentea* R. Brown *nom. illegit.* Fig. 30d–g.
Annual or perennial. Stems tufted, rounded-trigonous, 16–39cm; bladeless sheaths at base. Leaves c.½ length of stem, 1–2.5mm wide. Infl. of 3–6 sessile spikelets, involucral bracts 2–3, longest 3–9cm. Spikelets ovoid, blunt, 5–9 × 4–5.5mm. Glumes concave, spathulate, 2–2.6 × 0.9–1mm, midrib greenish, keeled, carried into broad, acute point so apex sub-trifid, sides whitish-hyaline, flecked brown. Hypogynous scales tightly enclosing flower, linear-lanceolate, acute, 1.8–2.3 × 0.5mm, ribbed, hyaline, flecked. Stamen 1 (additional filament sometimes present). Stigmas 2–3. Nut linear-obovoid, shouldered above, apiculate, c.1.2 × 0.4mm, straw-coloured.

Bhutan: C — Punakha (Punakha) and Tongsa (Tama) districts; **Terai** (Jalpaiguri to Siliguri); **Darjeeling** (Kalimpong, Mongpu, Pandam, Great Rangit, Riang, Rongbe); **Sikkim** (Phodong to Kabi, 1km N of Ranipul). Rice paddies; marsh by roadside, 250–2130m. Fr. March–September.

2. L. squarrosa (L.) Goetghebeur; *Rikliella squarrosa* (L.) J. Raynal; *Scirpus squarrosus* L.
Dwarf, tufted annual. Stems 3–9cm, slender. Stem leaves 1–3, basal, blade flat, very acute, 0.5–3.5cm; sheaths reddish. Infl. of 2–3, pseudolateral, squarrose spikelets; lowest bract erect, leaf-like, over 1cm. Spikelets ovoid, 2.5–6 × 2.5–3mm. Glumes oblong, narrowed into mucro, c.0.7 (excl. mucro) × 0.3mm, reddish-brown; mucro recurving, 0.5–0.6mm, green. Hypogynous scales absent. Filaments equalling nut; anthers c.0.2mm. Stigmas c.0.1mm. Nut c.0.5 × 0.3mm, faces oblong to narrowly obovate, ± truncate, cream to dark yellowish-brown, minutely papillose.
Darjeeling (junction of Great and Little Rangit). Sandy river shingle, 470m. Fr. August.

15. BLYSMUS Panzer ex Schultes

Rhizomatous perennials. Stems subterete, leafy below. Infl. a terminal spike of distichously arranged spikelets, subtended by a usually conspicuous bract. Spikelets with spirally inserted glumes. Flowers bisexual. Perianth of usually 6, barbed, hypogynous bristles. Stamens 3. Stigmas 2, style passing gradually into nut. Nut biconvex.

1. B. compressus (L.) Panzer ex Link. Fig. 30h–i.
Rhizomes extensively spreading. Stem 5–41cm, stout. Leaves on lower ¼ to ½ stem, widest near base, tapering to acute apex, flat, becoming inrolled, shorter than to equalling stem, 1.5–4.5mm wide, sometimes glaucous; lower sheaths and bladeless sheaths persistent, ribbed, brown. Infl. 1–2.5cm; lowest bract with glume-like base and green midrib extended into leaf-like point shorter than to exceeding spike. Spikelets 2–12, lowest occasionally slightly distant, linear-ellipsoid, lowest 6–9 × 2–3mm. Glumes ovate, blunt to sub-acute, 5–6 × 2–3.4mm, golden- to reddish-brown, shining, margins narrowly hyaline. Anthers with connective prolonged into acute point. Bristles 6, retrorsely barbed, slender, reddish-brown, about equalling stigmas. Stigmas c.3mm, exceeding style. Nut 2–2.5 × 1.2–1.4mm, flattened, obovate, abruptly contracted to shortly mucronate apex, dark brown with large, reflective, greyish epidermal cells.
Bhutan: C — Thimphu (Doteng, hill above Thimphu Hospital, above

Ragyo) and Bumthang (near Byakar Dzong, Khak Thang) districts; N —
Upper Mo Chu (Chebesa to Lingshi) and Upper Bumthang Chu (above
Lambrang) districts; **Sikkim** (Lachen, Chumegata, Kopup, Chamnago,
Lhonak); **Chumbi** (Yatung). Marshes, sandy soil and damp grassy ground by
stream, 2300–4570m. Fl. May; fr. June–September.

Probably merits further investigation in view of the wide range of altitudes at which it
occurs; however, var. *sikkimensis* C.B. Clarke does not seem worth recognising.

Commonly misidentified as a *Carex* of subgenus *Vignea*, which it superficially resembles.
The bisexual flowers and barbed bristles, however, easily distinguish it.

16. RHYNCHOSPORA Vahl

Tufted perennials with short rhizomes or sometimes annual. Stems trigonous to terete. Leaves mainly basal, some on stem, blades linear. Infl. composed
of a terminal and several lateral partial panicles subtended by leaf-like bracts,
or reduced to a dense terminal head. Spikelets with spirally inserted glumes;
lower 3–4 sterile, upper 2–6 fertile. Fertile flowers bisexual or lower female
and upper male (or sometimes sterile). Perianth of (0–)3–6 hypogynous bristles.
Stamens (1–)2–3. Stigmas 2; style swollen at base, with constriction between
base and nut. Nut biconvex, crowned by conspicuous swollen, persistent
style base.

1. Infl. loosely and narrowly paniculate with a terminal and several lateral
    partial infls. .... **1. R. rugosa**
+   Infl. a dense, capitate head .... **2. R. rubra**

1. R. rugosa (Vahl) Gale var. **rugosa**.
Perennial; rhizomes short. Stems tufted, to 62cm, with brown basal bladeless sheaths. Leaves on lower ½ stem, blades deeply channelled, margins
thickened, shorter than stem, 1–1.8mm wide, slightly greyish. Infl. a narrow,
elongate panicle with a terminal and 2–3 lateral partial infls.; lateral infls.
single or paired, corymbose, peduncled, subtended by leaf-like bracts. Spikelets
clustered, ellipsoid, acute, sessile, 3–4.4 × 1.5–2mm, dark brown. Sterile
glumes 3–4. Fertile glumes (1–)2(–4) convex, elliptic, acute, shortly (to 0.4mm)
mucronate, to 3.5 × 3mm, keeled, reddish-brown. Flowers bisexual. Bristles
6, antrorsely barbed, the longest slightly exceeding body of nut. Stamens 2–3.
Persistent style base triangular, c.0.6mm (c.half length of nut), greyish; nut
body obovate, truncate, stipitate, c.1.8 (excl. stipe and style base) × 1.5mm,
minutely rugulose, yellow-brown.

Bhutan: C — Thimphu (near Motithang Hotel). Wet grassy slope, 2660m. Fr. August.

var. **griffithii** (Boeckeler) D.M. Verma & V. Chandra; *R. griffithii* Boeckeler var. *levisetis* C.B. Clarke. Sikkim name: *jakcha naten*. Fig. 30j–k.
Differs from var. *rugosa* in its more slender habit (stems 20–56cm); spikelets longer and narrower (4–5 × 0.7–1mm); fertile glumes longer and narrower (3.7–4 × 1.2–1.8mm), acute not mucronate; persistent style base 0.7–1mm (more than half length of nut); nut body narrower c.1.5 × 1mm, orange-brown, shining; bristles exceeding style base, not barbed.
Bhutan: C — Punakha (below Gangtey Gompa towards Phubjikah) and Bumthang (above Dhur) districts; **Sikkim** (Lachen, Lachung). Bogs and wet meadows, 2130–3050m. Fl./fr. August–October.

var. **sikkimensis** (C.B. Clarke) D.M. Verma & V. Chandra; *R. sikkimensis* C.B. Clarke.
Differs from var. *griffithii* in its fatter (1.5–2mm) spikelets with 3–6 fertile glumes; nut (though perhaps immature) smaller (0.9 × 0.5mm); style base exceeding nut, oblong, tapering only at apex.
Sikkim (Lake Catsuperri). 1830m. Fr. January.

2. R. rubra (Loureiro) Makino; *R. wallichiana* Kunth. Fig. 30l–m.
Perennial; rhizome short. Stems tufted, trigonous, 35–54cm; basal sheaths sometimes decaying to fibres. Leaves mainly basal, stem leaf 1, sub-basal, blades shorter than stems, 2–2.5mm wide, midrib keeled below. Infl. a dense, hemispheric head, 1.2–1.5cm diameter; bracts 3 or more, leaf-like, unequal, longest 4–8cm, densely ciliate at base. Spikelets narrow, acute, sessile, pale reddish-brown. Glumes 5–6, lowest 3 small, sterile, 4th glume usually female, terminal 1–2 male (or sterile). Female glume convex, ovate, acute to minutely mucronate, c.3.7 × 1.6mm, keeled. Bristles to 6, unequal, longest c.½ length of nut. Stamens 3. Stigmas short; persistent style base short (under 0.4mm). Nut obovate, truncate, c.1.6 × 1.5mm, scabrid above, brown to blackish.
Terai (Dulkajhar). 150m. Fr. October.

R. corymbosa (L.) Britton (*R. aurea* Vahl) is recorded for Sikkim in F.B.I.; this probably refers to the Terai (specimen seen from Madreegora, Siliguri). It differs from both the above in being much taller (60–100cm) and stouter, with wider leaves (8–20mm); infl. larger composed of 2–5 corymbose partial infls. to 15cm across; style base very long (to 5mm) with 2 grooves; nut c.3 × 2mm.

17. SCLERIA Bergius

Rhizomatous perennials or annuals. Stems trigonous, sparsely leafy near base, sometimes with basal, bladeless sheaths. Leaves linear, with sheathing bases; sheaths sometimes winged, sometimes with contra-ligule (Fig. 30o) developed at apex on side opposite leaf blade. Infl. usually of one terminal and one or more lateral partial panicles, sometimes consisting of small, lateral, subsessile heads of spikelets or reduced to a spike. Flowers unisexual arranged in unisexual or bisexual spikelets. Spikelets usually with several sterile, basal glumes; glumes distichous. Female spikelets usually with a single, terminal fertile glume. Bisexual spikelets male above, with a single basal female glume. Female glumes usually persistent, sometimes falling with nut. Male glumes with 1–3 stamens. Stigmas 3. Nut with bony pericarp, commonly white and reticulately pitted, borne on a commonly 3-lobed disc adherent to base of nut and falling with it.

1. Infl. composed of ± sessile, few-flowered heads in axils of leaf-like bracts along length of stem; female spikelets falling entire at maturity (i.e. nut enclosed by glumes); plant annual .... **6. S. caricina**
+ Infl. with a terminal panicle and at least one lateral, peduncled partial panicle; nut falling from (persistent) glumes; plant annual or perennial   2

2. Annual; lower lateral partial panicles usually in groups of 2–3; lobes of disc short (c.0.5mm); panicles very slender .... **5. S. parvula**
+ Rhizomatous perennials; lateral partial panicles single at nodes; if lobes of disc short then panicle very branched 3

3. Spikelets all similar, bisexual; nut trigonous, disc not developed, reduced to 3 brown bands at base of nut (Fig. 30r) .... **4. S. lithosperma**
+ Male and female spikelets distinct; nut terete, 3-lobed disc present at base of nut 4

4. Panicles very branched (branches of partial panicles bearing lateral branchlets); lobes of disc obtuse, short (c.0.5mm) .... **1. S. terrestris**
+ Panicles simpler; lobes of disc acute or mucronulate 5

5. Lobes of disc long (1–2mm), narrowly triangular, acute; sheaths winged, sparsely pilose .... **2. S. levis**
+ Lobes of disc shorter (c.0.7mm), broadly triangular, mucronulate; sheaths angled, densely pilose .... **3. S. benthamii**

1. S. terrestris (L.) Fassett; *S. elata* Thwaites. Nep: *charpartan*. Fig. 30n–q.
    Rhizomatous perennial. Stems acutely trigonous, (0.6–)1.2–3m, stout, (2.5–)3.3–7mm diameter, usually glabrous. Leaves arranged evenly along stem,

blades acute, (0.5–)0.7–1.8cm wide, glabrous; sheaths narrowly 3-winged, inner face sometimes shortly hairy above, reddish-brown at base; contra-ligule shorter than wide, tongue-shaped, margin dark brown hyaline, glabrous. Infl. (14.5–)22–59 × (3–)5–14cm, with a terminal, and (2–)3–4 appressed partial panicles, bracts leaf-like. Partial panicles with stiff ascending branches, bearing lateral branchlets; bracteoles of branches and branchlets long, flexuous, fili-form, bearing single, sessile female spikelets near base and single or paired, pedicellate male spikelets above. Male spikelets lanceolate, 3.5–4 × 1–1.5mm; fertile glumes c.7, lanceolate, to c.3.5 × 2.2mm, dark reddish-brown. Female spikelets: fertile glume ovate, sharply acuminate, 3.5–4.5 × 3–3.9mm. Nut subglobose or slightly elongate, apiculate, base truncate, 2.8–4 (incl. disc) × 2.5–3mm, white or tinged dark purplish, reticulately pitted, shortly hairy or glabrous on reticulations; disc lobes 3, obtuse, short (c.0.5mm).

Bhutan: S — Chukka (Marichong), Sankosh (10km from Daga Dzong towards Goshi) and Gaylegphug (Gaylegphug) districts; C — Tongsa district (Tama); **Terai** (Dulkajhar); **Darjeeling** (Rishee, Mongpu, Tukvar, Garidoora, Singla, Kolbong, Rongsong, Rongpo). Old fields; rocky bank in warm, wet, broad-leaved forest, 120–1520m. Fr. May–October.

2. S. levis Retzius; *S. hebecarpa* Nees.

Like a slender form of *S. terrestris*, but differing as follows. Stem shorter (to 92cm), narrower (1.3–3mm diameter), often sparsely pilose. Leaves and bracts narrower (0.6–1.3cm wide), often sparsely pilose; leaf sheaths more prominently winged, contra-ligule densely pilose. Infl. smaller (8.5–45 × 2–8cm), narrower, lateral partial panicles 1–2, less branched (lateral branches of partial panicles not bearing branchlets). Nut ((2.4–)3.5–4 × 2.4–3mm) usually oblong-ovoid, shallowly reticulate and slightly hairy at first, becoming smooth, shining and glabrous; disc lobes acute, long (1–2mm).

Terai (Bamunpokri, Dulkajhar); **Darjeeling** (Mongpu, Rishee, Rangit, Silak, above Peshok). Roadside in open *Tectona/Shorea* plantation, 150–760m. Fr. July–October.

3. S. benthamii C.B. Clarke; *S. khasiana* C.B. Clarke non Boeckeler.

Differs from *S. levis* in having the sheaths and leaf undersurfaces densely pilose; sheaths angled, not winged; nuts more elongate (2.5–2.7 × 2.1–2.4mm), conspicuously reticulate, the ridges hairy, not becoming smooth; lobes of disc triangular, minutely apiculate, shorter (c.0.7mm).

Bhutan: C — Punakha district (near Chuzomsa). Dry grassland on steep hillside with scattered shrubs, 1200m. Fr. September.

4. S. lithosperma (L.) Swartz. Fig. 30r.

Slender perennial, differing from *S. levis* as follows. Rhizome slender, woody, creeping. Stems arising singly, very slender (1–1.7mm wide). Leaves 1–4mm wide; contra-ligule as long as wide, margin densely ciliate. Infl. extremely slender; branches of partial panicles linear, bearing distant, sessile spikelets; spikelets bisexual, with 1 female flower (above basal sterile glumes), and upper male flowers; female glumes broad, ovate, male glumes narrower. Nut obscurely trigonous, smooth and glabrous even at first, disc not developed, with 3 depressions at base with rugose, transverse, brown striations.

Darjeeling (Rangit). 610m. Fr. June.

5. S. parvula Steudel; *S. tessellata* sensu F.B.I. non Willdenow. Fig. 30s.

Annual. Stems scabrid on angles, 17–120(–166)cm, 1–2mm diameter. Leaves 1–2, sub-basal, blades 3.3–5mm wide, sparsely hairy on underside; sheaths hairy; contra-ligule broader than long, rather blunt, densely hairy. Infl. narrow, elongate, a terminal panicle and 2–3 groups of lateral partial panicles almost to base of stem. Terminal panicle 1.5–5.5 × 1.2–1.5cm, with several short, erect linear spikelet-bearing branches, bract erect, usually slightly exceeding infl.; lateral partial panicles in groups of 1–3, unequally peduncled, bracts leaf-like; bracteoles stiff, erect. Male spikelets oblong, shortly pedicelled, 3.5–4.5mm; sterile glumes 2, lanceolate, 3.3–4 × 1.2–1.9mm, sides hyaline flecked reddish, margins sometimes brown, midrib green; fertile glumes c.5. Female spikelets sessile, glumes c.3, narrowly ovate, very acute, 4.6–6 × 2.4mm, subequal, exceeding nut, keeled. Nut subglobose (sometimes slightly elongate), 2.5–3 × 2mm, white, reticulate-pitted (pits vertically elongate) except smooth band below apiculate apex, with wavy transverse bands of brownish hairs or glabrous; disc lobes, widely triangular, subacute, short (c.0.5mm).

Bhutan: C — Punakha (Samtengang, Lobesa, Gangtey Gompa to Phubjikah) and Bumthang (above Dhur) districts; **Darjeeling** (Darjeeling, Gareedora to Ruprool); **Sikkim** (Lachung). Grassy marshes; sphagnum bog; marsh around rice paddy; edge of lake, 1400–2870m. Fr. August–October.

S. biflora Roxb. is likely to occur. It is also annual, but differs in having triquetrous stems; female glumes ovate, shorter (c.3.2 × 2.3mm), with reddish-brown sides; nut squatter (c.2.2 × 2mm), pits ± square, apical band coloured dark purplish, disc lobes long-apiculate (c.1mm), with 2 pits visible behind each sinus.

6. S. caricina (R. Brown) Bentham; *Diplacrum caricinum* R. Brown. Fig. 30t.

Slender annual. Stems to 40cm, c.0.8mm diameter. Leaves 1–2, sub-basal. Bracts leaf-like regularly spaced along stem, to 4cm, 1.5–3mm wide, glabrous, sheaths glabrous, contra-ligule not developed. Infls. of small subsessile heads

(c.4–7mm diameter, fewer than 10 spikelets) in axils of bracts. Female spikelets falling entire at maturity, narrowly ovoid, c.2 × 1mm; glumes 2, oblong, apex sharply tridentate, c.1.7 × 1mm, hyaline flecked dark red, veins c.9, strong, green. Male spikelet c.1.7 × 0.7mm, sterile glume 1, narrowly lanceolate, c.1.5 × 1mm; fertile glumes c.6. Stamen 1. Nut ovoid-trigonous, angles thickened, acute, c.1 × 0.9mm, longitudinally ridged, cream, shining; disc lobes not developed.
Terai (Dulkajhar). 150m. Fr. October.

18. KOBRESIA Willdenow

Perennials. Rhizomes usually very short so stems densely tufted, occasionally stoloniferous and spreading. Leaves basal and sub-basal; blades flat with distinct keel on underside (Fig. 31b) or semicircular to V-shaped in section and lacking keel (Fig. 31i); bases sheathing; sheath bases often persistent, margins sometimes fibrillose. Infl. terminal, apparently simple (spike-like) or paniculate, sometimes unisexual, spikelets subtended by glume-like bracteoles (glumes in this account). Upper spikelets in infl. or branches commonly single-flowered and male (consisting of a glume + 3 stamens); lower spikelets single-flowered and female or androgynous. Prophyll of female spikelets modified as a 2-keeled perigynium, enclosing a single female + one or more male flowers (Fig. 32d) or a single female flower + a 'racheola' (a vestigial spikelet axis) (Fig. 32b). Prophylls compressed, so keels appearing to form (often ciliate) margins, open on adaxial face to varying degrees — sometimes to base, sometimes only at extreme apex so forming a *Carex*-like utricle. Stigmas 3 (2 in one non-Bhutanese species); nut trigonous, sometimes beaked. Perianth 0.

Identification requires microscopic examination, since spikelets are small and it is extremely difficult to see whether the prophylls are female-only or androgynous. Leaf sections are helpful. In species 18–22 care should be taken to collect all types of infl. and note how they are distributed between plants.

1. Infl. branched, with lateral spikes which may be closely appressed (Fig. 31a) 2
+ Infl. a simple spike (Fig. 31h) 10

2. Prophylls large, over 5mm; plants usually robust with leaves usually over 2mm wide 3
+ Prophylls under 3.5mm; plants usually slender, with narrow (usually under 1.5mm) leaves 8

3. Rhizomes spreading extensively .... **3. K. gammiei**
+ Rhizomes short, plants densely tufted 4

4. Lowest bract shorter than infl. 5
+ Lowest bract leaf-like, usually exceeding infl. 7

5. Infl. dense, spike-like with lateral spikes appressed 6
+ Infl. rather lax, with lateral spikes spreading .... **6. K. laxa**

6. Bases of sheaths shining brown, persistent; lower spikelets of spikes androgynous .... **1. K. pseuduncinoides**
+ Bases of sheaths pale, thin, not persistent; all spikelets single-flowered, lower ones female .... **2. K. uncinoides**

7. Lateral spikes long (lowest 2.5–6cm); prophyll over 8mm; female glumes over 6.5mm, narrowly ovate to oblong .... **4. K. curticeps**
+ Lateral spikes shorter (lowest to 2.5cm); prophyll under 5.5mm; female glumes under 5mm .... **5. K. sikkimensis**

8. Nut with long (equalling body), curved beak, conspicuously exserted from prophyll .... **9. K. curvirostris**
+ Nut not beaked, enclosed within prophyll 9

9. Lowest bract usually exceeding infl.; prophyll lanceolate, glabrous
8. K. fragilis
+ Lowest bract shorter than infl.; prophyll oblong, margins minutely ciliate .... **7. K. clarkeana**

10. Leaf blades distinctly keeled on undersurface, flat (grass-like) though sometimes becoming inrolled (Fig. 31b) 11
+ Leaf blades not keeled, semicircular (occasionally V-shaped) in section, often filiform (Fig. 31i) 14

FIG. 31. **Cyperaceae** VI (**Kobresia** habits and leaf types).
a–g, **K. pseuduncionides**, species showing paniculate infl. and flat, keeled leaf blade: a, habit (× 0.4); b, t.s. leaf (× 12.6); c, spikelet glume (× 3.6); d, whole spikelet (× 3.6); e, prophyll with male flowers removed (× 3.6); f, immature gynoecium (× 3.6); g, male flowers dissected out of prophyll (× 3.6). h–l, **K. woodii**, monoecious species with simple, spicate infls., male and female borne on same plant, leaf blade semicircular in section, not keeled: h, habit (× 0.4); i, t.s. leaf (× 21.6); j, glume of female spikelet (× 5.4); k, female spikelet (× 5.4); l, nut (× 5.4). Drawn by Mary Bates.

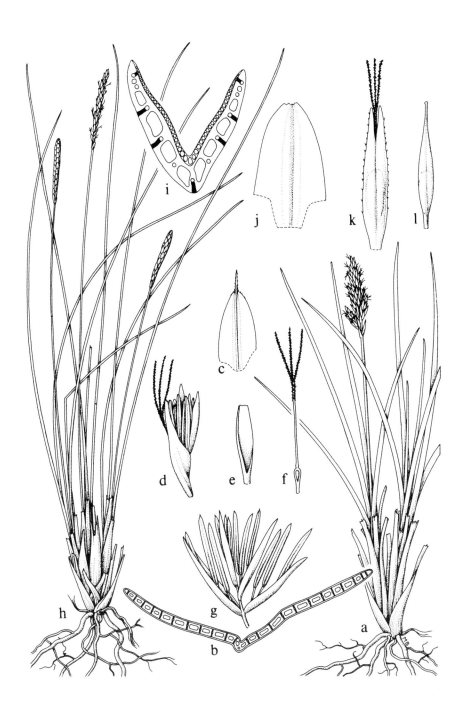

11. Spikelets androgynous, with 1 female and one or more male flowers enclosed in prophyll (Fig. 32d) 12
+ Spikelets all single-flowered (prophyll enclosing no male flowers); spikes commonly unisexual (Fig. 32b) 13

12. Spike linear, elongate; prophylls enclosing 1(–2) male flowers, margins ciliate .... **14. K. duthiei**
+ Spikes short, oblong, dense; prophylls enclosing 3(–4) male flowers, margins glabrous .... **11. K. humilis**

13. Prophylls open only near apex; female glumes rounded to acute
18. K. esenbeckii var. **esenbeckii**
+ Prophylls open to base; female glumes obtuse
18. K. esenbeckii var. **fissiglumis**

14. Spikelets androgynous (Fig. 32d) 15
+ Spikelets unisexual (Fig. 32b) 17

15. Spike elongate, lax; glumes with extremely broad hyaline margins
12. K. sargentiana
+ Spikes short, very dense; glume margins broadly hyaline or not 16

16. Sheath bases reddish-brown, not fibrillose; stem very stout (over 1.7mm); infl. club-shaped .... **10. K. schoenoides**
+ Sheath bases chocolate-brown, fibrillose; stem under 1.3mm wide; infl. linear .... **13. K. capillifolia**

17. Spikes androgynous i.e. male above, female below (Fig. 32a) 18
+ Spikes unisexual (if androgynous plants extremely dwarf) 19

18. Prophyll open only at apex; sheaths yellowish-brown; glumes mucronate .... **15. K. nepalensis**
+ Prophyll open to base; sheaths darker brown; glumes acute
16. K. stiebritziana

19. Plant extremely dwarf (under 4cm) forming dense mats; spikes androgynous or unisexual, male and female glumes blunt, under 4mm
17. K. pygmaea
+ Plant usually taller, if small then glumes over 4.5mm; spikes always unisexual 20

20. Culm over 20cm, stout; prophylls over 7mm .... **19. K. woodii**
+   Culm under 16cm, slender; prophylls under 6mm 21

21. Prophylls over 5mm, margins ciliate; male spikes linear, elongate (over 2cm); sheaths fibrillose; leaves very weakly filiform ...... **20. K. vaginosa**
+   Prophylls under 3.8mm, margins glabrous or minutely ciliate at extreme apex; male spikes ellipsoid, under 1.5cm; sheaths not fibrillose; leaves stiffly filiform tending to recurve 22

22. Female glumes appressed, usually under 4.2(–5)mm, rounded to subacute, midrib broad, c.⅓ glume width; male glumes to 3mm wide, subacute, midrib broad; nut beaked .... **21. K. vidua**
+   Female glumes obliquely spreading, over 4.5mm, very acute, midrib c.¼ glume width; male glumes under 2mm wide, very acute, midrib narrower; nut minutely apiculate, not beaked .... **22. K. prainii**

1. K. pseuduncinoides Noltie. Fig. 31a–g, Fig. 32e.

Densely tufted. Bases of sheaths dark yellowish-brown, shining, persistent, not fibrillose. Leaves sub-basal, blades flat, very acute, about equalling culm, 4.7–9mm wide, keeled, thick-textured, cross-veinlets prominent when dry. Culm erect, acutely trigonous, angles minutely scabrid, 24–38cm, stout (2.5–2.7mm wide). Infl. a dense spike-like panicle, 5.5–6.5 × 1.5–1.8cm; lateral spikes 8 or more, appressed; lowest bract with large (0.7–1.6cm), oblong, glume-like base, sides brown-hyaline, midrib broad, green, produced as filiform tip not exceeding infl. Lateral spikes with upper spikelets single-flowered, male, lower 5–10 spikelets androgynous with 1 female and 2–4 male flowers within prophyll; spikelet 'glumes' oblong-elliptic, acute to aristate, 6.2–6.5 (excl. arista) × 2.3–3.5mm, sides brown, midrib green, 1-veined, narrow, margins narrowly hyaline near apex. Prophyll oblong, truncate, open to base, 5.6–7 × 1–1.3mm, hyaline, flushed brown, keels minutely scabrid. Nut not seen.

Bhutan: N — Upper Kulong Chu district (Shingbe). Open marsh, 3810m. Fl. June.

2. K. uncinoides (Boott) C.B. Clarke. Fig. 32f.

Rhizome short, thick, woody. Bases of sheaths pale fawn, dull. Leaves basal and sub-basal, blades flat, very acute, shorter than culm, 2–4mm wide, keeled. Culm rigid, subterete, (2–)10.5–50cm, 1.2–2.2mm wide. Infl. a dense spike-like panicle, 2.5–8.3 × 0.8–1.7cm; lateral spikes 4–8, closely appressed, lowest slightly distant; lowest bract with oblong, glume-like, clasping base and aristate tip sometimes almost equalling infl.; long-peduncled basal spike also sometimes present. Lateral spikes androgynous or lower sometimes entirely female; spikelets all single-flowered. Female 'glumes' lanceolate to oblong-lanceolate, acute to obtuse, bases encircling axis, 5–8 (excl. awn) × 2.4–3.6mm, pale yellowish-brown, midrib green, developed as awn to 2(–4.5)mm. Prophyll utricle-like, linear-lanceolate, 5.5–8.3 × 1.4–1.5mm, ribbed, keels ciliate, open to less than half-way. Nut stipitate, oblong-trigonous, 3.1–3.5 × 0.8–1.4mm, brown, smooth. Racheola ciliate, about equalling to exceeding prophyll.

Bhutan: C — Thimphu (Pajoding, Barshong, Bimelang Tso to Dungtsho La, mountain E of Thimphu), Bumthang (above Gortsam), Tashigang (Merak, Chorten Kora) and Sakden (Orka La) districts; **N** — Upper Mo Chu (above Laya, Yale La, Soe to Lingshi) and Upper Bumthang Chu (above Lambrang) districts; **Sikkim** (Dzongri, Lachung, Lachen, Chola, Yakla, Megu, Sheravthang, Changu, Chakung Chu, etc.). Alpine pastures; damp hollows on hillside in open or among shrubs (incl. *Gaultheria* and *Rhododendron*), 3050–4880m. Fl./fr. June–October.

The record of *K. royleana* in F.E.H.1 should be referred to this species

3. K. gammiei C.B. Clarke; *K. williamsii* T. Koyama. Fig. 32g.
Rhizomes extensively spreading, stems loosely tufted, bases clothed with old leaves. Culm stiffly erect, subterete, 7–58cm, 1.2–1.5mm wide. Leaves basal and sub-basal, blades flat, shorter than culm, 1.5–3.5(–5)mm wide, keeled, sheaths long, inner faces membranous, orange-brown, produced upwards into short auricles; leaves of non-flowering shoots wider (to 6.5mm). Infl. densely paniculate, 2.6–5 × 0.8–1.5cm; lateral spikes 3–7, erect, lowest to 2.5cm, sometimes slightly distant; lowest bract with clasping glume-like base with retuse apex and long filiform tip shorter than infl.; long-peduncled sub-basal

---

FIG. 32. **Cyperaceae** VII (**Kobresia** prophylls).
o abaxial view of prophyll. + adaxial view of prophyll showing degree of openness.
* nut (+ racheola in female-only spikelets).
a, schematic diagram of **Kobresia** with androgynous, spicate infl. All spikelets single-sexed, the upper male, the lower female, as in species 15–18. b, detail of lower, female-only spikelet and subtending 'glume', the rachilla and nut concealed within prophyll. c, schematic diagram of **Kobresia** with spicate infl. with all spikelets androgynous, as in species 10–14. d, detail of spikelet and subtending 'glume', the prophyll enclosing 1 female and two male flowers. e, **K. pseuduncionoides**: androgynous spikelet from lower part of lateral spike. Male flowers removed from +. Nut immature in * (× 6). f, **K. uncinoides**: female spikelet from lower part of lateral spike (× 6). g, **K. gammiei**: female spikelet from middle part of lateral spike (× 6). h, **K. curticeps**: female spikelet from lower part of lateral spike (× 6). Drawn by Mary Bates.

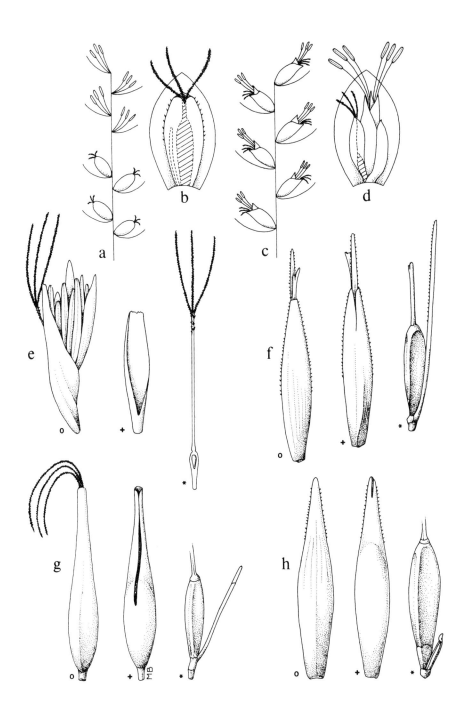

232. CYPERACEAE

spike sometimes also present. Lateral spikes with upper spikelets single-flowered, male, middle spikelets single-flowered, female and lower spikelets female or androgynous with 1 male and 1 female flower within prophyll. 'Glumes' elliptic to oblong, subacute to rounded, 5.5–8.5 × 3.2–4mm, midrib broad, green, 3-veined, sides pale reddish-brown. Prophyll utriculiform, lanceolate, gradually tapered above to oblique orifice, (5.6–)7–10.3 × 0.9–2.1mm, glabrous. Nut narrowly obovoid-trigonous, 3.3–3.8 × 1.7mm. Racheola (in female-only spikelets) linear, shorter than nut, 1-veined, glabrous.

Bhutan: C — Bumthang district (above Gortsam); **Sikkim** (Dzongri to Gamothang, Jamlinghang, Bikbari, Mon Lapcha, Thangshing to Lam Pokhri). Damp humus-rich slope in *Abies* forest; damp grassy ledge by waterfall; cliffs, 3660–4420m. Fl. May; fr. July–August.

Records of *K. sikkimensis* in F.E.H.1 refer to this species.

In the field very similar to *K. uncinoides* from which it can be told by its reddish-brown glumes and its spreading rhizomes.

4. K. curticeps (C.B. Clarke) Kükenthal; *Carex curticeps* C.B. Clarke. Fig. 32h.
Rhizomes short, stems tufted, bases clothed with remains of old leaf sheaths. Culm trigonous, leafy in lower half, 22–95cm, 1.1–2.2mm wide. Leaf blades flat, about equalling culm, 3–7mm wide, keeled, sheaths striate, pale brown, dull. Plants sometimes with female-only in addition to bisexual infls. Infl. a stiffly nodding, narrow panicle, 6.5–9.5 × 0.8–2.3cm; lateral spikes 3–12 ± appressed; lowest bract leaf-like, erect, exceeding infl. Terminal spike to 6.5cm, androgynous or entirely female; lateral spikes linear, lowest longest (to 6cm), upper spikelets single-flowered male (occasionally female), lower spikelets single-flowered, female (lowest 2 occasionally androgynous with 1 female and 2 male flowers within prophyll). Spikelet 'glumes' narrowly ovate or oblong, usually aristate (1–3.5mm), encircling spike axis, 5.5–10 × 2–3mm, midrib green, wide, sides tawny or hyaline. Prophyll utriculiform, linear-lanceolate, open only near apex, (6–)8–11.3 × 1–1.6mm, ribbed, faces hyaline, margins usually shortly hairy (occasionally glabrous). Nut shortly stipitate, linear-ellipsoid-trigonous, 5.5–6.5 × 1.3–1.7mm, pale brown. Racheola sometimes expanded and fimbriate at apex, to half length nut.

Bhutan: C — Thimphu district (summit of Dochu La); **Darjeeling** (Sandakphu); **Sikkim** (Singalila, Kyanglasha, Lungthung, Tsomgo, Potang La, Dzongri, Changu, Gnatong, Chakung Chu); **Chumbi** (Natula). Open stony or sandy slopes in damp conifer (incl. *Abies*) forest, 2740–4110m. Fl./fr. June–October.

5. K. sikkimensis Kükenthal.

Rhizomes short, stems densely tufted. Sheath bases pale fawn, persistent. Culm sharply trigonous, (40–)60cm, c.1.4mm wide. Leaf blades flat, about equalling culm, 2.3–3.3mm wide, keeled. Infl. a dense, curved panicle, 4.6–6.5 × 0.9–2cm. Lateral spikes 5–7, lowest to 2.5cm; spikes androgynous; spikelets all single-flowered, upper male, lower female; lowest bract leaf-like. Female 'glumes' lanceolate, acute, 4.6–5 × 2mm, midrib green, sides reddish-brown, margins narrowly hyaline; male 'glumes' with wider hyaline margins. Prophyll utriculiform, lanceolate, open only at hyaline apex, c.5.5 × 1.3mm, shortly hairy above, streaked brown. Nut narrowly obovoid-trigonous, apiculate (0.3mm), 3.2 × 1.1–2.4mm. Racheola shorter than nut, smooth.

Sikkim (locality unknown, CAL). [In Nepal on open grassy slopes, 3800m. May.]

6. K. laxa Nees. Fig. 33a.

Rhizome shortly creeping; stems loosely tufted. Culm 13–35cm, slender (c.0.9mm wide), with bladeless sheaths near base, gradually passing up into bladed leaves; sheaths pale to dark brown, dull, striate. Leaves sub-basal, flat, blades shorter than culm, 1.4–2.4mm wide, keeled. Infl. paniculate, 3.5–7.5 × 1–1.5cm; lateral spikes 9–16 divergent, overlapping, lowest spike sometimes branched; spikes narrowly lanceolate in outline, acute, reddish-brown; lowest bract with serrate, filiform blade shorter than infl. Spikelets all single-flowered. Lateral spikes in upper part of panicle androgynous, dense, primarily male (few basal female spikelets); spikes in lower part of panicle usually predominantly female, rather lax; entire infl. sometimes composed entirely of female or entirely of primarily male spikes. 'Glumes' oblong-lanceolate, acuminate to very acute apex, 3.9–5 × 1–1.4mm, orange- to reddish-brown, midrib green, margins broadly hyaline. Prophylls linear-lanceolate, apex very acute, notched, open to c.halfway, 5.5–6 × 0.5–0.7mm, keels ciliate. Stigmas stout. Nut linear-ellipsoid, c.2.5 × 0.4mm. Racheola exceeding nut, sometimes emerging from orifice, filiform, nerveless.

Sikkim (Lachen). Rocks, 3353–3658m. Fl./fr. June–July.

W Himalayan specimens have simpler, darker infls., with some basal androgynous spikelets and tend to be stouter and broader-leaved than the few specimens seen from Sikkim and Nepal.

7. K. clarkeana (Kükenthal) Kükenthal. Fig. 33b.

Similar to *K. laxa* but much more slender in all its parts.

Apparently tufted. Culm to 37cm, c.0.7mm wide. Leaf blades about equalling culm, 1.2–1.5mm wide; sheaths cream. Infl. a linear panicle, 4.5–6.5 × 0.3cm; lateral spikes c.7 appressed, upper overlapping, lower slightly distant;

lowest bract with filiform blade shorter than infl. Lateral spikes linear, lowest to 1.2cm, pale fawn, androgynous, all simple. Spikelets all single-flowered. Female 'glumes' ovate, subacute to acute, encircling axis, 2.5–2.7 × 1.6–2.4mm, pale fawn-hyaline speckled purplish, margins white-hyaline. Prophyll oblong, blunt, c.3 × 0.5mm, pale brown-hyaline, speckled purplish, keels minutely ciliate. Stigmas very slender. Nut linear-oblong, 2–2.5 × 0.5–0.7mm, pale fawn. Racheola about equalling prophyll, rather wide, 2-veined.

Sikkim (Cho-le-la); **Chumbi** (Chubithang). Marshes, 3810m. Fl./fr. June.

8. K. fragilis C.B. Clarke; *K. curvata* C.B. Clarke; *Carex curvata* Boott non Knaf. Fig. 33c.

Densely tufted. Culm curved to erect, trigonous, 5–45cm, 0.3–0.6mm wide. Leaves basal and sub-basal, V-shaped, becoming inrolled, blades shorter than to equalling culm, 0.9–1.5mm wide, not keeled; sheaths persisting as fibres, striate, pale brown, dull. Infl. a linear, spike-like panicle, often curved, 1–4.5 × 0.3–0.7cm; lateral spikes 2–5, short, just overlapping or lowest slightly distant, lower sometimes branched; lowest bract with hyaline clasping base and filiform blade equalling to greatly exceeding infl., bracts of lateral spikes sometimes also with long, filiform blades. Lateral spikes androgynous, with 1–4 single-flowered male and 2–7 single-flowered female spikelets. Female 'glumes' ovate, rounded to acute, mucronate, 2–3 × 1.4–1.8mm, hyaline becoming tinged yellowish-brown, midrib green. Prophyll utriculiform, curved, lanceolate, open only in upper part, 2–3.5 × 1–1.2mm, hyaline becoming pale brown, with 2 green, glabrous keels. Nut scarcely stipitate, narrowly ellipsoid-trigonous, 1.5–1.9 × 1mm, brown, shining. Racheola equalling to slightly exceeding prophyll, broad, 2-veined, ciliate.

Bhutan: C — Thimphu (Pajoding, above Ragyo, mountain E of Thimphu), Bumthang (above Gortsam) and Mongar (Sengor) districts; **Sikkim** (Tungu, Lachen, Tsomgo, Kopup, Karponang, Kyanglasha, Deosa, Nathula, Dzongri,

---

FIG. 33. **Cyperaceae** VIII (**Kobresia** prophylls).
o abaxial view of prophyll. + adaxial view of prophyll showing degree of openness. * nut (+ racheola in female-only spikelets).
a, **K. laxa**: female spikelet from lower part of lateral spike (× 6). b, **K. clarkeana**: female spikelet from lower part of lateral spike (× 12). c, **K. fragilis**: female spike from lower part of lateral spike (× 12). d, **K. curvirostris**: female spike from middle part of lateral spike. oo showing lateral view of prophyll (× 12). e, **K. schoenoides**: androgynous spikelet (× 6). f, **K. humilis**: androgynous spikelet. Nut immature in * (× 6). g, **K. sargentiana**: androgynous spikelet. Nut immature in * (× 6). Drawn by Mary Bates.

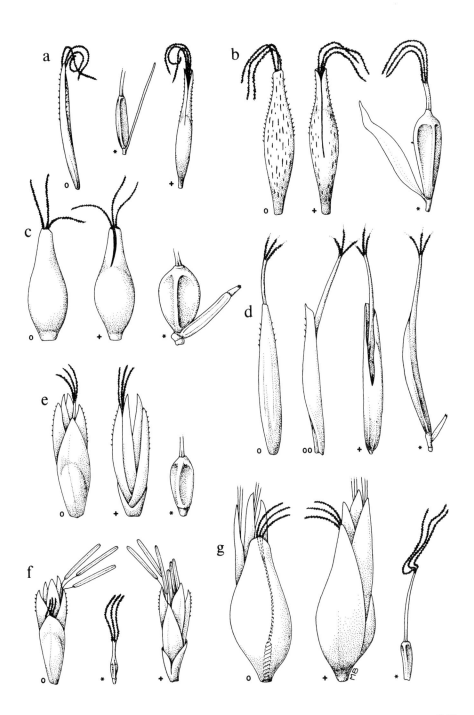

Mon Lapcha, Jamlinghang to Bikbari). Open grazed hillsides; pathside in rhododendron scrub, 2590–4270m. Fl./fr. June–September.

Variable in habit, specimens from grazed pasture having short, curved stems and infls., ones from ungrazed habitats taller with straight stems and infls.

9. K. curvirostris (C.B. Clarke) C.B. Clarke. Fig. 33d.

Slender, densely tufted. Culm curved or erect, trigonous, 5–27cm, c.0.5mm wide. Leaves basal and sub-basal, blades shorter than to exceeding culm, 0.7–1.5mm wide, keeled; sheaths persistent, dark brown, dull. Infl. a linear, spike-like panicle, outline fimbriate, 1–5 × 0.3–0.7cm; lateral spikes 2–9 short, overlapping, appressed; lowest bract with filiform blade shorter than or slightly exceeding infl. Upper spikelets single-flowered, male, middle spikelets single-flowered, female, lowest few spikelets androgynous with 1 female and 1 male flower within prophyll. Female 'glumes' oblong-ovate, acuminate, 1.7–2 (excl. mucro) × 1–1.4mm, pale brown-hyaline, midrib green, excurrent as mucro to 0.7mm. Prophyll curved, narrowly oblong, open to about halfway, 2.3–3.3 × 0.5–0.6mm, smooth, keels ciliate. Nut curved, linear, 1.7–2.3 × 0.4–0.5mm, pale brown, gradually narrowed into long (about equalling nut) beak protruding from prophyll. Racheola shorter than nut.

Bhutan: N — Upper Mo Chu district (above Kohina); **Sikkim** (Tungu). Wet rocky slopes and wet mossy cliffs, 3290–3960m. Fl./fr. July–September.

10. K. schoenoides (C.A. Meyer) Steudel; *K. deasyi* C.B. Clarke. Fig. 33e.

Densely tufted. Bases of leaf sheaths persistent, stiff, golden- to dark-brown, shining, non-fibrillose. Culm subterete, (4–)10–70cm, stout (1.7–2.5mm wide). Leaves basal, erect (sometimes curved), blades semicircular in section, channelled above, strongly inrolled on drying, shorter than to exceeding culm, c.2.5mm wide, not keeled. Infl. a dense, clavate spike, 1.5–3.7 × 0.6–0.8cm, with many spikelets borne directly on axis; lowest bract with glume-like clasping base about equalling lowest spikelet and very short filiform tip. Spikelets all androgynous, with 1 basal female and 3(–4) male flowers within prophyll. 'Glumes' oblong to lanceolate, usually rounded, 3.5–5 × 1.8–3mm, brown sometimes with narrow hyaline margins at apex; midrib broad, green, usually stopping short of apex, sometimes continued as short, serrate mucro. Prophyll lanceolate to narrowly elliptic, rounded, open to base, (3.2–)3.8–5.1 × 0.9–1.7mm, glabrous, scarcely keeled, brown above pale below. Nut ellipsoid- to narrowly obovoid-trigonous, apiculate, 1.7–2.8 × 1–1.2mm, finally grey.

Bhutan: C — Thimphu (below Darkey Pang Tso, Tremo La, between Pajoding and the lakes) and Sakden (Orka La) districts; **N** — Upper Mo Chu (Chawa Gassar to Seanchu Passa; Chebesa to Lingshi (F.E.H.2)) and Upper

Bumthang Chu (Kantanang) districts; **Sikkim** (Samding, Tungu, Chakung Chu, Chumegata, Lhonak, Chholhamu, Jelep La, Kopup, Natu La, Dikchu, Jemathang); **Chumbi** (Phari). Open boggy ground among rhododendrons; shady rocks; bogs; open grazed slopes; moist sedge moorland; edge of sandy pasture, 3800–5790m. Fl./fr. May–September.

11. K. humilis (C.A. Meyer) Sergievskaja; *K. royleana* (Nees) Boeckeler var. *humilis* (C.A. Meyer) Kükenthal. Fig. 33f.

Differs from dwarf forms of *K. schoenoides* chiefly vegetatively, lacking the prominent collar of persistent sheath bases and in the short, recurved leaves with flat blades, V-shaped in section, symmetrical about the keeled midrib.

Sikkim (Samding, Kareng-Chholhamu Plain). 4877m. Fr. August–September.

12. K. sargentiana Hemsley. Fig. 33g.

Densely tufted. Bases of sheaths pale brown, shining, fibrillose, persistent. Culm curved, 5(–13)cm. Leaves basal, blades stiffly curved, filiform, semicircular in section, channelled above, exceeding culm, c.1.2mm wide, not keeled. Infl. a dense spike, (2–)2.3(–2.5) × 0.4(–0.6)cm; most spikelets androgynous with 1 female and 3 male flowers within prophyll (upper few spikelets probably single-flowered and male); lowest bract glume-like. Prophyll broadly elliptic, open only in upper ⅓, 6.5(–7) × 2.5–2.7mm, hyaline, glabrous. 'Glumes' broadly ovate, rounded, encircling axis, (7.4–)7.8–8 × 5.5–6mm, midrib green 3–7-veined, sides orange-brown, margins very broadly (0.7–2mm) hyaline.

Chumbi (Tuna). Wet places, 4490m. Fl. May.

Very distinctive on account of its glumes, but poorly known; treated by Kükenthal as a var. of *K. robusta* Maximowicz, but probably better kept distinct.

13. K. capillifolia (Decaisne) C.B. Clarke. Fig. 34a.

Densely tufted, tufts surrounded by collar of shining, chocolate-brown sheaths with fibrillose margins. Culm stiffly erect, subterete, 9–36cm, 0.9–1.3mm wide. Leaves basal, blades stiffly erect, filiform, tubular in section, channelled above, shorter than to exceeding culm, c.0.8mm wide, not keeled. Infl. a narrow, dense spike of many spikelets, 1.1–2 × 0.1–0.6cm; lowest bract glume-like or with filiform tip shorter than spike. Spikelets all androgynous, with 1 female and (1–)3–4 male flowers within prophyll. 'Glumes' broadly oblong-lanceolate, rounded, encircling axis, 4.5–6 × 2.5–2.8mm, brown with broad hyaline margins and 3-veined, green midrib. Prophyll lanceolate, rounded, open to base, 2.7–4 × 0.8–1.2mm, keels weak, glabrous, hyaline tinged pale brown. Nut narrowly ellipsoid- to obovoid-trigonous, apiculate, 2–2.5 × 0.8–1.4mm, brown to dark brown, shining.

Bhutan: C — Thimphu district (hill above Thimphu Hospital, mountain E of Thimphu); **N** — Upper Mo Chu district (Lingshi); **Sikkim** (Chholhamoo, Gurudongmar, Giagong Plain). Dry, open, grassy mountain-top; meadow in *Abies* forest, 3480–3960(–5300)m. Fl./fr. July–August.

Our specimens tend towards var. *tibetica* (Maximowicz) Kükenthal which has longer, more stiffly erect stems and shorter, much denser spikes than the type. The glumes of our specimens are as in the type, being larger than in var. *tibetica*.

A specimen from Pajoding (*Wood* 5534) is probably an abnormal form of *K. capillifolia* from which it differs in having longer (to 4cm), laxer spikes, the filiform tip of the lowest bract slightly exceeding the spike and the lower glumes long-aristate.

14. K. duthiei C.B. Clarke. Fig. 34b.

Rhizome short, stems densely tufted. Culm stiffly erect, subterete, 4–29cm, slender (0.6–0.8mm wide), with basal bladeless sheaths. Leaves basal, blades flat, under half length to almost equalling culm, 1.4–2.3mm wide, keeled; sheaths striate, pale- to reddish-brown, dull; old leaves persistent. Infl. a dense, linear spike, 1–4.5 × 0.2–0.4cm, greenish when young, finally very pale brown; lowest bract usually glume-like, occasionally very shortly aristate. Spikelets all androgynous, with one male and one female flower within prophyll. Prophyll narrowly oblong to narrowly oblanceolate, open to less than halfway, 3.1–4.5 × 0.7–1mm, keels ciliate. 'Glumes' ovate, subacute, minutely ciliate at apex, 2.5–4.5 × 1.6–2.6mm, midrib broad, 3-veined, green, margins pale brown-hyaline. Nut stipitate, oblong, 2.1–2.8 × 0.6–0.9mm, pale brown, beak long (0.5–1.5(–2.5 in fr.)mm).

Bhutan: C — Thimphu district (above Pajoding, below Darkey Pang Tso, above Talukah Gompa); **N** — Upper Mo Chu district (Yale La); **Sikkim**

FIG. 34. **Cyperaceae** IX (**Kobresia** prophylls).
o abaxial view of prophyll. + adaxial view of prophyll showing degree of openness. * nut (+ racheola in female-only spikelets).
a, **K. capillifolia**: androgynous spikelet. Nut removed from + + to show two male flowers within prophyll (× 9). b, **K. duthiei**: androgynous spikelet. Single male flower and nut dissected out from prophyll in * (× 9). c, **K. nepalensis**: female spikelet from lower part of spike (× 9). d, **K. stiebritziana**: female spikelet from lower part of spike (× 9). e, **K. pygmaea**: female spikelet from lower part of spike (× 18). f, **K. esenbeckii**: female spikelet from lower part of spike. Lateral view shown in oo (× 9). g, **K. woodii**: female spikelet from female-only spike (× 9). h, **K. vaginosa**: female spikelet from female-only spike. Nut immature in * (× 9). i, **K. vidua**: female spikelet from female-only spike (× 18). j, **K. prainii**: female spikelet from female-only spike (× 18). Drawn by Mary Bates.

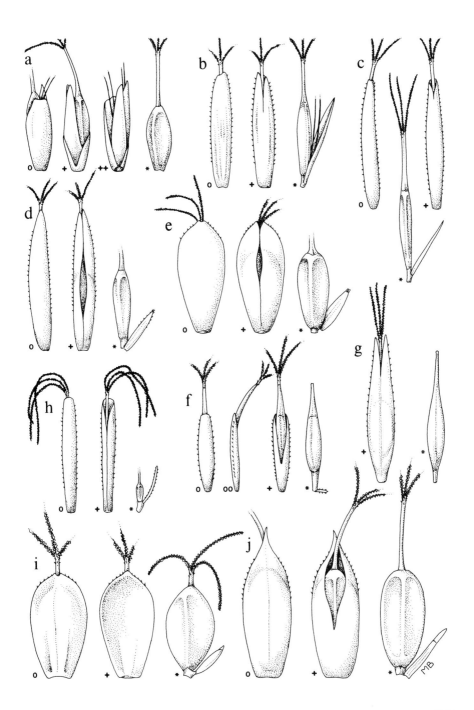

347

(Jelep La, Natu La, Bikbari, Dzongri to Prek Chu, Thangshing to Lam Pokhri, Jemathang to Goecha La). Bare mossy peat; wet, open flushes; exposed, dry, rocky ridge; north-facing grassy slopes, 3920–4600. Fl./fr. August–September.

Some specimens from Sikkim, namely Kopup (*Bor's Collector* 535, K) and Jelep La (*Bor's Collector* 640, K), differ from the above in being more robust (stems 29–44cm), with longer (3–5.2cm), laxer spikes, lower glumes long aristate and spikelets each with 2 male flowers, but are probably just large forms of *K. duthiei*.

15. K. nepalensis (Nees) Kükenthal; *Carex linearis* Boott. Fig. 34c.

Extremely densely tufted; tufts surrounded with collar of yellowish- to dark brown, slightly shining, strongly fibrillose sheath bases. Culm stiffly erect, subterete, 13–41cm, 0.7–1mm wide. Leaves basal, blades stiffly erect, filiform, semicircular in section, channelled above, about equalling culm, 0.4–0.7mm wide, not keeled. Infl. a linear spike, dense above, laxer below, 3.5–6.6 × 0.3–0.4cm, androgynous, with short, apical male section; lowest bract glume-like, sometimes slightly longer aristate. Spikelets all single-flowered, upper few male, majority female. Female 'glumes' ovate to lanceolate, subacute to blunt, usually mucronate, 3.3–5 (excl. mucro) × 2–2.4mm, midrib green, broad, with 1 strong central nerve and sometimes also 2 weak lateral ones, sides pale brown, margins hyaline especially near apex; midrib usually produced from below apex as serrate mucro ((0–)0.6–1.2mm). Prophyll utriculiform, curved, linear-lanceolate, open only near apex, 4.8–6.5 × 0.8–1.1mm, keels ciliate. Nut stipitate (stipe c.0.5mm), body oblong, 2.2–3.2 × 0.7–0.9mm, pale brown; beak 1.5–2.2mm, stout, tapering from broad base. Racheola linear, reaching about base of beak, 2-veined, ciliate.

Bhutan: C — Thimphu (below Darkey Pang Tso, above Phajoding) and Sakden (Orka La) districts; **N** — Upper Mo Chu district (Shodu); **Chumbi** (Yatung); **Sikkim** (Lachen, Chakung Chu, Natu La, Jelep La, very common in W Sikkim). Dry grassy slopes; exposed rocky and stony ridges; moraines; on mossy rocks and damp cliff-ledges, 3700–4270m. Fl./fr. June–October.

Var. *elachista* (C.B. Clarke) Kükenthal is only a dwarf growth form (stems 1.5–5cm; spikes 0.7–3cm). At first sight, however, it looks very different from *K. nepalensis* and more like *K. prainii* from which it differs in its dull, fibrillose sheaths, presence of terminal male flowers (though few and inconspicuous) and in its lanceolate, gradually tapering prophyll and very long-beaked, linear nut.

16. K. stiebritziana Handel-Mazzetti. Fig. 34d.

Superficially similar to *K. nepalensis* (and equally variable in stature), differing as follows: spike stouter, glumes narrower, acute, not mucronate, sides darker brown, margins more widely hyaline; prophyll open almost to base at maturity, with prominent midrib on abaxial face; racheola shorter (less than half length of nut); beak of nut shorter (0.5–1mm).

Bhutan: C — Thimphu district (between Pajoding and the lakes, below Laname Tso); **Sikkim** (Jelep La, E of Bikbari, Chuanrikiang, above Thangshing, Jemathang). Open gravelly and rocky slopes; sandy moraines; sedge-dominated, seasonally flooded moorland, 3900–4500m. Fl./fr. July–September.

17. K. pygmaea (C.B. Clarke) C.B. Clarke. Fig. 34e.

Extremely densely tufted, forming tough mats. Bases of leaf sheaths pale brown, dull, striate, persisting as fibres. Culm stiffly erect, subterete, 1–4(–6)cm, c.0.4mm wide. Leaves basal, blades erect, tubular-filiform, channelled above, about equalling culm, c.0.4mm wide, not keeled. Spikelets all single-flowered. Infl. a dense spike, 3–6.5 × 1–2mm; androgynous, male glumes apparently deciduous, so spikes often appearing entirely female by fruiting stage, female spikelets 3–6; lowest bract glume-like or sometimes shortly aristate. Female 'glumes' ovate, acute, encircling axis, 3–3.5 × 2.2–2.5mm, midrib broad, green, with 1 strong central nerve, sides orange-brown; male 'glumes' c.4 × 1.8mm, shape and colour as female. Prophyll narrowly oblong-elliptic, open to base, 2–2.5 × 1–1.2mm, keels glabrous or minutely ciliate at apex. Nut not stipitate, obovoid-trigonous, apiculate, c.1.7 × 1mm, shining, pale brown. Racheola very short (much shorter than nut), broad, not ciliate.

Bhutan: N — Upper Mo Chu district (Gangyuel to Lingshi, Chawa Gassar to Seanchu Passa; Yabu Thang to Laya (F.E.H.2)); **Sikkim** (Momay, Kongra Lama, Chaunrikiang, Jemathang, Llonak); **Chumbi** (Lingmatang, Chumolari, Phari). Exposed grassy ridge; on top of walls and rocks; bare turf by paths; boggy soil, 3350–4880m. Fl./fr. May–September.

18. K. esenbeckii (Kunth) Noltie var. **esenbeckii**; *K. trinervis* Boeckeler *nom. illegit.*; *K. hookeri* Boeckeler; *K. hookeri* var. *dioica* C.B. Clarke; *K. angusta* C.B. Clarke. Fig. 34f.

Rhizomes short, stems extremely densely tufted. Bases of old leaf sheaths persistent, pale- to dark brown, slightly shining, sometimes forming a collar to half length of stem, eventually becoming fibrous. Culm trigonous, scabrid on angles, 4–33(–38)cm, slender (0.3–0.7mm wide). Leaf blades stiffly erect, flat, margins serrate, about equalling culm, (1–)1.4–3.5mm wide, keeled. Infl. spicate, unisexual or androgynous; plants sometimes bearing male and female, or male, female and androgynous spikes or plants apparently bearing spikes of only one sex (i.e. dioecious); all spikelets single-flowered. Female and androgynous spikes linear, 1.7–4.5 × 0.15–0.4cm, androgynous spikes primarily composed of rather lax female spikelets, with few terminal male flowers; female 'glumes' oblong to narrowly ovate, rounded to acute, especially lower

349

glumes often mucronate (mucro to 0.5mm), 2.2–4 (excl. mucro) × 1–2mm, midrib narrow, green, sides and margins orange-brown. Male spikes very dense, (1.7–)3.5–5 × 0.4cm; male 'glumes' oblong, blunt, with wide hyaline margins, (5–)7–8.5 × 1.6–2mm. Prophyll oblong, blunt, initially open only near apex, splitting to base when nut matures, 1.9–3.5 × 0.4–0.7mm, pale brown-hyaline, keels ciliate. Nut stipitate (0.4–0.7mm), finally curved, long exserted from prophyll at maturity, giving spike fimbriate outline, body linear-ellipsoid, 2.5–3.4 × 0.5–0.7mm; beak straight or curved, (1–)1.4–2mm.

Bhutan: C — Thimphu (Darkey Pang Tso, Laname Tso, between Pajoding and the lakes) and Bumthang (above Gortsam) districts; **N** — Upper Mo Chu (Laya) and Upper Kulong Chu (Shingbe) districts; **Darjeeling** (Sandakphu, Phalut Top, Tonglu Top); **Sikkim** (very common in W Sikkim, Lachen, Lachung, Singalila, Namdee, Changu, Lhonak, Tsomgo, Tuko La, Natu La, Jelep La, etc.); **Chumbi** (Gantsa). Wet rocks and cliff-ledges; grazed slopes; lake shore; on trees, 3500–4270(–4850)m. Fl./fr. May–October.

var. **fissiglumis** (C.B. Clarke) Noltie.

Differs from small forms of *K. esenbeckii* with unisexual spikes in having prophylls open to base even when young, keels smooth; female glumes usually oblong, truncate, with distinct hyaline apex.

Sikkim (W of Chemathang, Bikbari to Chaunrikiang, Thegu). On rocks and cliff-ledges, 4250–4300m.

Distinctive extremely dwarf, apparently truly dioecious, forms from glacial debris at extreme altitudes (4550–4850m) in Sikkim are almost certainly growth forms (as occur in other high-altitude species of *Carex* and *Kobresia*) and are not worth taxonomic recognition.

19. K. woodii Noltie. Fig. 31h–l, Fig. 34g.

Densely tufted perennial. Upper parts of leaf sheaths straw-coloured, lower parts chocolate brown with darker margins, slightly shining, persistent, not fibrillose. Leaf blades semicircular to V-shaped in section, slightly exceeding culms, c.1.5mm diameter, scarcely keeled. Culm ± terete, 24–28cm, c.1mm diameter. Infls. spicate, linear, unisexual, male and female borne on same plant; all spikelets single-flowered. Female infl. 6 × 0.6cm, lower spikelets slightly distant; female glumes oblong-ovate, blunt, 6.5 × 4mm, brown, margins narrowly hyaline, midrib wide, green, 1-veined; lowest bract glume-like, shortly aristate. Male infl. c.4 × 0.3cm; male glumes oblong-oblanceolate, rounded, c.8 × 2.5mm, margins narrowly hyaline. Prophyll linear oblong, open to base, 7–9 × 1mm, keels minutely scabrid, apex brown. Nut linear: stipe c.1mm, body c.2.5 × 0.7mm, beak c.1.5mm; style c.2.5mm; stigmas 3, c.3mm.

Bhutan: C — Thimphu district (below Pajoding). Grassland, possibly seasonally burnt, in upper forest zone. 3300m.

20. K. vaginosa C.B. Clarke. Fig. 34h.

Forming dense swards. Sheath bases yellowish-brown, dull, fibrillose. Culm slender, 3.5–10cm. Leaves flaccid, equalling to exceeding culm, filiform, c.0.5mm wide. Infl. spicate, spikes unisexual (occasionally gynaecandrous), some individuals bearing both male and female spikes, some apparently bearing spikes of only one sex (i.e. dioecious); all spikelets single-flowered. Female spike linear, 2–3.5 × c.0.2cm, glumes appressed, lower rather distant; female glumes lanceolate, blunt to subacute, 5–6.2 × 1.7–2mm, midrib narrow, sides orange-brown, margins broadly hyaline. Male spikes linear, 1.5–3 × c.0.2cm; male glumes oblong-lanceolate, 6–8 × 1.2–2mm, colour as female. Prophyll linear, apex hyaline, open to base, 4.7–5.7 × 0.7mm, margins brown, keels minutely ciliate. Racheola to half length of prophyll.

Sikkim (E of Bikbari, Momay). Dry slopes, where sometimes dominant, 4300–4877m. Fl./fr. July–September.

21. K. vidua (Boott ex C.B. Clarke) Kükenthal; *Carex vidua* Boott ex C.B. Clarke. Fig. 34i.

Forming dense swards. Sheath bases brown to chocolate brown, shining, persistent, non-fibrillose. Culm stiffly erect, 1.5–3.5cm at flowering, female elongating in fruit (to 14cm), c.0.9mm wide. Leaves basal, blades semicircular in section, margins serrate, shorter than to exceeding culm, filiform, 0.5–1.1mm wide, not keeled. Infls. spicate, usually unisexual, occasionally androgynous, some individuals bearing spikes of both sexes, some apparently only one (i.e. dioecious); spikelets single-flowered; lowest bract glume-like. Female spikes linear, (0.9–)1.7–2.7 × 0.2–0.4cm, glumes appressed; female glumes oblong-ovate, rounded, 2–5 × 1.3–2.4mm, midrib very broad (c.⅓ width of glume), green, usually stopping below apex, sides brown, margins narrowly hyaline. Male spikes narrowly ellipsoid, 0.7–1.7 × 0.2–0.3(–0.6)cm; male glumes oblong or oblanceolate, 5.5–7 × 1.6–3mm, midrib broad, sides orange-brown. Prophyll utriculiform, narrowly elliptic, open only at apex, 2.2–4 × 0.6–1.5mm, ciliate on keels especially when immature. Nut stipitate, body elliptic, 1.6–1.8 × 0.9–1.1mm, pale brown, beak stout, green, 0.4–1mm. Racheola lanceolate, about half length nut, not veined, not ciliate.

Bhutan: N — Upper Mo Chu (N side of Shingche La), Upper Bumthang Chu (Lambrang Chu) and Upper Kulong Chu (Shingbe) districts; **Sikkim** (Lachen, above Thangshing, Thangshing to Lam Pokhri); **Chumbi** (Lingmatang, Chumbithang). Dry soil on plain; on rocks; dry grassy, rocky or sandy slopes, 3350–4530m. Fl./fr. May–September.

22. K. prainii Kükenthal; *K. utriculata* C.B. Clarke. Fig. 34j.
Differs from *K. vidua* as follows: female glumes spreading obliquely, longer (4.5–6mm), narrowly lanceolate, very acute, midrib narrower (c.¼ width of glume); male glumes longer (5.6–7.5mm); nut longer (2.2–2.4mm), oblong-elliptic, minutely apiculate, not beaked.

Sikkim (Naku La, Muguthang); **Chumbi** (Phari, Yatung, Lingmatang; Kyoo La (Kükenthal, 1909)). On dry slopes and rocks, 3350–5270m. Fl./fr. May–September.

19. CAREX L.

Perennial herbs. Rhizomes short (stems tufted), or elongate and creeping. Leaves borne mainly on vegetative shoots, and/or mainly at base of culm, blades usually linear, bases sheathing. Culms (flower stems) commonly trigonous, upper part leafless or leafy, sometimes bearing bladeless sheaths at base. Plants monoecious (in Bhutan). Flowers unisexual each subtended by a glume; male flowers with usually 3 stamens; female flowers each of a single pistil enclosed within a utricle, stigmas 2 or 3, ovary developing into a biconvex or trigonous nut. Utricle commonly extended upwards into a beak, ending in an aperture through which stigmas project. Flowers arranged in spikes which may be entirely female, entirely male, or mixed (gynaecandrous when female above male; androgynous when male above female); spikes arranged in panicles or racemes or sometimes single. Bracts subtending spikes or partial infls. commonly leaf-like; spikes (or peduncles) usually subtended by a glume-like or utricle-like 'cladoprophyll', which may be hidden by, or appear as 'auricles' to, the subtending bract.

No satisfactory infrageneric treatment has been produced for the Himalayan species and none has therefore been used in the following account. Closely related species are

---

FIG. 35. **Cyperaceae X (Carex** habits and inflorescence types).
a, **C. nubigena**: habit (× ⅓), showing inflorescence of sessile spikes characteristic of subgenus *Vignea* (spp. 1–7). b, **C. rara**: habit (× ⅓), showing infl. of a single spike characteristic of subgenus *Primocarex* (spp. 19–21). c, **C. filicina**: habit (× ¼), showing paniculate inflorescence characteristic of Section *Vigneastra* (spp. 23–30), and basal bladeless sheaths. d, **C. myosurus**: habit (× ¼), showing paniculate infl. of large spikes and fibrillose leaf sheaths. e, **C. decora**: habit in male phase (× ¼), showing infl. of fascicles of spikes characteristic of Section *Oligostachyae* (spp. 36–41). f, **C. laeta**: habit (× ¼), showing infl. of terminal male and lateral female spikes, and old leaf-sheath bases persisting as fibres. Drawn by Mary Bates.

placed together, but Sectional names are only given in the case of certain well-defined groups with characteristic but complex infl. morphology.

Notes on key:

1. Only specimens with ripe fruit are identifiable unambiguously.

2. In attempting to determine the number of stigmas it is usually necessary to examine many utricles, since these fragile structures often break off. It is usually easier to say whether or not a specimen has 2 stigmas, since this is correlated with a lenticular (biconvex) or plano-convex utricle. (There is much more variation in utricle shape in the species with 3 stigmas.)

3. It is necessary to examine spikes very closely to determine disposition of the sexes, since by fruiting the conspicuous anthers are usually shed. However, it is nearly always possible to see the remains of filaments in male flowers.

4. *Androgynous* spikes are male at apex and female at base. *Gynaecandrous* spikes are female at apex and male at base. In many species, however, the disposition of sexes within spikes (especially the terminal one) is variable and not entirely reliable as a taxonomic character.

5. In many species the characters from the bases of the lower leaf sheaths are important. In some species the inner, membranous face splits to form a characteristic ladder-like structure of *fibrils* ('sheaths fibrillose') (Fig. 35d). The degree and form of persistence of these sheaths is also important. Some persist as membranous fragments, others remain as a collar of *fibres* (Fig. 35f). *Bladeless sheaths* (Fig. 35c) are linear to oblong, non-herbaceous structures borne at the base of flowering (sometimes on non-flowering) stems in some species and are often characteristically coloured.

6. Measurements for utricle length include the beak. Measurements for glumes include any point or mucro except for a few stated exceptions where the awn is exceptionally long.

1. Stigmas 2, utricles biconvex or plano-convex 2
+   Stigmas 3, utricles usually trigonous (though hard to see in species with small nuts and/or swollen utricles) 19

2. Spikes all sessile, infl. a dense capitate head or a spike-like panicle (interrupted to various degrees) 3
+   At least the lower spikes distinctly peduncled 9

3. Infl. capitate or an interrupted spike-like panicle with 5 or fewer lateral spikes 4
+   Infl. of more than 5 spikes, densely cylindrical (at least in upper part)   5

4. Infl. densely capitate, spikes all similar, predominantly female with a few apical male flowers .... **1. C. pseudofoetida** subsp. **afghanica**
+   Infl. a short, interrupted spike-like panicle; terminal spike male at base, others entirely female .... **2. C. echinata**

5. Lowest bract filiform, short (about equalling lowest spike) 6
+   Lowest bract leaf-like, equalling or exceeding infl. 7

6. Infl. green in fruit; leaves wide (over 5mm) .... **6. C. foliosa**
+   Infl. brown in fruit; leaves narrow (under 5mm) .... **3. C. diandra**

7. Infl. brown in fruit; stem leafy throughout, leaves greatly exceeding
    stem .... **5. C. thomsonii**
+   Infl. green in fruit; stem leafless in upper part, leaves about equalling
    stem 8

8. Spikes narrow at maturity (under 4mm wide); terminal spike male at
    base or entirely male, other spikes entirely female; margins of beak
    minutely serrate .... **7. C. remota**
+   Spikes ovoid at maturity (over 5mm wide); all spikes male at apex
    (hard to see when utricles mature); margins of beak smooth
                                                              4. C. nubigena

9. Spikes all androgynous 10
+   Terminal spike male (sometimes with female flowers at top or in centre),
    others female 12

10. Spikes stout, peduncles borne singly at infl. nodes (though spikes
    sometimes shortly branched at base); utricles over 5mm .. **8. C. longipes**
+   Spikes small, slender, more than one peduncle per infl. node or spikes
    borne on branched partial panicles; utricles under 4mm 11

11. Utricle abruptly contracted into long (about equalling body of utricle),
    slender beak; stigma lobes very slender, exceeding utricle
                                                              9. C. longicruris
+   Utricles gradually narrowed into short (c.half length of body of utricle)
    beak; stigma lobes stouter (tending to recurve), shorter than utricle
                                                                 10. C. lenta

12. Female glumes blunt to acute, or minutely mucronate (in which case
    mucro smooth), blackish 13
+   Female glumes with midrib long-excurrent as serrate point, if acute
    then pale reddish-brown 16

13. Utricles widely ellipsoid (over 1.4mm wide), beakless, becoming fus-
    cous-purple; female spikes usually under 2cm; stem under 21cm
                                                              11. C. orbicularis

+ Utricles narrower, sometimes beaked, not becoming purplish; female
  spikes over 2cm; stems usually over 20cm 14

14. Utricles beakless; male spike long-peduncled, remote
    14. C. nigra subsp. **drukyulensis**
+ Utricles beaked; male spike overlapping with female 15

15. Utricles abruptly contracted into beak, under 2.9mm; beak notched;
    glumes dark purplish-brown .... **12. C. notha**
+ Utricles gradually narrowed into beak, usually over 3mm; beak not
  notched; glumes black .... **13. C. fucata**

16. Utricles distinctly beaked; stigmas reddish, very long, persistent
    15. C. rubro-brunnea
+ Utricles beakless; stigmas short, not persistent 17

17. Faces of utricles densely brown-glandular, margins green   **16. C. phacota**
+ Utricles greenish 18

18. Female spikes short (under 2.5cm), fat, pendent on slender peduncles
    17. C. pruinosa
+ Female spikes usually over 4cm, linear-cylindric, peduncles short,
  ascending .... **18. C. teres**

19. Infl. a single, terminal spike 20
+ Infl. of several to many spikes 23

20. Bract exceeding spike; utricles hairy .... **22. C. radicalis**
+ Lowest bract glume-like, much shorter than spike; utricles glabrous . 21

21. Utricles ellipsoid, small (under 2.5mm), spreading at maturity
    19. C. rara
+ Utricles lanceolate, large (over 4mm), strongly deflexed at maturity . 22

22. Utricle with bristle projecting through aperture, so stigmas appearing
    lateral .... **20. C. microglochin**
+ Utricle without bristle projecting through aperture, stigmas appearing
  terminal .... **21. C. parva**

23. Infl. a branching panicle with spikes borne directly on major primary
    branches which may be branched again (Fig. 35c, d) 24

+ Infl. with spikes borne singly or in groups of 2–11 directly from nodes on flower stem (peduncles occasionally bearing more than one spike) (Fig. 35e, f) 34

24. Spikes small (mostly under 1cm) (Fig. 35c) 25
+ Spikes large (over 1cm) (Fig. 35d) 32

25. Veins at bases of leaf sheaths darkened, contrasting with paler background; utricles glabrous, faces each with 7 or more close, parallel, evenly thickened veins, rather fat, abruptly contracted into beak; glumes cream to pale brown 26
+ Veins at bases of leaf sheaths not contrasting with usually brownish background; utricles hispid or glabrous, with fewer veins per face (if swollen, veins unequally thickened); glumes brown to cinnamon-coloured (pale usually only when immature) 27

26. Leaves thin-textured, usually over 1cm wide; filiform tips of bracteoles short (under 2.5mm); male section of spikes short (under 8mm) **23. C. stramentita**
+ Leaves coriaceous, under 1cm wide; filiform tips of bracteoles long, conspicuous; male section of spikes commonly over 1cm .. **24. C. indica**

27. Utricle beak with oblique aperture, not bidentate; spikes distant and stiffly divaricate .... **25. C. filicina**
+ Utricle beak sharply bidentate (if oblique then utricle swollen and whitish); spikes not stiffly divaricate 28

28. Male section of spikes very small (usually under 4mm), male glumes under 2.5mm 29
+ Male section of spikes conspicuous (over 5mm), male glumes over 3mm 30

29. Utricle glabrous, swollen, whitish in life, beak ± straight **26. C. cruciata** var. **argocarpa**
+ Utricle hispid, not swollen, brownish-green, beak curved **30. C. continua**

30. Basal sheaths not fibrillose; branchlets of partial infls. densely congested; utricles exceeding glumes, beak long (c.1mm) curved **27. C. condensata**
+ Basal sheaths fibrillose; branchlets of partial infls. lax; utricles about equalling glumes, beak short (c.0.5mm), straight 31

31. Branches of partial infls. erect; male glumes c.4mm, acute; female glumes not or minutely mucronate; fruits greenish-brown, fr. July–November .... **28. C. vesiculosa**
+ Branches of partial infls. filiform, flexuous; male glumes under 3.5mm, mucronate; female glumes long-mucronate; fruits suffused chestnut; fr. April–May .... **29. C. burttii**

32. Utricles swollen, whitish when immature, finally red and fleshy, glabrous; secondary branches not developed so spikes stiff and congested on panicle branches .... **34. C. baccans**
+ Utricles not swollen, brownish-green, not fleshy, usually hispid 33

33. Utricles over 2.6mm, ribbed, glabrous to sparsely hispid, gradually tapered into beak; spikes commonly drooping .... **32. C. myosurus**
+ Utricles c.2.3mm, not ribbed, densely hairy, abruptly contracted into beak; spikes erect .... **33. C. composita**

34. Infl. nodes each bearing 3 or more peduncles, each peduncle usually bearing a single (occasionally more) spike 35
+ Peduncles single or in pairs at infl. nodes, each peduncle sometimes bearing more than one spike 41

35. Spikes ellipsoid .... **35. C. polycephala**
+ Spikes linear 36

36. Stem leafy throughout, leaves gradually passing into bracts
37. C. insignis
+ Leaves basal and sub-basal, distinct from bracts 37

37. Utricles hispid 38
+ Utricles glabrous (or with a few hairs on margins of beak) 39

38. Bases of leaf sheaths golden-brown .... **38. C. daltonii**
+ Bases of leaf sheaths dark chestnut-brown .... **39. C. crassipes**

39. Plant stout, leaves over 0.5cm wide; utricles over 4mm, beak over 1.6mm .... **36. C. decora**
+ Plant slender, leaves under 0.3cm wide; utricles under 3.5mm, beak under 1mm 40

40. Female glumes with broad hyaline apex, midrib excurrent; utricles chestnut-brown, over 2.9mm; spikes 2–4 per fascicle ... **40. C. anomoea**
+ Female glumes scarcely hyaline at apex, midrib not excurrent; utricles olive-brown, under 2.7mm; spikes 3–11 per fascicle .... **41. C. pulchra**

41. Utricles glabrous on faces (margins of beak sometimes hispid), if hairy
   above then terminal spike gynaecandrous and other spikes female ...   42
+ Utricles hairy (sometimes minutely) on at least one face   68

42. Spikes very dark, ± ellipsoid (except sometimes terminal one), glumes
   fuscous-purple or blackish, not aristate   43
+ Spikes greenish or olive- to orange-brown, if dark then female ones
   linear to cylindric, or glumes aristate   55

43. Terminal spike gynaecandrous, others female   44
+ Terminal spike male or androgynous (if gynaecandrous then other
   spikes small and drooping)   52

44. Rhizomes spreading   45
+ Rhizomes short, stems densely tufted   46

45. Lower spikes large (over 1cm), nodding; stems and rhizomes stout
                                                        48. C. atrofusca
+ Lower spikes smaller under 0.9cm, sessile, erect; stems and rhizomes
   slender .... **42. C. gracilenta**

46. Utricles hispid (sometimes minutely) above
                                          47. C. obscura var. **brachycarpa**
+ Utricles completely glabrous   47

47. Utricle with distinct (over 0.5mm), deeply notched beak
                                                      44. C. psychrophila
+ Utricle not beaked, or with extremely short (under 0.2mm) beak   48

48. Terminal spike distant, lower spikes pendent; leaves c.1.4mm wide
                                                    49. C. montis-everestii
+ Terminal spike overlapping with adjacent one; leaves over 2mm wide   49

49. Aperture of beak notched; blade of lowest bract filiform, much shorter
   than infl.   50
+ Aperture of beak entire; blade of lowest bract leaf-like, usually equalling to greatly exceeding infl.   51

50. Robust plant; stems over 19cm; spikes over 1.6cm; utricles over
   3.4mm .... **45. C. atrata** var. **atrata**
+ Small plant; stems under 18cm; spikes under 1.3cm; utricles under
   2.5mm .... **45. C. atrata** var. **glacialis**

51. Spikes under 0.9cm, oblong in outline, lowest erect; utricle obovoid; female glumes under 1.7mm .... **46. C. lehmannii**
+ Spikes over 1.2cm, narrowly elliptic in outline, lowest commonly pendent; utricle oblong-elliptic; female glumes over 2.2mm ... **43. C. duthiei**

52. Lower spikes drooping 53
+ Lower spikes erect or sessile 54

53. Stems densely tufted; leaves c.1.4mm wide; female glumes shining, acute; utricle beakless, papery, shining .... **49. C. montis-everestii**
+ Stems not tufted; leaves over 2.5mm wide; female glumes dull, acuminate; utricle shortly beaked, semi-coriaceous, dull .... **48. C. atrofusca**

54. Upper spike male, female spikes under 1cm wide; glumes under 8mm
50. C. moorcroftii
+ Spikes all androgynous, globose, very large (over 1cm wide); female glumes over 8mm .... **51. C. praeclara**

55. Peduncles bearing a cluster of spikes 56
+ Peduncles each bearing a single spike 57

56. Spikes elongate (over 2cm), gynaecandrous and female; utricle beak c.1.5mm .... **57. C. fastigiata**
+ Spikes short (under 1cm), very densely clustered (superficially appearing as a single spike), male and androgynous; utricle beak c.0.5mm
31. C. oligostachya

57. Lower spikes female 58
+ Lower spikes androgynous, or if female then terminal spike androgynous and branched 66

58. Utricles large (over 4.8mm), inflated, shining; beak over 1.5mm, aperture very sharply bidentate .... **52. C. obscuriceps**
+ Utricles not as above 59

59. Female spikes greenish (glumes hyaline or pale brown) 60
+ Female spikes (utricles and/or glumes) darker 64

60. Spikes crowded near apex of stem 61
+ At least lower spikes distant 62

61. Female spikes sessile, subspherical, under 0.5cm .... **53. C. viridula**
+ Female spikes shortly peduncled, cylindrical, over 2cm
54. C. alopecuroides

62. Female spikes short (under 2.6cm), utricles gradually tapered to aperture, not beaked; female glumes with excurrent midrib .. **55. C. jackiana**
+ Female spikes long (over 2.5cm), utricles long-beaked (over 2mm); glumes subacute, midrib not excurrent 63

63. Leaves narrow, gradually tapered to apex; bases of leaf sheaths chestnut to reddish-brown .... **56. C. fusiformis** subsp. **fusiformis**
+ Leaves short, broad, abruptly contracted at apex; bases of leaf sheaths reddish-purple .... **56. C. fusiformis** subsp. **finitima**

64. Margins of beak setose; female spikes linear, very slender **59. C. setosa**
+ Beak smooth; female spikes cylindric 65

65. Leaves wide (usually over 1cm); lowest bract very shortly sheathing; utricle shortly (under 0.9mm) beaked, curved .... **60. C. olivacea**
+ Leaves under 6.6mm; lowest bract long (over 1.8cm) sheathing; beak long, commonly swollen in middle, utricle not becoming curved
61. C. oedorrhampha

66. Utricles large (c.7mm incl. beak), beak long, serrate; female glumes aristate .... **62. C. desponsa**
+ Utricles under 4mm, beak under 1.2mm, minutely hispid or smooth; female glumes subacute to acute 67

67. Terminal spike usually branched at base; female glumes small (under 2.5mm), pale brown, subacute .... **63. C. munda**
+ Terminal spike unbranched; female glumes over 3.2mm, reddish brown, very acute .... **65. C. inclinis**

68. All spikes androgynous, or terminal one(s) sometimes male 69
+ Terminal spike(s) male, others female 72

69. Female glumes greenish white, wide, blunt to subacute; utricles strongly trigonous; all spikes similar .... **64. C. speciosa**
+ Female glumes reddish-brown, very acute to aristate; utricles not strongly trigonous; terminal spikes commonly all male, lower androgynous 70

70. Rhizome condensed; bases of sheaths pale cream; spikes very narrow, rather lax .... **66. C. fragilis**
+ Rhizomes extensively spreading; bases of sheaths reddish-brown; spikes dense 71

71. Female glumes acute or minutely mucronate; utricles minutely hairy
65. C. inclinis
+ Female glumes long-aristate (awn over 1.2mm); utricles densely hairy
67. C. setigera

72. Bases of stems bearing reduced scale-like leaves, middle and upper part of stems leafy; auricles hairy .... **73. C. hebecarpa**
+ Bases of stems leafy; auricles not hairy 73

73. Rhizomes slender, spreading 74
+ Rhizomes short, woody, stems densely tufted 75

74. Female glumes and sheaths reddish; sheaths fibrillose
68. C. schlagintweitiana subsp. **deformis**
+ Female glumes and sheaths chocolate brown; sheaths not fibrillose
70. C. griersonii

75. Spikes subsessile 76
+ Most spikes on long, slender peduncles 77

76. Female glumes becoming tinged orange-brown, midrib not to minutely excurrent .... **69. C. inanis**
+ Female glumes white hyaline, midrib long-excurrent
71. C. breviculmis subsp. **royleana**

77. Utricle abruptly contracted into distinct (over 0.6mm), setose beak
59. C. setosa
+ Utricle not or extremely shortly (to 0.2mm) beaked 78

78. Male spike usually more than one (up to 7); utricles large (over 3.3mm), narrowly elliptic, tapered to deeply cleft aperture, upper part becoming fuscous .... **72. C. haematostoma**
+ Male spike single; utricles under 3.4mm, obovoid, aperture entire, greenish-brown .... **58. C. laeta** subsp. **laeta**

1. C. pseudofoetida Kükenthal subsp. **afghanica** Kukkonen. Fig. 36a.
Rhizome c.2mm diameter, branched, creeping extensively, covered with scales with dark brown sheaths and short, paler, strongly ribbed blades. Bases

of leaf sheaths pale, not persisting as fibres. Leaves sub-basal, blades deeply channelled, becoming inrolled, shorter than stem, 0.5–2mm wide, stiff. Culm 3–20.5cm, rather stout (1–1.5mm diameter). Infl. a densely congested head, ovoid to oblong (depending on posture of lower spikes), 0.7–1.4 × 0.4–0.8cm; spikes 3, androgynous, 5–8mm, utricles erect; lowest bract glume-like, encircling and tightly clasping infl., suborbicular, shorter than lowest spike, brown with hyaline margins; midrib broad, triangular, green, 1–3(–?8)-veined, shortly excurrent. Utricles very shortly stipitate, narrowly elliptic, plano-convex, c.3.1 × 1.3mm, smooth, glabrous, becoming pale golden; beak c.0.5mm, aperture hyaline, shallowly notched, margins smooth; stigmas 2. Female glumes suborbicular to ovate, subacute, 3.2–3.5 × 2.6mm, brown with wide (to 0.5mm) hyaline margins and green midrib.

Bhutan: C — Thimphu district (Shodu to Barshong (F.E.H.2); included with some trepidation since a voucher (BM) for other record in same reference is *Juncus thomsonii*!); **Sikkim** (Kareng-Chholhamu Plain); **Chumbi** (Yatung). Wet places; bed of backwater of river (short turf on slopes with moving groundwater in E Nepal), 3050–4780m. Fr. June–August.

Further work is needed on the distinction between this species and *C. stenophylla* Wahlenb.

2. C. echinata Murray. Fig. 36b.

Rhizomes short, stems tufted. Bases of leaf sheaths pale fawn; bladeless sheaths present at base. Leaves inserted on lower ⅓ of culm, blades shorter than to just exceeding culm, 0.8–1.4mm wide. Culm 9–25cm, stiffly erect. Infl. 1.7–2.1cm, spikes 3–5, evenly spaced, slightly overlapping, lowest internode under 0.7cm; lowest bract glume-like, shortly mucronate, shorter than lowest spike, hyaline. Terminal spike androgynous, male and female sections equal, male glumes appressed and inconspicuous after anthesis, lower spikes entirely female; utricles spreading widely at maturity. Utricles narrowly to widely triangular-ovate, plano-convex, margins thickened, 3–3.3 × 1–1.6mm, glabrous, ribbed, green, body becoming golden-brown; beak c.1mm, aperture notched, margins minutely serrate; stigmas 2. Female glumes ovate, acute, c.2 × 1.5mm, hyaline, brownish near centre, midrib green.

Bhutan: C — Thimphu (below and to W of Pajoding Monastery) and Bumthang (2km N of Byakar Dzong) districts; **Chumbi** (Yatung). Wet flush; sedge-dominated marshy meadow, 2750–3660m. Fr. June–July.

3. C. diandra Schrank; *C. teretiuscula* Goodenough. Fig. 36c.

Rhizomes short; stems densely tufted. Bases of lower sheaths and basal, bladeless sheaths chocolate brown, dull, persisting as fibres. Leaves inserted on lower ¼ of culm, blades shorter than culm, 2.2–4.3mm wide. Culm 47–67cm. Infl. dense, narrowly cylindric, spike-like, 2.5–7cm, brown, lateral spikes borne directly on axis in upper part, on very short appressed branches near base, spikes and branches continuous or lower separated by up to 1.5cm; lowest bract filiform, about equalling lowest branch. Spikes androgynous, small (c.5mm), predominantly female, male section minute (c.4-flowered), utricles spreading at maturity. Utricles shortly stipitate, ovate, cordate, knobbly, biconvex, 2.1–3 × 1–1.6mm, glabrous, with thickened ribs on both faces, brown; beak 0.3–0.9mm, aperture shallowly notched, margins serrulate; stigmas 2. Female glumes ovate, acute or minutely mucronate, 1.8–2.5 × 1.3–1.8mm, brown with wide hyaline margins and green midrib.

Bhutan: C — Thimphu district (Dotena to Thimphu, below Umsho, Pangri Zampa, hill above Thimphu Hospital). Damp flushes and streamsides; marshy ground in terraced paddy-field, 2300–3000m. Fr. May–June.

The Bhutan plant belongs to f. *major* (Koch) Kükenthal, differing from the typical plant in its wider leaves and larger, compound infl.

4. C. nubigena D. Don ex Tilloch & Taylor (incl. *C. pleistogyna* V. Kreczetowicz). Sikkim name: *jyakcha*. Fig. 35a, Fig. 36d.

Rhizomes short, woody; stems densely tufted. Bases of leaf sheaths straw-coloured, persisting as fibres; short, brown bladeless sheaths present. Leaves inserted along lower ⅓ of culm, blades shorter than to equalling culm, 1.2–1.8mm wide. Culm slender, not sharply trigonous, 7–40cm. Infl. narrowly cylindric, spike-like, 1.5–4cm, greenish; lateral spikes sessile, upper densely congested at fruiting, lower sometimes slightly distant; lowest bract leaf-like, 2–3 × infl., with hyaline margins at base clasping stem. Spikes androgynous, shortly cylindric to ovoid, lowest 0.5–1cm, utricles very dense, ascending; male section very small, scarcely visible at fruiting. Utricles plano-convex, lanceolate, gradually tapered into beak, 2.9–5 × 1–1.6mm, glabrous, ribbed, pale in centre with green margins and beak; nut not filling utricle; beak 1–1.5mm, aperture slightly notched; stigmas 2. Female glumes oblong-lanceolate to -ovate, acuminate, 2–3 × 1.2–1.8mm, hyaline (occasionally flushed brown near midrib), with green midrib usually extended into scabrid point (0–2mm).

FIG. 36. Cyperaceae XI (**Carex** utricles: spp. with 2 stigmas).
a, **C. pseudofoetida** subsp. **afghanica**. b, **C. echinata**. c, **C. diandra**. d, **C. nubigena**. e, **C. thomsonii**. f, **C. foliosa**. g, **C. remota** subsp. **rochebrunii**. h, **C. longipes**. i, **C. longicruris**. j, **C. lenta**. k, **C. orbicularis**. l, **C. notha**. m, **C. fucata**. n, **C. nigra** subsp. **drukyulensis**. o, **C. rubro-brunnea**. p, **C. phacota**. q, **C. pruinosa**. r, **C. teres**. s, **C. cf. teres** (*Clarke* 35636). All × 12. Drawn by Mary Bates.

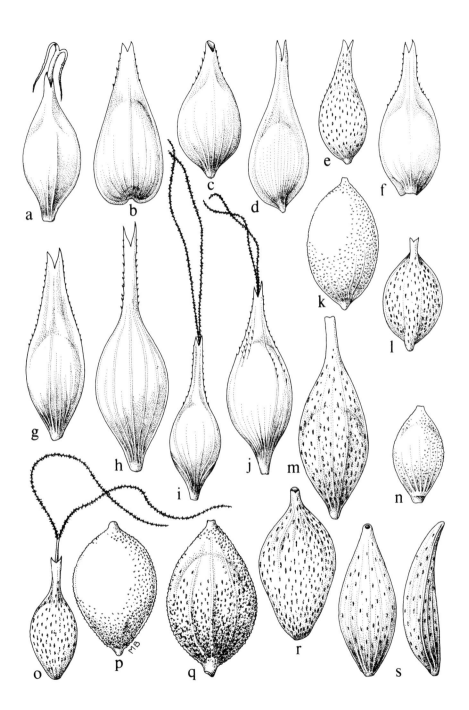

Bhutan: S — Chukka district (Chukka Colony to Taktichu); **C** — Ha (Ha), Thimphu (Paga, Namselling, Thimphu, etc.), Punakha (Mishichen to Khosa (F.E.H.2)); Punakha/Tongsa (Ritang to Rukubi), Bumthang (Yuto La, Gyetsa, above Dhur) and Sakden (Orka La, Mera) districts — under-recorded, extremely common; **Darjeeling** (Tiger Hill, Phalut, Jalapahar, Ghum to Sukia Pokhri); **Sikkim** (Tungu, Lachen, Lachung, Kopup, Karponang, Mintok Khola to Prek Chu, etc.); **Chumbi**. Dry roadsides and waste places; beside streams, rivers and ditches; banks of rice paddies, 1200–4270m. Fl. April–July; fr. May–October.

5. C. thomsonii Boott. Fig. 36e.

Differs from *C. nubigena* in its wider (2.2–3.5mm) leaves which are evenly disposed along entire culm (which is hidden by the sheaths); culms (13–31cm) stouter, rigid, persistent, becoming blackish and sometimes producing new plants from apex after flowering; infl. narrower (linear-cylindric), longer (4–10cm), more rigid, brownish, with spikes more numerous (20–30) and more regularly spaced, the lower less than 0.6cm apart; lowest bract only just exceeding infl.; spikes smaller and narrower, the axes persistent; utricles smaller (2.5–2.8 × 1.3–1.6mm), marked (as are glumes) with minute dark red dots; nut filling utricle; scarcely beaked (under 0.5mm), margins of beak minutely hispid.

Bhutan: C — Punakha district (Samtengang to Chusom; Chusom to Mishina, Punakha to Bhotoka (F.E.H.2)); **Darjeeling** (Little Rangit Valley). Among stones in river bed, 305–1500m. Fl. April; fr. April–August.

6. C. foliosa D. Don ex Tilloch & Taylor; *C. muricata* L. var. *foliosa* (D. Don ex Tilloch & Taylor) C.B. Clarke. Fig. 36f.

Differs from *C. nubigena* in its wider (5–6.4mm) leaves; taller (c.56cm) and stouter, ridged, acutely trigonous culms; lowest bract filiform, much shorter than infl.; infl. longer (8.5–10.5cm); spikes longer (lowest 1.2–1.5cm), the lower compound with 2 or more, sessile, lateral, subsidiary spikes; utricles smaller (3–3.5 × c.1.2mm), not ribbed, with nut filling utricle, margins of beak and upper part of utricle hispid.

Bhutan: S — Chukka district (S of Chimakothi); **C** — Bumthang district (above Dhur); **Sikkim** (Lachen). Open wet areas (e.g. paths and clearings) in broad-leaved and oak/blue pine forest, 2150–2870m. Fr. August–September.

7. C. remota L. subsp. **rochebrunii** (Franchet & Savatier) Kükenthal; *C. monopleura* V. Kreczetowicz. Fig. 36g.

Rhizomes creeping, stems densely tufted. Bases of lower sheaths pale brown, strongly ribbed; dark brown, dull, fibrous, bladeless sheaths present at base. Leaves inserted on lower ⅓ of culm, blades shorter than to slightly exceeding culm, 1.5–2.4mm wide. Culm 14–33(–42)cm. Infl. linear, spike-like, 3.5–7.5cm; lowest lateral spike 8–14.5 × 2–3mm. Spikes (4–)7–10, sessile, erect, closely appressed to axis, usually overlapping, lower sometimes more distant, lowest internode under 1cm (occasionally to 2.5cm); topmost 1–3 spikes entirely male (in which case very small) or gynaecandrous, other spikes entirely female or with a few basal male flowers, utricles erect; lowest bract leaf-like, greatly exceeding infl., not sheathing. Utricles lanceolate, plano-convex, (2.9–)3.5–4.5 × 0.9–1.3mm, flattened margins widened upwards into short (1.4–1.7mm), bidentate beak, strongly veined on curved face, obscurely veined on flat face, greenish-straw-coloured, margins of beak minutely serrate; stigmas 2. Female glumes narrowly lanceolate to ovate, subacute to very acute, 2.7–3.3 × 1–1.8mm, hyaline with green midrib.

Bhutan: C — Bumthang district (above Gortsam); **Darjeeling** (Sandakphu, Senchal Lake, Tonglu, Ryang, Batasia to Palmajua, Nayathang to Phalut, Tiger Hill); **Sikkim** (Karponang, Kalapokri, above Tsoka). Wet, muddy places in fir and oak forest, 2400–3750m. Fl. May–June; fr. June–August.

Some of the specimens collected in 'Sikkim' (Tonglu) by Hooker and included by Kükenthal under var. *remotiformis* (Komarov) Kükenthal are merely starved forms of subsp. *rochebrunii* with etiolated infls.

subsp. **stewartii** Kukkonen.

Differs from subsp. *rochebrunii* as follows: culms more slender, taller (32–39cm); infl. much longer (6–13.5cm), zigzag; spikes remote (lowest floral internode 1.5–4.5cm), smaller and wider (lowest 6–8mm long); utricles smaller and wider (2.3–2.5 × 0.9–1.1mm), beak shorter (c.1mm).

Bhutan: C — Thimphu (hill above Thimphu Hospital, Dotena, above Motithang) and Bumthang (above Gortsam) districts; **Sikkim** (Domang, Lachen). Wet humus in oak and spruce/fir/bamboo forest; bare earth banks in deciduous forest; semi-shaded grazed marsh by stream, 2700–3150m.

Sometimes occurring with subsp. *rochebrunii*; intermediates occur.

8. C. longipes D. Don ex Tilloch & Taylor. Sikkim name: *namo*. Fig. 36h.

Rhizomes short, woody; stems tufted. Bases of leaf sheaths pale brown, persisting as stiff brown fibres. Leaves mainly basal, sometimes 1 on lower part of culm, blades c.⅓ to ½ length of culm, 1.7–2.6mm wide. Culm 27–61cm; angles rounded. Infl. of 4–7 spikes, upper 2–3 sessile, congested, lower usually simple (occasionally compound) on long, erect, filiform peduncles, along upper half of culm (sometimes also a long-peduncled sub-basal spike); bracts with blades shorter than infl., bases with close-fitting sheaths. Spikes 2–3.2cm, androgynous, all similar, slender, utricles ascending; male section short (0.5–1.5cm). Utricles narrowly ellipsoid, biconvex, gradually narrowed into beak, margins thickened, 5–6.2 × 1.4–1.6mm, glabrous, faces prominently ribbed, olive-green; beak 1.9–2.1mm, aperture hyaline, deeply bifid, margins scabrous; stigmas 2, very long and thin. Female glumes narrowly elliptic, 2.7–3.6 (excl. awn) × 1–1.8mm, straw-coloured with very wide hyaline margins, midrib wide, green, keeled, produced into long (0.7–3mm) excurrent, scabrous awn. Male glumes narrowly elliptic, c.3.1 × 1.6mm, brownish-hyaline, with green midrib produced into slightly (c.0.9mm) excurrent point.

Bhutan: S — Chukka district (below Chimakothi); **C** — Thimphu (near Motithang Hotel, below Dotena, S of Chapcha, etc.), Tongsa (Chendebi) and Bumthang (Dhur Chu) districts — under-recorded, common; **Darjeeling** (Ghum to Kurseong, Ramam, Lloyd Botanic Garden; Lopchu (F.E.H.1)); **Sikkim** (Lachen, Domang, Yoksum). Grassy meadows and stream banks; under open oak forest; flushes, 1500–3050m. Fl. June–August; fr. May–August.

Rather variable in habit: forms with tall stems and spikes on long, slender, nodding peduncles look rather different from more compact forms with short, erect peduncles.

9. C. longicruris Nees. Fig. 36i.

Stems tufted. Bases of leaf sheaths dark brown, fibrillose. Leaves borne on lower third of culm, blades c.½ length to equalling culm, 3–4mm wide. Culm to 150cm. Infl. long, slender, drooping, with terminal and up to 12 lateral partial infls., each with up to 6 filiform, unequal peduncles, the longer branched bearing several spikes; bracts with very short blades (shorter than partial infls.). Spikes androgynous, all similar, 1.2–2cm, very slender, utricles ascending; male section short (4–6mm). Utricles stipitate, narrowly ellipsoid, biconvex, abruptly contracted into long beak, margins sparsely scabrous, c.3.9 × 1mm, glabrous, faces prominently ribbed, yellowish-green, shining; stipe to 0.7mm; beak c.1.5mm (equalling body of utricle), linear, aperture bifid, margins scabrous; stigmas 2, very slender, exceeding utricle. Female glumes lanceolate, very acute, 2.5–3 × 1mm, hyaline with brown margins, midrib green, sometimes shortly excurrent. Male glumes lanceolate, acute, c.3.5 × 0.7mm, brownish-hyaline, with green midrib.

Bhutan: S — Chukka district (between Bunakha and Chimakothi); **C** — Thimphu district (Thimphu to Dochu La (F.E.H.2)). Moist cliffs in broad-leaved forest, 2000–2450m. Fr. September.

Work is required on *C. longicruris* and the following species, and the closely related Khasian *C. teinogyna* Boott.

10. C. lenta D. Don; *C. brunnea* sensu F.B.I. non Thunberg. Fig. 36j.
Differs from *C. longicruris* mainly in the utricles: gradually narrowed into short beak (c.half length of body of utricle), sometimes shortly hispid on nerves of faces; stigmas stouter and shorter than utricle; midrib of female glumes not excurrent.
Bhutan: S — Chukka district (Chukka Colony to Taktichu); **C** — Punakha district (c.5km above Punakha). Dry grassy bank; shaded gully by stream on clay soil in cultivated area, 1000–1780m. Fr. July–October.

The Punakha and part of the Chukka gathering represent typical material with utricles c.4 × 1.5mm; however, part of the latter consists of a much smaller form with shorter, narrower leaves, more delicate infls. and smaller utricles (c.2.5 × 0.7mm). Specimens identical to this from Yunnan were described as *C. mosoynensis* Franchet, but as the two seem to grow intermixed, they are probably only forms and Kükenthal was probably correct in sinking the latter under '*C. brunnea*'.

11. C. orbicularis Boott; *C. rigida* sensu F.B.I. non Goodenough. Fig. 36k.
Rhizomes extensively spreading, branched, relatively stout. Bases of leaf sheaths pale fawn, ribbed, cross-veinlets prominent; bladeless sheaths of vegetative shoots short, purplish to blackish, shining. Leaves persistent, basal, blades shorter than culm, 2–4(–5.5)mm wide. Culm 3–21cm, acutely trigonous. Infl. of 2–4 erect spikes on short (lowest 0.2–2cm) peduncles, upper overlapping, lowest internode 0.7–2.3(–4)cm (occasionally also a sub-basal long-peduncled spike); terminal spike male, 1.2–2.3cm; lower 1–3 spikes female (upper occasionally male at apex) 0.7–2(–3)cm, dense (lower flowers occasionally slightly distant), utricles suberect; bracts with narrow, elongate blackened 'auricles', lowest with stiff blade shorter than infl., narrow (0.6–1.9mm), margins scabrid, base not sheathing. Utricles shortly stipitate, widely elliptic, compressed-biconvex, margins sometimes with few, minute bristles, 2.4–3.6 × 1.4–1.8mm, minutely papillose, nerveless, glabrous, upper part becoming fuscous-purple; beak minute (under 0.2mm), aperture entire; stigmas 2. Female glumes oblong, apex rounded or occasionally with midrib shortly (to 0.5mm) excurrent, 2.1–3.5 × 0.9–1.2mm, fuscous-purple, midrib green. Male glumes 2–3.8 × 0.7–1.5mm, apex sometimes with narrow hyaline margin.
Sikkim (Lachen, Naku La); **Chumbi** (Phari). Probably under-recorded, almost certainly present in Bhutan. Short turf with moving groundwater; streamsides, 3660–5270m. Fl. May; fr. July–August.

12. C. notha Kunth. Fig. 36l.

Rhizomes relatively stout, spreading, deeply buried. Bases of leaf sheaths fawn, persistent, not becoming fibrous; vegetative (and occasionally flowering) shoots with bladeless sheaths. Leaves sub-basal with up to 2 on lower half of culm, blades shorter than to slightly exceeding culm, 2.2–4.2mm wide. Culm (9–)24–47cm, acutely trigonous. Infl. of 4–5 erect, overlapping spikes; peduncles short, slender; terminal spike male, 1.7–2.5cm; lower 3–4 spikes female, narrowly cylindric, 1.5–4 × c.0.4cm, utricles suberect, dense throughout; lowest bract leaf-like, equalling to slightly exceeding infl., upper filiform with conspicuous black auricles. Utricles compressed biconvex, elliptic, abruptly contracted into beak, margins sometimes with a few minute white bristles, 2.1–2.9 × 1–1.6mm, obscurely (to conspicuously) 2–4-ribbed on each face, surface minutely, densely whitish-papillose when immature, olive brown with linear, pale reddish-brown glands at maturity; beak 0.3–0.6mm, aperture notched, darkened; stigmas 2. Female glumes narrowly elliptic, blunt, 1.8–2.8 × 0.9–1mm, fuscous-purple, midrib green. Male glumes oblanceolate, rounded, 2.6–4 × 1–1.2mm.

Bhutan: S — Chukka district (above Lobnakha); **C** — Thimphu district (Drugye Dzong, Taba, above Motithang Hotel, above Ragyo, hill above Thimphu Hospital); **Chumbi** (Yatung, Chubithang). Marshes, streamsides and damp grassy slopes in open or fir forest or *Quercus semecarpifolia* scrub, 2250–3810m. Fl./fr. April–July.

13. C. fucata Boott ex C.B. Clarke; *C. sikkimensis* C.B. Clarke. Fig. 36m.

Differs from *C. notha* in its longer, fatter, darker spikes (male (3–)4.2–6cm; female 3–5 × c.0.6cm); larger (3–)3.5–4.6 × 1–1.5mm utricles, lanceolate in outline, gradually tapered into beak, aperture ± entire; female glumes longer, narrower (3.3–4.5 × c.1.2mm), midrib sometimes minutely excurrent; lowest bract greatly exceeding infl. Terminal spike variable in distribution of sexes, gynaecandrous, male, or male with female flowers scattered throughout.

Bhutan: N — Upper Kulong Chu district (Lao); **Sikkim** (Lachen, Dzongri, Toong, Singalila, Phedang, Mamaichu Lake). Waterlogged meadow; river bed; streamside, 2740–3960m. Fr. August–October.

14. C. nigra (L.) Reichard subsp. **drukyulensis** Noltie. Fig. 36n.

Differs from *C. notha* in having narrower (1.6–2.6mm) leaves; longer, laxer infl., sometimes with a sub-basal spike on a very long (12cm) peduncle, male spike distant on long (over 1cm) peduncle, upper female spikes male at apex, lowest bract shorter than infl. Utricles smaller (2–2.4 × 1–1.2mm), grey green at maturity, minutely papillose, beakless, prominently ribbed (5–9 per face).

Bhutan: C — Thimphu (6km N of Thimphu Dzong) and Bumthang

(Byakar, Chunkar) districts. Wet meadow; flush; bog, 2450–2900m. Fl. April; fr. June.

15. C. rubro-brunnea C.B. Clarke. Fig. 36o.

Rhizomes slender, ?creeping. Bases of leaf sheaths pale fawn, cross-veinlets prominent, margins reticulate-fibrillose, bases persistent, not becoming fibrous; 1 or more triangular bladeless sheaths sometimes present. Leaves sub-basal, blades about equalling culm, 2.3–2.7(–3)mm wide. Culm 17–30(–60)cm. Infl. of 5–6 erect, sessile, congested spikes; terminal spike male (sometimes with some female flowers near middle), linear, 4–6.5cm; female spikes 4–5, linear-cylindric, 4.2–7.2cm, dense, utricles suberect; lowest bract with blade exceeding infl., base not sheathing. Utricles compressed-biconvex, elliptic, abruptly contracted into beak, 2.5–2.8 × 1.1mm, glabrous, not ribbed, olive-brown (darkening with age), with minute linear orange-brown glands; beak 0.6–1mm, slender, aperture deeply notched; stigmas 2, extremely long, red-brown, persistent. Female glumes linear-acuminate, 2–3.5 × 0.6mm, consisting largely of green, 1–3-veined midrib, with narrow hyaline margins (white near centre, brown on outside).

Bhutan: C — Tashigang district (Tashi Yangtse Dzong). Streamside, 1680m. Fr. April.

16. C. phacota Sprengel. Fig. 36p.

Rhizomes short; stems densely tufted. Bladeless sheaths dark to light cinnamon-coloured, passing gradually upwards (through intermediates) into bladed leaves, of which 2 on lower half of culm; inner face of sheaths thin, pale brown, splitting into reticulate fibrils, persisting as broad membranous margins to lower parts of leaves. Leaf blades flat, exceeding culms, 3.3–7mm wide. Culms 41–79cm, acutely trigonous. Infl. of 3–5, erect, overlapping (lowest internode 1.8–4.5cm) spikes, upper sessile, lower on slender, erect peduncles (lowest 1.5–3.3cm); terminal spike male, 2.5–5.8cm; lower 2–4 spikes female (male at apex in some Indian specimens), 2–5.7cm; lowest bract leaf-like, greatly exceeding infl., base not sheathing. Utricles compressed-biconvex, elliptic to suborbicular, tapered to subacute, beakless apex, 2.5–3 × 1.4–2mm, not ribbed, densely covered with small glands, dark brown except for thickened, green margins; aperture entire; stigmas 2. Female glumes persistent, 2.3–3.5 (incl. awn) × 0.7mm, consisting mainly of 3-veined midrib produced into long (0.7–2mm), very scabrid awn, margins oblong, truncate, hyaline with reddish-brown flecks. Male glumes c.3.6 × 0.8mm, very shortly (c.0.3mm) aristate.

Terai (Sivoke); ?**Sikkim.**

A Hooker specimen at K is labelled Lachung (1830–2130m), but this seems unlikely.

17. C. pruinosa Boott. Fig. 36q.

Vegetatively similar to *C. phacota* though more slender; apparently not tufted; sheaths not reticulate-fibrillose. Leaf blades narrower (2–3.3mm wide). Culms shorter (32–53cm). Infl. of 3–4 spikes, terminal male c.2.5cm, lower 2–3 female 1.8–2.4cm, nodding, on slender (lowest 1.5–3cm) peduncles. Utricles 3.2–3.8 × 1.8–2.1mm, 3-ribbed on one face, olive, suffused pale reddish-brown, covered with glands. Female glumes lanceolate, gradually tapered to shortly (0.3–0.6mm) mucronate apex, c.3.5 × 1.4mm, midrib green, 3-ribbed, margins reddish-brown, hyaline. Male glumes oblanceolate, subacute, not mucronate, c.2.5 × 1.4mm.

Bhutan: C — Punakha (Tinlegang (F.E.H.2)) and Tongsa (Tongsa) districts. Damp flush by footpath, 1950–2350m. Fr. May.

The single Bhutanese specimen seen differs from Khasian ones which have longer (4.5–5.5cm) female spikes, male at apex. Ours resembles some western Chinese examples which tend towards subsp. *maximowiczii* (Miquel) Kükenthal of E Asia.

18. C. teres Boott; *C. praelonga* C.B. Clarke. Fig. 36r.

Rhizome thick, woody, ?creeping. Bladeless sheaths red-brown, margins dark brown, passing gradually upwards (through intermediates) into leaves, inner face of sheaths thin, pale brown, reticulately fibrillose, persisting as wide margins to leaf bases. Leaf blades flat, apex trigonous, equalling to exceeding culm, (4.7–)5.9–8.8mm wide. Culm 36–64cm, stout, triquetrous. Spikes 5–11, finally spreading, relatively congested, peduncles short (lowest 1–2.5cm), ascending; terminal 1(–3) spikes male, gynaecandrous or with female flowers in middle, (1.4–)4.8–8.5cm; lower 4–10 spikes female, 4.5–9.5cm, utricles suberect; lowest bract leaf-like, blade just exceeding infl., base scarcely sheathing, upper bracts filiform, with fuscous 'auricles'. Utricles compressed-biconvex, narrowly elliptic to elliptic, abruptly contracted into beak at maturity, 2.5–3.3 × 1.1–2.1mm, usually strongly 3–5-nerved, sometimes smooth, yellowish- to brownish-olive with large or small purplish glands; beak 0.3–0.5mm, aperture transverse, entire, commonly darkened; stigmas 2. Female glumes 2.5–2.9(–3.3) (incl. awn) × 0.8–1.3mm, oblong to narrowly elliptic, apex truncate to emarginate (commonly asymmetrically) with long (0.5–1mm) excurrent minutely scabrid awn, midrib 3-veined, green, margins fuscous purple. Male glumes oblanceolate-acuminate, subacute to very shortly mucronate, 4.2–4.4 × 1–1.2mm.

Bhutan: C — Tashigang district (Shapang); **Darjeeling** (Senchal, Tonglu, Rungbool Marsh near Darjeeling, below Tiger Hill); **Sikkim** (near Dzongri, Dikeeling). Grassy marsh beside streams and ponds; streamside in oak forest, 1980–3960m. Fl./fr. April–July.

Two Darjeeling specimens (Sandakphu, 2743m, *Clarke* 35667, K; Tonglu, 2438m, *Clarke* 35636, K) differ from *C. teres* in having longer (to 12.5cm) spikes; utricles (Fig. 36s) tending to recurve at maturity, larger (3.5 × 1.7mm), fatter, pyriform-biconvex, more strongly ribbed; female glumes longer and narrower (3.5 × 0.6mm) with fuscous margins gradually tapered into longer (1.7mm) awn. The specimens have no basal parts and further collections are highly desirable; they most probably represent a monstrous form.

19. C. rara Boott; *C. capillacea* Boott. Fig. 35b, Fig. 37a.

Rhizomes very short, stems densely tufted. Bases of leaf sheaths pale brown, not becoming fibrous; bladeless sheaths long, ribbed, brown. Leaves sub-basal, blades shorter than culm, 0.3–0.7mm wide, usually rather stiff and erect. Culms 6–43cm, very slender. Infl. a single androgynous spike, 4.2–15mm, male section linear (2.2–8mm), female section 1.7–7.5mm, utricles 7–15, spreading at maturity; lowest bract glume-like, equalling to slightly exceeding lowest utricle, brown, midrib green, shortly excurrent. Utricles obscurely trigonous, widely to narrowly ellipsoid, slightly contracted to beak, 1.7–2.5 × 0.7–1.3mm, glabrous, weakly to strongly ribbed, pale green to olive-brown, occasionally with minute red dots; beak 0.2–0.5mm, aperture brown, scarcely notched; stigmas 3. Female glumes oblong-ovate, blunt, 1.2–2 × 0.8–1.2mm, pale to reddish-brown with green midrib. Male glumes lanceolate, acute, c.2.5 × 0.8mm.

Bhutan: S — Chukka district (above Lobnakha); **C** — Thimphu (Drugye Dzong, hill above Thimphu Hospital, Taba; Dotanang to Thimphu, Thimphu to Dochu La (F.E.H.2)), Punakha (Ritang to Rukubi (F.E.H.2)), Tongsa (Tongsa, Tongsa to Uto La Road) and Bumthang (Bumthang, Byakar) districts; **Darjeeling** (Tonglu to Phalut, Phalut; Sandakphu to Garibans (F.E.H.1)); **Sikkim** (Changu, Lachen, Yakla, Laghep, Phedang); **Chumbi** (Yatung). Damp flush; wet meadow; beside streams and irrigation channels, 2250–3500m. Fl./fr. April–July.

C. capillacea is supposed to differ from *C. rara* in being smaller and more delicate in all parts, with utricles reflexed and dotted red, and female glumes deciduous. However, the red dots occur in only a few specimens, while the posture of utricles and persistence of glumes depend on condition and maturity of specimen. The slender forms do tend to have smaller (under 2mm) utricles, but correlation between utricle size and stature is by no means absolute. I therefore take '*capillacea*' to be a growth form (perhaps from higher altitudes) and certainly not worthy of subspecific (as treated by Kükenthal) or even varietal rank. Admittedly the Sikkim plants tend to be more slender than the Bhutan ones, but this is likely to be an artefact of collecting; I have seen no high-altitude specimens from Bhutan.

20. C. microglochin Wahlenberg. Fig. 37b.

Rhizomes slender, creeping; stems not tufted. Bases of leaf sheaths reddish-brown, ribbed; bladeless sheaths present. Leaves sub-basal, blades deeply channelled, inrolled, blunt, about equalling culms, 0.3–0.8mm wide, rather stiff. Culms 1.5–9.5(–11.5)cm. Infl. a single androgynous spike, ebracteate, 6–10mm, male section 5.3–5.5mm, female section 1.5–3mm, utricles 5–14, finally deflexed. Utricles narrowly lanceolate, gradually tapering to hyaline, oblique aperture from which protrudes a bristle, 4.1–5.5 (incl. bristle) × c.0.5mm, pale green, suffused brown especially near apex, glabrous, smooth; bristle arising below nut, stout, tapering to acute tip; stigmas 3, appearing lateral. Female glumes rather blunt, 2.7–3.5 × 1.6–2mm, lower ovate, upper narrower, dark buff with paler midrib and narrow hyaline margin at apex. Male glumes oblong to oblanceolate, subacute, c.3.4 × 1.4mm.

Bhutan: C — Thimphu district (below Darkey Pang Tso); **N** — Upper Mo Chu (Shingche La to Chebesa; Chebesa to Lingshi (F.E.H.2)) and Upper Bumthang Chu (above Lambrang) districts; **Sikkim** (above Lambi; Gamothang (F.E.H.1)). Flushes and meadows beside streams and rivers, 3600–4100m. Fr. May–September.

21. C. parva Nees. Fig. 37c.

Differs from *C. microglochin* in being larger and stouter in all its parts (leaf blades 0.9–1.5mm wide; stems 12.5–50cm; infl. 10–15mm); infl. with a basal glume-like bract with aristate tip to 8mm; utricles longer (6.8–8mm), ribbed, bristle not projecting from aperture so stigmas appearing terminal; female glumes dark chestnut, larger (6–7mm) lower aristate, upper acute.

Bhutan: N — Upper Mo Chu (Shingche La to Chebesa (F.E.H.2)) and Upper Bumthang Chu (above Lambrang) districts; **Sikkim** (Lachen, Kopup). Flush in valley bottom, 3660–4780m. Fr. July–August.

22. C. radicalis Boott. Fig. 37d.

Rhizome short, woody, stems densely tufted. Bases of leaf sheaths reddish-brown, persisting as dark brown fibres. Leaves basal, blades 14–35cm, 1.2–2.5mm wide. Culms extremely short, hidden in leaf bases, what appear to be culms are in fact peduncles bearing single, terminal spikes; peduncles 6–30cm, filiform, acutely trigonous. Spikes androgynous, 0.5–0.7cm, male part

---

FIG. 37. **Cyperaceae** XII (**Carex** utricles: spp. with 3 stigmas). a, **C. rara**. b, **C. microglochin**. c, **C. parva**. d, **C. radicalis**. e, **C. stramentita**. f, **C. filicina**. g, **C. cruciata** var. **argocarpa**. h, **C. condensata**. i, **C. vesiculosa**. j, **C. burttii**. k, **C. continua**. l, **C. oligostachya**. m, **C. myosurus**. n, **C. composita**. o, **C. baccans**. p, **C. polycephala**. q, **C. decora**. r, **C. insignis**. s, **C. daltonii**. All × 12. Drawn by Mary Bates.

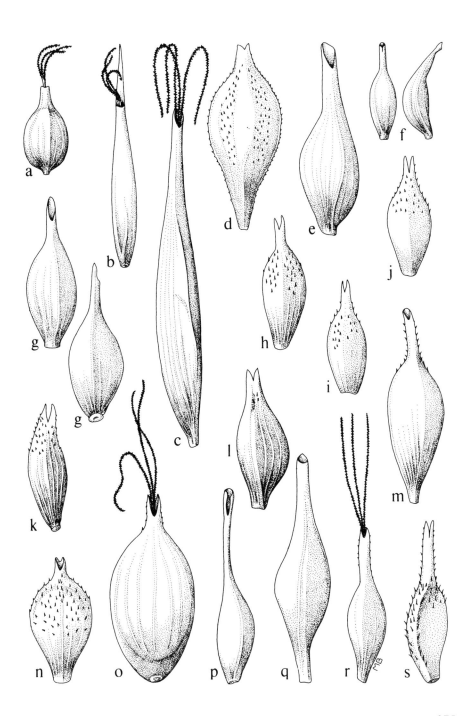

a

b

c

d

e

f

g

g

h

i

j

k

l

m

n

o

p

q

r

s

3.5–4 × 0.7–1.2mm, hidden by utricles at maturity, utricles 3–5, suberect; spike bract ovate, clasping, whitish, midrib excurrent as long, serrate point equalling to 4 × length of spike. Utricles narrowly obovoid-trigonous, knobbly, gradually narrowed into beak, long-attenuate to base, 2.5–5 × 1–2mm, densely covered with short hairs, 3-ribbed on two faces, green to pale olive-brown; beak short (0.5–0.8mm), aperture shallowly notched; stigmas 3. Female glumes ovate, 2–2.5 × 1.2–1.8mm, greenish-white, lower with midrib excurrent as short point, upper subacute. Male glumes ovate, blunt to subacute, 1.5–2 × 1.4–1.6mm, margins broad, whitish.

Bhutan: C — Thimphu district (hill above Thimphu Hospital); **Sikkim** (Lachen). Meadow in clearing in fir forest; bare soil by path in shady, deciduous forest, 2900–3480m. Fr. July–August.

Species 23–30 belonging to Section **Vigneastra** Tuckerman (Sect. *Indicae* Clarke) form an easily recognised, but taxonomically difficult, group with highly branched, paniculate infls. with all spikes small and androgynous. Infl. composed of partial panicles inserted singly or in unequal pairs at nodes (except terminal node where 3 or more) and subtended by leaf-like bracts. Branching of the partial panicles is characteristic, with spikes borne directly on axis in upper part and on secondary branches (sometimes again branched) in lower part. Spikes and secondary branches emerge from a ribbed, utricle-like 'cladoprophyll' with a free, flanged apex and subtended by small bracteoles commonly with filiform tips. Relative lengths of male and female portions of spikes characteristic; measurements given for lowest spike borne directly on axis of lowest partial panicle.

It is not always possible to be certain of the identity of a given specimen: characters are not always well developed, due to growth conditions or perhaps other factors.

23. C. stramentita Boott ex Boeckeler. Fig. 37e.

A very distinctive species with wide leaves, cream spikes and large, very strongly trigonous, chunky, utricles.

Rhizomes stout, creeping, covered with dark brown scales. Bases of leaf sheaths persisting as dark brown fibres, when dry pale, striped with conspicuous dark veins. Leaves basal and 1 on lower part of stem, blades usually shorter than culm at maturity, 8–14.8mm wide, rather thin-textured, pale green. Infl. nodes 2–3, lower partial panicles single, broadly triangular in outline, widely spaced at maturity (very dense in most commonly collected, conspicuous, immature, phase); bracts exceeding infl.; bracteoles with scabrid point c.2.5mm. Spikes primarily male, male section 3.5–8mm, female section 1.4–3.7mm, utricles spreading at maturity. Utricles not curved, acutely trigonous, fairly abruptly contracted into beak, 3.5–4.5 × 1.2–1.6mm, faces narrowly rhomboid, glabrous, conspicuously veined; beak 0.7–1.5mm, apex hyaline, aperture oblique, not notched, margins sometimes scabrid; stigmas 3. Female glumes ovate, acute, 1.4–3.8 × 1.2–1.6mm, cream, ribbed, with scabrid point 0.5–1.5mm. Male glumes oblong-lanceolate, very shortly mucronate, 2.8–4.5 × 1.2–2mm, ribbed, cream.

Bhutan: S — Sankosh (Balu Khola) and Gaylegphug (Taklai Khola) districts; **C** — Tongsa district (above Dakpai); **Duars** (Buxa); **Darjeeling** (Sittong, Sureil, Mongpu, Barnesbeg, Sukna, Tista, Rongpo Road, Kurseong, Choonbuttee, Punding, Rangit Valley, Sivoke); **Sikkim** (Pakhyong, Burmiak). Slopes and river banks in subtropical (incl. Sal) forest; rocky slope under scrub in dry valley, 150–1800m. Fl. March–May; fr. April–June.

Records from Phuntsholing/Chukka (Phuntsholing to Chimakothi), Chukka/Thimphu (Chimakothi to Thimphu) and Tongsa (Tongsa to Tratang) districts determined by Ohwi (F.E.H.2) are probably reliable, but Koyama (E.F.N.) misunderstood the species and most of his Nepalese records refer to *C. cruciata* s.l.

24. C. indica L.

Similar to *C. stramentita* in its leaf bases and pale glumes but differs in having narrower (under 10mm) leaf blades, a more open (even when immature) infl., with narrower, more divaricate spikes, which have a very long (to 1cm or more) male section; utricles less strongly trigonous, with thickened, more prominent veins; bracteoles and female glumes with longer and more conspicuous, flexuous, filiform aristate tips.

A single immature specimen from **Terai** (Dulkajhar, 150m, *Clarke* 36996, K) probably belongs to this species.

25. C. filicina Nees. Fig. 35c, Fig. 37f.

Plants stout; rhizomes woody, creeping, under 4mm diameter. Bases of leaf sheaths cream or reddish-purple, persistent, not becoming fibrous, margins sometimes fibrillose; bladeless sheaths sometimes present, reddish-purple, margins dark brown. Leaves basal + 1–3 on culm, blades about equalling culm, 2–13mm wide. Culm 26–114cm. Infl. 11–88cm, nodes 2–6(–7), partial panicles usually in unequal pairs; bracts usually shorter than infl.; filiform tips of bracteoles inconspicuous (usually under 3mm). Partial panicles open, rigid, triangular in outline, axis hispid, spikes decreasing markedly in length upwards so very variable in size within infl., borne directly on axis in upper half and on secondary branches in lower half. Spikes primarily female, utricles lax, evenly spaced, spreading to deflexed at maturity; female section 5–20.2mm; male section 2.2–6.5mm. Utricles curved, narrowly ellipsoid-trigonous, gradually narrowed into beak, 2–3.3(–3.8) × 0.7–1mm, glabrous, strongly ribbed, olive-green; beak deflexed, 0.6–1.6(–2)mm, apex hyaline, aperture oblique, not notched; stigmas 3. Female glumes ovate (occasionally lanceolate), acute (occasionally minutely mucronate), 1–1.9 × 0.8–1.2mm, sometimes minutely hispid, pale brown with darker flecks to rusty-brown (occasionally straw-coloured). Male glumes lanceolate, 1.7–3.2 × 0.6–1.2mm.

Bhutan: S — Chukka district (Jumudag to Tala, Chimakothi to Phuntsholing); **C** — Thimphu (hill above Thimphu Hospital, Dotena to Barshong, S of Chapcha), Punakha (above Tinlegang), Tongsa (near Shemgang, Longte Chu below Chendebi), Bumthang (below Dhur, near Swiss Farm), and Mongar (Sengor) districts — under-recorded, common; **Darjeeling** (Senchal, Ramam, Batasia to Palmajua, Peshok, etc.); **Sikkim** (Lachen, Bakkim, Karponang, Gangtok, Chungthang, Tsoka, etc.). Wet slopes and cliffs in open or in shady (incl. *Castanopsis* and *Quercus semecarpifolia*) forest, 610–3660m. Fl. April–August; fr. May–February.

Very variable, especially in stature; plants from exposed habitats (e.g. footpaths) are reduced in all parts but hardly worthy of taxonomic recognition. Boott (1862) described them as *C. filicina* var. *minor* and I would agree that his t. 317 is merely a reduced form of *C. filicina*. However, his t. 318 (also labelled '*C. filicina (minor)*', described in the text as 'in some respects connecting *C. filicina* with *C. cruciata* of Nees', is different; similar intermediate specimens between *C. filicina* and *C. cruciata* var. *argocarpa* have been seen in E Nepal and Sikkim (Bakkim) and require further investigation.

Most of our material has short (under 1.6mm) beaks; specimens from Kurseong (*Hooker* s.n., K) and Chukka district (*Kanai et al.* 86, BM) differ in having very large (to 3.8mm) utricles with very long (c.2mm) beaks; Boott (1862) illustrated the former (t. 320) and referred them to '*C. cruciata* sensu Nees'.

Specimens from Darjeeling (Tonglo, Dilpa to Dhodre, Ramam) possibly represent an undescribed species (some of these were originally proposed by Clarke (on labels) as a new species but he later changed his mind and included them under *C. condensata*; one was determined as *C.* aff. *ceylanica* by E. Nelmes). They agree with *C. filicina* in having glabrous utricles with long deflexed beaks and oblique apertures, but differ in having denser infls., longer, very conspicuous male sections of the spikes and dark purplish-red glumes. Similar specimens seen from Yunnan: further work required.

C. cruciata s.l.

The following 5 taxa represent a taxonomic nightmare and the modern tendency has been to unite the first two under *C. cruciata*. However, despite great variability and the presence of intermediate specimens, two taxa seem to occur in E Himalaya, the extreme forms of which are easily distinguished morphologically and also, at least partly, ecologically.

26. C. cruciata Wahlenberg var. **argocarpa** C.B. Clarke. Nep: *harkatey*. Fig. 37g.

Rhizomes stout, woody, stems clothed at apex with remains of old leaves; bases of sheaths pale brown, not fibrillose, not persisting as fibres. Leaves sub-

basal and on lower part of culm, blades exceeding infl., 6–9mm wide. Culm
67–92cm. Infl. 23–49cm, narrowly cylindric, nodes 4–6; partial panicles inserted
singly, narrowly pyramidal, branchlets and spikes evenly spaced. Bracts leaf-
like, exceeding infl.; bracteoles minutely filiform (under 2mm). Spikes short,
longest (i.e. near base of partial panicles) to 1cm, predominantly female (longest with up to 10 utricles); male portion 2.5–5mm, utricles spreading at
maturity; utricles appearing swollen at maturity, ellipsoid-trigonous, abruptly
contracted into ± straight beak, angles thickened, 2.5–3.7 × 0.8–1.4 mm,
glabrous, 2–3 thickened ribs per face, whitish in life; beak to 1mm, sometimes
minutely hispid, aperture bifid or sometimes slightly oblique; stigmas 3. Female
glumes ovate, minutely mucronate, 1.5–2.5 × 1.2–1.6mm, pale orange streaked
dark red. Male glumes lanceolate, minutely mucronate, 2–3.2mm.

Bhutan: S — Samchi (Dhoan Khola River), Phuntsholing (Gedu to
Kharbandi), Chukka (Chukka Colony to Taktichu) and Gaylegphug (Betni,
Rani Camp to Tama) districts; **C** — Tongsa (Tama) and Mongar (Lhuntse)
districts — probably under-recorded; **Darjeeling** (Mongpu, Punkabari,
Kurseong, Rongbe, Sittong, Sivock, etc.); **Sikkim** (Bakkim to Yoksum,
Gangtok). Damp rocky banks and wet cliffs in subtropical and oak forest,
400–2250m. Fl. May; fr. August–January.

27. C. condensata Nees. Nep: *partay ghans*. Fig. 37h.
Vegetatively similar to *C. cruciata* var. *argocarpa*. Infl. very dense, spike
bearing branchlets of partial infls. densely congested. Spikes primarily male
(male portion 4.5–9.5mm; female portion 1.4–3mm, utricles fewer than 6);
utricles narrowly ellipsoid-trigonous, gradually tapered into curved beak, 3–4
× 0.8–0.9mm, exceeding glumes, hispid (occasionally glabrous), olive-brown;
beak 0.9 –1.3mm, aperture deeply cleft. Female glumes ovate, 2.1–3 ×
0.9–1.6 mm, usually with aristate tip 0.7–1.5mm. Male glumes 2.5–3.5(–4) ×
1–1.4mm, usually aristate. Glumes commonly orange-brown to dark tawny
when mature, initially cream.

Bhutan: S — Chukka district (above Lobnakha, below Chimakothi);
C — Thimphu (Hinglai La, near Motithang Hotel, hill above Thimphu
Hospital), Punakha (Dochu La, E of Chusutsa), Tongsa (Tongsa, Mangde
Chu W of Tongsa) and Tashigang (Chorten Kora, Bamri Chu) districts; **N** —
Upper Mo Chu district (Gasa, Goen Shari); **Darjeeling** (Rangit, Soke, Sureil
Road); **Sikkim** (Lachen, above Yoksum, Kabi). Stony banks and hillsides,
wet or dry, often in disturbed places in open, among bushes or forest,
210–3350m. Fl. April–August; fr. June–November.

28. C. vesiculosa Boott. Fig. 37i.
Very similar to *C. condensata*, differing as follows: stems more slender, leaf
sheaths fibrillose, bases often reddish, leafless sheaths present; leaf blades

narrower, usually under 4mm (2.7–5.8mm); infl. laxer, partial infls. cylindric with spike bearing branchlets distant; bracteoles usually with conspicuous filiform tips; utricle smaller (c.2.3 × 0.9mm), shorter than to equalling female glume, distinctly shouldered into short (c.0.5mm), straight beak; female glumes lanceolate, shortly mucronate; male glumes longer (usually c.4mm), acute.

Bhutan: S — Chukka district (c.1km S of Awaka, c.3km S of Chimakothi); **Darjeeling** (Darjeeling). Roadside cliffs and damp slopes in open, or forest, 1700–2010m. Fr. August–November.

Var. *paniculata* C.B. Clarke described from Darjeeling is a large form doubtfully distinct from *C. condensata*.

29. C. burttii Noltie; *C. vesiculosa* sensu F.B.I. p.p. (Bhutan plants); *C. vesiculosa* Boott f. *pallida* Kükenthal; *C. continua* sensu F.E.H.2. Fig. 37j.

Vegetatively similar to *C. vesiculosa*, having narrow leaves and fibrillose leaf sheaths. Stem bases slender, with characteristic purplish bladeless sheaths. Differing from *C. vesiculosa* as follows: axis of partial panicles filiform, flexuous, branchlets scarcely developed so spikes clustered; glumes whitish at first becoming suffused chestnut, female ovate, aristate, male smaller (to 3.5mm), mucronate; utricles becoming suffused with chestnut; fruiting earlier.

Bhutan: C — Tongsa district (Tashiling to Tongsa, Kinga Rapten to Tongsa, Tashiling to Charikhachor); **Sikkim** (Ramtek Gompa to Murtam, Tumlong, Kabi). On rocks and wet cliffs; partly shaded rocky slopes, 1070–2200m. Fl./fr. April–May.

30. C. continua C.B. Clarke. ?Lep: *mangsel*. Fig. 37k.

Characterised by its infl. of very small, evenly spaced spikes. Similar to *C. cruciata* in its spike-form with very short male section (c.3mm); utricles and glumes as in *C. condensata* but smaller (utricles 2.3–3.2 × 0.7–0.9mm; female glumes 1.5–2 × 1.2–1.6mm, mucronate tip 0.5–1.1mm; male glumes c.2.4 × 1mm); bracteoles conspicuously filiform (to 8mm).

Darjeeling (Tista, Rangit, Bakrikat, Riang). Habitat not recorded, 150–610m. Fr. August–November.

31. C. oligostachya Nees; *C. rhizomatosa* Steudel. Fig. 37l.

Rhizomes thick, woody. Bases of leaf sheaths persisting on rhizomes as stiff, brown fibres. Leaves basal and along culm, gradually passing into bracts, blades shorter than culm, 2.5–2.8mm wide. Culm 22–35cm. Infl. a terminal cluster of spikes and 2–5 lateral partial infls. on upper half of stem; peduncles borne singly or in pairs, slender, erect each bearing a dense cluster of spikes (appearing as a single spike), apical spike of each cluster primarily male (to 9mm), lateral 2–3 very small (to 5mm), androgynous, utricles spreading; bracts

leaf-like, about equalling partial infls. Utricles ellipsoid-trigonous, 3–3.5 ×
1.2mm, gradually narrowed into short (c.0.5mm) bifid beak; body prominently
ribbed, olive green, minutely hispid on upper part of ribs and margins of beak;
stigmas 3. Female glumes ovate, retuse, c.2 × 1.2mm, midrib excurrent as
short, scabrid mucro, orange-brown with narrow hyaline margins. Male glumes
lanceolate, mucronate, 3–3.4 × 1.4mm, orange brown.
  Terai (Kolabari, Buxa Reserve). Seasonally burnt grassland. Fl. February.

32. C. myosurus Nees, *C. spiculata* Boott (incl. var. *nobilis* (Boott) C.B.
Clarke). Lep: *mungshiell*. Fig. 35d, Fig. 37m.
  Rhizomes short, thick, woody; stems tufted. Bases of leaf sheaths reddish
to brownish-purple, dull, reticulately fibrillose, persisting as fibres. Leaves
inserted along lower half of culm, blades coarse, 0.5–1.5cm wide. Culm
79–165cm, stout, angles rounded. Infl. 9–64cm, a slender to massive panicle,
nodes 5–11, bracts long, leaf-like (usually exceeding infl.), sheaths long (lowest
6–12cm), close-fitting; primary panicle branches borne singly, erect, bearing
spikes or spikes and secondary branches (each bearing up to 10 spikes). Spikes
androgynous, usually all similar, usually drooping, sometimes stiffer and ±
erect, (1.2–)2–6cm, male section very short to equalling female section; utricles
suberect. Utricles narrowly ellipsoid-trigonous, 2.6–4 × 1–1.4mm, gradually
narrowed into short (c.0.5mm), deeply bifid beak, serrulate on strongest ribs
and margins of beak, surface of upper part glabrous to shortly hairy, body
prominently veined, olive green suffused brown to varying degrees; stigmas 3.
Female glumes (lanceolate to) oblong, acute, (1.8–)2.5–3.6 × (0.8–)1–1.3mm,
olive-brown to dark brown with hyaline tip; midrib scabrous, sometimes very
shortly (to 0.6mm) excurrent. Male glumes narrowly oblanceolate, c.4.6 ×
1mm, tapered at apex to very shortly excurrent midrib, reddish-brown.
  Bhutan: S — Samchi (Dhoroka), Sarbhang (near Sarbhang High School)
and Deothang (c.15km N of Deothang) districts; unlocalised Griffith specimen
(K); **Darjeeling** (Selim, Soke, Kurseong, Mongpu, Sureil, Senchal, etc.); **Sikkim**
(Yoksum, Kulhait, Ratong River, Sittong). Banks in forest (incl. oak–rhododendron); bank among disturbed, secondary scrub; damp slope in open, wet
forest, 300–2440m. Fl. August–October; fr. September–July.

Very variable in degree of branching of infl., colour of glumes, shape and hairiness of
utricle and degree of notching of aperture. None of the following varieties seems to be
worth recognising: var. *eminens* (Boott) Boeckeler supposedly differs in its shorter,
wider utricles and darker glumes, var. *floribunda* (Boeckeler) Kükenthal in its massively
branched panicle, var. *ratongensis* C.B. Clarke in its small almost simple raceme and
narrow leaves. Similarly *C. spiculata* Boott (reduced to a subspecies by Kükenthal) and
its var. *nobilis* (Boott) C.B. Clarke supposedly have more rigid panicles and paler
female glumes but seem to grade imperceptibly into typical material.

33. C. composita Boott. Fig. 37n.

Differs from forms of *C. myosurus* with narrow panicles (i.e. secondary branches not developed) and stiffly erect spikes in its smaller (c.2.3 × 1.3mm), elliptic-obovate-trigonous utricles which are very abruptly contracted into the beak, more densely hairy and not ribbed.

Bhutan: S — Phuntsholing (below Suntlakha), Chukka (1km S of Gedu) and Deothang (Samdrup Jongkhar (M.F.B.)) districts; unlocalised Griffith record in F.B.I. Steep, wet cliff in broad-leaved forest; dry bank by roadside, 1690–2000m. Fr. July–November.

34. C. baccans Nees. Nep: *harkato*. Fig. 37o.

Similar to *C. myosurus* from which it differs in its more congested infl. — secondary panicle branches extremely short so all spikes appearing sessile, spikes stiffer, more erect; male portions of spikes smaller (not more than ⅓ total) relative to female; utricles spreading horizontally, 3–4 × 1.4–2mm, inflated, oblong-ellipsoid, very abruptly contracted into beak, greenish-white, translucent when young, fleshy and red at maturity, shining, more strongly ribbed, completely glabrous or with a few short setae on margins of beak.

Bhutan: S — Samchi (Sangura), Chukka (W of Chukka Colony) and Gaylegphug (Sham Khara) districts; **C** — Punakha district (Waecha to Nobding); **Sikkim** (Mamring, Gangtok); **Darjeeling** (Rungbee, Badamtam, Mongpu, Numsong, Gopaldora, Sittong, Sureil; Tenzing Norgay Road, Darjeeling (F.E.H.1)). Dry grassy banks by roadside, 910–2130m. Fr. July–December.

35. C. polycephala Boott. Name at Tongsa: *samsoi*. Fig. 37p.

Rhizomes short, stems tufted. Bases of leaf sheaths pale brown, ribbed, dull, not persisting as fibres; bladeless sheaths short. Leaves mostly in stout, elongate (sheaths long) rosettes, blades equalling or exceeding culms, 5–9.6mm wide, upper surface scabrous. Culms 28–66cm, stout, acutely trigonous. Infl. a usually narrow panicle, nodes 4–6, peduncles in groups of 2–8, relatively slender, erect, each bearing a single spike or sometimes again branched with up to 4 spikes. Bracts with blades exceeding infl., sheaths slightly inflated, pale. Spikes androgynous, all similar, narrowly ellipsoid, 1.6–3 × 0.6–0.9cm, primarily female, male section to 0.7cm, longer ones sometimes branched at base, utricles spreading to ascending. Utricles narrowly ellipsoid-trigonous, gradually narrowed into beak, margins conspicuously thickened, 3.4–4.1 × 0.9–1mm, glabrous, olive-green; beak 1.7–2.2mm, aperture oblique, scarcely notched; stigmas 3. Female glumes lanceolate, acute, 3.6–4.5 × 1–1.6mm, hyaline or straw-coloured with green midrib. Male glumes linear-lanceolate, acute, c.6 × 1mm, straw-coloured.

Bhutan: S — Deothang district (c.2km from Tashigang-Samdrup Jongkhar Road towards Pemagatsel); **C** — Thimphu (Thimphu to Dochu La), Punakha (c.6km E of Nobding, above Lometsawa; Ritang to Rukubi (F.E.H.2)), Tongsa (Changkha; Tashiling to Charikhachor (F.E.H.2)) and Sakden (Tashigang to Sakden) districts; **Darjeeling** (Tonglu, Sandakphu, Rungirun, Sureil); **Sikkim** (Karponang, Dikeeling, Kulhait, Kalapokri). Roadside embankment in moist, mixed forest; rocks by waterfall in small ravine, 1520–3660m. Fr. April–July.

Species 36–41 form a distinctive group (Section **Oligostachyae** C.B. Clarke (Sect. *Frigidae* Fries Subsect. *Decorae* Kükenthal)), characterised by having fascicles of peduncled spikes subtended by sheathing bracts. Within each fascicle distribution of sexes is complex: shortest spikes (on shortest peduncles) tend to be female, longer peduncles bearing longer spikes (commonly branched) have an increasing tendency to maleness (androgynous — except for *C. pulchra* — or wholly male), the tendency towards entirely male spikes increasing towards the top of the infl.

36. C. decora Boott. Fig. 35e, Fig. 37q.

Rhizome slender, creeping. Leaves mainly in stout, basal rosettes, usually one on lower part of culm, blades shorter than culm, 0.5–1.2cm wide, flat, rather stiff. Bases of old leaves reddish, dull, persistent, not becoming fibrous. Culm 54–77cm, stout, erect; spike-bearing nodes 5–7, bracts narrow, leaf-like, blades slightly exceeding infl. Fascicles with 3–9 unequal peduncles, erect at flowering, comprising 1–4 female spikes (2.5–3.5cm), peduncles short at flowering, lengthening to 10cm in fruit, 1–2 longer-peduncled, androgynous spikes (7–9.5cm, with few basal female flowers) and 1–4 single or paired male spikes (5–6cm) on longest peduncles, uppermost fascicle usually of male spikes only. Female spikes and parts of spikes rather lax, glumes scarcely overlapping. Utricles strongly trigonous, very narrowly obovoid, 4–6 × 1–1.5mm (larger lower in spike), abruptly contracted into long (1.6–2mm), scarcely notched beak, margins of beak (and occasionally body) with few, short hairs, faces prominently 2 ribbed, olive-brown. Female glumes oblong, acute, (3–)5–7.5 × 1.9–3mm, orange-brown to chestnut, tip and margins widely hyaline, midrib strongly ribbed, green, excurrent from below apex, sometimes projecting by up to 1.5mm beyond tip of glume; stigmas 3. Male glumes narrowly oblanceolate, 6–7.5 × 1.3–3mm.

Bhutan: C — Thimphu (Thimphu to Dochu La, above Ragyo, Dotena to Barshong, above Motithang), Tongsa (Pele La, Tongsa, Tongsa to Dorji Gompa), Bumthang (above Lami Gompa) and Mongar (above Sengor) districts; **N** — Upper Mo Chu district (Gasa to Chamsa); **Darjeeling** (Sandakphu, Kanglan, Tonglu, Phalut, Sedonchen); **Sikkim** (Bakkim, Chola, Phulalong, Laghep, Tsomgo, etc.). Banks in wet forest (*Abies/Rhododendron* and *Tsuga*); shady streamside in meadows, 2400–3660m. Fl. May–July; fr. July–November.

Apparently strongly protandrous and thus very variable in appearance, most collections are of the (conspicuous) male phase, which is very different from the fruiting stage when the male spikes have withered and peduncles of female spikes greatly elongated. More collections are required from Bhutan as some of the specimens are atypical, with very thin spikes and dark glumes.

A specimen collected by C.B. Clarke (*Clarke* 35673, K) from Sandakphu, to which he refers in F.B.I. and which he originally intended to describe as a species and then as a form of *C. decora* ('f. *eximia*' on duplicate from his own herbarium) certainly seems worthy of specific status, but further collections are required. In infl. form it is intermediate between *C. decora* and *C. myosurus*, but differs from both species in its enormous glumes and (glabrous) utricles (7–8mm) with long, curved, glabrous beaks and the beak asymmetrically and extensively (1–1.5mm) produced beyond the aperture.

37. C. insignis Boott. Fig. 37r.

Differs from *C. decora* as follows: rhizomes stouter; leaves evenly disposed along longer (61–93cm) culm which is thus almost completely hidden by the sheaths, blades narrower (4–6.8mm); infl. longer (6–14 nodes); peduncles less unequal; spikes more predominantly female — those on longest 2–3 peduncles of each fascicle androgynous, often branched, other spikes entirely female (topmost fascicle entirely of male spikes); utricles smaller (3.1–4 × 1–1.1mm), with shorter (c.1mm), deflexed beak, body sometimes sparsely hairy, faces with more than 2 prominent ribs; female glumes smaller (2.2–2.7 × 1.4–1.8mm) with much less conspicuous (narrower) hyaline tips and margins; male glumes smaller (c.3.5 × 0.8mm).

Bhutan: S — Chukka (between Jumudag and Chasilakha) and Deothang (near Wamrung (M.F.B.)) districts; **Darjeeling** (Mongpu, Kurseong, Rungtong, Lloyd Botanic Garden; Butia Basti (F.E.H.1)); **Sikkim** (Penlong, Tari). Warm broad-leaved (incl. oak) forest, 1220–2130(–3960)m. Fl. July; fr. September–February.

38. C. daltonii Boott. Fig. 37s.

Vegetatively similar to *C. decora* but differs as follows: sheath bases golden brown, shining; fascicles with more (11–18) peduncles which are more slender and longer (to 16cm), similar in disposition of the sexes; utricles smaller (3–4.1 × 0.5–1mm), more gradually narrowed into shorter (1.2–1.8mm), deeply bifid beak; body and beak densely hispid, chestnut at maturity; female glumes smaller (2.3–2.8 × 1.4–2mm), straw-coloured to dark brown with much narrower hyaline borders and a much longer awn (2.2–2.4mm).

Sikkim (Chungtam). Habitat unknown, 2440–3350m. Fr. May.

39. C. crassipes Boeckeler.

Specimens collected by Griffith in Bhutan (locality unknown since field number lost, *KD* 2664, *HEIC* 6086 K, BM, P) were thought (though not unhesitatingly) by Boott to be dwarf specimens of *C. daltonii*, but were described as *C. crassipes* by Boeckeler. The specimens are unfortunately immature, but agree with *C. daltonii* in having hispid utricles with a bifid beak, but differ in being much smaller plants (17–28cm) with shining dark chestnut bases to the leaf sheaths, narrower leaf blades (3.5–4.6mm), smaller infls. with fewer spikes and glumes with shorter (c.1mm) awns. It is almost certainly a distinct species, and it would be extremely interesting to obtain modern collections.

40. C. anomoea Handel-Mazzetti; *C. inaequalis* Boott ex C.B. Clarke, *nom. illegit.* Fig. 38a.

Tufted. Bases of leaf sheaths cream, persisting as fibres. Leaves basal with a few on lower part of culm, blades shorter than culms, 2–2.5mm wide. Culm 22–54cm, angles rounded. Infl. long, slender, flexuous, nodes 2–5. Fascicles with 2–4 unequal, slender, erect, scabrous peduncles to 6cm; bracts with blades shorter than infl., sheaths membranous, upper part pale brown. Spikes on short peduncles female, those on longer peduncles sometimes branched and androgynous, terminal fascicle sometimes with 1–2 male spikes. Female spikes ± linear, 1–1.9cm, utricles ± erect. Utricles narrowly ellipsoid-trigonous, 2.9–3.5 × 0.6–1mm, gradually narrowed upwards into short, curved (0.7–1mm) beak, chestnut and shining at maturity, beak margins with a few hairs; stigmas 3. Female glumes oblong-elliptic, subacute, 2.4–2.8 × 1–1.4mm, midrib shortly excurrent (c.0.6mm) from below apex, reddish-brown with hyaline apex and margins. Male glumes narrowly oblanceolate, c.5 × 1.5mm, narrowed above into shortly excurrent midrib.

Sikkim (Lachen). Habitat unknown (wet banks in conifer forest in Nepal), 2740–3050m. Fr. July(–August).

41. C. pulchra Boott. Fig. 38b.

Rhizomes short, stems tufted. Bases of leaf sheaths dull, fawn or pale orange-brown, persisting as fibres. Leaves in basal rosettes, with 1 on lower part of culm, blades shorter than to equalling culm, 1.3–2.7mm wide, rather stiff, hispid on upper surface (at least when young). Culm 19–51cm; infl. slender, flexuous or nodding, nodes 2–4. Lowest node bearing a single peduncle, upper with fascicles of 3–11 slender, erect, scabrous peduncles to 2cm, bract blades not exceeding infl. Most spikes entirely female, longer peduncles sometimes bearing 2 spikes, longest peduncled spike in each fascicle sometimes gynaecandrous, or in terminal fascicle sometimes entirely male. Female spikes linear, 1.2–3cm, rather lax, utricles ± erect. Utricles narrowly ellipsoid-

232. CYPERACEAE

trigonous, 2.2–2.7 × 0.5–0.8mm, abruptly contracted into beak (0.5–0.9mm), glabrous, olive-brown with reddish-brown flecks, eventually suffused dark brown, apex hyaline, aperture oblique, scarcely notched; stigmas 3. Female glumes oblong-obovate, subacute, 1.8–2.8 × 0.8–1.2mm, golden brown, sometimes with a narrow hyaline border. Male glumes narrowly lanceolate, very acute, c.3.7 × 1.2mm, midrib keeled.

Bhutan: C — Thimphu (N of Chapcha, hill above Thimphu Hospital), Bumthang (Byakar, Dhur Chu) and Tashigang (Kaling) districts; **Darjeeling** (Darjeeling); **Sikkim** (Lachung, Lachen, Karponang, Zemu Valley). Marshy and peaty meadows; banks in evergreen oak forest, 2000–3350m. Fr. July–October.

A large, robust plant from Darjeeling (Sukia Pokhri to Manibhanjang, *ESIK* 1136, E) with stems to 74cm and leaves to 3.5mm wide appears to belong to this species.

42. C. gracilenta Boott ex Boeckeler; *C. alpina* Liljeblad [non Schrank] var. *gracilenta* (Boott ex Boeckeler) C.B.Clarke; *C. infuscata* sensu E.F.N. p.p., non Nees. Fig. 38c.

Rhizomes slender, extensively spreading. Bases of leaf sheaths pale fawn, striate, persisting as fibres; small, dark purplish bladeless sheaths sometimes present. Leaves sub-basal, blades shorter than to equalling culm, 0.8–2mm wide. Culm 4–47cm. Infl. narrow, of 2–4 erect, overlapping (occasionally congested) spikes, lowest shortly peduncled; terminal spike gynaecandrous, 8–13mm, other spikes female, 4.5–9 × 2–4mm, utricles suberect; lowest bract with blade shorter than infl., base not sheathing, with clasping membranous 'auricles' darkening with age. Utricles turgid, rhomboid- to ellipsoid-trigonous, gradually narrowed into beak, 2.2–2.5 × 1.2–1.5mm, pale green becoming olive-brown with dark markings, minutely papillose above; beak 0.2–0.5mm, sometimes minutely serrate, aperture notched; stigmas 3. Female glumes widely rhombic-elliptic, tapered into shortly excurrent point, 1.7–2.6 × 1.2–1.8mm, dark chestnut brown, midrib broad, green. Male glumes linear to narrowly elliptic.

Bhutan: C — Ha (hillside opposite Ha Bazaar), Thimphu (above Pajoding, Bimelang Tso to Dungtsho La, Dochu La) and Bumthang (above Gortsam)

FIG. 38. **Cyperaceae** XIII (**Carex** utricles: spp. with 3 stigmas).
a, **C. anomoea**. b, **C. pulchra**. c, **C. gracilenta**. d, **C. duthiei**. e, **C. psychrophila**. f, **C. atrata** var. **atrata**. g, **C. atrata** var. **glacialis**. h, **C. lehmannii**. i, **C. obscura** var. **brachycarpa**. j, **C. atrofusca** var. **atrofusca**. k, **C. montis-everestii**. l, **C. moorcroftii**. m, **C. praeclara**. n, **C. obscuriceps**. o, **C. viridula**. p, **C. alopecuroides** var. **alopecuroides**. q, **C. jackiana**. r, **C. fusiformis** subsp. **fusiformis**. All × 12. Drawn by Mary Bates.

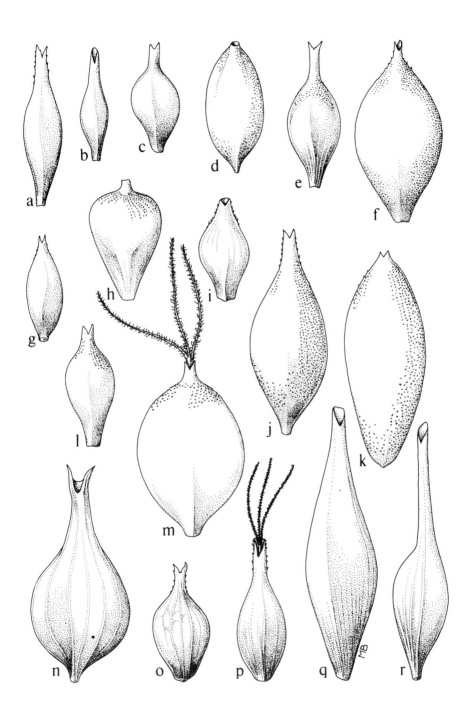

districts; **N** — Upper Bumthang Chu district (below Ju La); **Sikkim** (above Tangu, Kopup, Lachen, Eumtso La, Dzongri, Thangshing, Phedang). Grazed yak pasture; grassy bank on forest margin; wet grass beside streams in open or among dwarf shrubs, 3200–4320m. Fl. June; fr. June–September.

Habit very variable from extremely dwarf, with weak, recurved stems when grazed, to robust, with erect stems in ungrazed or shaded sites. Clarke records *C. alpina* (by implication the W Himalayan *C. infuscata* Nees) for Sikkim (Yeumtong) in F.B.I. but no specimens seen.

A form with beakless utricles has been seen from Sikkim (Bikbari, 3950m). This is analogous to *C. pseudobicolor* Boeckeler which seems to be merely a beakless form of the W Himalayan *C. infuscata* as originally thought by Boott (1858), who described it as *C. alpina* var. *erostrata* Boott.

43. C. duthiei C.B. Clarke; *C. atrata* L. subsp. *pullata* (Boott) Kükenthal. Plate 12, Fig. 38d.

Rhizomes short, stems densely tufted. Extreme bases of leaf sheaths and bladeless sheaths striped brownish- to blackish-red, shining. Leaves sub-basal and with 1–2 on lower half of culm, blades about half length to equalling culm, 2–5(–6.5)mm wide. Culm (12–)21–87cm, acutely trigonous. Infl. of (2–)3–4(–6) drooping spikes, upper 2–3 shortly peduncled, crowded, lowest often distant on longer peduncle; spikes narrowly cylindric, tapering to apex, 1.2–3 × 0.6–0.8cm, terminal gynaecandrous, others female; lowest bract with blade developed to varying degrees (short and setaceous to leaf-like and exceeding infl.), 'auricles' short, clasping, dark, those of upper bracts very conspicuous. Utricles oblong-elliptic-trigonous, (2.2–)2.4–3.9 × 1–1.9mm, glabrous, minutely papillose, pale green (darkening later), beakless (occasionally with very short beak to 0.2mm), aperture entire, blackened; stigmas 3. Female glumes narrowly lanceolate, acute to acuminate, (2.2–)3.2–4.8 × 1.2–1.5(–1.8)mm, purplish-black, minutely papillose, midrib sometimes excurrent as scabrid awn.

Bhutan: C — Thimphu (Barshong, above Pajoding, below Darkey Pang Tso), Bumthang (Oroling to Ura La) and Sakden (Orka La) districts; **N** — Upper Mo Chu district (above Laya); **Darjeeling** (Sandakphu; Phalut to Sandakphu (F.E.H.1)); **Sikkim** (Dzongri, Changu, Lhonak, Lachen, Bikbari, etc.); **Chumbi**. Wet grassy slopes; among grass and shrubs; marshes; streamsides; stony grassland, 3660–4570m. Fl. June–July; fr. July–October.

Very variable in stature, size of spikes and utricles, degree of development of bract blade and leaf width.

A specimen with short (11–24cm), slender, weak stems and dense, terminal clusters of 2–3 small (c.1 × 0.5cm) spikes and small (c.2.2 × 1.1mm), beakless utricles from

Sikkim (Nathu La, 4267m, *Bor's Collector* 250, K) is probably a form of this species: similar one seen from Nepal (*de Haas* 2481, BM).

44. C. psychrophila Nees. Fig. 38e.
Very similar to *C. duthiei* differing chiefly in the utricle (c.3 × 0.8–1.2mm) which is narrowly ellipsoid- to rhomboid-trigonous, abruptly contracted into a prominent beak, body prominently ribbed; beak 0.5–0.7mm, apex fuscous-purple, aperture deeply notched; female glumes smaller (2–2.5 × 1–1.4mm). Spikes in Sikkim specimens smaller (under 1.5 × 0.5cm).
Sikkim (Yeumtong, Phalut). Habitat not recorded, 3350–3660m. Fr. June–September.

45. C. atrata L. var. **atrata**. Fig. 38f.
Similar to *C. duthiei* in the intensely coloured basal sheaths, but leaf blades wider (4.3–8.1mm); culm (19–112cm), stouter (?more erect); lowest bract with filiform blade not exceeding infl., clasping black 'auricles' conspicuous; spikes 3–5, larger (1.6–3.3 × 0.6–0.8cm), all terminal, shortly peduncled, suberect; utricles larger (3.4–4.5 × 1.6–2.3mm), more widely elliptic, fuscous purple, abruptly contracted into minute (0.2–0.4mm), bidentate beak, very thin-textured, nut less than half-filling utricle; female glumes 4.2–4.7 × 1.3–1.5mm, exceeding utricles.
Bhutan: N — Upper Mo Chu district (E bank of Tharizam Chu); **Sikkim** (Lhonak, Lachen, Samdong, Lhasal, Bijan); **Chumbi** (Chomo Lhari, Upper Khangbu, below Ghara La). Grassy hillsides among scrub, 3960–5182(–5790)m. Fl. June; fr. July–September.

var. **glacialis** Boott. Fig. 38g.
Differs in its smaller stature (culms 4–18cm); smaller spikes (0.8–1.3 × 0.5–0.7cm) and smaller (to 2.5mm), thicker-textured utricles with relatively larger nuts.
Chumbi (E of Phari); **Sikkim** (Momay, Kinchinjhow, Kankola, Yeumtong) [also E. Nepal, where on scree beside stream]. 4570–5180m. Fr. August–September.

Placed under *C. duthiei* by Clarke (F.B.I.) and Kükenthal (1909), but I revert to Boott's (1862) treatment. It differs from *C. duthiei* in its short, stiff leaves; densely clustered, erect spikes; lowest bract with short, setaceous blade and conspicuous clasping auricles; utricles distinctly beaked, aperture bidentate. Possibly worthy of specific rank.

46. C. lehmannii Drejer. Fig. 38h.
Rhizomes short, stems tufted. Bases of leaf sheaths pale fawn, becoming fibrous; small, dark purplish-red, bladeless sheaths sometimes present. Leaf

blades erect, almost equalling culms, 2.5–4mm wide. Culms erect, 15–53cm, very slender (0.5–1.1mm diameter), trigonous, angles smooth. Infl. of 3–4 congested spikes, on slender, erect peduncles (decreasing in length upwards), sometimes also with a very long-peduncled spike much lower down culm; spikes squat, cylindric, 6–9 × 3–5mm, utricles spreading, terminal gynaecandrous, others female; lowest bract leaf-like, exceeding infl., narrow (1.3–2.1mm wide), not sheathing, 'auricles' short, blunt, becoming blackened. Utricles turgid (nut filling utricle), narrowly obovoid-trigonous, tapering to apex (or upper part sometimes rather flattened), abruptly contracted into beak, 1.9–2.6 × 1–1.8mm, glabrous, whitish-green becoming golden-brown, beak minute (c.0.1mm), purplish, aperture entire; stigmas 3. Female glumes ovate, blunt, 1.2–1.7 × 1–1.4mm, much smaller than mature utricles, dark blackish-purple.

Bhutan: C — Thimphu district (Shodu to Barshong); **N** — Upper Mo Chu district (Laya, E bank of Tharizam Chu); **Sikkim** (Yeumting); **Chumbi** (Yatung). Grassy hillsides among dwarf rhododendrons and shrubs; mossy boulders in *Abies* forest; streamsides and wet sand in stream beds, 3050–4150m. Fl. June; fr. June–October.

47. C. obscura Nees var. **brachycarpa** C.B. Clarke. Fig. 38i.

Rhizomes short, stems densely tufted. Bases of leaf sheaths and long, bladeless sheaths dark purplish, shining. Leaf blades equalling culms, 3.2–5.5mm wide. Culms erect, 15–92cm, stout (1.4–2.3mm diameter), trigonous, angles minutely serrate. Infl. of 4–6(–7) erect, congested, subsessile spikes (occasionally with a long peduncled spike from lower down culm); spikes very dark, cylindric, sometimes (when apical flowers not fully developed) tapering to apex, 1.2–1.9 × 0.4–0.5cm, terminal gynaecandrous, others female; lowest bract not sheathing, blade greatly exceeding infl., 2.4–4.3mm wide. Utricles narrowly rhomboid-trigonous, gradually tapered to beakless apex, 1.7–2.5 × 0.9–1.3mm, shoulders hispid (sometimes only minutely), aperture entire, darkened; stigmas 3. Female glumes widely ovate to orbicular, blunt, 1.5–2.6 × 1.2–2mm, dark purplish, midrib paler streaked greenish.

Bhutan: C — Thimphu (above Motithang, above Ragyo, Gidakom Valley) and Bumthang (W side of Ura La, Yuto La, above Gortsam) districts; **Darjeeling** (unlocalised); **Sikkim** (Dzongri, Jamlinghang, Thangshing, Kopup, Patang La, Yeumtong, Kankola, Lachen, Zemu Valley, etc.); **Chumbi** (Yatung). Edge of streams and wet slopes in conifer forest (*Picea/Tsuga* and *Abies*), 2700–3960m. Fl. June; fr. June–October.

48. C. atrofusca Schkuhr var. **atrofusca** (incl. var. *minor* Boott); *C. ustulata* sensu F.B.I., non Wahlenberg. Fig. 38j.

Rhizomes extensively spreading, covered with pale brown scales. Bases of leaf sheaths cream to pale brown, dull, striate. Leaves persistent, basal, with

0–3 on lower part of culm, blades much shorter than culms (2–12cm), 2.5–4.7mm wide, widest near base, slightly glaucous. Culm 6–37cm. Infl. of 2–5 blackish spikes on drooping peduncles near apex of stem (sometimes also a long peduncled, sub-basal spike); lower spikes female (occasionally male at apex), ellipsoid, 1–2 × 0.6–1cm, dense, utricles suberect, terminal spike entirely male (when narrower than others) to entirely female; bract blades filiform, shorter than sheath and adjacent spike, sheaths long, loose, mouths oblique, hyaline, darkened. Utricles lanceolate in outline, gradually tapered into beak, 2.5–4.5 × 1–1.7mm, thin-textured, nut not nearly filling utricle, glabrous, minutely papillose, purplish-black (pale at base), beak short (under 1mm), aperture notched, angles commonly minutely scabrid; stigmas 3. Female glumes lanceolate, acuminate, 2.8–4.5 × 0.7–1.2mm (about equalling utricle), finally uniformly purplish-black (midrib paler when young), surface dull, minutely papillose. Male glumes oblanceolate (sometimes narrowly), 3.5–5 × 1–1.6mm.

Bhutan: C — Ha (Kang La to Ha) and Thimphu (Shodug, beyond Pajoding) districts; **N** — Upper Mo Chu (Laya, Shingche La, Yale La, Tharizam Chu) and Upper Kulong Chu (Shingbe) districts; **Sikkim** (Chemathang to Thangshing, Bikbari, Zemu Valley, Lhonak, Momay, Chumegata); **Chumbi**. In shallow streams and marshes; open grassland and grassy stream banks; wet, rocky places; among *Rhododendron* and *Potentilla* scrub, 3810–5180(–5790)m. Fl. May–August; fr. July–October.

var. **angustifructus** Kükenthal.
Differs in being larger in all its parts: culms 21–40cm; female spikes 1.4–2.4 × 0.8–1cm; utricles 4.6–6 × 1.2–1.8mm.
Bhutan: N — Upper Bumthang Chu district (Waitang); **Sikkim** (Tangu, Yeumtong, Lhonak, ?Lachen, Zemu Valley, Tang-ka-la, above Lambi). Gravel by river, 3660–5270m.

This species belongs to a difficult group which requires much further work. Further collections are needed, especially of large plants superficially similar to the above but with longer spikes. Specimens from Upper Mo Chu district (below Chhew La, 4160m, *Sinclair & Long* 5445, E, K) with utricles much larger than the glumes and with curious cleft, but beakless apertures and from Bumthang/Mongar district (Rudong La, 3658m, *Ludlow & Sherriff* 289, BM) with very finely acuminate glumes with green midribs and utricles with longer, slender, deeply bifid beaks belong to this group but probably represent undescribed species. Both *C. cruenta* Nees and *C. nivalis* Boott have been recorded from our area but all specimens seen that have been so named have been re-determined as *C. atrofusca* var. *angustifructus*.

49. C. montis-everestii Kükenthal. Fig. 38k.

Differs from *C. atrofusca* in its densely tufted habit, reddish-brown leaf-sheath bases; extremely narrow leaf blades (c.1.4mm wide); female spikes shorter and more broadly cylindric (c.1.2 × 0.9cm), utricles more widely spreading; female glumes acute, shining; utricles beakless, papery, shining.
Sikkim (Lhonak). Moraines, 4570m. Fr. August.

50. C. moorcroftii Falconer ex Boott. Fig. 38l.

Rhizomes (c.2mm diameter) extensively spreading; stems densely tufted. Bases of leaf sheaths pale reddish-brown, striate, dull, persistent, not becoming fibrous, forming a long, underground 'collar' around stem bases. Leaves basal, blades erect, less than half to ⅔ length of culm, 2.6–4.2mm wide, widest near base, very stiff. Culm curved, 7–24cm, acutely trigonous. Infl. of 3–5(–6) spikes, upper densely congested forming a rather irregular head to 1.5cm wide, lowest on short, erect peduncle (occasionally one c.⅓ down culm on longer peduncle); terminal spike male, 1–1.6cm, clavate at anthesis, becoming narrowly cylindric, pale; lower spikes female, 0.9–1.8 × 0.5–1cm, decreasing in size upwards, very dense, utricles suberect; lowest bract of terminal cluster with broad blackened 'auricles', margins hyaline, clasping (sometimes encircling) stem, blade usually filiform, shorter than adjacent spike; bract of distant spike (when present) leaf-like. Utricles ellipsoid-trigonous, abruptly contracted into beak, 2.6–4 × 1.2–2mm, glabrous, minutely papillose, pale olive-brown, becoming fuscous purple, base and margins remaining green; beak c.0.2mm, aperture bifid; stigmas 3. Female glumes narrowly to broadly ovate, acute or acuminate, occasionally suborbicular, blunt, 2.3–5.2 × 2–2.7mm, dark purplish-brown, midrib paler, margins narrowly to broadly hyaline near apex. Male glumes oblanceolate, subacute, reddish-brown, margins very broadly hyaline near apex.

Sikkim (Gurudongmar, Kambay, Yume Chu, Chholhamoo); **Chumbi** (Phari). Glacial sand beside rivers, 4270–5430m. Fl. May–June; fr. July–August.

A group of specimens from similar localities and extreme altitudes resemble this species in their congested heads, bracts and dark glumes with hyaline margins, but are much smaller in all parts, especially the extremely narrow (c.0.6mm) leaves and the slender, grooved, subterete stems. These require further investigation and may be related to the Tibetan *C. ivanovae* Egorova.

51. C. praeclara Nelmes. Fig. 38m.

Similar to *C. moorcroftii* but differing in its infl. forming a wider (2.5–3.5cm), denser head, composed of 4–5 widely ovoid spikes; spikes androgynous, 1.2–2 × 0.9–1.5cm, all similar; only immature utricles seen,

apparently wider and with longer (0.3–0.5mm) beak; female glumes much larger 8–9.5 × 2.6–3.2mm.
Sikkim (Lhonak, Jonsong La Valley). Habitat not recorded, 4880–5030(–5790)m. Fr. August.

52. C. obscuriceps Kükenthal; *C. lurida* C.B. Clarke non Wahlenberg. Fig. 38n.
Rhizomes ?short; bases of leaf sheaths cream to dark brown, dull, not persistent. Leaves inserted near base of culm, blades half length to equalling culms, 3.5–6.9mm wide, tips rather blunt, blades and sheaths with conspicuous, short, transverse veinlets. Culm 14–78cm. Infl. of 1–2(–3) terminal male spikes and (1–)2–4 female spikes, rather distant on upper ⅓ of culm; male spike linear, 1.9–4.4cm (secondary ones if present smaller); female spikes cylindric, 1.7–5.7cm, utricles spreading horizontally, peduncles erect (lowest 1.1–8cm); bracts exceeding infl., lowest with sheathing base. Utricles inflated, ovoid-trigonous, gradually narrowed upwards into beak, 4.8–5.5 × 2–2.5mm, glabrous, strongly ribbed, straw-coloured, shining, translucent, reticulate pattern of cells visible at low magnification; beak 1.5–2mm, aperture very sharply bidentate; stigmas 3. Female glumes narrowly lanceolate, 2.9–4.5 × 1–1.9mm, midrib green, 3-veined, excurrent (at least in lower glumes of each spike) as scabrid point 0.9–2.1mm, margins narrow, purplish-brown, hyaline. Male glumes oblong-lanceolate, subacute, 5.1–5.7 × 1.3–1.6mm, purplish-brown with green midrib.
Bhutan: S — Chukka district (above Lobnakha); **C** — Thimphu (Chapcha, Taba, hill above Thimphu Hospital, below Umsho La, 6km N of Thimphu Dzong) and Bumthang (Byakar, Gyetsa to Yuto La, above Dhur) districts; **Sikkim** (Lachen). Marshes and streamsides in open or open blue pine or evergreen oak forest; ponds, 2300–3660m. Fl./fr. May–August.

53. C. viridula Michaux; *C. flava* sensu F.B.I. non L. Fig. 38o.
Densely tufted; bases of leaf sheaths pale, ribbed, dull. Leaves sub-basal, blades deeply channelled, less than half length of culm, 0.6–1.5mm wide. Culms 9.5–17cm. Infl. a congested head; male spike solitary, terminal, sessile, linear, 0.5–1.3cm; female spikes 3–4, sessile, subspherical to oblong-obovoid, 0.4–0.5cm; lowest bract leaf-like, spreading to deflexed, base scarcely sheathing, 2.6–7.5cm, greatly exceeding infl. Utricles knobbly, obovoid-trigonous, 2.1–3 × 0.8–1.3mm, abruptly shouldered into minutely serrate beak (0.4–1mm), strongly ribbed, yellowish-green; stigmas 3. Female glumes ovate, subacute, 1.5–2.3 × 1–1.2mm, tawny-brown with green midrib. Male glumes oblong-lanceolate, subacute to blunt, c.3.8 × 1.1mm, pale brown with hyaline margins and green midrib.
Bhutan: C — Thimphu district (hill above Thimphu Hospital, Paro Bridge,

above Ginekah). Open marsh by stream; mud by pool; marshy meadows, 2290–2800m. Fr. June–September.

This plant comes closest to the N European and N American var. **pulchella** (Lonnroth) B. Schmid, having very small spikes, small, shortly beaked utricles and narrow leaves.

54. C. alopecuroides D. Don ex Tilloch & Taylor var. **alopecuroides**; *C. japonica* sensu F.B.I. p.p. & F.E.H.1, non Thunberg. Lep: *mongsher*. Fig. 38p.

Tufted, spreading by long creeping rhizomes covered with cream, ribbed scales; bases of leaf-sheaths cream, not persistent, margins splitting into fibrils; bladeless sheaths long. Leaves sub-basal and sheathing lower ⅓ of culm, often with upper culm leaf, blades more than half length of culm, 2.6–5.5mm wide. Culm 26–45cm, acutely trigonous. Infl. of 1 terminal male spike and 3–5 female spikes, all erect, congested, shortly peduncled (lowest peduncle 0.8–3.5cm); male spike linear, 2.3–5cm; female spikes narrowly cylindric, 2.2–7cm, utricles finally spreading; bracts leaf-like, greatly exceeding infl., not sheathing. Utricles slightly swollen, narrowly ellipsoid-trigonous, gradually tapering to apex, margins thickened, 2.5–3.5 × 0.9–1mm, glabrous, strongly ribbed, straw-coloured; beak 0.6–1mm, minutely serrate, aperture weakly bidentate, hyaline, often reflexed; stigmas 3. Female glumes narrowly lanceolate, 2.1–2.8 × 0.6–1mm, hyaline, midrib green, acuminate into excurrent, scabrid point (0.6–1mm). Male glumes narrowly oblong to lanceolate, subacute to finely acuminate with scabrid tip to 1.1mm, 3.4–4 × 0.6–1.4mm, hyaline with green midrib.

Bhutan: C — Thimphu (hill above Thimphu Hospital) and Tongsa (Tama) districts; **Sikkim** (Gangtok, Lachung, Tumlong, Baghghora); **Darjeeling** (Rimbick to Ramam, Ghum Forest, Rangit, Tiger Hill, Tonglu, Labdah, Rungnoo Valley). Marsh beside stream; forest, 1220–2400m. Fr. April–May.

var. **chlorostachya** C.B. Clarke.

Differs in having wider leaf blades (5.5–11mm), larger (4–4.5 × 1.1mm) utricles more gradually tapered into a longer (c.1.5mm) beak, apex herbaceous, erect, deeply bifid.

Bhutan: C — Thimphu district (Drugye Dzong); **Darjeeling** (Rangit, Batasia to Palmajua). Damp ground among rocks and cliffs, 1530–3050m. Fr. April–May.

Records from **Darjeeling** (Senchal to Takdah (F.E.H.1)) and **Bhutan: S** — Chukka/ Phuntsholing district (Chimakothi to Phuntsholing (F.E.H.2)) under *C. japonica* subsp. *chlorostachys* and *C. doniana* must be referred to *C. alopecuroides* s.l.

55. C. jackiana Boott. Fig. 38q.

Rhizomes creeping, stems tufted. Bases of leaf sheaths pale brown, persistent, not becoming fibrous. Leaves basal, blades shorter than culm, 4.4–6mm wide, rather thin-textured. Culm 38–62cm, winged. Infl. of 3–5 spikes on upper ⅔ stem, upper congested, ± sessile, lower distant (lowest internode 11–27cm) on long peduncles (lowest 4–18cm); terminal spike male, 1.2–3cm; lower 2–4 female (or upper with some apical male flowers), 1.1–2.6cm, lax, with relatively few, erect utricles; lowest bract with long, sheathing base and leaf-like blade shorter than infl. Utricles strongly trigonous, 5.5–6 × 1.2–1.7mm, faces lanceolate, gradually tapering to apex, glabrous, ribs prominent, many, pale green, aperture hyaline, oblique, notched; stigmas 3. Female glumes oblong to obtrullate, narrowed to subacute apex, folded around base of utricle, 4.4–7.5 (incl. awn) × c.2mm, hyaline; midrib green, 3-ribbed, excurrent to 1–2.5mm. Male glumes narrowly elliptic, subacute, c.4.9 × 1.6mm, midrib 2-ribbed, not excurrent.

Bhutan: C — Tongsa district (below Tongsa Rest House). Grassy slope, 2350m. Fr. May.

56. C. fusiformis Nees subsp. **fusiformis**; *C. finitima* sensu F.B.I. p.p. & E.F.N. Fig. 38r.

Rhizome slender, ?short; plant tufted. Bases of leaf sheaths cream, usually becoming chestnut brown, not becoming fibrous; bladeless sheaths short. Leaves mostly basal, or with one on lower part of culm, blades gradually tapered to trigonous apex, half length to almost equalling culm, 2.9–6.2(–6.5)mm wide. Culm (21–)34–89cm, angles rounded. Infl. of 1 male (occasionally with a few terminal female flowers) and 4–7 female spikes, topmost (1–)2–3 spikes subsessile, clustered, others overlapping, inserted singly along upper half of culm on slender, erect peduncles; male spike linear, 2–3.8cm, female spikes narrow, 2.5–8.5(–11)cm, lax, utricles erect; bract blades exceeding infl., very finely tapered at apex, sheath of lowest 2–4cm. Utricles narrowly ellipsoid-trigonous, margins thickened, (4.7–)5–6.2(–7.2) × 1.1–1.4mm, glabrous, pale green, shining, gradually narrowed upwards into long (2–3.5mm) beak with hyaline tip, aperture oblique becoming torn; stigmas 3. Female glumes oblong-oblanceolate, fimbriate-truncate to subacute, (2.5–)3.2–4.1 × 1.2–1.8mm, keeled, hyaline or with pale brown markings towards centre, midrib green, if excurrent then point not projecting beyond apex. Male glumes oblong, acute, c.5.4 × 1.2mm.

Bhutan: C — Thimphu (Drugye Dzong, above Taba, Thimphu to Dochu La, Dechencholing to Punakha) and Bumthang (W side of Ura La) districts; **Darjeeling** (Tonglu, Sandakphu); **Sikkim** (Lachen, Kulhait, Karponang, Kalapokri). Wet slopes and streamsides in conifer (*Picea*/*Tsuga* and *Pinus*/ *Picea*) forest; cliffs, 2250–3500m. Fl. May; fr. April–September.

subsp. **finitima** (Boott) Noltie.
Differs from subsp. *fusiformis* as follows: leaf-sheath bases more intensely coloured (reddish-purple), shining; leaf blades broader (5.5–7.2mm), shorter (11–22cm), more abruptly contracted to apex; bract blades not so finely tapered, sheaths longer (lowest 3–7.5cm); male glumes smaller (c.4.6 × 1.8mm), with broad, blunt apex; utricle beak under 2mm.
Sikkim/Darjeeling (?Tonglu).

57. C. fastigiata Franchet. Fig. 39a.
Rhizome stout, short, woody; plant tufted. Bases of leaf sheaths cream, becoming dark brown, not becoming fibrous. Leaves mostly basal, or with one on lower part of culm, blades to half length of culm, c.3.5mm wide. Culm to 44cm. Spikes in clusters of c.4, terminal sessile, lateral clusters to 5, long-peduncled on upper half of culm, sometimes also with sub-basal one; peduncles drooping, slender; bracts shorter than peduncles. Spikes slender, all ± similar, 2–4.5cm, terminal (and sometimes lower ones) of each cluster gynaecandrous, lower ones sometimes female, utricles erect. Utricles narrowly ellipsoid-trigonous, gradually narrowed into beak, margins thickened, 4.5–5 × 1mm, glabrous, pale green; beak c.1.5mm, apex hyaline, aperture oblique, margins minutely scabrous; stigmas 3. Female glumes oblong-elliptic, subacute, c.3 × 1mm, pale brown with broad hyaline margins. Male glumes pale whitish-brown.
Bhutan: C — Bumthang district (above Gortsam); **N** — Upper Bumthang Chu district (Lambrang). Muddy gravel on track through fir forest; yak pasture, 3520–4000m. Fr. August.

Some specimens from Sikkim (Lachen, 3048 & 3353m, *Hooker* s.n., K) differ from *C. fusiformis* and *C. fastigiata* in being much stouter with wider (7.8–12.7mm) leaves, long female spikes (9–11cm) and long (3–4.5mm), aristate female glumes; the terminal spike is entirely male. They resemble the Nepal specimen (*Stainton, Sykes & Williams* 6062) identified by Koyama as *C. fastigiata* and probably represent an undescribed taxon close to *C. fastigiata*, but further collections are required.

58. C. laeta Boott subsp. **laeta**. Fig. 35f, Fig. 39b.
Rhizome short, thick, woody, stems densely tufted. Bases of leaf sheaths pale brown, persistent as dull, reddish-brown fibres. Leaves all basal, blades

FIG. 39. **Cyperaceae** XIV (**Carex** utricles: spp. with 3 stigmas).
a, **C. fastigiata**. b, **C. laeta** subsp. **laeta**. c, **C. laeta** subsp. **gelongii**. d, **C. setosa**. e, **C. olivacea**. f, **C. oedorrhampha**. g, **C. desponsa**. h, **C. munda**. i, **C. speciosa** subsp. **speciosa**. j, **C. speciosa** subsp. **dilatata**. k, **C. inclinis**. l, **C. fragilis**. m, **C. setigera**. n, **C. schlagintweitiana** subsp. **deformis**. o, **C. inanis**. p, **C. griersonii**. q, **C. breviculmis** subsp. **royleana**. r, **C. haematostoma**. s, **C. hebecarpa**. All × 12. Drawn by Mary Bates.

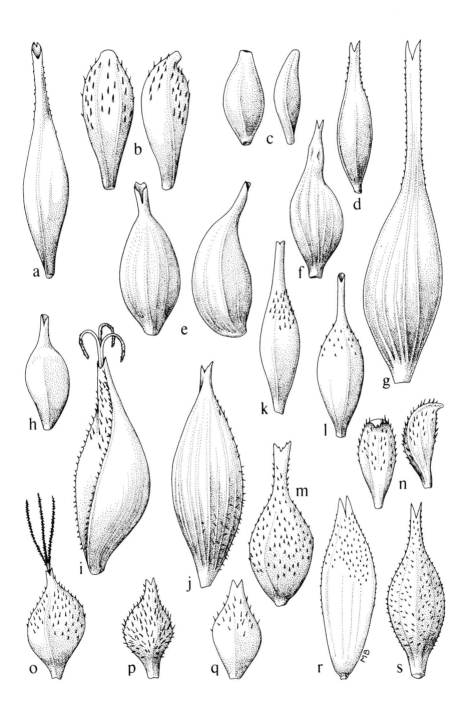

much shorter than culms, 1–1.6mm wide. Culm 8–41cm, slender, flexuous, rounded-trigonous. Infl. of 2–4 nodding, slender spikes, the upper on slender peduncles slightly exceeding spikes, lowest commonly sub-basal on long (to 26cm), filiform peduncle resembling a culm; terminal spike male, 0.7–1.3cm; lower 1–3 female, 1–1.7cm, rather lax, utricles suberect; bracts with ± filiform, serrulate blade, shorter than adjacent spike, sheath loose, mouth oblique and margins hyaline. Utricles knobbly, narrowly obovoid-trigonous, margins thickened, with short bristles, attenuate to base, 3–3.4 × 1–1.3mm, faces (or outer only) shortly hairy, abruptly shouldered to extremely short (c.0.2mm), deflexed beak, pale green tinged orange-brown, aperture entire; stigmas 3. Female glumes widely oblong-obovate, truncate (sometimes ovate, acute), base truncate, 2.2–3.3 × 1.5–2.6mm, folded round utricle, chestnut-brown with narrow hyaline margins, midrib green, 3-ribbed, very shortly (0.2–0.5mm) excurrent as scabrid point.

Bhutan: N — Upper Mo Chu district (above Laya, below Chhew La); **Sikkim** (Tsomgo, Tungu); **Chumbi** (Yatung). Damp turf among boulders; wet rocks; bare hillsides, 3660–4315m. Fr. June–September.

var. **major** Boott.

Female spikes longer (1.5–2.8cm), denser, on longer, more erect peduncles; utricles larger (3.5–4mm); female glumes broader, chocolate brown.

Chumbi (Gantsa, foot of Natu La); **Sikkim** (Tungu). 3660–3960m. Fr. June–July.

subsp. **gelongii** Noltie. Fig. 39c.

Differs from subsp. *laeta* in its more slender habit, smaller in all its parts; utricles 1.7–2.5mm, glabrous, elliptic in outline, distinctly beaked, beak 0.3–0.7mm, not deflexed.

Bhutan: C — Thimphu district (below Darkey Pang Tso N of Paro); **N** — Upper Bumthang Chu district (above Lambrang); **Chumbi** (Yatung). Wet, flushed areas beside streams, 3200–4950m. Fr. June–August.

59. C. setosa Boott. Fig. 39d.

Rhizomes short, woody; stems densely tufted. Bases of leaf sheaths pale brown, persisting as stiff, dark brown fibres. Leaves basal and with 2 on lower part of culm, sometimes also with an intermediate bract-like leaf, blades flat, shorter than culm, 1.5–2.4mm wide. Culm (8–)11–74cm, rather slender. Infl. lax; male spike 1, terminal, (0.8–)1.6–2.9cm; female spikes (2–)3–5, (1.1–)2.2–5.3cm, very slender, lowest flowers distant, lowest spike occasionally with small basal branch; peduncles long, slender, nodding, lowest (0.9–)3.5–8cm; bracts with stiff blades shorter than adjacent spike, sheaths long, apex oblique, brown-hyaline, rather widely open. Utricles linear-ellipsoid-trigonous, distinctly shouldered, margins thickened, sparsely setose, 2.8–3.5 × 0.7–1mm, body glabrous or sometimes shortly hairy near margins, shining, pale green becoming suffused chestnut; beak 0.6–1.2mm, conspicuously notched, margins setose; stigmas 3. Female glumes oblong-ovate, usually blunt (upper ones in spike more acute), 2.2–3 × 1–1.6mm, reddish-brown to chestnut, with extremely narrow hyaline margins, midrib green, scabrid, very shortly excurrent (0.2–0.9mm). Male glumes narrowly oblanceolate, 6–7 × 1.2–2mm, paler, acute to aristate.

Bhutan: C — Thimphu district (above Ragyo N of Paro); **Sikkim** (Lachen, Namdee, Jelep La, Potang La, Tsomgo, Kopup, Phalut, Chamnago). Shady rocks; streamsides; wet meadows, 3000–4270m. Fl. May–June; fr. June–September.

Close to *C. laeta* (especially var. *major*) differing in its utricles: beak long, aperture notched, hairs longer, restricted to the margins and the usually longer, more slender spikes. A specimen from Sakden district (Orka La, 3962m, *Kingdon Ward* 13731, BM) is intermediate between *C. setosa* and *C. laeta* var. *major*: it has stems to 36cm, female spikes chocolate brown, to 3.5cm, female glumes narrowly oblong 3.7–4.2 × 1.2–1.6mm, utricles to 4mm, beak short (0.3–0.5mm), bristly on margin. The whole group (i.e. Sect. Digitatae) is clearly in need of revision in the Sino-Himalaya.

60. C. olivacea Boott. Fig. 39e.
Rhizomes relatively slender, spreading; bases of leaf sheaths cream, dull, not splitting into fibrils. Leaves basal, with 1 on culm, blades rather stiff, gradually narrowed to very acute apex, exceeding culm, (0.7–)1.2–1.8cm wide. Culm (30–)44–91cm, stout. Infl. of 1 terminal male and 3–9 female (the uppermost sometimes male at apex) spikes, all erect, very shortly peduncled, upper congested, sometimes with a distant female spike on long (to 9cm) peduncle; male spike 3–6.7cm; female spikes cylindric, 2.2–8(–16)cm, utricles spreading to deflexed at maturity; bracts leaf-like, blades broad, greatly exceeding infl., bases very shortly sheathing. Utricles curved, turgid, ovoid-trigonous, abruptly contracted into beak, 2.9–4 × 1.3–1.8mm, glabrous, strongly ribbed, dark green becoming purplish-brown; beak 0.5–0.9mm, fuscous-purple, aperture oblique, scarcely notched; stigmas 3. Female glumes narrowly lanceolate, 2.9–5.2 × 0.6–1mm, dark reddish-purple with green midrib excurrent into long (1.5–3.5mm) smooth or scabrid awn. Male glumes oblong, 4.3–6.5 × 1–1.1mm truncate, or emarginate midrib very shortly excurrent, sometimes subacute, purplish with green midrib, margins hyaline especially at apex when young.

Bhutan: S — Samchi district (Chamurchi); C — Thimphu (Thimphu to Dochu La, Jujo near Paro, above Motithang, Simtokha to Talukah) and

Bumthang (above Dhur) districts; **N** — Upper Kulong Chu district (Tobrang); **Duars** (Rechi La); **Sikkim** (unlocalised Hooker specimen). Marshes and streamsides in clearings or forest (blue pine and mixed), (305–)2000–3050m. Fl. April–May; fr. June–September.

It is rather odd that the Sikkim and Samchi records are from subtropical altitudes (305–914m) whereas the other specimens are all from over 2000m.

61. C. oedorrhampha Nelmes; *C. tumida* Boott non Beilschmied. Fig. 39f.

Rhizomes short, stems tufted; base of leaf sheaths dull, cream becoming reddish-brown, margins fibrillose; bladeless sheaths present. Leaves mostly basal, usually with 1 or more on lower part of culm, blades very acute, equalling to exceeding infl., 2.4–6.6mm wide. Culm 45–57cm, sharply trigonous. Infl. of 1 terminal male spike and 2–4 female spikes, all erect, upper ones crowded, lower 1–2 female spikes distant; male spike linear, 4–6.5cm; female spikes narrowly cylindric, 4–8.3cm, utricles suberect even at maturity; peduncles short, visible part of lowest one 1–4.5cm; bracts exceeding infl., bases sheathing. Utricles slightly inflated (nut not filling utricle), ellipsoid-trigonous, fairly abruptly contracted into beak, 3.3–3.6 × 1–1.5mm, rather thin-textured, distinctly ribbed, olive-brown becoming purplish-brown; beak 1–1.2mm, lower part stout, upper part tapering, sometimes appearing swollen in middle, aperture scarcely notched; stigmas 3. Female glumes linear-lanceolate, 3–3.3 × 0.5mm, hyaline with broad, 3-veined, green midrib excurrent into long (1.4–2mm) scabrid point. Male glumes linear, 5–12 × 0.5–0.8mm, hyaline, with green midrib excurrent into very long (1.6–6mm) scabrid point.

Bhutan: C — Tashigang district (Tashi Yangtsi Chu); **Darjeeling** (Kurseong, Takdah). Streamside in oak forest, 1600–1980m. Fr. April–May.

62. C. desponsa Boott. Fig. 39g.

Rhizomes short, stems tufted. Bases of leaf sheaths and bladeless sheaths of vegetative shoots reddish-purple. Leaves all basal, blades exceeding culms, (4–)5.1–5.5mm wide, cross-veinlets prominent, rather thin-textured. Culms (30–)45–54(–90)cm. Infl. of (5–)6(–7) erect, non-overlapping spikes on upper ⅔ of culm, upper sessile, lower on peduncles (lowest 4.3–7.5cm); terminal spike male, 2.3–2.6cm; lower spikes androgynous, predominantly female, male section from merely a few flowers to almost half total length, 2–5 × 0.5–0.7cm, utricles dense, erect; bracts leaf-like, sheaths long, upper part of inner face brownish-membranous. Utricles lanceolate in outline, gradually tapered into long beak, c.7 × 1.8mm, glabrous, shining, two sides greenish with prominent thickened ribs, one side pinkish-brown, not prominently ribbed; beak 2.5–3mm, margins conspicuously serrate, aperture hyaline, deeply notched; stigmas 3.

Female glumes elliptic to oblong, 2.5–4 (excl. awn) × 1.6mm, apex gradually narrowed or truncate (sometimes asymmetric), hyaline, midrib green, strongly 3-ribbed, excurrent as long (1.6–4mm) scabrid awn. Male glumes narrowly elliptic, acute to shortly mucronate, c.6 × 1.3mm, brown.

Bhutan: C — Tongsa district (1km S of Tongsa). Bank in *Quercus* forest, 2250m. Fr. May.

63. C. munda Boott. Fig. 39h.

Rhizome creeping, but usually condensed; stems inserted in a single row. Bases of leaf sheaths cream or fawn, bladeless sheaths short. Leaves mainly in elongate, non-flowering rosettes, 2–3 on lower part of culms, blades half length to equalling culms, 2.1–3.3mm wide, often with wrinkled patches on blade. Culm 21–44cm, minutely scabrous on angles. Infl. of 2–6 spikes on long, filiform peduncles, lowest usually single, upper 2–3 per bract; bracts leaf-like, blade not exceeding infl., sheaths long, close-fitting. Terminal spike androgynous with conspicuous male section (0.6–1cm), often branched (with 1–2 subsidiary spikes at base), other spikes usually with a few terminal male flowers, or all female. Predominantly female spikes very narrow, 1.2–2.5cm, utricles erect, lax. Utricles narrowly ellipsoid-trigonous, gradually tapered into beak, 2.2–3 × 1–1.1mm, glabrous, prominently nerved on one face, olive-green; beak curved, 0.3–0.5mm, aperture truncate; stigmas 3. Female glumes oblong to narrowly obovate, subacute, 1.8–2.5 × 1–1.6mm, pale brown, midrib green, margins hyaline. Male glumes oblanceolate, blunt, c.2.6 × 0.8mm, pale brown, margins hyaline.

Bhutan: C — Thimphu (Dotena to Barshong, E side of Chile La, above Pajoding, above Ragyo) and Mongar (above Sengor) districts; **N** — Upper Kulong Chu district (Shingbe); **Darjeeling** (Sandakphu; Phalut (F.E.H.1)); **Sikkim** (Lachen, Dzongri, Laghep, Nathu La, Kopup, Lungthung, etc.); **Chumbi** (Gantsa). On rocks, banks, pathsides and wet screes in *Abies/ Rhododendron* forest; wet, shaded alpine turf; in rhododendron scrub, 3050–4270m. Fl. June; fr. June–October.

64. C. speciosa Kunth. subsp. **speciosa**. Fig. 39i.

Rhizomes short, stems tufted. Bases of leaf sheaths pale, persisting as brown fibres. Leaves basal, blades exceeding culm, 3–5.8mm wide. Culm 19–46cm, rounded-trigonous, relatively stout. Infl. of 2–4 spikes, on short (scarcely emerging from sheaths), stiffly erect peduncles, evenly disposed along stem (not overlapping); spikes all similar, androgynous, 2.5–5.2cm, whitish-green, concolorous, female section c.⅔ total length, dense, utricles erect; bracts leaf-like, blades exceeding infl., bases sheathing. Utricles very strongly trigonous, ellipsoid, tapered above to shallowly notched (hardly beaked) aperture,

4.2–5 × 1.7–2mm, far exceeding glumes, adaxial face and angles shortly hairy, conspicuously ribbed, ribs more than 15 per face, equally thickened; stigmas 3. Female glumes encircling axis, ovate, apex rounded, 2.4–3mm long, greenish-white. Male glumes subacute, 2.5–2.6mm.

Darjeeling (Sureil, Tista, Mumchele Keelong). Habitat not recorded, 305–1520m. Fr. June–August.

subsp. **dilatata** Noltie. Fig. 39j.

Differs from subsp. *speciosa* in its wider (1–1.6cm) leaves, shorter than culms; taller culms (56–73cm); spikes 5–6, longer (3–5.6cm); utricles smaller (4–4.5mm), only just exceeding glumes, ribs fewer than 10 per face, unequally thickened; female glumes longer (3–3.5mm), subacute.

Bhutan: C — Punakha district (above Lomitsawa); **Darjeeling** (Ramam). Steep, shady slopes in moist broad-leaved forest, 1830–2200m. Fr. August.

subsp. **pinetorum** Noltie.

Differs from subsp. *speciosa* in its more slender habit; utricles smaller (2.2–4mm), shortly hairy on all faces, scarcely ribbed, less strongly trigonous; glumes acute.

Bhutan: S — Chukka district (above Lobnakha); **C** — Thimphu district (below Motithang Hotel, hill above Thimphu Hospital, ridge W of Thimphu Valley, below Sharadango Gompa). Dry blue pine/oak forest, 2500–2750m.

65. C. inclinis C.B. Clarke. Fig. 39k.

Rhizomes slender, extensively branched and spreading. Bases of leaf sheaths purplish-red, shining, not decaying into fibres. Leaves mainly basal, blades greatly exceeding (to twice length) culms, 2.2–3.3mm wide. Culms 9–19cm, hidden among leaves. Infl. of 4–8 spikes, on slender, erect peduncles on upper part of stem, inserted singly or in pairs, blades of bracts not usually exceeding infl., sheaths long, closely fitting. Spikes all androgynous, or terminal 1–2 all male; androgynous spikes 1.4–3cm, male part shorter than to equalling female, utricles relatively lax, spreading at maturity. Utricles ellipsoid, gradually tapered into slightly deflexed beak, 3.4–4 × 1mm, glabrous or minutely hispid, olive-brown; beak 0.8–1.2mm apex hyaline, aperture ± truncate; stigmas 3. Female glumes lanceolate, gradually narrowed to very acute apex, 3.2–4.5 × 1.4–2mm, pale- to reddish-brown, borders hyaline. Male glumes lanceolate, very acute, 5–6.2 × 2mm, midrib pale, keeled.

Bhutan: C — Thimphu (Tzatogang to Dotanang, Nala to Tzatogang, above Talukah Monastery, Dechencholing to Punakha, above Changri Monastery) and Bumthang (above Gortsam) districts; **Darjeeling** (Sandakphu, Tonglu, Batasi to Palmajua); **Sikkim** (Lachen). Wet banks in conifer (incl.

Abies, *Tsuga* and *Picea*/*Pinus*) forest, occasionally on rocky slopes, 2300–3960m. Fl. May; fr. May–August.

66. C. fragilis Boott. Fig. 39l.
Differs from *C. inclinis* in its very contracted (but creeping) rhizomes, stems inserted in a single row, lack of colouring in sheath bases and glumes and longer male sections to spikes.
Sikkim (Lachen, Lachung). 2740–3350m.

Recent specimens from Sikkim (unlocalised) and Yunnan confirm this to be a distinct species; previously only known from Hooker's type collections.

67. C. setigera D. Don; *C. wallichiana* sensu F.B.I. p.p. (Sikkim plant). Fig. 39m.
Rhizomes stout (c.3mm diameter), extensively spreading, stems tufted. Basal leaf sheaths dark purplish-brown splitting into rather stout, reticulate fibrils; bladeless sheaths short. Leaves inserted along lower half of culm, blades longer than culms, 1.9–4.6mm wide. Culms 21–47cm, rounded-trigonous. Infl. of 4–7 erect, overlapping spikes, upper sessile, lowest on peduncle 0–3.2cm; upper 1–4 spikes male, 3.5–4.7cm, terminal occasionally with some female basal flowers, other male spikes smaller; lower 3–6 spikes androgynous, 2.2–6.3cm, female in lower ½–⅔, very dense, utricles ± spreading; bracts leaf-like, blades greatly exceeding infl., sheaths rather loose. Utricles narrowly ellipsoid-trigonous, gradually tapered upwards into beak, margins thickened, 3–3.8 × 1–1.6mm, densely covered with short, stiff hairs, olive-brown; beak 0.6–1mm, aperture membranous, brownish, bidentate; stigmas 3. Female glumes oblong to lanceolate, acuminate, 3–4.2 (excl. awn) × 1–2mm, orange-brown with hyaline margin near apex, midrib keeled, projecting as long (1.2–3.5mm) scabrid awn. Male glumes narrowly oblanceolate, c.6 × 1.6mm, dark chestnut midrib paler, excurrent as scabrid awn.
Bhutan: C — Thimphu (Thimphu to Dochu La, above Changri Monastery), Punakha (E side of Dochu La), Tongsa (Tongsa to Tratang) and Mongar (W side of Donga La) districts; **N** — Upper Mo Chu district (Gasa to Chamsa); **Darjeeling** (Tonglu; Ghum to Lopchu (F.E.H.1)); **Sikkim** (Chungtam). Wet forest (incl. *Tsuga*); rocks in evergreen riverine forest, 2130–3000m. Fl. April; fr. April–June.

68. C. schlagintweitiana Boeckeler subsp. **deformis** Noltie. Fig. 39n.
Rhizomes slender, spreading, clothed with purplish-red, fibrous scales. Bases of leaf sheaths pale reddish-fawn. Leaves sub-basal, blades exceeding culms, 1.6–2.5mm, wide, rather weak. Culms 17–25cm, very slender. Infl. of 3–4 delicate, erect spikes near top of stem, upper overlapping; peduncles erect,

slender, lowest 0.4–1.5cm; terminal spike male, 1.4–2.1cm; lower 2–3 spikes female (upper sometimes with a few terminal male flowers), 1–1.8 × c.0.3cm, moderately dense, utricles suberect; bracts with narrow leaf-like blades — lowest exceeding infl., bases sheathing. Utricles narrowly obovoid, apex deflexed, 1.4–1.7 × c.0.7mm, densely hairy, becoming marked reddish-brown; beak extremely short (under 0.2mm), aperture entire; stigmas 3. Female glumes lanceolate, acuminate into short (under 0.5mm) awn, 2.3–3.3 (total length) × 0.6–1.2mm, reddish-brown, margins hyaline, midrib green, keeled. Male glumes linear-lanceolate, c.3.5 × 0.6mm.

Bhutan: C — Thimphu district (above Talukah Monastery, hill above Thimphu Hospital, below Ragyo); **Chumbi** (Pipitang, Yatung). Rocky hillsides and screes; open forest (spruce, fir, oak/pine), 3050–3960m. Fr. May–June.

69. C. inanis C.B. Clarke. Fig. 39o.

Rhizomes short, woody, stems densely tufted. Basal leaf sheaths splitting into rather weak fibrils, bases persisting as stiff, erect, purplish-brown fibres. Leaves inserted on lower half of culm, blades usually longer than culms, 1–2.2mm wide. Culm 6–42cm (very variable depending on grazing, etc.), rounded-trigonous. Infl. of 4–7 erect, subsessile spikes crowded at apex of culm (sometimes also with a more distant, peduncled spike); terminal spike male, 0.8–2.2cm; lower 3–6 spikes female or with very short terminal male section, 1.2–2.3cm, densely cylindric, sometimes with 1–2 short, basal branches, utricles ± spreading; bracts greatly exceeding infl., sheaths close-fitting. Utricles knobbly, broadly ellipsoid- to narrowly-obovoid-trigonous, abruptly contracted into short beak, (1.5–)1.9–2.4 × 1–1.4mm, upper part densely covered with short, whitish, spreading hairs, olive-brown; beak 0.2–0.4mm, weakly bidentate; stigmas 3. Female glumes oblong to lanceolate, acute, sometimes minutely emarginate, 1.7–2.7 × 0.9–1.7mm, midrib not to slightly projecting beyond tip, orange-brown, margins hyaline, midrib green. Male glumes narrowly oblanceolate, midrib strongly keeled, produced into shortly excurrent tip.

Bhutan: S — Chukka district (Chukka Colony to Taktichu, W of Gedu); **C** — Ha (Shenkoona to Ha), Thimphu (Paga, near Taksang Monastery, W side of Dochu La, Dotena to Barshong, below Lobnakha), Tongsa (above Rukubji), Bumthang (above Dhur) and Mongar (Sengor) districts — probably under-recorded; **Darjeeling** (Ghum to Lopchu, Birch Hill, Lloyd Botanic Garden, Senchal); **Sikkim** (Kangling, Karponang, Lachen, Tumbok, Islumbo, Domang, Yoksum to Mintok Khola, etc.); **Chumbi** (Yatung). Grassy hillsides, pastures and rock-clefts; wet shady cliff, 1500–4270m. Fl. June; fr. July–November.

70. C. griersonii Noltie. Fig. 39p.

Differs from *C. inanis* in its more slender habit and its slender, extensively spreading rhizomes; female spikes smaller (0.5–1.4 × 0.3–0.5cm), more compact; female glumes fuscous-purple.

Bhutan: C — Ha district (opposite Ha Bazaar). Rock crevices and around rocks in dry, open grassland, 2600m. Fr. June.

71. C. breviculmis R. Brown subsp. **royleana** (Nees) Kükenthal. Fig. 39q.

Rhizomes short, stems densely tufted. Bases of leaf sheaths pale brown, persisting as dark brown fibres. Leaves basal and 1–2 on lower ⅓ of culm, blades shorter than to equalling culms, 1.9–2.8mm wide. Culms (3–)10–27cm, acutely trigonous. Infl. of 3–4 erect, overlapping, subsessile spikes (occasionally also a long-peduncled sub-basal one); terminal spike male, 0.6–1.1cm; lower 2–3 female, 0.7–1.5cm, densely cylindric, utricles suberect; lowest bract with blade equalling to slightly exceeding infl., base not sheathing. Utricles knobbly, ellipsoid-trigonous, gradually tapered to apex, 2.1–2.5 × 0.9–1.1mm, surface covered with short, scattered hairs, pale green, scarcely beaked, aperture shallowly notched; nut with annulus at apex; stigmas 3. Female glumes oblong or obovate, truncate, emarginate or narrowed at (sometimes asymmetric) apex, 1.7–2.3 (excl. awn) × 1–1.9mm, silver-hyaline sometimes with brown markings, midrib green, prolonged into long, (1–2.4mm), serrate, point. Male glumes narrowly oblanceolate, gradually tapered upwards into awn.

Bhutan: C — Thimphu district (Thimphu, ridge W of Thimphu Valley, hill above Thimphu Hospital; Mishina to Thimphu, Dotanang to Thimphu (F.E.H.2)); **Chumbi** (Yatung). Open grassy ridge; bare ground among open pine/oak scrub; wet places, 2250–3050m. Fr. May–July.

72. C. haematostoma Nees; *C. nakaoana* T. Koyama; *C. bhutanica* T. Koyama. Fig. 39r.

Rhizomes short; stems densely tufted. Bases of leaf sheaths yellowish-brown, margins becoming fibrous, old leaves and sheaths persistent. Leaves basal and 1–3 on lower part of culm, blades shorter than culm, 1.7–4.5mm wide, becoming inrolled. Culm 21–108cm, rather smooth. Infl. 6–30cm; male spikes single or in unequal pairs, 1–3(–7), terminal, crowded, cylindric to clavate, 1.2–4cm; female spikes single or in pairs, narrowly cylindric, sometimes branched at base, 2–5(–8), 1.2–3.5(–4) × 0.2–0.3cm, lowest flowers slightly distant, uppermost sometimes male at apex, peduncles short and erect or long and filiform (so spikes nodding), lower spikes not to slightly overlapping; blade of lowest bract slender, usually not exceeding infl., sheath with rather wide, oblique mouth. Utricle biconvex, narrowly elliptic, not beaked, 3.3–5.5 × 0.7–1.7mm, aperture deeply cleft, margins and faces of upper part densely scabrid, ± ribbed, green below, becoming fuscous-purple in upper part; stigmas

3. Female glumes narrowly to broadly rhombic to obovate, acuminate, 2.5–3.9 × 1–1.6mm, minutely scabrid to varying degrees, purplish, margins hyaline at apex, midrib green, scabrid, usually shortly (0–1.5mm) excurrent. Male glumes 3.2–4.2 × 1–2.4mm.

Bhutan: C — Ha (Ha La to Kyu La) and Thimphu (above Pajoding, below Darkey Pang Tso) districts; **N** — Upper Mo Chu district (below Gangyuel, E bank of Tharizam Chu, above Laya, Soe-Lingshi-Yale La, Shingche La); **Sikkim** (Tangu, Thangshing, Dzongri, Lachen, Yeumtong, Samdung, Jelep La, etc.); **Chumbi**. Open, grazed grassland, block-scree, etc. chiefly above treeline; banks in *Betula utilis* woodland, 3200–5030m. Fl. June; fr. June–October.

Occurring over a wide altitudinal range and correspondingly variable (in stature, and size and hairiness of glumes and utricles): large plants have been described as *C. bhutanica*; smaller, more slender ones as *C. nakaoana*. Field observations confirm that the variation is continuous and not worth treating taxonomically. However, a dwarf form (8–20cm) from extreme altitudes and habitats (e.g. glacial gravel) has a very distinct appearance, having slender, stiffly erect spikes and smaller in all its parts. Its distribution is as follows:

Bhutan: C — Thimphu district (below Laname Tso); **N** — Upper Mo Chu district (Yale La); **Sikkim** (Phalung, Naku Chu Valley, Llonak, Lunrip, Samiti, Chaunrikiang); **Chumbi** (Phari, Chomo Lhari). 4267–5182m.

Some specimens of this form have been determined as *C. nakaoana*, but a photograph of the type of that species reveals it to be typical (if smallish) *C. haematostoma*.

An interesting sedge resembling *C. haematostoma* is known from two gatherings (Marsyandi Valley, Nepal, *McBeath* 1547 (E) and Dotha, Chumbi, 3810m, *Bor & Ram* 20462 (K)). Further collections are required, but it is probably related to *C. sempervirens* Villars subsp. *tristis* (M. Bieberstein) Kükenthal, from which it differs in its much larger spikes. Leaves coarse, blades c.half length culm, 6–7.2mm wide; sheath bases cream, persistent. Culm 21–31cm, acutely trigonous. Infl. of 4–6 spikes, upper 2 male, others female (some male at apex), upper 3 spikes shortly peduncled, congested, lower 2 spikes distant on long (3–5cm), erect peduncles; bracts leaf-like, shorter than infl., bases sheathing. Female spikes cylindric, dense, 1.8–2.7 × c.0.5cm; male spikes c.2.5cm. Immature utricles narrowly lanceolate, 5mm, gradually tapered into long beak, body shortly hairy, margins strongly setose; beak 1.5mm, aperture notched, becoming fuscous. Female glumes linear-lanceolate, subacute, 6.5–7 × 1.6–1.8mm, fuscous with wide hyaline margins at apex.

73. C. hebecarpa C.A. Meyer. Fig. 39s.

Rhizome short, woody, stems tufted. Culms to 50cm, hispid on the sharp angles, bases wiry, purplish, bearing bladeless sheaths. Leaves evenly disposed along stem from middle upwards, blades c.4mm wide; auricles conspicuous, shortly reddish-hairy. Infl. a terminal cluster of 1 male and 2 or more female spikes, lower 1–3 spikes entirely female, borne singly, shortly peduncled, bracts leaf-like, exceeding infl. Male spike c.1.5cm, whitish. Female spikes 1–2.5cm, utricles erect, slightly overlapping. Utricles strongly trigonous, ellipsoid, gradually narrowed into short (c.0.7mm) beak, c.3.5 × 1.5mm, faces densely hispid; stigmas 3, short. Female glumes ovate, c.1.5 × 1.2mm, whitish, margins sometimes becoming brown, midrib green, shortly excurrent.

Bhutan: C — Thimphu district (valley above Leprosy Hospital, Gidakom). Dry deciduous woodland, 2200m. Fr. August.

C. hebecarpa var. *maubertiana* (Boott) Franchet has been recorded for Sikkim by Kükenthal (1909) but I have not seen the specimen; it is scarcely worth recognising being slightly stouter, with denser female spikes and wider, more congested leaves than the typical variety.

Doubtfully recorded and imperfectly known species:

C. eleusinoides Turczaninow ex Kunth.
   Recorded from Sikkim (Na Tong, *Dungboo*, s.n., 1878. CAL, K) in Rao & Verma (1980). The specimens are very immature and impossible to name, but are perhaps dwarfed forms of *C. fucata* or else an undescribed species.

C. fedia Nees.
   Recorded for Sikkim in F.B.I. (as *C. wallichiana* Prescott) but specimen re-determined as *C. setigera*; there is, however, a correctly named Kurz specimen at CAL labelled 'Sikkim'.

C. kingiana C.B. Clarke non Léveillé & Vaniot.
   Described from Pheedong, Sikkim, ex herb. King — specimen not found at CAL; description inadequate. Ghildyal & Bhattacharyya (1985) performed a less than useful service in giving it a new name (*C. sahnii*) without augmenting the description or citing any specimens or the location of the type.

C. sclerocarpa Franchet.
   This Chinese (Sichuan) species has been recorded from Chukka/Phuntsholing district (Chimakothi to Phuntsholing, 300–2250m (F.E.H.2)) — specimen not seen and rather unlikely; possibly refers to *C. fusiformis*.

Family 233. POACEAE (GRAMINEAE) will form Volume 3 Part 2 of the
Flora of Bhutan

Family 234. ARECACEAE (PALMAE)

Armed or unarmed shrubs, dwarf to tall trees or slender climbers. Stems solitary or clustered, usually unbranched, pleonanthic (flowering in successive years) or hapaxanthic (flowering once and dying). Leaves spirally inserted, occasionally distichous, often forming a crown. Sheaths initially tubular, sometimes forming a smooth apical cylinder (crownshaft), sometimes spiny or bearing spiny whips (flagella). Leaves plicate in bud, commonly splitting to give a 1(–2)× pinnate or palmate blade, leaflets reduplicate (Λ-shaped) or induplicate (V-shaped); rachis sometimes extended into an armed whip (cirrus). Hermaphrodite, monoecious, dioecious or polygamous (i.e. with bisexual and unisexual flowers). Infl. axillary, lateral or apparently terminal, sometimes borne in axils of fallen leaves so infrafoliar, a simple spike to many times compound, subtended by large or small, deciduous or persistent, 2-keeled prophyll (spathe), bracts and bracteoles commonly tubular, axis sometimes developed into armed flagellum. Flowers borne on rachillae, single or variously grouped (e.g. in triads), unisexual or bisexual, commonly sessile, calyx 3-lobed, lobes often fused at base; corolla 3-lobed, lobes often fused at base. Stamens (3–)6 to many; filaments free or fused into collar, bases free or united to base of corolla; anthers basifixed or dorsifixed; staminodes usually developed to some degree in female flowers. Ovary of 3 free or partly fused carpels, syncarpous with 3 locules or pseudomonomerous with a single fertile ovule; each carpel or locule with a single ovule; separate or fused styles developed or not; pistillode developed to varying degrees in male flowers, often trifid; ovary sometimes covered with scales. Fruit indehiscent, usually 1-seeded, mesocarp dry, fleshy (fruit drupe-like) or fibrous (fruit nut-like). Seed with large endosperm.

Recent collections of this family are woefully lacking and old specimens are fragmentary and poorly annotated. The Sikkim species, however, are relatively well known, having been well-described by Anderson (1871), Griffith (1850), Hooker (F.B.I.) and Beccari (1890, 1908, 1911, 1913, 1931): their accounts have been used extensively in the following account. The generic descriptions rely heavily on Uhl & Dransfield (1987). It is highly

---

Fig. 40. **Arecaceae** I (habits).
a, **Caryota** cf. **urens**: hapaxanthic palm in fruit showing bipinnate leaves (× ¹⁄₂₀₀).
b, **Areca catechu**: showing crown shaft and infrafoliar infls. (× ¹⁄₈₀). c, **Calamus acanthospathus**: pleonanthic rattan showing flagella and lateral, flagellate infls. (× ¹⁄₁₀₀).
d, **Plectocomia himalayana**: hapaxanthic rattan with terminal infl. (× ¹⁄₅₀).
e, **Trachycarpus** cf. **khasianus**: fan-leaved palm (× ¹⁄₁₂₀). f, **Wallichia densiflora**: stemless palm with praemorse leaflets (× ¹⁄₃₀). Drawn by Glenn Rodrigues.

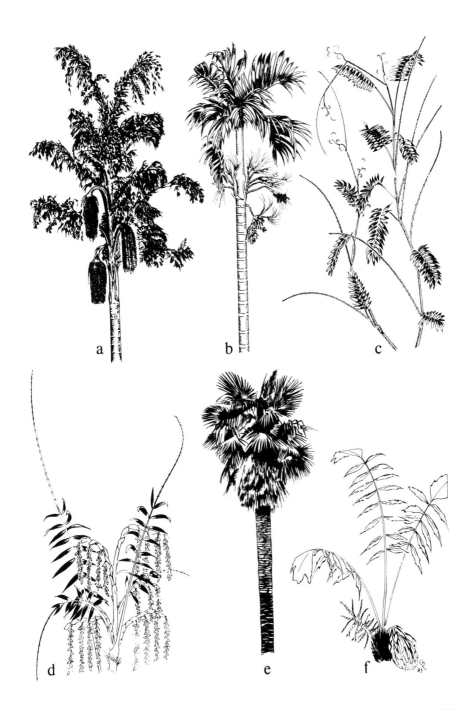

a

b

c

d

e

f

desirable that new collections and ethnobotanical observations be made on this interesting and useful family in Bhutan.

The Shachop name *lai-ree* supposedly refers to a species of *Corypha*; it no doubt refers to some other species of fan-leaved palm.

1. Leaves fan-shaped 2
+ Leaves pinnate 4

2. Leaves divided to base into wedge-shaped, reduplicate segments; dwarf palm, infl. with few (to 5) drooping rachillae .... **3. Licuala**
+ Leaves partially divided into acuminate, induplicate segments; trunk tall, bare for most of length; infl. much branched with many stiff, spreading rachillae 3

3. Margins of petiole crenulate; dioecious (NB female flowers with staminodes); carpels free at apex .... **1. Trachycarpus**
+ Margins of petiole spiny; hermaphrodite (flowers bisexual); carpels free below but united into common style .... **2. Livistona**

4. Leaves twice pinnate, leaflets wedge-shaped; tall hapaxanthic tree **10. Caryota**
+ Leaves simply pinnate 5

5. Leaf sheaths spiny, often also with spiny flagella from sheaths or spiny cirri from leaf rachises; slender climbers 6
+ Leaf sheaths and other parts not spiny; not climbing, habit various (stemless or with slender to stout trunk) 8

6. Leaflets irregularly fascicled, with long, filiform apices; sheaths with needle-like spines on wavy lamellae (Fig. 41p); leaf sheath lacking a knee; hapaxanthic; rachillae hidden by oblong bracts .... **7. Plectocomia**
+ Leaflets regularly spaced, inserted singly, apices not or very shortly filiform; spines on sheaths not as above; leaf sheath usually with a conspicuous knee; pleonanthic; rachillae not hidden by bracts 7

7. Flowers sessile; female corolla (visible even in fr.) scarcely exceeding calyx (Fig. 41f); prophyll small .... **5. Calamus**
+ Flowers shortly stalked; female corolla c.twice length of calyx (Fig. 41o); prophyll large, spiny, enclosing infl. in bud . **6. Daemonorops**

8. Tips of leaflets acute or acuminate, margins smooth 9
+ Tips of leaflets praemorse (irregularly toothed), margins usually with praemorse steps 12

9. Trunk almost smooth with distant annulae, crown-shaft prominent; infl. infrafoliar 10
+   Trunk rough and closely annulate or not developed, crown shaft not prominent; infl. interfoliar 11

10. Stems slender (under 5cm diameter); infl. simple, spike-like, female flowers distributed all along rachillae .... **11. Pinanga**
+   Stems stouter (over 20cm diameter); infl. branched to 3 orders; female flowers only at base of rachillae .... **12. Areca**

11. Leaflets induplicate, lower modified as spines; dioecious; trunk short or lacking if tall then fruits small .... **4. Phoenix**
+   Leaflets reduplicate, lower not spine-like; monoecious; trunk tall; fruits massive .... **13. Cocos**

12. Leaves distichous on a trunk, leaflets inserted in groups
8.2. Wallichia disticha
+   Stemless, leaves not distichous, leaflets inserted singly 13

13. Sepals of male flowers free to base; leaflets narrowly oblong, symmetrical about midrib, margins scarcely stepped but toothed especially near apex .... **9. Arenga**
+   Sepals of male flowers tubular below; leaflets oblong to rhombic, asymmetric, margins conspicuously stepped ..... **8.1. Wallichia densiflora**

1. TRACHYCARPUS H. Wendland

Pleonanthic; trunks usually solitary, moderately tall, usually clothed with persistent petiole bases at least in upper part. Leaf sheaths becoming fibrous. Petiole long, margins lacking spines. Leaves fan-shaped, induplicate, partially divided into single-fold segments, bifid at apex. Dioecious, or sometimes polygamous; infls. interfoliar, branched to 4 orders; peduncles bearing 1–3 coriaceous bracts tubular at base; prophyll tubular at base; rachillae bearing spirally arranged flowers, solitary or in 2s or 3s, on a small tubercle, each subtended by a small bracteole. Male and female flowers similar: calyx with 3 lobes united at base, not overlapping; corolla lobes imbricate, exceeding calyx. Male flowers: stamens 6, filaments fleshy, anthers basifixed, latrorse, pistillodes 3, small. Female flowers: carpels 3 each with a single, sub-basal ovule, style short, staminodes 6. Fruit purplish-black with pale bloom, circular to oblong in outline, kidney-shaped in section, mesocarp thin; seed grooved.

411

234. ARECACEAE (PALMAE)

1. T. fortunei (Hooker) H. Wendland.
Trunk to 7m, slender, clothed with fibrous remains of leaf sheaths at least beneath crown. Crown dense, subspherical. Leaf sheaths short. Petioles to 0.9m, margins crenulate. Leaves to 1.5m diameter, suborbicular, with c.40 segments, divided irregularly to halfway or more, segments under 2cm wide, transverse veinlets scarcely visible. Dioecious. Infls. 1–1.5m, branched to 3 orders, spreading, branches very stiff, bracts coriaceous, becoming orange-brown, longest to 30cm, concealing rachis. Male flowers not seen. Female flowers borne singly, yellowish-brown, globose-ovoid, calyx lobes c.1.5 × 2mm, broadly ovate, subacute, corolla lobes c.2 × 3mm, broadly ovate, subacute, carpels densely white-hairy, usually 2 aborting; staminodes — filaments c.0.8mm, flat, triangular; antherodes c.0.5mm, lobes divergent, cordate. Fruit c.6 × 9mm, irregularly reniform, blackish, glaucous.

Darjeeling (Lloyd Botanic Garden, Windamere Hotel etc.); **Sikkim** (Gangtok). Planted in gardens.

Native to N Burma and SW China but commonly cultivated for ornament in temperate regions.

Further work is needed on the genus in our area. The commonly planted species appears to be *T. fortunei*; however, *T. martianus* (Wall. ex Martius) H. Wendland (Nep: *kasru*; Lep: *talaerkop*) has been recorded as being planted (e.g. in the Lloyd Botanic Garden) and there are old records from Darjeeling: Rissom; Rungbong (Beccari, 1931) when it was apparently native in Upper Hill Forest (1200–1980m). *T. martianus* differs chiefly in its fruit being longer than wide and having a naked trunk. It was described from C Nepal, but the Khasian *T. khasyanus* (Griff.) H. Wendland has usually been taken to be synonymous with it. A plant seen in Chukka District (Lobnakha, 2700m) (Fig. 40e) agreed with Khasian specimens in having a naked trunk and with leaves glaucous beneath and scurfy when young. A pair of very distinctive trees apparently belonging to this genus are planted at the Windamere Hotel Darjeeling. They have elongate fruits and very wide (over 3cm) leaf segments, but clearly belong to a different species.

2. LIVISTONA R. Brown

Vegetatively similar to *Trachycarpus* but larger and petioles often with spiny margins. Differing chiefly in having bisexual flowers. Filaments connate in fleshy ring, anthers medifixed; carpels 3, free below, united by a common, elongate style; seed not grooved.

1. L. jenkinsiana Griff. Lep: *talainyom, purbong*.
Trunk annulate, 6–9m, grey. Petiole to 2m, margins with deflexed spines (decreasing in size upwards). Leaves to 0.9–1.8m diameter, divided to about halfway, segments 70–80, bidentate, thin in texture, transverse veinlets promi-

nent when dry. Flowers in clusters of up to 8–10. Mature buds ovoid, 3.5–4 × 2.5–3mm; calyx divided to about halfway, lobes broadly ovate, obtuse; corolla about twice length calyx, divided to about halfway, lobes triangular, acute. Fruit globose, 2.2–2.8cm diameter, bluish.

Darjeeling (Kungbheek Jhora, Tista, Sivoke Hills; Mahanadi Valley (Gamble, 1922)), near Sittong (Anderson, 1871)). Lower Hill Forest. 910m.

According to Gamble, leaves used by Lepchas for thatching and as umbrellas.

L. chinensis (Jacquin) Martius. Cultivated for ornament at Kurseong and characterised by its large leaves with elegantly drooping leaf-tips.

Photograph seen of another species planted in the Terai at Sarbhang: possibly **L. rotundifolia** (Lamarck) Martius which has a tall (to 15m), prominently ringed trunk; crown elongate, petioles spiny only near base, leaves suborbicular, divided only to c.⅓.

3. LICUALA Thunberg

Pleonanthic; acaulous or trunks short, solitary or clustered, clothed with persistent petiole bases. Leaf sheaths becoming fibrous. Petioles unarmed or with spines on margins. Leaf blade palmate, divided to varying degrees into cuneate, truncate-dentate, 1–several-fold, reduplicate segments. Usually hermaphrodite. Infl. interfoliar, spicate to branched to 3 orders, prophyll and peduncle bracts similar, tubular. Rachillae with spirally arranged usually bisexual flowers borne singly or in 2s or 3s, subtended by small bracteoles. Calyx sometimes stalk-like below, tubular with 3 short lobes; corolla lobes exceeding calyx. Stamens 6, filaments ± free or all or 3 fused into a collar to various degrees; anthers small, latrorse; carpels 3, free below, united into slender style above, ovules basal. Fruit globose to spindle-shaped, often brightly coloured, mesocarp usually thinly fleshy, fibrous.

No specimens seen: descriptions taken from literature (Uhl & Dransfield, Beccari and Griffith).

1. L. peltata Roxb. ex Buchanan-Hamilton. Lep: *tale lama kuri.*
Trunks clumped, stout, 1.5–4.5m. Petiole 2.4–3.6m, spiny on margins, spines recurved. Leaf blades 1.2–1.8m diameter, circular in outline, divided to base into 12–30, subequal segments. Infls. several, branched to 1 order, peduncle exceeding leaves, tightly sheathed with bracts. Rachillae 3–5, lowest 25–30cm, pendulous, flowers numerous, solitary on short tubercles. Flowers

in bud 15–17 × 7mm; calyx shortly toothed; corolla twice length of calyx, lobed to halfway, lobes triangular, acute. Stamens 6, filaments distinct, only lower parts (adnate to corolla) connate; ovary hairy; style exceeding ovary, stigma obscurely 3-lobed. Fruit c.2cm, oblong-ovoid, orange, bearing abortive carpels at apex.

Darjeeling (valleys near Tista River (Gamble, 1922)). Lower Hill Forest, to 1830m.

4. PHOENIX L.

Pleonanthic; trunks solitary or clustered, very short to tall, often clothed with persistent petiole bases. Leaf sheaths fibrous. Leaf blades imparipinnate, leaflets borne singly or in fascicles, induplicate, lower ones modified as spines. Dioecious, infls. interfoliar, branched to 1 order, borne on a flattened scape, subtended by prophyll; prophyll tubular at base, often deciduous; flowers subtended by a small bracteole. Male flowers: calyx cup-like, 3-pointed, corolla lobes 3, exceeding calyx, fused only at base, stamens 6, anthers dehiscing latrorsely, attached basally. Female flowers globose, persistent: corolla lobes very wide, strongly overlapping, staminodes small, carpels 3, connate, each with a single, basal ovule and a single, recurved stigma. Fruit oblong-ovoid, mesocarp fleshy; seed elongate, grooved.

1. Leaflets inserted singly and regularly, spreading in a single plane, usually over 1.5cm wide .... **3. P. rupicola**
+   Leaflets in interrupted fascicles, spreading in more than one plane, usually under 1.5cm wide 2

2. Tree, trunk to 15m, petiole bases not persistent .... **4. P. sylvestris**
+   Dwarf tree, trunk to 3m, clothed with persistent petiole bases 3

3. Trunk slender, to 3m; infl. exserted from prophyll on long scape (over 10cm); female rachillae slender, fruits not crowded .... **1. P. loureiri**
+   Stem bulbous, extremely short; infl. scarcely exceeding prophyll; female rachillae stout with fruits densely crowded .... **2. P. acaulis**

1. P. loureiri Kunth; *P. humilis* Royle ex Beccari. Nep: *takal.*

Trunk to 3m, rough with persistent leaf bases. Leaf sheaths reddish-brown, reticulately fibrous. Leaflets fascicled in irregularly distant groups of 2–3, spreading in more than one plane, longest (16–)26–29cm, widest 0.4–1.6cm wide, midrib distinct but not keeled on lower surface, glabrous; basal spines to 5cm. Scape elongate (over 10cm), flattened; prophyll reddish-brown, base tubular, sheathing, blade becoming lacerate. Male infl. 10–13cm, club-shaped;

rachillae to 8.5cm, slender (c.0.9mm), tortuous, irregularly inserted on axis, stiffly erect. Male flowers subtended by minute bracteole, calyx c.1mm, cup-shaped ± truncate with 3 short points, membranous; corolla lobes c.5 × 2.5mm, oblong, convex, apex rounded, hooded; anthers c.3mm, filaments c.1mm. Fruiting female infl. similar but larger (14–21cm) and stouter; rachillae longer (to 19cm), stouter (2–4mm wide), fruits well-separated; calyx c.2.5mm deep, subglobose; corolla lobes c.2.5 × 4mm, broadly transversely elliptic, apex slightly retuse, strongly overlapping; carpels c.2mm. Fruit c.1.5 × 1cm, oblong-ellipsoid; calyx persistent, c.4mm diameter.

Duars (Dhupguri, Salgara Terai); **Darjeeling** (above Rangit, Badamtam, Barnesbeg to Singla Bazaar, above Sukna, Darogadhara; Pashok, Sivoke (Gamble, 1922)). Dry Sal forest and grassland on lower hills and in terai, 300–760m. Fl. November; fr. January–May.

There is confusion over the name and identification of this species. The widely used *P. humilis* unfortunately appears not to have been validated until 1890, and so *P. loureiri* has precedence if it applies to the same plant.

2. P. acaulis Roxb. Nep: *thakal*.

Leaves similar to *P. loureiri* but differing in its very short, bulbous stem; scapes very short (under 10cm), infls. scarcely exceeding prophylls. Female fruiting specimens very distinct in having short (under 10cm), very stout rachillae, with fruits densely crowded and larger, persistent flowers (c.6mm diameter).

Terai (Singi Jhora; Siliguri (Beccari,1890)). Terai forest.

Most old records of this species refer to *P. loureiri* as probably does Anderson's (1871) information that '*P. acaulis*' is called *schap* by the Lepchas who eat its astringent fruits.

3. P. rupicola T. Anderson. Nep: *tarika*; Lep: *schiap*.

Differs from *P. loureiri* as follows: leaves (to 2m) with leaflets closely and regularly inserted, alternate to subopposite, in a single plane, usually wider (1.4–2.6cm), midrib keeled on underside, often tomentose, petiole spines usually under 5cm; rachillae longer (female to 28cm) in horizontal fascicles.

Bhutan: S — Samchi (Chepuwa Khola SE of Samchi, Khagra Valley near Gokti), Sankosh (2km W of Pinkhua), Gaylegphug (near Tatapani) and Deothang districts (Deothang (Griffith teste Anderson, 1871)); **C** — Tongsa district (Tama); **Darjeeling** (Sivoke, Birick, Nimbong; Tista and Mahanadi Valleys (Gamble, 1922)). Ravines and shaded cliffs in subtropical (Lower Hill) forest; steep rocky hillside, 360–1220m. Fl. May–June.

Interior of stem eaten uncooked by Lepchas (Gamble, 1922).

4. P. sylvestris (L.) Roxb. Lep: *kubong, rotong*; Eng: *wild date*.

Trunk to 15m; roughly annulate but petiole bases not persistent; leaflets fascicled, spreading in more than one plane.

Darjeeling (below Barnesbeg, Mongpu, Lal). Planted occasionally in the terai and lower valleys up to 1220m.

Fruits eaten; sap used to make toddy and sugar in India.

The following three genera comprise the commercially important rattans (Nep: *bet*). Canes are used for basket and furniture making, fibres as a rope, the bitter shoots eaten in curries and the fruits of some species chewed as a stimulant. According to Basu (1985, pp. 87 and 215), cultivation trials of *C. guruba, C. latifolius, C. acanthospathus, C. tenuis, C. flagellum, C. leptospadix* and *Daemonorops jenkinsiana* have been made in W Bengal (Buxa and Sukna), plantings being under *Shorea robusta* and *Lagerstroemia speciosa*. In Bhutan rattans are collected from the wild and therefore probably under threat; it is probably no coincidence that on a recent visit to Darjeeling the only species observed were species with useless canes.

5. CALAMUS L.

Pleonanthic. Stems slender, erect or scandent, single or clustered. Petiole kneed at junction with sheath in climbing species. Leaves paripinnate, leaflets inserted regularly or in groups, terminal pair sometimes partly fused, midrib conspicuous, usually with 2 or more thickened parallel costae, costae and margins often setose. Rachis developed into an armed, climbing whip (cirrus) or not. Leaf sheath usually heavily armed, inner face forming an ocrea, sometimes splitting and persisting as auricles; armed whip-like flagellum (modified infl.) often produced on sheath. Dioecious, infls. lateral, commonly armed, much branched (male commonly to 3 orders, female to 2 orders), subtended by a tubular (often splitting later), prophyll, primary axis sometimes ending

---

FIG. 41. **Arecaceae** II (rattans).

a–b, **Calamus erectus** var. **schizospathus**: a, leaf base showing spiny petiole and ocreate leaf sheath (× ⅕); b, armature of leaf sheath. c, **C. flagellum**: armature of leaf sheath. d, **C. leptospadix**: armature of leaf sheath. e–f, **C. tenuis**: e, armature of leaf sheath; f, fruit (× 2). g–i, **C. acanthospathus**: g, armature of leaf sheath; h, leaf base showing flagellum (× ¼); i, flagellate infl. (× ½). j, **C. guruba**: armature of leaf sheath. k, **C. latifolius**: armature of leaf sheath. l, **C. inermis**: leaf sheath. m–o, **Daemonorops jenkinsiana**: m, armature of leaf sheath; n, infl. showing persistent spathes (× ⅕); o, fruit (× 2). p–q, **Plectocomia himalayana**: p, armature of leaf sheath; q, leaf apex showing cirrus (× ⅕). Armature of leaf sheaths all × ½. Drawn by Glenn Rodrigues.

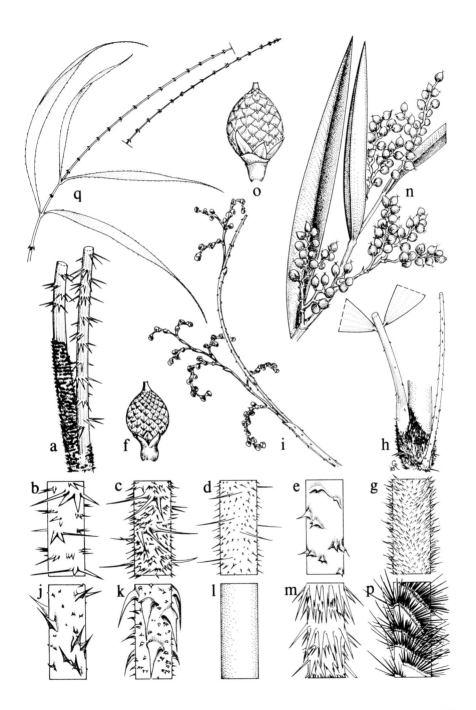

in an armed flagellum; partial infls. subtended by prophyll-like bracts. Rachillae long or condensed, flowers commonly distichous; flowers usually sessile, calyx deeply 3-lobed, tubular at base, corolla 3-lobed, tubular at base. Male flowers borne singly subtended by cup-like bracteole and funnel-shaped bract, corolla exceeding calyx, stamens 6, anthers dorsifixed, usually latrorsely dehiscing, filaments stout, free. Female flowers accompanied by a cup-like bracteole and a sterile staminate flower, together subtended by a funnel-shaped bract, corolla about equalling calyx, ovary partly 3-loculed, with 3 basal ovules, scaly on outside, stigmas 3, stout, style usually short, persistent, ovary surrounded at base by a corona of 6 fused filaments bearing sterile anthers. Fruit covered by rhombic, downward pointing ± shining scales, bearing a single seed, apiculate with persistent style.

The genus is very poorly collected in Bhutan (due presumably to collecting difficulties arising from their large size and prickly nature); additional species could well occur. Care should be taken to collect representative specimens (i.e. male and female infl., mature leaves and stems with leaf sheaths together with adequate notes on habit and on the parts not shown on the small fragments possible to preserve on herbarium sheets).

The Sikkim and Darjeeling names of rattans given in Gamble (1922) and Cowan & Cowan (1929) have been corrected from a manuscript by Prain at E.

The following uses are given for our species in Basu (1985): alpenstocks (*C. inermis, C. latifolius*); baskets (*C. acanthospathus, C. guruba, C. latifolius, C. leptospadix, C. tenuis, D. jenkinsiana, P. himalayana*); chair bottoms (*C. guruba, C. tenuis*); furniture frames (*C. acanthospathus*), umbrella handles (*C. acanthospathus, C. latifolius*).

The following Shachop names have been recorded for species of *Calamus*, but no vouchers seen so application is uncertain: *menji* (supposedly for *C. tenuis*), *fashi* ('bigger than *menji*, with branch'), *day* ('bigger than *menji*, without branch').

1. Leaflets under 2cm wide 2
+   Leaflets over (2.5–)3cm wide 4

2. Sheaths with bristly, spiny auricles, largest spines on sheaths not decur-
    rent at base; rachis rusty-red tomentose; infl. flagelliform, rachillae
    scorpioid .... **3. C. leptospadix**
+   Sheaths with auricles not developed or long, papery and deciduous,
    largest spines on sheaths with decurrent bases; rachis lacking rusty-red
    hairs; infl. flagelliform or not, rachillae not scorpioid 3

3. Infl. flagelliform, bracts of partial infl. blade-like above, not armed, exceeding partial infls.; flowers minute (under 2.5mm) ...... **6. C. guruba**
+   Infl. not flagelliform; partial infl. bracts tubular, armed, much shorter than partial infls.; flowers larger (over 3mm) .... **4. C. tenuis**

4. Leaves cirrate 5
+   Leaves ecirrate 6

5. Leaf sheaths unarmed; rachillae long, lax, axis developed between flowers so zigzag .... **8. C. inermis**
+   Leaf sheaths heavily armed, the larger spines with decurrent bases; rachillae short, dense, not obviously zigzag .... **7. C. latifolius**

6. Leaf sheaths not flagelliferous, sheath auricles long, bristly, persistent; rachis with straight, flattened spines, usually in half-whorls   **1. C. erectus**
+   Leaf sheaths flagelliferous, auricles very short or deciduous; rachis with a single row of singly-inserted hooks 7

7. Leaflets with conspicuous midrib and weaker lateral veins; rachillae long, straight or slightly flexuous; bracts funnel-shaped, overlapping
    2. C. flagellum
+   Leaflets with 5–7 equal costae; rachilla short, recurved; bracts very small, not overlapping .... **5. C. acanthospathus**

1. C. erectus Roxb. var. **schizospathus** (Griff.) Beccari. Sha: *kheershing*; Nep: *pekri* (*phekri, phekori*) *bet*; Lep: *rhom*; *rong*. Fig. 41a–b.
    Stems 2.5–6m, erect, tufted. Sheaths 2.5–3.5cm wide, eflagellate, spiny above; ocrea 16–24cm, densely covered with transverse lamellae bearing blackish-brown bristles; auricles 4–9cm. Spines on sheaths, lower part of rachis and petiole in half-whorls of 3 or more, long (to 3cm), straight, flattened; upper part of rachis fawn-tomentose with singly inserted, straight, flattened spines. Petiole to 80cm, not kneed. Leaves to 1.5(–5)m, ecirrate; leaflets to 35 each side, subopposite, linear-lanceolate, longest 40–74cm, widest (2.5–)3–4.4cm wide, midrib sparsely setose on both surfaces, lateral veins weaker, many, margins setose especially near finely acuminate apex, terminal pinnae wider, fused at base. Infls. branched to 1 or 2 orders, 34–100cm, eflagellate, covered with short, pale brown felt, pendent; prophylls and infl. bracts tubular, splitting on one side, rapidly lacerate. Male rachillae to 30cm, stout, flexuous, bracts overlapping, funnel-shaped, 3.5–5mm long, 6–7mm diameter, flowers sessile on cup-like bracteole; calyx c.6mm, lobed to c.⅓; corolla lobes 6.5–8 × 2.5–3.7mm; anthers 4–7mm, filaments c.7mm. Female infl. similar to male,

usually less compound; bracts longer, more tubular (c.7.5mm long); flower —
calyx c.5mm, lobed to c.halfway; corolla lobes c.2mm, scarcely exceeding calyx.
Fruit 2.8–3.8 × 2.5cm, ellipsoid, scales 0.6–0.9cm wide, orange-brown with
dark brown intramarginal line, grooved.

Bhutan: S — Samchi (near Gokti), Sarbhang (Noonpani to Tori Bari),
Gaylegphug (Lodarai Khola) and ?Deothang (Rizoma) districts; **C** — Tongsa
district (Pertimi to Tintibi Bridge); **Terai** (Chamokdangi, Jogikhola); **Darjeeling**
(Mongpu, Chunbati, below Rimbik, Bamunpokri, Sivoke, Ryang, Lal bhanjan,
Rongbe, Rongsong, Rangpo to Tista). Subtropical (Lower Hill) forest,
305–1220m. Fl. November–April; fr. April–August.

According to Gamble (1922) the commonest rattan in Lower Hill Forest but canes
useless. Fruit chewed as a stimulant in Bhutan.

2. C. flagellum Griff. Lep: *rhim, rheem*. Fig. 41c.

Differs from *C. erectus* as follows: scandent; sheaths flagellate, spines more
slender and more scattered, auricles short and deciduous; spines on petiole
and lower rachis inserted singly; upper part of rachis with singly borne, short,
downward pointing, hooked spines; infl. flagellate.

Bhutan: S — Samchi district (Chamurchi Forest (M.F.B.)); **Darjeeling**
(Rongbe, Dulkajhar, Kurseong, Great Rangit, Lal, 1km above Mongpu).
Middle Plains and Lower Hill Forest, 610–1360m. Fl. April–July; fr. June.

Sterile specimens from Samchi (Buduni, 380m, Nep: *phegre bet*), Sarbhang (Burborte
Khola near Phipsoo, 290m) and Gaylegphug (Taklai Khola, 360m) districts probably
belong to this species which they resemble in their leaves, and sheath and petiole
armature. However, they lack flagella but this may be because they are immature.

Seeds eaten according to Hooker; canes soft, only used for tying (Gamble, 1922).

3. C. leptospadix Griff. Nep: *kukhre bet* (Bhutan), *phekori* (Darjeeling); Lep:
lat. Fig. 41d.

Stems 2–6m, slender, scandent. Sheaths 1–1.5cm wide, usually with slender
flagella, reddish-brown tomentose, bearing long, straight, flattened (to 23 ×
1.7mm), singly inserted spines and shorter, bristle-like ones; ocrea 4.5–7.5cm,
densely covered with dark rusty brown hairs and spines; auricles 1–2.5cm,
spiny. Petiole kneed, 25–42cm with long, straight, singly inserted spines. Rachis
densely rusty-hairy towards apex, with a single row of spines on underside,
the distal ones deflexed. Leaf blade 85–100cm, ecirrate; leaflets many (to 35
each side), closely set, alternate, linear, longest to 26cm, widest 0.8–1.2cm
wide, glossy, midrib and 2 lateral costae sparsely setose above and beneath,
margins setulose especially near shortly (to 2cm) filiform apex, terminal pinnae
free at base. Infls. branched to 2 orders, very slender, flagellate, erect, partial

infls. few, distant; prophylls and infl. bracts short, armed, tubular, splitting on one side, not lacerate; partial infls. dense, the tubular rachilla bracts overlapping. Male rachillae, to 3cm, dense (bracts overlapping), decurved; bracts 2.5–3mm diameter, ovate, open, apiculus ciliate; bracteole almost equalling bract, flowers sessile; calyx c.3mm, lobed to halfway; corolla lobes c.3.5 × 1.2mm, oblong, shortly and widely acuminate; anthers c.2mm, filaments c.3mm. Female rachillae c.1cm, stout, erect, dense; bracts open, c.3mm diameter, bracteoles of female flower c.2.5mm diameter, cup-shaped. Fruit c.1cm diameter, subspherical, apiculate; scales c.2.5mm diameter, pale brown with dark brown intramarginal stripe, grooved.

Bhutan: S — Samchi (Deo Pani Khola), Sarbhang (Kami Khola, Lam Pati 10km E of Sarbhang) and Gaylegphug (Taklai Khola E of Gaylegphug) districts; **Duars** (Buxa Reserve); **Darjeeling** (Tista, Great Rangit, Rishap, Sivoke). Subtropical forest; damp places along rivers forming thickets, 305–910m. Fl. February–April.

Canes thin and useless according to Gamble (1922).

4. C. tenuis Roxb. Fig. 41e–f.

Climber. Sheaths to 2cm wide, usually with slender flagella, spines rather few, flattened, deflexed, with decurrent bases; auricles very short, papery. Petiole to 15cm. Rachis with a single row of stout, deflexed hooks on underside. Leaf blade to 1m, ecirrate; leaflets many, closely set, alternate, linear-lanceolate, longest to 30cm, widest to 1.9cm wide, midrib and 2 weak lateral costae sparsely setose above, setae long (to 1cm), dark, margins setulose especially near acuminate apex, terminal pinnae free to base. Infl. (only female seen) branched to 2 orders; prophyll long, tubular, not lacerate, bearing deflexed hooks; partial infls. distant bearing many spreading rachillae subtended by tubular-funnel-shaped bracts. Rachillae dense, to 7cm; bracts funnel-shaped, c.1.5mm long, c.2.3mm diameter; bracteole of female flower very small, rather open; calyx c.3.5mm, shortly stalked in fruit; corolla lobes 1.8 × 1mm. Fruit small, ellipsoid c.0.5 (+ 0.2) × 0.4mm, scales 1.5mm diameter, pale yellowish-brown.

Duars (Muraghat; Titalya to Kishenganj, Sath Bhaia Jhar (Gamble, 1922)). Marshy places in terai. Fr. January.

No recent specimens seen.

5. C. acanthospathus Griff. Dz: *tsim*; Sha: *minji*; Nep: *pukka bet, gouri bet*; Lep: *rue, rhu*. Fig. 40c, Fig. 41g–i.

Stems to 4m or more, scandent. Sheaths 2–4cm wide, usually flagellate (flagella to 5.6m), densely covered with spines; longer spines to 1.5cm, confluent at base into short, horizontal groups, straight, flattened, erose, scurfy, smaller bristle-like spines also present; mouth obliquely truncate, auricles very short (c.2mm), bristly. Petiole short (0–21cm), scarcely kneed. Rachis pale fawn felted, with single row of singly inserted, deflexed hooks on underside, upper side spiny especially near base. Leaf blade 0.9–1.6m, commonly imparipinnate, ecirrate; leaflets usually alternate (upper sometimes subopposite), rather few (to 8 each side), distant, narrowly elliptic, plicate, widest to 3.5–5.6cm wide, longest to 21–41cm; costae 5–7, subequal usually glabrous, occasionally setulose above; margins setulose especially near acuminate apex; terminal pinnae free at base. Infls. (only female seen) branched to 2(–3) orders, flagellate, partial infls. distant, short, recurved; prophylls and infl. bracts tubular, not lacerate. Female rachillae to 8cm, stout, slightly zigzag, recurved; flowers rather distant, bracts shortly tubular (c.1mm), collar-like, bracteoles of female flowers, saucer-shaped, 3–4mm diameter; flowers sessile, calyx c.3.5 × 4mm, widely cylindric, lobed to halfway; corolla lobes equalling calyx, c.3 × 2mm. Fruit c.2 × 1.5cm, ellipsoid, apiculate; scales 0.4cm wide, orange-brown with reddish-brown intramarginal line, erose-ciliate especially at apex.

Bhutan: S — Sarbhang district (2.5km below Getchu on Chirang road); **C** — Tongsa (Shemgang) and Mongar (above Shersing Thang) districts; **Darjeeling** (Pugraingbong, Karmi). Dense broad-leaved (Lower and Middle Hill) forest, 910–1900m. Fl. March–July; fr. February.

Used in Bhutan for its cane and as rope; apex of young stems eaten. According to Gamble (1922) the best species of cane and used for walking sticks, bridges and chair making, but implies that it was, even then, over-collected.

6. C. guruba Buchanan-Hamilton ex Martius. Nep: *dute bet, dudheya bet*. Fig. 41j.

Slender climber, to 4m. Sheaths 0.6–1.5cm wide, with relatively few, singly inserted spines, spines straight, flattened, horizontal or upward-pointing, bases decurrent; flagella long, slender; auricles long (6–7cm), reddish-brown, papery, becoming torn, then deciduous. Petiole minutely kneed, short (c.11cm) unarmed or with few long, straight spines. Rachis with dark rusty brown hairs and strongly deflexed spines. Leaf blade 68–76cm, with many (to 30 each side), linear, closely placed pinnae, terminal 2 free, ecirrate; pinnae alternate, linear, longest 17–28cm, widest 1.2–1.7cm, with midrib and 2 lateral costae sparsely setulose above, margins setulose especially at filiform apex. Male infl. flagellate, branched to 2 orders, axis spiny, prophyll and infl. bracts very long (exceeding partial infl. they subtend), tubular only at base, upper part blade-like; partial infl. stiffly erect, with stiff secondary branchlets bearing rachillae. Rachillae 0.5–1.5cm, zigzag; bracts open, ovate, c.1mm wide; bracteoles smaller. Calyx

c.1.5 × 1mm, very shallowly lobed. Corolla lobes c.2.3 × 0.7mm. Anthers c.0.8mm. Female infl. branched to 1 order; rachillae longer (to 7.5cm in fruit). Fruit subspherical, body c.5.5mm, beak c.1mm; scales c.1mm wide, yellowish with reddish-brown intramarginal line, margin broadly hyaline.
Terai (Jalpaiguri, Marabari, Sath Bhaia Jhar, ?Gazal). Fl. November–April; fr. December.

Hamilton's spelling (on specimen from his own herbarium at E) is 'gurula' from the Bengali vernacular *gurul beti*.

7. C. latifolius Roxb. Nep: *putle bet*; Lep: *gorot, rubee*. Fig. 41k.
Climber. Sheath 1.2–3.5cm wide, reddish-brown, armed with large and small spines — large singly inserted, to 2.5cm, deflexed, flattened, bases decurrent, mouth oblique, auricles minute, papery. Petiole kneed. Leaves to 2.5m, cirrate with downward-pointing spines and hooks on rachis and cirrus; pinnae relatively few (to 15 each side), longest to 19–37cm, widest 4–7cm, elliptic with 5 or more glabrous costae, alternate, sometimes paired each side of rachis, distant, margins setulose especially near acuminate apex. Male infl. branched to 2–3 orders; prophyll tubular, small (2.5–4cm), armed, partial infl. bracts narrowly funnel-shaped, the larger armed; secondary branches bearing evenly spaced rachillae; rachillae short (to 2.5–6cm), very dense, recurved; bracts 1.5–2.7mm long, rather open (very shortly funnel-shaped to ± ovate), 3.5–4.5mm wide, margins ciliate, bracteoles smaller, lunate. Calyx c.5 × 1.8mm, lobed to less than ⅓; corolla lobes (immature) c.3.5 × 1.5mm. Anthers (immature) c.2mm.
Terai (Chamokdangi, Jogighora, Sivoke); **Duars** (Buxa Reserve). Damp ground; Mixed Plains and Lower Hill Forest. Fl. April.

There appears (from Prain's manuscript) to be confusion over Gamble's local names: the above are taken from a Cave specimen. Canes good but not much used due to rarity of plant (Gamble, 1922).

8. C. inermis T. Anderson. Nep: *dangri bet*; Lep: *brool*. Fig. 41(1).
Similar to *C. latifolius* in its cirrate leaves but differs in its narrower (to 3.5cm), parallel-sided leaflets and unarmed leaf sheaths. Only female infl. seen: stout, unarmed, rachillae zigzag, bracts long and funnel-shaped.
Sikkim/Darjeeling. Hot, damp valleys, 300–610m.

Included under *C. latifolius* in F.B.I.

Not seen recently despite repeated attempts to refind it (Basu, 1985, p. 88); formerly used for making walking sticks (Gamble, 1922).

6. DAEMONOROPS Blume

Similar to *Calamus* from any of which our single species can be distinguished as follows: infl. initially enclosed by large, spiny, boat-shaped, deciduous bracts; calyx very shallowly lobed; corolla lobes of female flower (visible in fruit) twice as long as calyx.

1. D. jenkinsiana (Griff.) Martius. Nep: *dhangru*; *dudia bet* (Kurseong — Burkill in Beccari, 1911); *garra, cheka bet* (Jalpaiguri teste Beccari, 1911). Fig. 41m–o.

Stout climber. Sheaths c.3.5cm wide, with horizontal to oblique rows of large spines alternating with bands of smaller, bristle-like spines; large spines to 4 × 0.4cm, flat, straight, dark brown, downward-pointing, margins setose at base; mouth obliquely truncate, auricles extremely narrow, papery. Petioles stout flattened, spiny on underside, not or slightly kneed. Leaves pinnate, upper cirrate, leaflets alternate, very dense, longest to 32cm, widest to 2mm, rather thick-textured, midrib prominent, lateral veins weaker, one pair with long (c.6mm) setae on upper surface; margins evenly setulose throughout. Rachis with deflexed hooks. Infl. (only female seen) branched to 2 orders, erect, stout, relatively short (to 30cm); prophyll very spiny, deciduous; rachillae to 7cm (in fr.) axis internodes c.6mm, zigzag; bracts collar-like, asymmetric, short, c.1mm, diads distinctly peduncled, bracteoles c.3mm diameter, shallow. Female flowers: calyx c.2mm with 3 ciliate points; corolla lobes c.4mm, narrowly triangular; filaments oblong-ovate, apiculate, fused only at base. Fruit 1.3–1.7cm, subspherical, shortly (c.1mm) apiculate; scales 3–4mm wide, grooved, pale orange-brown, with reddish-brown intramarginal stripe, margins hyaline, apex triangular, brown.

Terai/Duars (Chekopara, Buxa Reserve, Dulkajhar, Apalchand Forest Jalpaiguri Duars; Singari Puhar near Sivoke (Gamble, 1922)); '**Sikkim**' (unlocalised Cave specimen most probably from Terai); **Darjeeling** (Kurseong (Beccari, 1911)). Mixed Plains Forest, beside water-courses. Fr. January–February.

According to Gamble (1922) the commonest species of rattan in the Mixed Plains Forest; canes very long, rather soft, used to make baskets.

7. PLECTOCOMIA Martius ex Blume

Vegetatively similar to scandent species of *Calamus* with cirrate leaves, but leaf sheaths lacking knees, plants hapaxanthic and our species easily distinguished by its long, filiform leaflet tips. Infl. differing greatly from *Calamus*, terminal, the partial infls. drooping and the short rachillae enclosed in overlapping ± oblong, coriaceous bracts.

1. P. himalayana Griff. Nep: *gowri bet* (Bhutan); *tokri bet, tara bet* (Darjeeling); Lep: *rinul.* Fig. 40d, Fig. 41p–q.

Stems 3–15m or more, scandent. Leaf sheaths tomentose (orange-brown in lower part, pale fawn above), heavily armed with oblique or wavy-horizontal lamellae bearing unequal (longest to 2.5cm), slender, straight spines; mouth oblique, auricles not developed. Petiole very short, not kneed. Rachis flattened, fawn-tomentose, lower part sometimes spiny, upper part developed into long cirrus (to 2m) heavily armed with grapnels. Leaves to 3.5m; pinnae relatively few (to 25), linear-lanceolate, gradually drawn into long (to 8cm) filiform apex, longest 30–45cm, widest 2–3.6cm, irregularly inserted — singly or in groups of 2–3, margins spinulose, with many, glabrous, parallel veins. Male infl. branched to 2 orders, main axis stout, covered by overlapping, tubular bracts of partial infls.; partial infls. pendent, covered with overlapping bracts of rachillae; rachilla bracts 4.5–5.5 × 1.7–2cm, oblong to obcuneate, acuminate, coriaceous, abaxial surface finely tomentose. Male rachillae to 2.5cm, slender, zigzag, axis rusty-tomentose, bracteoles minute, linear. Male flowers: calyx lobes c.2 (+ 0.5) × 3mm, broadly rhombic, apiculate, ribbed, reddish-brown; corolla lobes c.7 × 2.5mm, oblong, acuminate; anthers c.3mm, exceeding filaments. Female infl. similar; rachilla bracts shorter and wider (c.3.5 × 2.5cm); rachillae shorter (to 1.5cm), stouter. Female corolla lobes longer (to 9mm). Fruit to 1cm, spherical; scales c.0.5mm wide, reddish-brown, yellowish at base, margins erose.

Bhutan: S — Gaylegphug district (W bank of Chabley Khola 54km along Tongsa road); **C** — Tongsa (Shamgong to Pertimi 9km from Shamgong) and Mongar (Shersing Thang to Namning) districts; **Darjeeling** (descent N of Darjeeling, Birch Hill, Jorebungalow to Mongpu, Darjeeling to Lebong; Senchal Range, Tukdah, Rungbee (Madulid, 1981)). Warm broad-leaved (Upper and Middle Hill) forest, 1220–2130m. Fl. August–December; fr. April.

The bitter stem pith is eaten in Bhutan. According to Gamble (1922) only used for tying fences and rough baskets.

8. WALLICHIA Roxb.

Hapaxanthic; stems clustered or solitary, often very short. Leaves spiral or distichous; sheaths disintegrating into fibres, petiole distinct. Leaves imparipinnate, leaflets induplicate, lateral unicostate, rather irregularly lanceolate to rhombic, margins praemorse (i.e. with smaller veins projecting irregularly as teeth); terminal lobe usually larger and several costate. Monoecious, male and female infls. separate on same plant; infls. interfoliar, branched to 1 order; peduncles stout bearing many coriaceous bracts tubular at base; prophyll small, tubular at base. Male infl. slender, densely flowered. Male flowers: calyx tubular, corolla tubular at base, stipitate, with 3 long lobes; stamens (3–)6(–15); anthers basal, latrorse, filaments fused at base and adnate to corolla tube. Female infl. stouter, erect. Female flowers globose, subtended by small bract and 3 bracteoles; calyx with 3 imbricating lobes; corolla lobes scarcely exceeding calyx; ovary 2–3-loculed, each with a single, sub-basal ovule; stigma sessile, small; staminodes 0–3. Fruit 2–3 seeded, mesocarp thin, fleshy.

1. Stem not or scarcely developed, leaves spiral .... **1. W. densiflora**
+ Stem developed, leaves distichous .... **2. W. disticha**

1. W. densiflora (Martius) Martius. Nep: *takoru, thakal*; Lep: *uh, oho, u-pe*. Fig. 40f.

Shrubby, stemless palm. Petiole and rachis rusty-felted. Leaves 1–3m, leaflets dark green above, silvery sometimes with rust-coloured bands below, lateral leaflets inserted singly, alternate or lower subopposite, longest 18–50 × 5–11cm, irregular, roughly oblong to rhombic, strongly asymmetric about midrib, base cuneate, edge obliquely stepped, one side more so than other, steps praemorse. Terminal leaflet 17.5–38.5 × 14–25.5cm, broadly fan-shaped, 3–4-costate, cuneate, usually 3-lobed, lobes praemorse. Peduncle bracts to 37 × 5cm, coriaceous, narrowly elliptic, acute, rusty tomentose on outside. Male rachillae slender, densely flowered. Male flowers: calyx c.5.5 × 2.3mm, cylindric, truncate, corolla lobes c.8.5 × 1mm, narrowly oblong, hooded; stamens 6; anthers c.5mm exceeding filaments. Female infl. with 7–10, stout, stiffly erect rachillae, rachillae 16–20cm in fl. (to 34cm in fr.), axis accrescent in fr. Female flowers depressed globose, almost sunk into axis; calyx lobes c.1.5 × 2mm, broad, imbricating, fused at base; corolla lobes scarcely exceeding calyx, lobes triangular-ovate c.1.5 × 1.5mm, not overlapping; ovary ellipsoid-prismatic, acute; staminodes 0. Fruit 5–6 × 4–5mm, greenish-brown, narrowly ovoid, usually 2-seeded.

Bhutan: S — Samchi (Khagra Valley near Gokti; Dhumdhum Hill (M.F.B.)), Phuntsholing (tributary of Torsa River 2km N of Phuntsholing) and Sankosh (2km W of Pinkhua) districts; **Duars** (Mal); **Darjeeling** (Darjeeling, Copper Mine Ghat, Tista, Chunbati, Mongpu, Rongbe); **Sikkim** (Singtam). Subtropical (Lower and Middle Hill) forest, especially on rocks under dense forest, 200–1220m. Fl. May; fr. February–March.

According to Gamble (1922) leaves used for feeding ponies and leaf rachis used by Nepalis for making combs.

2. W. disticha T. Anderson. Nep: *thakal*; Lep: *katong*.
Differs from *W. densiflora* in having a well-developed trunk 3–4.5m, clothed with persistent petiole bases; leaves 2.4–3m, arranged distichously; lower lateral leaflets inserted in groups of 4–5, more elongate (longest 53–74cm), more shortly and abruptly 'stepped', steps with longer teeth.
     Terai (Leesh Forest); **Darjeeling** (Choklong, ridge above Sivoke, Lopchu (planted); between Mahanadi and Tista Rivers (Anderson, 1871)). Lower Hill Forest, to 1650m.

According to Gamble (1922) Lepchas eat the pith from near the apex of the trunk; however, Anderson (1871) states that the Lepchas make no use of the tree and dread touching it because of its reputed irritating properties.

9. ARENGA Labillardière

     Differs from *Wallichia* chiefly in its male flowers which have calyx lobed almost to base, lobes imbricate; stamens usually numerous.

1. A. westerhoutii Griff.
     Leaves to 2m, oblong-elliptic in outline, curved. Rachis rusty-brown-scurfy, with darker scales below. Lateral leaflets c.40, singly inserted, alternate, narrowly lanceolate (upper ones c.31 × 3.5cm), symmetric about midrib, glaucous beneath, margins praemorse near apex, sides scarcely stepped. Flowers not seen.
     Bhutan: S — Phuntsholing district (below Suntlakha); **C** — Tongsa district (5km SE of Shamgong). Steep slope in dense, warm mixed forest, 1570–2000m.

Only stemless (young?) plants seen in Bhutan: should eventually develop a tall trunk.

A. nana (Griff.) H.E. Moore (*Didymosperma nana* (Griff.) H. Wendland & Drude ex Hook.f.) is an Assamese species recorded (possibly correctly) from Narfong (M.F.B.). Vegetatively it resembles a dwarf *Wallichia densiflora* but differs in its flowers and in having only 5–7 rather distant leaflets per leaf and a usually simple, stout, erect infl.

10. CARYOTA L.

     Hapaxanthic; trunks clustered or solitary, usually tall, becoming bare. Leaves twice pinnate, leaflets induplicate, lateral leaflets asymmetrically and irregularly triangular, veins radiating from base, distal margin praemorse; terminal leaflet symmetric, usually several-lobed. Monoecious, infls. of one type, interfoliar, branched to 1 order, developing sequentially from apex downwards; peduncles stout bearing many coriaceous bracts tubular at base; prophyll small, tubular at base; rachillae bearing protandrous triads of flowers — 2 males flanking a female, subtended by a minute bract and small bracteoles. Male flowers: calyx with 3 broad, imbricate lobes; corolla with 3 long lobes fused at base; stamens 6–c.100; anthers basal, latrorse. Female flowers: calyx with 3 broad, imbricating lobes; corolla tubular in lower part, lobes exceeding calyx; ovary 3-loculed, 1–2 with a single, sub-basal ovule; stigma shortly 2–3-lobed; staminodes 0–6. Fruit globose, 1–2-seeded, mesocarp fleshy.

1. C. cf. **urens** L. Nep: *rungbong, rangbhang*; Lep: *simong, sa-mon, ruri bari, somong-kung*. Fig. 40a.

Trunk to 15m, grey, smooth with conspicuous annular leaf scars; crown elongate. Leaf sheaths long. Petiole very short. Leaves very large (to 6 × 3.6m), drooping at apex, twice pinnate bearing long secondary rachises, lateral leaflets to 24cm, irregularly triangular, cuneate, apex oblique, irregularly lobed, praemorse, green on both sides. Rachillae long, to 65cm in fl., to 150cm in fr. Male flowers: calyx lobes c.3 × 4.5mm, coriaceous and swollen with narrow hyaline, ciliate, margin; corolla lobes c.10 × 3mm, oblong, acute; stamens 20(–45); anthers c.7.5mm, filaments c.1mm, connate. Female flowers: calyx lobes like male; corolla lobes c.6.5 × 3mm, lanceolate. Fruit c.1.3cm diameter, reddish.

Bhutan: C — Punakha district (Lometsawa); **Sikkim** (19km W of Singtam); **Darjeeling** (Chenga Forest, Mahanadi, ridge above Sivoke, Rongbe, Lopchu, Mongpu, Barnesbeg). Proabably native in Lower Hill Forest, but usually planted, 305–2000m. Fl. February.

Seeds and young shoots eaten, according to Hooker; fruits chewed by children (Rai & Rai, 1994); inner part of pith eaten green by Lepchas according to Gamble (1922). In India the leaves are used as a source of fibre (for ropes, brooms, etc.).

An immature *Caryota* specimen (stemless palm to 2m) from Taklai Khola E of Gaylegphug presumably belongs to this species.

The name *C. urens* applies strictly to plants from S India and Ceylon and according to Dransfield (pers. comm.) the E Himalayan plant almost certainly belongs to a different species. Our plant might be one of the Chinese species such as *C. ochlandra* Hance, but collections of flowering material are required.

11. PINANGA Blume

Pleonanthic. Stems solitary or clustered, often slender, smooth with distant annular scars, sometimes acaulous. Sheaths tubular, upper usually forming a crownshaft. Leaves pinnate or undivided (with or without apical notch); leaflets reduplicate, regularly or irregularly arranged, several costate, $\pm$ flat or several-folded, especially apical ones usually lobed. Infl. monoecious, of one type, usually infrafoliar, pendent, unbranched or branched to 1 order; peduncle flattened, usually short; prophyll membranous, deciduous. Flowers sessile, subtended by minute triangular bracteoles, arranged in triads, a small, pro-togynous female flanked by 2, larger, deciduous males. Male flower: calyx small, cup-like, 3-pointed; corolla asymmetric, lobed to base, lobes 3, fleshy; stamens (6–)12–30, filaments short, anthers basally attached, dehiscing latror-sely. Female flowers globose, symmetric, calyx lobed to base, similar to corolla, lobes imbricate; staminodes absent; ovary single-loculed, ovule single, basal; style short, stout; stigma irregularly lobed, papillose. Fruit often passing from pink to crimson to black, globose to ellipsoid, mesocarp fleshy, thin; seed large.

1. P. gracilis (Roxb.) Blume. Lep: *khur, karr patung, kar kuri*.
Stems slender (0.9–1.5cm diameter), erect, brown. Sheaths long, tight, striate when dry with conspicuous midrib. Petiole short. Leaf blade paripinnate, oblong-ovate in outline, leaflets 2–5 per side, $\pm$ paired, oblong, flat, costae prominent on upper surface, lower surface paler, terminal pair truncate-dentate, others falcately acuminate, penultimate ones 33.5–55 × 6–8.5cm, 5–7-costate. Infls. infrafoliar, spike-like, 9–16cm (to 26cm in fr.), pendent, emerging from sheathing prophyll; peduncle short (to 3cm); triads spirally arranged, subtended by minute bracteoles. Male flowers pink, deciduous: calyx saucer-shaped, 3-pointed, c.2mm deep, 6mm wide; corolla lobed to base, lobes c.7.5 × 8mm, $\pm$ triangular; stamens over 20. Female flowers: globose, calyx and corolla lobes similar, persistent and accrescent, transversely elliptic, 1.5–3 × 3.5–4.5mm, margins ciliate. Fruit 1.2–1.8 × 0.7–1cm, ellipsoid, shortly beaked.
Bhutan: S — Sarbhang district (Burborte Khola near Phipsoo); **Terai** (Sivoke); **Darjeeling** (Katanbari, Great Rangit, Tista, Birick, Jogikhola, Darjeeling, ?Urpital). Ravine in subtropical (Lower Hill) forest; Mixed Plains Forest of Terai, 290–910m. Fl. November–April.

Flower buds eaten in Bhutan.

12. ARECA L.

Vegetatively similar to *Pinanga*. Differing chiefly in the infl. which is much branched (to 3 orders); triads confined to basal parts of infl., upper parts of rachillae bearing single or paired male flowers or naked near apex; male flowers small; female flowers larger; stigmas 3, fleshy, recurved.

1. A. catechu L. Bhutanese: *doma*; Nep: *supari, gua*; Eng: *betel nut*. Fig. 40b.
Trunk to 15m or more, c.20cm diameter, greenish when young, finally grey, with conspicuous, distant annular scars. Crownshaft prominent. Leaves to 1.8m, pinnate; leaflets narrow, many. Infls. very stiff, spreading to erect in fl., drooping in fr., rachillae tortuous. Male flowers: calyx minute; corolla lobes c.3.5 × 1.5mm, lanceolate, ribbed; stamens 6; pistillodes 3. Female flowers: calyx lobed to base, lobes c.10 × 9mm, oblong-ovate, imbricate, fibrous; corolla lobes similar but smaller. Fruits large, mesocarp fibrous; seed (nut) ellipsoid.
Bhutan: S — Phuntsholing district (Phuntsholing); **Terai** (common); **Darjeeling** (below Barnesbeg, Punkabari); **Sikkim** (Singtam). Cultivated up to 700m.

Cultivated for its seeds (betel nuts) which are extensively chewed for their mildly stimulatory effect.

13. COCOS L.

Pleonanthic; trunks tall, naked, annular. Leaves in a large crown. Sheaths fibrous. Leaves pinnate, leaflets reduplicate, single-fold, narrow, arranged in one plane. Monoecious. Infl. of one type, branched to at least 2 orders, interfoliar, protandrous; prophyll small, tubular below, persistent, becoming fibrous; peduncle stout, bearing a large bract, enclosing infl. in bud, becoming woody. Infl. axis bearing spirally inserted, stout rachillae. Rachillae with few sub-basal triads a female flanked by 2 male flowers, upper part with single or paired male flowers. Male flowers narrowly ovoid, calyx lobes triangular, keeled, corolla lobes exceeding calyx, boat-shaped, acute; stamens 6, anthers medifixed, versatile, latrorse; pistillode small, 3-lobed. Female flowers large, ovoid, calyx lobes rounded, imbricate, corolla lobes similar, larger; ovary 3-loculed with 1–3 sub-basal ovules (of which only 1 usually develops), greatly fibrously thickened above, stigmas 3, sunken; ovary subtended by staminodal ring (lacking sterile anthers). Fruit large, ellipsoid, obscurely 3-angled, green becoming brown; mesocarp very thick, fibrous; endocarp woody with 3 basal pores; seed with liquid-filled cavity.

1. C. nucifera L. Nep: *naryal*; Eng: *coconut*.
Description as for genus.
Bhutan: S — Samchi (Chengmari) and Phuntsholing (Phuntsholing) dis-

tricts; **Terai** (common); **Darjeeling** (below Barnesbeg, Punkabari); **Sikkim** (Singtam to Rangpo). Planted up to 700m.

Perhaps originating in SW Pacific area but widely cultivated pantropically. A very versatile species of which almost every part is used in areas where it is grown more extensively (e.g. fruits as a source of food (endosperm), oil and copra, and for fibres from the husk; immature infls. as a source of toddy; leaves for thatching; wood, etc.). From its rarity in our area it is evidently not used much, possibly only for the fruits.

Other species and genera are probably occasionally cultivated for use or ornament, especially in the Terai.

Family 235. PANDANACEAE

Small sparingly branched pachycaul trees or shrubs, often with aerial or prop roots. Leaves usually 3-ranked, ensiform, margins and midrib spiny, coriaceous. Dioecious; flowers lacking perianths, ± coalescent in dense spadices borne singly or in racemes, subtended by spathes. Male flowers with filaments free or fused, pistillode sometimes present. Female flowers with pistil consisting of a single or several fused carpels, monocarpellary pistils sometimes partly fused in groups; ovules 1–many per carpel; staminodes present or absent; stigmas sessile. Fruit an aggregate of berries or drupes; drupes 1–many seeded.

Nakao & Nishioka (1984) record the use of the apical meristem of a *Pandanus* sp. (presumably *P. furcatus*) as a vegetable at Pangbang, Shemgang (Tongsa district).

1. PANDANUS L.

Trees or shrubs. Filaments of stamens free or fused below, pistillodes absent in male 'flowers'. Carpels free or fused into groups, each with a single ovule; staminodes absent in female flowers. Fruit an aggregate (syncarp) of 1-seeded drupes.

1. Tree usually over 3m; syncarps finally reddish-orange, usually several, large (over 15cm); drupes over 3.5cm; anthers c.5mm ..... **1. P. furcatus**
+ Shrub under 2m; syncarps finally yellow, borne singly, under 10cm; drupes under 3cm; anthers c.2mm .... **2. P. unguifer**

No satisfactory account can currently be given of the *screw-pines* (Eng) of Sikkim and Bhutan, due to the very few, inadequate, poorly annotated (e.g. lacking notes on stature and fruit colour) specimens seen. The position has merely been confused by St John's 'monographic study' (1972) of the Indian members of Section *Rykia* with its two new

431

taxa from the area based on very few specimens, and minute supposed differences. Its key is unusable and the lack of references and synonymy astonishing. A conservative view has therefore been taken (following F.B.I. and Cowan & Cowan, 1929) that two species occur in the area — a tree and a shrub — but much further fieldwork is required.

1. P. furcatus Roxb.; *P. nepalensis* St John; *P. sikkimensis* sensu F.E.H.3. Sha: *pairumnang shing*; Nep: *tarika*; Lep: *bor-kung*. Fig. 42a–c.
Tree (3–)4–9m, dichotomously branched above, girth at breast height 15–30cm, with numerous thick aerial roots from lower part. Leaves drooping, ensiform, finely acuminate to long, spiny caudate apex (to 10cm), abruptly contracted near base, to 2–3(–6)m, to 7cm wide, cross-veinlets sometimes prominent when dry, midrib spiny on underside, margins with closely (c.1cm apart in lower part) and regularly set forward-pointing spines; spines scarcely curved, 1–2mm on shorter (upper) edge. Male infl. (not seen, description from literature) a pendulous, compound spike, 10–15 × 2–3cm; stamens in groups of c.10, filaments fused for lower ⅔, anthers 5mm, mucronate; spathes broadly lanceolate, shortly acuminate, to 1 × 0.1m, coriaceous, golden-yellow. Female infl. of 1–8 syncarps in a dense, pendent spike; spathes oblong, abruptly acuminate, to 32 × 4.5cm; style persistent, flattened, usually bifid, 0.5–0.7cm at fl.; fruiting syncarps ellipsoid to oblong, 15–28(–45) × 10–13cm, reddish-orange when ripe; drupes narrowly obovoid, irregularly hexagonal in section, 3.5–5 × (0.8–)1.5–2.2cm, free apical part shallowly domed, c.0.8cm deep (excl. style), obscurely 6-faceted, faces smooth.
Bhutan: S — Samchi (Tamangdhanra Forest), Phuntsholing (tributary of Torsa River 2km N of Phuntsholing) and Gaylegphug (near Gaylegphug) districts; **C** — Tongsa district (Shemgang); **Darjeeling** (Selimpahar, Pankabari, above Sukna, Barnesbeg, above Mongpu, Rangpo to Tista); **Sikkim** (N of Jorethang, 14km W of Singtam, Singtam to Ranipul). Common in Lower Hill Forest (Cowan & Cowan, 1929). Subtropical forest, 200–1520m. Fr. August–May.

Fruits eaten by children (Rai & Rai, 1994).

2. P. unguifer Hook.f.; *P. minor* Buchanan-Hamilton ex Solms; *P. sikkimensis* St John. Fig. 42d–g.
Differs from *P. furcatus* in being a clump-forming shrub; stems (0.3–1(–2?)m) simple, slender, decumbent. Leaves lorate, more abruptly acumi-

---

FIG. 42. **Pandanaceae.**
a–c, **Pandanus furcatus**: a, habit (× ¹⁄₁₀); b, leaf apex (× ⅓); c, drupe showing bifid style (× ½). d–g, **P. unguifer**: d, leaf apex (× ⅓); e, immature female infl. (× ⅓); f, drupe showing simple style (× ½); g, male infl. (× ⅓). Drawn by Glenn Rodrigues.

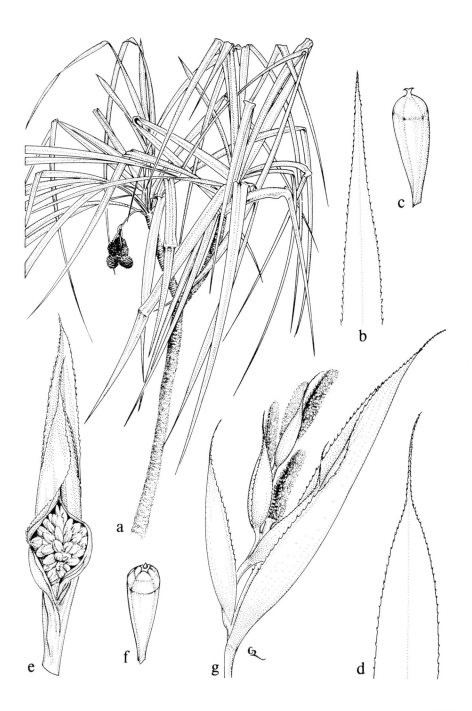

433

nate, more gradually narrowed to base, to 60–90 × 3.5–6cm, teeth longer (3–5mm on upper edge), more curved, more widely and irregularly spaced (often over 2cm apart near base). Male infl. a compound, terminal spike to 17 × 6cm; anthers 1.5–2mm; lowest spathe 18 × 7cm. Female infl. solitary, spreading, subterminal, surrounded by involucre of spathes, lowest spathe 18 × 5cm; fruiting syncarp 6.5–14 × 5.5–9cm, yellow when ripe; drupes 2–2.7 × 0.8–1.5cm, free apical part steeply domed with 6 sharply defined faces marked with horizontal ridges, persistent style commonly simple, concave.

Bhutan: S — Gaylegphug district (Rang Khola 4km NE of Surey, Karai Khola, above Aie bridge); **Darjeeling** (Mongpu, N of Kurseong, Farsing; common in Mixed Plains Forest (Cowan & Cowan, 1929)). Beside watercourses in subtropical broad-leaved forest, in terai and lower hills, 150–1550m. Fl. June–August.

The name *P. minor* was not validated until 'February 1878' (presumably the latter part of February since the last part of the previous volume of *Linnaea* was also published that month). It seems that *P. unguifer* dating from 1 ii 1878, therefore, has priority. Both species were based, at least partly, on the same Hooker material from Khasia. St John (1972) claims that *P. minor* does not belong to Sect. *Rykia* though without stating where he believes it to belong (presumably Sect. *Acrostigma* as he misdetermined a Khasian specimen of *P. foetidus* at E as *P. minor*).

St John described his *P. sikkimensis* from old and inadequate material and could not therefore describe its habit. From the description and the locality of the holotype (Mongpu) it seems to be referable to *P. unguifer*. The Japanese collections from Tista which (though confusing the references, illustrations, localities and specimens) he refers to the same species can be seen from the photograph (Hara, 1968b, plate 8) quite clearly to belong to *P. furcatus*, having a racemose female infl.

It is possible that **P. foetidus** Roxb. occurs in the Terai — it is not recorded for the area in F.B.I. but Hooker (1878) states that he gathered it 'in Sikkim'. This shrubby species belongs to Sect. *Acrostigma*, differing in having drupes with apices gradually tapering into a long, sharp, simple style and stamens not united into a column.

BIBLIOGRAPHY

Bibliography for records and notes in the present volume; cited in the text by author and date. Numbers in parentheses refer to the existing *Flora of Bhutan* Bibliography.

Anderson, T. (1871). An enumeration of the palms of Sikkim. *J. Linn. Soc., Bot.* 11: 4–14. (1).

Bailey, L.H. (1929). The case of *Ophiopogon* and *Liriope*. *Gentes Herb.* 2: 1–37.

Baker, J.G. (1893). A synopsis of the genera and species of Musaceae. *Ann. Bot.* 7: 189–222.

Basu, S.K. (1985). The present status of rattan palms in India — an overview. In: Wong, K.M. & Manokaran, N. (eds) *Proceedings of the Rattan Seminar, 2–4 October 1984, Kuala Lumpur, Malaysia*, pp. 77–94. Kepong.

Beccari, O. (1890). Rivista monografica delle specie del genere *Phoenix* Linn. *Malesia* 3: 345–416.

Beccari, O. (1908). Asiatic Palms — Lepidocaryeae. Part 1 — The species of *Calamus*. *Ann. Royal Bot. Gard. (Calcutta)* 11: 1–518; t. 1–231.

Beccari, O. (1911). Asiatic Palms — Lepidocaryeae. Part 2 — The species of *Daemonorops*. *Ann. Royal Bot. Gard. (Calcutta)* 12: 1–237; t. 1–109.

Beccari, O. (1913). Asiatic Palms — Lepidocaryeae. Supplement to Part 1 — The species of *Calamus*. *Ann. Royal Bot. Gard. (Calcutta)* 12: 1–142; t. 1–83.

Beccari, O. (1931). Asiatic Palms — Corypheae. *Ann. Royal Bot. Gard. (Calcutta)* 13: 1–356; t. 1–99.

Boott, F. (1858). *Illustrations of the genus Carex, 1*. London.

Boott, F. (1862). *Illustrations of the genus Carex, 3*. London.

Buchenau, F. (1885). Die Juncaceen aus Indien, insbesondere die aus dem Himalaya. *Bot. Jahrb.* 6: 187–232.

Chase, M.W. *et al.* (1993). Phylogenetics of seed plants: an analysis of nucleotide sequences from the plastid gene rbcL. *Ann. Missouri Bot. Gard.* 80: 528–580.

Cheesman, E.E. (1948). Classification of the bananas III: Critical notes on species. *Kew Bull.* 1948: 11–28.

Clarke, C.B. (1874). *Commelynaceae et Cyrtandraceae Bengalenses*. Calcutta.

Clarke, C.B. (1881). Commelinaceae. In: de Candolle, A. & C. (eds) *Monographiae Phanerogamarum* 3, pp. 113–324. Paris.

Clarke, C.B. (1884). On the Indian species of *Cyperus*; with remarks on some others that specially illustrate the subdivisions of the genus. *J. Linn. Soc., Bot.* 21: 1–202.

Clarke, C.B. (1885). Botanic notes from Darjeeling to Tonglo and Sundukphoo. *J. Linn. Soc., Bot.* 21: 384–386. (26).

Cowan, A.M. & Cowan, J.M. (1929). *The Trees of Northern Bengal.* Calcutta. (34).

Dahlgren, R.M.T., Clifford, H.T. & Yeo, P.F. (1985). *The Families of the Monocotyledons.* Berlin, etc.

Dasgupta, S. & Deb, D.B. (1984). Taxonomic revision of the genus *Lilium* L. in India and adjoining region. *Candollea* 39: 487–506.

Dasgupta, S. & Deb, D.B. (1986). Taxonomic revision of the genus *Gagea* Salisb. (Liliaceae) in India and adjoining regions. *J. Bombay Nat. Hist. Soc.* 83: 78–97.

Diels, L. (1912). Plantae Chinenses Forrestianae. *Notes Royal Bot. Gard. Edinburgh* 5: 161–308.

Fyson, P.F. (1923). *The Indian Species of Eriocaulon* [reprinted from *J. Indian Bot. Soc.*, vols 2 & 3]. Madras.

Gamble, J.S. (1922). *A Manual of Indian Timbers — reprint of second edition with some corrections and additions.* London.

Ghildyal, N. & Bhattacharyya, U.C. (1985). A new name for *Carex kingiana* Clarke (Cyperaceae). *J. Econ. Taxon. Bot.* 7: 447.

Goetghebeur, P. (1989). Studies in Cyperaceae 9. Problems in the lecto-typification and infrageneric taxonomy of *Cyperus* L. *Bull. Soc. Roy. Bot. Belgique* 122: 103–114.

Graebner, P. (1900). Sparganiaceae. In: Engler, A. (ed.) *Das Pflanzenreich 2.IV.10 Sparganiaceae.* Leipzig.

Griffith, W. (1847). *Journals of Travels in Assam, Burma, Bootan, Affghanistan and the Neighbouring Countries.* Ed. J. McClelland. Calcutta. (63).

Griffith, W. (1848). *Itinerary Notes of Plants Collected in the Khasyah and Bootan Mountains, 1837–8, in Affghanistan and Neighbouring Countries, 1839–41.* Ed. J. McClelland. Calcutta. (66).

Griffith, W. (1850). *Palms of British East India.* Calcutta.

Habib-Ur-Rehman, Siddiqui, F. & Khokhar, I. (1992). Anticancer agents of *Arisaema jacquemontii* Blume. *Pakistan J. Sci. Industr. Res.* 35: 406–408.

Hara, H. (1961). New or noteworthy flowering plants from Eastern Himalaya (1). *J. Jap. Bot.* 36: 75–80.

Hara, H. (1965). New or noteworthy flowering plants from Eastern Himalaya (3). *J. Jap. Bot.* 40: 19–22.

Hara, H. (1968a). New or noteworthy flowering plants from Eastern Himalaya (6). *J. Jap. Bot.* 43: 44–48.

Hara, H. (ed.) (1968b). *Photo-album of Plants of Eastern Himalaya.* Tokyo. (70).

Hara, H. (1969). Variations in *Paris polyphylla* Smith, with reference to other Asiatic species. *J. Fac. Sci. Univ. Tokyo* 10: 141–180.

Hara, H. (1984). The genus *Disporum* (Liliaceae) of the Himalayas. *Bot. Helv.* 94: 255–259.

Hara, H. (1987). Notes towards a revision of the Asiatic species of the genus *Smilacina. J. Fac. Sci. Univ. Tokyo* 14: 137–159.

Hara, H. (1988). A revision of the Asiatic species of the genus *Disporum* (Liliaceae). In: Ohba, H. & Malla, S.B. (eds) *The Himalayan Plants, Volume 1*, pp. 163–209. Tokyo.

Hong, D.Y. (1974). Revisio Commelinacearum Sinicarum. *Acta Phytotax. Sin.* 12: 459–488.

Hooker, J.D. (1854). *Himalayan Journals or Notes of a Naturalist in Bengal, the Sikkim and Nepal Himalayas, the Khasia Mountains etc.* London. (78).

Hooker, J.D. (1878). *Pandanus unguifer. Bot. Mag.* 104: t. 6347.

Jacobsen, N. (1980). The *Cryptocoryne albida* group of mainland Asia (Araceae). *Misc. Pap. Landbouwhoogeschool Wageningen* 19: 183–204.

Jessop, J.P. (1976). A revision of *Peliosanthes* (Liliaceae). *Blumea* 23: 141–159.

Jonker, F.P. (1938). A monograph of the Burmanniaceae. *Meded. Bot. Mus. Herb. Rijks Univ. Utrecht* 51: 1–279.

Kammathy, R.V. (1983). Rare and endemic species of Indian Commelinaceae. In: Jain, S.K. & Rao, R.R. (eds) *An Assessment of Threatened Plants of India*, pp. 213–221. Howrah.

Kern, J.H. (1974). Cyperaceae. In: van Steenis, C.G.G.J. (ed.) *Flora Malesiana* 7, pp. 435–753. Leyden.

King, G. (undated manuscript). Four forms of soboliferous Musas with pendulous inflorescences in Sikkim. Copy of unpublished ms made by J.D. Hooker at Kew.

Kränzlin, F. (1912). Cannaceae. In: Engler, A. (ed.) *Das Pflanzenreich 56.IV.47.* Leipzig.

Kükenthal, G. (1909). Cyperaceae-Caricoideae. In: Engler, A. (ed.) *Das Pflanzenreich 38.IV.20.* Leipzig.

Kükenthal, G. (1936). Cyperaceae-Scirpoideae-Cypereae. In: Engler, A. (ed.) *Das Pflanzenreich 101.IV.20.* Leipzig.

Lama, P.C. (1989). A preliminary report on the ethnobotanical importance of the Sukhia Pokhri region of Darjeeling Himalaya. *J. Bengal Nat. Hist. Soc.* 8: 56–62.

Larsen, K. (1964). Studies on Zingiberaceae IV; *Caulokaempferia*, a new genus. *Bot. Tidsskr.* 60: 165–179.

Madulid, D.A. (1981). A monograph of *Plectocomia* (Palmae: Lepidocaryoideae). *Kalikasan* 10: 1–94.

Matthew, K.M. (1967). A preliminary list of plants from Kurseong. *Bull. Bot. Surv. India* 8: 158–168. (95).

Mikage, M. & Suzuki, M. (1993). A morphological study on the rhizomes of

Polygonatum cirrhifolium and *P. verticillatum* (Liliaceae) from far west Nepal. *J. Phytogeogr. Taxon.* 41: 15–19.

Mukherjee, A. (1988). *The Flowering Plants of Darjiling.* Delhi & Lucknow. (218).

Murata, J. (1990). Three subspecies of *Arisaema flavum* (Forssk.) Schott (Araceae). *J. Jap. Bot.* 65: 65–73.

Naithani, H.B. (1990). *Flowering Plants of India, Nepal and Bhutan not recorded in Sir J.D. Hooker's Flora of British India.* Dehra Dun.

Nakao, S. & Nishioka, K. (1984). *Flowers of Bhutan.* Tokyo. (165).

Parker, C. (1992). *Weeds of Bhutan.* Simtokha.

Pradhan, U.C. (1990). *Himalayan Cobra-lilies (Arisaema): their Botany and Culture.* Kalimpong. (226).

Prain, D. & Burkill, I.H. (1936). An account of the genus *Dioscorea* in the east. Part 1 — the species which twine to the left. *Ann. Royal Bot. Gard. (Calcutta)* 14(1): 1–210; t. 1–85.

Prain, D. & Burkill, I.H. (1939). An account of the genus *Dioscorea* in the east. Part 2 — the species which twine to the right. *Ann. Royal Bot. Gard. (Calcutta)* 14(2): 211–528; t. 86–150.

Rai, T. & Rai, L. (1994). *Trees of the Sikkim Himalaya.* New Delhi.

Rao, A.S. & Verma, D.M. (1980). Notes on Cyperaceae of Assam. *Bull. Bot. Surv. India* 22: 80–90.

Rao, R.S. (1964a). Indian species of Commelinaceae — miscellaneous notes. *Notes Royal Bot. Gard. Edinburgh* 25: 179–189.

Rao, R.S. (1964b). A botanical tour in the Sikkim state, eastern Himalayas. *Bull. Bot. Surv. India* 5: 165–205. (104).

Rao, R.S. (1966). Indian species of Commelinaceae — Miscellaneous notes II. *Blumea* 14: 345–354.

Rao, R.S. & Deori, N.C. (1980). *Belosynapsis ciliata* (Bl.) R. Rao (Commelinaceae), a new record from India and its identity and distribution in SE Asia. *Indian J. Bot.* 3: 1–5.

Rao, R.S., Kammathy, R.V. & Raghavan, R.S. (1968). Cytotaxonomic studies on Indian Commelinaceae. *J. Linn. Soc., Bot.* 60: 357–372.

Roscoe, W. (1828). *Monandrian Plants of the Order Scitamineae.* Liverpool.

Satyavati, G.V., Raina, M.K. & Sharma, M. (eds) (1976). *Medicinal Plants of India.* New Delhi.

Sikdar, J.K. (1981). Notes on some plant records for Bengal. *J. Bombay Nat. Hist. Soc.* 78: 419–421.

Sikdar, J.K. & Samanta, D.N. (1984). Herbaceous flora (excluding Cyperaceae, Poaceae and Pteridophytes) of Jalpaiguri District, West Bengal — a check-list. *J. Econ. Taxon. Bot.* 4: 525–538. (174).

Simmonds, N.W. (1957). Botanical results of the banana collecting expedition, 1954–5. *Kew Bull.* 1956: 463–489.

Smith, W.W. (1911). Some additions to the flora of the Eastern Himalaya. *Rec. Bot. Surv. India* 4: 261–272. (111).

Smith, W.W. (1913). The alpine and subalpine vegetation of south-east Sikkim. *Rec. Bot. Surv. India* 4: 323–431. (112).

Smith, W.W. & Cave, G.H. (1911). The vegetation of the Zemu and Llonakh valleys of Sikkim. *Rec. Bot. Surv. India* 4: 141–260. (113).

Srivastava, R.C. (1985). Notes on distribution and citation of some species of genus *Hedychium* Koen. Zingiberaceae. *J. Econ. Taxon. Bot.* 7: 500–503.

St John, H. (1972). The Indian species of *Pandanus* (section *Rykia*). *Bot. Mag. (Tokyo)* 85: 241–262.

Takhtajan, A. (1983). A revision of *Daiswa* (Trilliaceae). *Brittonia* 35: 255–270.

Uhl, N.W. & Dransfield, J. (1987). *Genera Palmarum*. Lawrence.

Van Bruggen, H.W.E. (1985). Monograph of the genus *Aponogeton* (Aponogetonaceae). *Biblioth. Bot.* 137: 1–76.

Wang, F.T. & Tang, T. (eds) (1980). Liliaceae (1). *Flora Reipublicae Popularis Sinicae* Vol. 14. Beijing.

Wiegleb, G. (1990). A redescription of *Potamogeton distinctus* including remarks on the taxonomy of the *P. nodosus* group. *Pl. Syst. Evol.* 169: 245–259.

Wight, R. (1853). *Icones Plantarum Indiae Orientalis: or Figures of Indian Plants*. Vol. 6. Madras.

Xu, J.M. (1990). Key to the Alliums of China. Translated by P. Hanelt & C.L. Long from account in *Flora Reipublicae Popularis Sinicae*. *Herbertia* 46: 140–164.

Taxonomic papers relevant to this volume have been published in the *Notes relating to the Flora of Bhutan* series in the *Edinburgh Journal of Botany* as follows:

XVII	Zingiberaceae by R.M. Smith 48: 13–15 (1991)
XIX	Cyperaceae (*Kobresia*) by H.J. Noltie 50: 39–50 (1993)
XX	Liliaceae (*Lloydia*) by H.J. Noltie 50: 51–57 (1993)
XXI	Cyperaceae (*Carex*) by H.J. Noltie 50: 185–206 (1993)
XXII	Asparagaceae (*Asparagus filicinus*) and Convallariaceae (*Maianthemum oleraceum*) by H.J. Noltie 50: 207–210 (1993)
XXIV	Juncaceae (*Juncus*) by H.J. Noltie 51: 129–143 (1994)
XXVI	Smilacaceae (*Smilax*) by H.J. Noltie 51: 147–163 (1994)
XXVIII	Eriocaulaceae (*Eriocaulon*), Musaceae (*Musa*), Cyperaceae (*Actinoscirpus*) by H.J. Noltie 51: 169–174 (1994)

INDEX OF BOTANICAL NAMES

440

var. *foliosa* (D. Don ex Tilloch & Taylor)
C.B. Clarke, 366
myosurus Nees, 381
var. *eminens* (Boot) Boeckeler, 381
var. *floribunda* (Boeckeler) Kükenthal,
381
var. *ratongensis* C.B. Clarke, 381
nakaoana T. Koyama, 405
nigra (L.) Reichard
ssp. drukyulensis Noltie, 370
nivalis Boott, 391
notha Kunth, 370
nubigena D. Don ex Tilloch & Taylor, 364
obscura Nees
var. brachycarpa C.B. Clarke, 390
obscuriceps Kükenthal, 393
oedorrhampha Nelmes, 400
oligostachya Nees, 380
olivacea Boott, 399
orbicularis Boott, 369
parva Nees, 374
phacota Sprengel, 371
pleistogyna V. Kreczetowicz, 364
polycephala Boott, 382
praeclara Nelmes, 392
praelonga C.B. Clarke, 372
pruinosa Boott, 372
ssp. maximowiczii (Miquel) Kükenthal,
372
pseudobicolor Boeckeler, 388
pseudofoetida Kükenthal
ssp. afghanica Kukkonen, 362
psychrophila Nees, 389
pulchra Boott, 385
radicalis Boott, 374
rara Boott, 373
remota L.
var. *remotiformis* (Komarov)
Kükenthal, 367
ssp. rochebrunii (Franchet & Savatier)
Kükenthal, 366
ssp. stewartii Kukkonen, 367
rhizomatosa Steudel, 380
rigida sensu F.B.I., 369
rubro-brunnea C.B. Clarke, 371
sahnii Ghildyal & Bhattacharyya, 407
schlagintweitiana Boeckeler
ssp. deformis Noltie, 403
sclerocarpa Franchet, 407
setigera D. Don, 403
setosa Boott, 398
sempervirens Villars
ssp. tristis (M. Bieberstein) Kükenthal,
406

speciosa Kunth
ssp. dilatata Noltie, 402
ssp. pinetorum Noltie, 402
ssp. speciosa, 401
sikkimensis C.B. Clarke, 370
spiculata Boott, 381
var. *nobilis* (Boott) C.B. Clarke, 381
stenophylla Wahlenberg, 363
stramentita Boott ex Boeckeler, 376
teinogyna Boott, 369
teres Boott, 372
teretisucula Goodenough, 363
thomsonii Boott, 366
tumida Boott, 400
ustulata sensu F.B.I., 390
vesiculosa Boott, 379
f. *pallida* Kükenthal, 380
Var. *paniculata* C.B. Clarke, 380
vesiculosa sensu F.B.I. p.p., 380
vidua Boott ex C.B. Clarke, 351
viridula Michaux, 393
var. *pulchella* (Lonnroth) B. Schmid, 394
wallichiana sensu F.B.I. p.p., 403
Caryota L., 427
ochlandra Hance, 428
cf. urens L., 428
Caulokaempferia K. Larsen, 196
secunda (Wall.) K. Larsen, 198
sikkimensis (King ex Baker) K. Larsen, 196
Cautleya (Bentham) Hook.f., 193
cathcartii Baker, 193
gracilis (Smith) Dandy, 193
lutea (Royle) Hooker, 193
robusta Baker, 194
spicata (Smith) Baker, 193
Chlorophytum Ker Gawler, 73
arundinaceum Baker, 74
breviscapum sensu F.B.I. p.p., 75
comosum (Thunberg) Jacques
'Variegatum', 75
khasianum Hook.f., 74
nepalense (Lindley) Baker, 74
undulatum Wall. ex Hook.f., 74
Clinogyne dichotoma Salisbury, 215
Clintonia Rafinesque, 94
udensis Trautvetter & C.A. Meyer
ssp. alpina (Kunth ex Baker) Hara, 94
Clivia Lindley
miniata (Lindley) Bosse, 87
Cocos L., 430
nucifera L., 430
COLCHICACEAE, 92
Colocasia Schott, 136
affinis Schott, 137

INDEX OF COMMON NAMES

INDEX OF COMMON NAMES

456